Problems on Algebra

William McCallum, Eric Connally, Deborah Hughes Hallett et al., 2010

Problems on Algebraic Expressions

Section 1.3: Equivalent Expressions

A person's monthly income is I, her monthly rent is R, and her monthly food expense is F. In Problems 1–4, do the two expressions have the same value? If not, say which is larger, or that there is not enough information to decide. Explain your reasoning in terms of income and expenses.

1. $I - R - F$ and $I - (R + F)$

2. $12(R + F)$ and $12R + 12F$

3. $I - R - F + 100$ and $I - R - (F + 100)$

4. $\dfrac{R + F}{I}$ and $\dfrac{I - R - F}{I}$

Chapter 1 Review

In Problems 5–10, both a and x are positive. What is the effect of increasing a on the value of the expression? Does the value increase? Decrease? Remain unchanged?

5. $ax + 1$

6. $a + x - (2 + a)$

7. $x - a$

8. $\dfrac{x}{a} + 1$

9. $x + \dfrac{1}{a}$

10. $ax - \dfrac{1}{a}$

Section 5.1: Linear Functions

11. If the tickets for a concert cost p dollars each, the number of people who will attend is $2500 - 80p$. Which of the following best describes the meaning of the 80 in this expression?

    (i) The price of an individual ticket.

    (ii) The slope of the graph of attendance against ticket price.

    (iii) The price at which no one will go to the concert.

    (iv) The number of people who will decide not to go if the price is raised by one dollar.

Section 5.2: Working with Linear Expressions

12. A car trip costs \$1.50 per fifteen miles for gas, 30¢ per mile for other expenses, and \$20 for car rental. The total cost for a trip of d miles is given by

$$\text{Total cost} = 1.5\left(\frac{d}{15}\right) + 0.3d + 20.$$

    (a) Explain what each of the three terms in the expression represents in terms of the trip.

    (b) What are the units for cost and distance?

    (c) Is the expression for cost linear?

13. Greta's Gas Company charges residential customers \$8 per month even if they use no gas, plus 82¢ per therm used. (A therm is a quantity of gas.) In addition, the company is authorized to add a rate adjustment, or surcharge, per therm. The total cost of g therms of gas is given by

$$\text{Total cost} = 8 + 0.82g + 0.109g.$$

    (a) Which term represents the rate adjustment? What is the rate adjustment in cents per therm?

    (b) Is the expression for the total cost linear?

Section 9.1: Quadratic Functions

14. A peanut, dropped at time $t = 0$ from an upper floor of the Empire State Building, has height in feet above the ground t seconds later given by

$$h(t) = -16t^2 + 1024.$$

    What does the factored form

$$h(t) = -16(t - 8)(t + 8)$$

    tell us about when the peanut hits the ground?

Section 10.3: Working with the Exponent

Prices are increasing at 5% per year. What is wrong with the statements in Problems 15–23? Correct the formula in the statement.

15. A \$6 item costs $\$(6 \cdot 1.05)^7$ in 7 years' time.

16. A \$3 item costs $\$3(0.05)^{10}$ in ten years' time.

17. The percent increase in prices over a 25-year period is $(1.05)^{25}$.

18. If time t is measured in months, then the price of a \$100 item at the end of one year is $\$100(1.05)^{12t}$.

19. If the rate at which prices increase is doubled, then the price of a \$20 object in 7 years' time is $\$20(2.10)^7$.

20. If time t is measured in decades (10 years), then the price of a \$45 item in t decades is $\$45(1.05)^{0.1t}$.

21. Prices change by $10 \cdot 5\% = 50\%$ over a decade.

22. Prices change by $(5/12)\%$ in one month.

23. A \$250 million town budget is trimmed by 1% but then increases with inflation as prices go up. Ten years later, the budget is $\$250(1.04)^{10}$ million.

WHAT IS OUR APPROACH TO ALGEBRA?

McCallum | Connally | Hughes-Hallett | et al.

The fundamental approach of *Algebra: Form and Function* is to foster strategic competence and conceptual understanding in algebra, in addition to procedural fluency. Fluency—the ability to carry out procedures such as expanding, factoring, and solving equations—is important, but too often students do not see the purpose or structure of the procedures. Strategic competence and conceptual understanding in algebra mean being able to read algebraic expressions and equations in real-life contexts, not just manipulate them, and being able to make choices of which form or which operation will best suit the context. They also mean being able to translate back and forth between symbolic representations and graphical, numerical, and verbal ones. As with our previous books, the heart of this book is in innovative problems that get students to think. We foster strategic competence, conceptual understanding and procedural fluency by

- Interpretation problems, which ask students to think about the form and function of the expressions and equations, not just blindly manipulate them (Sample sheet, 1-4, 12, 16)

- Problems that foster algebraic foresight, helping students see their way through an algebraic calculation and formulate strategies for problem-solving (Sample Sheet, 26-33)

- Modeling problems that train students in symbolic representation of real-world contexts, and in graphical numerical interpretation of expressions and equations (Sections 2.5, 3.4, 6.5)

- Real expressions and equations, presenting students with the expressions and equations they will encounter in later study and work (Sample sheet, 17–25)

- Drill problems to develop basic techniques (Skills chapter and exercises throughout text).

We have also continued our tradition of clear, concise exposition. Many problems that students have in algebra result from confusion about fundamental concepts, such as the difference between an expression and equation. We have written the text to be readable by students and make clear the basic ideas of algebra.

This book supports later courses using our Precalculus and Calculus texts, which are based on the Rule of Four: that concepts should be presented symbolically, graphically, numerically, and verbally. As is appropriate for an Algebra text, it focuses mainly on the symbolic and verbal strands, while not neglecting the other two. The text assumes that students have access to graphing calculators, although could quite easily be used without them by avoiding certain exercises.

In addition to problem types for which the Consortium is already known, the sample sheets on the following pages illustrate the new types of problems we have written for *Algebra: Form and Function* to foster interpretation and foresight in working with expressions and equations.

Problems on Algebraic Equations

Section 1.2: Equations

In Problems 24–30, does the equation have a solution? Explain how you know without solving it.

24. $2x - 3 = 7$ 25. $x^2 + 3 = 7$ 26. $4 = 5 + x^2$

27. $\dfrac{x+3}{2x+5} = 1$ 28. $\dfrac{x+3}{5+x} = 1$ 29. $\dfrac{x+3}{2x+6} = 1$

30. $2 + 5x = 6 + 5x$

Section 1.4: Equivalent Equations

31. Which of the following equations have the same solution? Give reasons for your answer that do not depend on solving the equations.

   I. $x + 3 = 5x - 4$ II. $x - 3 = 5x + 4$
   III. $2x + 8 = 5x - 3$ IV. $10x + 6 = 2x - 8$
   V. $10x - 8 = 2x + 6$ VI. $0.3 + \dfrac{x}{10} = \dfrac{1}{2}x - 0.4$

In Problems 32–35, the solution depends on the constant a. Assuming a is positive, what is the effect of increasing a on the solution? Does it increase, decrease, or remain unchanged? Give a reason for your answer that can be understood without solving the equation.

32. $x - a = 0$ 33. $ax = 1$
34. $ax = a$ 35. $\dfrac{x}{a} = 1$

Chapter 1 Review

36. You plan to drive 300 miles at 55 miles per hour, stopping for a two-hour rest. You want to know t, the number of hours the journey is going to take. Which of the following equations would you use?

   (A) $55t = 190$ **(B)** $55 + 2t = 300$
   (C) $55(t + 2) = 300$ **(D)** $55(t - 2) = 300$

Section 5.3: Solving Linear Equations

In Problems 37–44, decide for what value(s) of the constant A (if any) the equation has

(a) The solution $x = 0$ (b) A positive solution in x
(c) No solution in x.

37. $3x = A$ 38. $Ax = 3$
39. $3x + 5 = A$ 40. $3x + A = 5$
41. $3x + A = 5x + A$ 42. $Ax + 3 = Ax + 5$
43. $\dfrac{7}{x} = A$ 44. $\dfrac{A}{x} = 5$

Chapter 5 Review

Without solving them, say whether the equations in Problems 45–56 have a positive solution, a negative solution, a zero solution, or no solution. Give a reason for your answer.

45. $7x = 5$ 46. $3x + 5 = 7$
47. $5x + 3 = 7$ 48. $5 - 3x = 7$
49. $3 - 5x = 7$ 50. $9x = 4x + 6$
51. $9x = 6 - 4x$ 52. $9 - 6x = 4x - 9$
53. $8x + 11 = 2x + 3$ 54. $11 - 2x = 8 - 4x$
55. $8x + 3 = 8x + 11$ 56. $8x + 3 = 2x + 11x$

Section 9.3: Solving Quadratic Equations by Completing the Square

In Problems 57–60, for what values of the constant A (if any) does the equation have no solution? Give a reason for your answer.

57. $3(x - 2)^2 = A$ 58. $(x - A)^2 = 10$
59. $A(x - 2)^2 + 5 = 0$ 60. $5(x - 3)^2 + A = 10$

Section 9.4: Solving Quadratic Equations by Factoring

Without solving them, say whether the equations in Problems 61–68 have two solutions, one solution, or no solution. Give a reason for your answer.

61. $3(x - 3)(x + 2) = 0$ 62. $(x - 2)(x - 2) = 0$
63. $(x + 5)(x + 5) = -10$ 64. $(x + 2)^2 = 17$
65. $(x - 3)^2 = 0$ 66. $3(x + 2)^2 + 5 = 1$
67. $-2(x - 1)^2 + 7 = 5$ 68. $2(x - 3)^2 + 10 = 10$

Section 10.4: Solving Exponential Equations

69. Match each statement (a)–(b) with the solutions to one or more of the equations (I)–(VI).

   I. $10(1.2)^t = 5$ II. $10 = 5(1.2)^t$
   III. $10 + 5(1.2)^t = 0$ IV. $5 + 10(1.2)^t = 0$
   V. $10(0.8)^t = 5$ VI. $5(0.8)^t = 10$

   (a) The time an exponentially growing quantity takes to grow from 5 to 10 grams.
   (b) The time an exponentially decaying quantity takes to drop from 10 to 5 grams.

70. Assume $0 < r < 1$ and x is positive. Without solving equations (I)–(IV) for x, decide which one has

   (a) The largest solution
   (b) The smallest solution
   (c) No solution

   I. $3(1 + r)^x = 7$ II. $3(1 + 2r)^x = 7$
   III. $3(1 + 0.01r)^x = 7$ IV. $3(1 - r)^x = 7$

Without solving them, say whether the equations in Problems 71–80 have a positive solution, a negative solution, a zero solution, or no solution. Give a reason for your answer.

71. $25 \cdot 3^z = 15$ 72. $13 \cdot 5^{t+1} = 5^{2t}$
73. $(0.1)^x = 2$ 74. $5(0.5)^y = 1$
75. $5 = -(0.7)^t$ 76. $28 = 7(0.4)^z$
77. $7 = 28(0.4)^z$ 78. $0.01(0.3)^t = 0.1$
79. $10^t = 7 \cdot 5^t$ 80. $4^t \cdot 3^t = 5$

ALGEBRA: FORM AND FUNCTION

ALGEBRA: FORM AND FUNCTION

Produced by the Calculus Consortium and initially funded by a National Science Foundation Grant.

William G. McCallum
University of Arizona

Eric Connally
Harvard University Extension

Deborah Hughes-Hallett
University of Arizona

Philip Cheifetz
Nassau Community College

Ann Davidian
Gen. Douglas MacArthur HS

Patti Frazer Lock
St. Lawrence University

David Lovelock
University of Arizona

Ellen Schmierer
Nassau Community College

Pat Shure
University of Michigan

Carl Swenson
Seattle University

Elliot J. Marks

with the assistance of

Andrew M. Gleason
Harvard University

Pallavi Jayawant
Bates College

David Lomen
University of Arizona

Karen Rhea
University of Michigan

John Wiley & Sons, Inc.

PUBLISHER	Laurie Rosatone
PROJECT EDITOR	Shannon Corliss
MARKETING MANAGER	Sarah Davis
FREELANCE DEVELOPMENTAL EDITOR	Anne Scanlan-Rohrer
SENIOR PRODUCTION EDITOR	Ken Santor
EDITORIAL ASSISTANT	Beth Pearson
MARKETING ASSISTANT	Diana Smith
COVER DESIGNER	Madelyn Lesure
COVER PHOTO	©Patrick Zephyr/Patrick Zephyr Nature Photography

This book was set in Times Roman by the Consortium using TeX, Mathematica, and the package AsTeX, which was written by Alex Kasman. It was printed and bound by R. R. Donnelley/Jefferson City. The cover was printed by R. R. Donnelley/Jefferson City. The process was managed by Elliot Marks.

This book is printed on acid-free paper.

The paper in this book was manufactured by a mill whose forest management programs include sustained yield harvesting of its timberlands. Sustained yield harvesting principles ensure that the numbers of trees cut each year does not exceed the amount of new growth.

This material is based upon work supported by the National Science Foundation under Grant No. DUE-9352905. Opinions expressed are those of the authors and not necessarily those of the Foundation.

ISBN paperbound 978-0470-52143-4
ISBN casebound 978-0471-70708-0
ISBN binder-ready 978-0470-55664-1

Printed in the United States of America

10 9 8 7 6 5 4 3 2 1

PREFACE

Algebra is fundamental to science, engineering, and business. Its efficient use of symbols to represent complex ideas has enabled extraordinary advances throughout the natural and social sciences. To be successful in any quantitative field, students need to master both symbolic manipulation and algebraic reasoning.

Balance: Manipulation and Interpretation

The fact that algebra can be encapsulated in rules sometimes encourages students to try and learn the subject merely as a set of rules. However, both manipulative skill and understanding are required for fluency. Inadequate practice in manipulation leads to frustration; inadequate attention to understanding leads to misconceptions, which easily become firmly rooted. Therefore we include both drill and conceptual exercises to develop skill and understanding together.

By balancing practice in manipulation and opportunities to see the big picture, we offer a way for teachers to help students achieve fluency. Our approach is designed to give students confidence with manipulations as well as a solid understanding of algebraic principles, which help them remember the many different manipulations they need to master.

Laying the Foundation: Expressions and Equations

We start the book by revisiting the two fundamental ideas that underpin algebra: expressions and equations. The distinction between the two is fundamental to understanding algebra—and to choosing the appropriate manipulation. After introducing each type of function, we study the types of expressions and the types of equations it generates. We pay attention to the meaning and purpose of expressions and equations in various contexts. On these foundations we proceed to study how each type of function is used in mathematical modeling.

Achieving Algebraic Power: Strategic Competence

We help students use algebra effectively by giving them practice in identifying the manipulation needed for a particular purpose. For each type of function, we give problems about recognizing algebraic forms and understanding the purpose of each form.

Functions and Modeling

Students who have a grasp of both the basic skills and the basic ideas of algebra, and a strategic sense of how to deploy them, discover a new confidence in applying their knowledge to the natural and social sciences. Mastery of algebra enables students to attack the multi-step modeling problems that we supply for each type of function.

Technology and Pedagogy

The classes who used the preliminary edition of this book were taught in a variety of pedagogical styles and a mix of lecture and discussion. The students in these classes used a range of technologies, from none to computer algebra systems. The emphasis on understanding enables this book to be used successfully with all these groups.

Student Background

We expect students who use this book to have completed high school algebra. Familiarity with basic manipulations and functional forms enables students to build on their knowledge and achieve fluency.

After completing this course, students will be well-prepared for precalculus, calculus, and other subsequent courses in mathematics and other disciplines. The focus on interpreting algebraic form, supported by graphical and numerical representations, enables students to obtain a deeper understanding of the material. Our goal is to help bring students' understanding of mathematics to a higher level. Whether students go on to use "reform" texts, or more traditional ones, this knowledge will form a solid foundation for their future studies.

Content

This content represents our vision of how algebra can be taught.

Chapter 1: The Key Concepts of Algebra

In this chapter we look at the the basic ideas of expression and equation, and at the difference between them. We discuss the underlying principles for transforming expressions, and we construct, read, and analyze examples of expressions and equations.

Chapter 2: Rules for Expressions and the Reasons for Them

This chapter reviews the rules for manipulating expressions that flow from the basic rules of arithmetic, particularly the distributive law, which provides the underlying rationale for expanding, factoring, combining like terms, and many of the manipulations of algebraic fractions.

Chapter 3: Rules for Equations and the Reasons for Them

This chapter reviews the rules for transforming equations and inequalities, laying the foundation for the more complicated methods of solving equations that are covered in later chapters. It also covers equations involving absolute values.

Chapter 4: Functions, Expressions, and Equations

In this chapter we consider functions defined by algebraic expressions and how equations arise from functions. We consider other ways of describing functions—graphs, tables, and verbal descriptions—that are useful in analyzing functions. We look at the average rate of change and conclude the chapter with a discussion of proportionality as an example of modeling with functions.

Chapter 5: Linear Functions, Expressions, and Equations

In this chapter we introduce functions that represent change at a constant rate. We consider different forms for linear expressions and what each form reveals about the function it expresses. We see how linear equations in one and two variable arise in the context of linear functions. We use linear functions to model data and applications. We conclude the chapter with a discussion of systems of linear equations.

Chapter 6: Rules for Exponents and the Reasons for Them

This chapter reviews the rules for exponents, including fractional exponents, and rules for manipulating expressions involving radicals.

Chapter 7: Power Functions, Expressions, and Equations

Power functions express relationships in which one quantity is proportional to a power of another. We relate the basic graphical properties of a power function to the properties of the exponent and use the laws of exponents to put functions in a form where the exponent can be clearly recognized. We consider equations involving power functions and conclude with applications that can be modeled by power functions.

Chapter 8: More on Functions

In this chapter we use what we have learned about analyzing algebraic expressions to study functions in more depth. We consider the possible inputs and outputs of functions (domain and range), see how functions can be built up from, and decomposed into, simpler functions, and consider how to construct inverse functions by reversing the operations from which they are made up.

Chapter 9: Quadratic Functions, Expressions, and Equations

We start this chapter by looking at quadratic functions and their graphs. We then consider the different forms of quadratic expressions—standard, factored, and vertex form—and show how each form reveals a different property of the function it defines. We consider two important techniques for solving quadratic equations: completing the square and factoring. The first technique leads to the quadratic formula.

Chapter 10: Exponential Functions, Expressions, and Equations

In this chapter we consider exponential functions such as 2^x and 3^x, in which the base is a constant and the variable is in the exponent. We show how to interpret different forms of exponential functions. For example, we see how to interpret the base to give the growth rate and how to interpret exponents in terms of growth over different time periods. We then look at exponential equations. Although they cannot be solved using the basic operations introduced so far, we show how to find qualitative information about solutions and how to estimate solutions to exponential equations graphically and numerically. We conclude with a section on modeling with exponential functions.

Chapter 11: Logarithms

In this chapter we develop the properties of logarithms from the properties of exponents and use them to solve exponential equations. We explain that logarithms do for exponential equations what taking roots does for equations involving powers: They provide a way of isolating the variable so that the equation can be solved. We consider applications of logarithms to modeling, and conclude with a section on natural logarithms and logarithms to other bases.

Chapter 12: Polynomials

In this chapter, as in the chapter on quadratics, we consider the form of polynomial expressions, including the factored form, and study what form reveals about different properties of polynomial functions. We conclude with a section on the long-run behavior of polynomials. If desired, Chapters 12 and 13 could be taught immediately after Chapter 9.

Chapter 13: Rational Functions

In this chapter we look at the graphical and numerical behavior of rational functions on both large and small scales. We examine the factored form and the quotient form of a rational function, and consider horizontal, vertical, and slant asymptotes.

Chapter 14: Summation Notation

This brief chapter of one section introduces subscripted variables and summation notation in preparation for the following three chapters.

Chapter 15: Sequences and Series

In this chapter we consider arithmetic and geometric sequences and series, and their applications. We also look briefly at recursively defined sequences.

Chapter 16: Matrices and Vectors

This chapter introduces matrices and vectors from an algebraic point of view, using concrete examples to motivate matrix multiplication and multiplication of vectors by matrices. We cover the use of matrices in describing linear equations and discuss the purpose of echelon form, concluding with an introduction to row reduction.

Chapter 17: Probability and Statistics

In this chapter we discuss the mean and standard deviation as ways of describing data sets, and provide a brief introduction to concepts of probability, including a discussion of conditional probability and independence.

Changes since the Preliminary Edition

In response to reviewer comments and suggestions, we have made the following changes since the preliminary edition.
- Added four new chapters on summation notation, sequences and series, matrices, and probability and statistics.
- Split the initial chapter reviewing basic skills into three shorter chapters on rules and the reasons for them.
- Added Solving Drill sections at the end of the chapters on linear, power, and quadratic functions. These sections provide practice solving linear, power, and quadratic equations.
- Added material on radical expressions to Chapter 6, the chapter on the exponent rules.
- Added material on solving inequalities, as well as absolute value equations and inequalities, to Chapter 3, the chapter on rules for equations.
- Added more exercises and problems throughout.

Supplementary Materials

The following supplementary materials are available for the First Edition:
- **Instructor's Manual** (ISBN 978-0470-57088-3) contains information on planning and creating lessons and organizing in-class activities. There are focus points as well as suggested exercises, problems and enrichment problems to be assigned to students. This can serve as a guide and check list for teachers who are using the text for the first time.
- **Instructor's Solution Manual** (ISBN 978-0470-57258-0) with complete solutions to all problems.
- **Student's Solution Manual** (ISBN 978-0471-71336-4) with complete solutions to half the odd numbered problems.
- **Student's Study Guide** (ISBN 978-0471-71334-0) with key ideas, additional worked examples with corresponding exercises, and study skills.

ConcepTests

ConcepTests (ISBN 978-0470-59253-3), modeled on the pioneering work of Harvard physicist Eric Mazur, are questions designed to promote active learning during class, particularly (but not exclusively) in large lectures. Our evaluation data show students taught with ConcepTests outperformed students taught by traditional lecture methods 73% versus 17% on conceptual questions, and 63% versus 54% on computational problems. A new supplement to *Algebra: Form and Function*, containing ConcepTests by section, is available from your Wiley representative.

About the Calculus Consortium

The Calculus Consortium was formed in 1988 in response to the call for change at the "Lean and Lively Calculus" and "Calculus for a New Century" conferences. These conferences urged mathematicians to re-design the content and pedagogy used in calculus. The Consortium brought together mathematics faculty from Harvard, Stanford, the University of Arizona, Southern Mississippi, Colgate, Haverford, Suffolk Community College and Chelmsford High School to address the issue. Finding surprising agreement among their diverse institutions, the Consortium was awarded funding from the National Science Foundation to design a new calculus course. A subsequent NSF grant supported the development of a precalculus and multivariable calculus curriculum.

The Consortium's work has produced innovative course materials. Five books have been published. The first edition Calculus was the most widely used of any first edition calculus text ever; the precalculus book is currently the most widely used college text in the field. Books by the Consortium have been translated into Spanish, Portuguese, French, Chinese, Japanese, and Korean. They have been used in Australia, South Africa, Turkey, and Germany.

During the 1990s, the Consortium gave more than 100 workshops for college and high school faculty, in addition to numerous talks. These workshops drew a large number of mathematicians into the discussion on the teaching of mathematics. Rare before the 1990s, such discussions are now part of the everyday discourse of almost every university mathematics department. By playing a major role in shaping the national debate, the Consortium's philosophy has had widespread influence on the teaching of mathematics throughout the US and around the world.

During the 1990s, about 15 additional mathematics faculty joined the Consortium. The proceeds from royalties earned under NSF funding were put into a non-profit foundation, which supported efforts to improve the teaching of mathematics.

Since its inception, the Calculus Consortium has consisted of members of high school, college, and university faculty, all working together toward a common goal. The collegiality of such a disparate group of instructors is one of the strengths of the Consortium.

Acknowledgments

We would like to thank the many people who made this book possible. First, we would like to thank the National Science Foundation for their trust and their support; we are particularly grateful to Jim Lightbourne and Spud Bradley.

Working with Laurie Rosatone, Jessica Jacobs, Ken Santor, Shannon Corliss, Madelyn Lesure, Anne Scanlan-Rohrer, and Sarah Davis at John Wiley is a pleasure. We appreciate their patience and imagination.

Without testing of the preliminary edition in classes across the country, this book would not have been possible. We thank Victor Akasta, Jacob Amidon, Catherine Aust, Irene Gaither, Tony Giaquinto, Berri Hsiao, Pallivi Ketkar, Richard Lucas, Katherine Nyman, Daniel Pinzon, Katherine Pinzon, Abolhassan Taghavy, Chris Wetzel, Charles Widener, Andrew Wilson, Alan Yang, and all of their students for their willingness to experiment and for their many helpful suggestions.

Many people have contributed significantly to this text. They include: Lauren Akers, Charlotte Bonner, Pierre Cressant, Tina Deemer, Laurie Delitsky, Valentina Dimitrova, Carolyn Edmund, Melanie Fulton, Tony Giaquinto, Gregory Hartman, Ian Hoover, Robert Indik, Pallavi Jayawant, Selin Kalaycioglu, Alyssa Keri, Donna Krawczyk, Lincoln Littlefield, Malcolm Littlefield, David Lomen, Kyle Marshall, Abby McCallum, Gowri Meda, Hideo Nagahashi, Judy Nguyen, Igor Padure, Alex Perlis, Maria Robinson, Elaine Rudel, Seung Hye Song, Naomi Stephen, Elias Toubassi, Laurie Varecka, Mariamma Varghese, Joe Vignolini, and Katherine Yoshiwara.

Special thanks are owed to Faye Riddle for administering the project, to Alex Kasman for his software support, and to Kyle Niedzwiecki for his help with the computers.

William G. McCallum	Ann Davidian	Ellen Schmierer
Eric Connally	Patti Frazer Lock	Pat Shure
Deborah Hughes-Hallett	David Lovelock	Carl Swenson
Philip Cheifetz	Elliot J. Marks	

To Students: How to Learn from this Book

- This book may be different from other mathematics textbooks that you have used, so it may be helpful to know about some of the differences in advance. At every stage, this book emphasizes the *meaning* (in algebraic, practical, graphical or numerical terms) of the symbols you are using. There is more emphasis on the interpretation of expressions and equations than you may expect. You will often be asked to explain your ideas in words.

    Why does the book have this emphasis? Because *understanding* is the key to being able to remember and use your knowledge in other courses and other fields. Much of the book is designed to help you gain such an understanding.

- The book contains the main ideas of algebra written in plain English. It was meant to be read by students like yourself. Success in using this book will depend on reading, questioning, and thinking hard about the ideas presented. It will be helpful to read the text in detail, not just the worked examples.

- There are few examples in the text that are exactly like the homework problems, so homework problems can't be done by searching for similar-looking "worked out" examples. Success with the homework will come by grappling with the ideas of algebra.

- Many of the problems in the book are open-ended. This means that there is more than one correct approach and more than one correct solution. Sometimes, solving a problem relies on common sense ideas that are not stated in the problem explicitly but which you know from everyday life.

- The following quote from a student may help you understand how some students feel. "I find this course more interesting, yet more difficult. Some math books are like cookbooks, with recipes on how to do the problems. This math requires more thinking, and I do get frustrated at times. It requires you to figure out problems on your own. But, then again, life doesn't come with a cookbook."

- This book attempts to give equal weight to three skills you need to use algebra successfully: interpreting form and structure, choosing the right form for a given application, and transforming an expression or equation into the right form. There are many situations where it is useful to look at the symbols and develop a strategy before going ahead and "doing the math."

- Students using this book have found discussing these problems in small groups helpful. There are a great many problems which are not cut-and-dried; it can help to attack them with the other perspectives your classmates can provide. Sometimes your teacher may organize the class into groups to work together on solving some of the problems. It might also be helpful to work with other students when doing your homework or preparing for exams.

- You are probably wondering what you'll get from the book. The answer is, if you put in a solid effort, you will get a real understanding of algebra as well as a real sense of how mathematics is used in the age of technology.

WileyPLUS

This online teaching and learning environment integrates the entire digital textbook with the most effective instructor and student resources to fit every learning style. With WileyPLUS:

- Students achieve concept mastery in a rich, structured environment that's available 24/7.
- Instructors personalize and manage their course more effectively with assessment, assignments, grade tracking, and more.

WileyPLUS can complement your current textbook or replace the printed text altogether.

For Students: Personalize the Learning Experience

Different learning styles, different levels of proficiency, different levels of preparation—each student is unique. WileyPLUS empowers each student to take advantage of his or her individual strengths:

- Students receive timely access to resources that address their demonstrated need. Practice and homework questions linked to study objectives allow students and instructors to have a better understanding of areas of mastery and areas needing improvement, including the ability to link to "self study" resources based on performance.
- Integrated multimedia resources provide multiple study-paths to fit each student's learning preferences and encourage more active learning.
- WileyPLUS includes many opportunities for self-assessment linked to the relevant portions of the text. Students can take control of their own learning and practice until they master the material.

For Instructors: Personalize the Teaching Experience

WileyPLUS empowers you with the tools and resources you need to make your teaching even more effective:

- You can customize your classroom presentation with a wealth of resources and functionality ranging from PowerPoint slides to a database of rich visuals. You can even add your own materials to your WileyPLUS course.
- With WileyPLUS you can identify those students who are falling behind and intervene accordingly, without having to wait for them to come to office hours.
- Diagnostics:
  - WileyPLUS simplifies and automates such tasks as assessing student performance, making assignments, scoring student work, keeping grades, and more.
  - WileyPLUS questions are evaluated by Maple. This means questions are graded accurately the first time, saving you time and instilling confidence in your students.
  - WileyPLUS includes a "Show Work" feature. This virtual whiteboard allows students to show their work in WileyPLUS and submit it electronically in their homework assignment. Now you can take a closer look at how students came up with their answers and see where your students may be struggling.
  - Algorithmic graphing problems in WileyPLUS allow students to draw lines and curves on graphs to answer graphing problems, simulating paper-and-pencil homework. This provides students with extensive graphing practice to improve their course grade.

Table of Contents

Chapter 1

The Key Concepts of Algebra

CONTENTS

1.1 EXPRESSIONS

An *algebraic expression* is a way of representing a calculation, using letters to stand for numbers. For example, the expression

$$\pi r^2 h,$$

which gives the volume of a cylinder of radius r and height h, describes the following calculation:

- square the radius
- multiply by the height
- multiply by π.

Notice that the verbal description is longer than the expression and that the expression enables us to see in compact form all the features of the calculation at once.

Example 1	(a) Describe a method for calculating a 20% tip on a restaurant bill and use it to calculate the tip on a bill of $8.95 and a bill of $23.70. (b) Choosing the letter B to stand for the bill amount, represent your method in symbols.
Solution	(a) Taking 20% of a number is the same as multiplying it by 0.2, so

$$\text{Tip on } \$8.95 = 0.2 \times 8.95 = 1.79 \text{ dollars}$$
$$\text{Tip on } \$23.70 = 0.2 \times 23.70 = 4.74 \text{ dollars.}$$

(b) The tip on a bill of B dollars is $0.2 \times B$ dollars. Usually in algebra we leave out the multiplication sign or represent it with a dot, so we write

$$0.2B \quad \text{or} \quad 0.2 \cdot B.$$

We call the letters r, h, and B *variables*, because they can stand for various different numbers.

Evaluating Algebraic Expressions

If we give particular values to the variables, then we can find the corresponding value of the expression. We call this *evaluating* the expression.

Example 2	Juan's total cost for 5 bags of chips at $c each and 10 bottles of soda at $s each is given by

$$\text{Total cost} = 5c + 10s.$$

If a bag of chips costs $2.99 and a bottle of soda costs $1.29, find the total cost.

Solution	We have $c = 2.99$ and $s = 1.29$, so

$$\text{Total cost} = 5c + 10s = 5 \cdot 2.99 + 10 \cdot 1.29 = 27.85 \text{ dollars.}$$

Usually, changing the value of the variables changes the value of the expression.

Example 3   Evaluate $3x - 4y$ and $4x^2 + 9x + 7y$ using

(a) $x = 2, y = -5$ (b) $x = -2, y = 3$.

Solution   (a) If $x = 2$ and $y = -5$, then

$$3x - 4y = 3 \cdot 2 - 4 \cdot (-5) \qquad = 6 + 20 \qquad = 26$$
$$4x^2 + 9x + 7y = 4 \cdot 2^2 + 9 \cdot 2 + 7(-5) = 16 + 18 - 35 = -1.$$

(b) If $x = -2$ and $y = 3$, then

$$3x - 4y = 3 \cdot (-2) - 4 \cdot 3 \qquad = -6 - 12 \qquad = -18$$
$$4x^2 + 9x + 7y = 4 \cdot (-2)^2 + 9 \cdot (-2) + 7 \cdot 3 = 16 - 18 + 21 = 19.$$

How Do We Read Algebraic Expressions?

One way to read an expression is to give a verbal description of the calculation that it represents.

Example 4   The surface area of a cylinder of radius r and height h is given by the formula

$$\text{Surface area} = 2\pi r^2 + 2\pi rh.$$

Describe in words how the surface area is computed.

Solution   First square the radius and multiply the result by 2π, then take the product of the radius and the height and multiply the result by 2π; then add the results of these two calculations.

Because algebraic notation is so compact, similar expressions can have quite different values. Paying attention to the way operations are grouped using parentheses is particularly important.

Example 5   In each of the following pairs, the second expression differs from the first by the introduction of parentheses. Explain what difference this makes to the calculation and choose values of the variables that illustrate the difference.

(a) $2x^2$ and $(2x)^2$ (b) $2l + w$ and $2(l + w)$ (c) $3 - x + y$ and $3 - (x + y)$

Solution   In each case the parentheses change the order in which we do the calculation, which changes the value of the expression.

(a) In the expression $2x^2$, we square the number x and multiply the result by 2. In $(2x)^2$, we multiply the number x by 2 then square the product. For example, if $x = 3$, the first calculation gives $2 \cdot 9 = 18$, whereas the second gives $6^2 = 36$.

(b) In $2l + w$, we begin with two numbers, l and w. We double l and add w to it. In $2(l + w)$, we add l to w, then double the sum. If $l = 3$ and $w = 4$, then the first calculation gives $6 + 4 = 10$, whereas the second gives $2 \cdot 7 = 14$.

(c) In $3 - x + y$, we subtract x from 3, then add y. In $3 - (x + y)$ we add x and y, then subtract the sum from 3. If $x = 1$ and $y = 2$, then the first calculation gives $2 + 2 = 4$, whereas the second gives $3 - 3 = 0$.

Breaking Expressions into Simpler Pieces

Longer expressions can be built up from simpler ones. To read a complicated expression, we can break the expression into a sum or product of parts and analyze each part. The parts of a sum are called *terms*, and the parts of a product are called *factors*.

Example 6   Describe how each expression breaks down into parts.

(a) $\dfrac{1}{2}h(a + b)$ (b) $3(x - y) + 4(x + y)$ (c) $\dfrac{R + S}{RS}.$

Solution   (a) This expression is the product of three factors, $1/2$, h, and $a + b$. The last factor is the sum of two terms, a and b.
(b) This expression is the sum of two terms, $3(x - y)$ and $4(x + y)$. The first term is the product of two factors, 3 and $x - y$, and the second is the product of 4 and $x + y$. The factors $x - y$ and $x + y$ are each sums of two terms.
(c) This expression is a quotient of two expressions. The numerator is the sum of two terms, R and S, and the denominator is the product of two factors, R and S.

Comparing the Value of Two Expressions

Thinking of algebraic expressions as calculations with actual numbers can help us interpret them and compare their value.

Example 7   Suppose p and q represent the price in dollars of two brands of MP3 player, where $p > q$. Which expression in each pair is larger? Interpret your answer in terms of prices.

(a) $p + q$ and $2p$ (b) $p + 0.05p$ and $q + 0.05q$ (c) $500 - p$ and $500 - q$.

Solution   (a) The first expression is the sum of p and q, whereas the second is the sum of p with itself. Since $p > q$, we have $2p = p + p > p + q$, so the second expression is larger. This says that the price of two of the expensive brand is more than the price of one of each brand.
(b) We already know $p > q$, which means $0.05p > 0.05q$ as well, so the first expression is larger. This says that the cost after a 5% tax is greater for the expensive brand than for the cheap one.
(c) Subtracting a larger number leaves a smaller result, so $500 - p < 500 - q$. This says that you have less money left over from $500 after buying the more expensive brand.

Finding the Hidden Meaning in Expressions

Just as English sentences can have a hidden meaning that emerges on a second or third reading, algebraic expressions can have hidden structure. Comparing different expressions of the same type helps reveal this structure.

Example 8   Guess possible values of x and y that make each expression have the form

$$\frac{(x + y) + xy}{2}.$$

(a) $\dfrac{(3 + 4) + 3 \cdot 4}{2}$ (b) $\dfrac{10 + 21}{2}$ (c) $\dfrac{(2r + 3s) + 6rs}{2}$

Solution   (a) If $x = 3$ and $y = 4$, then

$$\frac{(x+y)+xy}{2} = \frac{(3+4)+3\cdot4}{2}.$$

(b) Here we want

$$\frac{(x+y)+xy}{2} = \frac{10+21}{2},$$

so we want $x + y = 10$ and $xy = 21$. Possible values are $x = 7$ and $y = 3$.

(c) Here we want $x + y = 2r + 3s$ and $xy = 6rs$. Since $(2r)(3s) = 6rs$, we can choose $x = 2r$ and $y = 3s$. Notice that in this case the value we have given to x and y involves other variables, rather than specific numbers.

How Do We Construct Algebraic Expressions?

In Example 4 we converted an algebraic expression into a verbal description of a calculation. Constructing algebraic expressions goes the other way. Given a verbal description of a calculation, we choose variables to represent the quantities involved and construct an algebraic expression from the description.

Example 9   A corporate bond has a face value of p dollars. The interest each year is 5% of the face value. After t years the total interest is the product of the number of years, t, and the interest received each year. The payout is the sum of the face value and the total interest.

(a) Write an expression for the total interest after t years.
(b) Write an expression for the payout after t years.

Solution   (a) The variables are the number of years, t, and the face value, p. Because $5\% = 0.05$, the interest each year is $0.05 \cdot$ (Face value), so the total interest is given by

$$\text{Total interest} = (\text{Number of years}) \cdot 0.05 \cdot (\text{Face value}) = t0.05p = 0.05tp.$$

(Although it is not mathematically necessary to rewrite $t0.05p$ as $0.05tp$, we do so because it is customary to place constants in front of variables in a product.)

(b) The payout is the sum of the face value and the total interest and is given by

$$\text{Payout} = p + 0.05tp.$$

Example 10   A student's grade in a course depends on a homework grade, h, three test grades, t_1, t_2, and t_3, and a final exam grade, f. The course grade is the sum of 10% of the homework grade, 60% of the average of the three test grades, and 30% of the final exam grade. Write an expression for the course grade.

Solution   The grade is the addition of three terms:

$$\text{Course grade} = 10\% \text{ of homework} + 60\% \text{ of test average} + 30\% \text{ of final}$$

$$= 0.10h + 0.60\left(\frac{t_1 + t_2 + t_3}{3}\right) + 0.30f.$$

Exercises for Section 1.1

EXERCISES

In Exercises **1–7**, evaluate the expression using the values given.

1. $3x^2 - 2y^3$, $x = 3$, $y = -1$
2. $-16t^2 + 64t + 128$, $t = 3$
3. $(0.5z + 0.1w)/t$, $z = 10$, $w = -100$, $t = -10$
4. $(1/4)(x + 3)^2 - 1$, $x = -7$
5. $(a + b)^2$, $a = -5$, $b = 3$
6. $(1/2)h(B + b)$, $h = 10$, $B = 6$, $b = 8$
7. $((b - x)^2 + 3y)/2 - 6/(x - 1)$, $b = -1$, $x = 2$, $y = 1$

In Exercises **8–11**, describe the calculation given by each expression and explain how they are different.

8. $p + q/3$ and $(p + q)/3$
9. $(2/3)/x$ and $2/(3/x)$
10. $a - (b - x)$ and $a - b - x$
11. $3a + 4a^2$ and $12a^3$

In Exercises **12–15**, describe the sequence of operations that produces the expression.

12. $2(u + 1)$
13. $2u + 1$
14. $1 - 3(B/2 + 4)$
15. $3 - 2(s + 5)$

Suppose P and Q give the sizes of two different animal populations, where $Q > P$. In Exercises **16–19**, which of the given pair of expressions is larger? Briefly explain your reasoning in terms of the two populations.

16. $P + Q$ and $2P$
17. $\dfrac{P}{P + Q}$ and $\dfrac{P + Q}{2}$
18. $(Q - P)/2$ and $Q - P/2$
19. $P + 50t$ and $Q + 50t$

In Exercises **20–23**, write an expression for the sequence of operations.

20. Subtract x from 1, double, add 3.
21. Subtract 1 from x, double, add 3.
22. Add 3 to x, subtract the result from 1, double.
23. Add 3 to x, double, subtract 1 from the result.

In Exercises **24–27**, write an expression for the sales tax on a car.

24. Tax rate is 7%, price is $\$p$.
25. Tax rate is r, price is $20,000.
26. Tax rate is 6%, price is $1000 off the sticker price, $\$p$.
27. Tax rate is r, price is 10% off the sticker price, $\$p$.

28. A caterer for a party buys 75 cans of soda and 15 bags of chips. Write an expression for the total cost if soda costs s dollars per can and chips cost c dollars per bag.

29. Apples are 99 cents a pound, and pears are $1.25 a pound. Write an expression for the total cost, in dollars, of a pounds of apples and p pounds of pears.

30. (a) Write an expression for the total cost of buying 8 apples at $\$a$ each and 5 pears at $\$p$ each.
    (b) Find the total cost if apples cost $0.40 each and pears cost $0.75 each.

31. (a) Write an expression for the total cost of buying s sweatshirts at $50 each and t t-shirts at $10 each.
    (b) Find the total cost if the purchase includes 3 sweatshirts and 7 t-shirts.

32. (a) Pick two numbers, find their sum and product, and then find the average of their sum and product.
    (b) Using the variables x and y to stand for the two numbers, write an algebraic expression that represents this calculation.

PROBLEMS

33. To buy their mother a gift costing $\$c$, each child contributes $\dfrac{c}{5}$.
    (a) How many children are there?
    (b) If the gift costs $200, how much does each child contribute?
    (c) Find the value of c if each child contributes $50.
    (d) Write an expression for the amount contributed by each child if there are 3 children.

34. The number of people who attend a concert is $160 - p$ when the price of a ticket is $\$p$.
    (a) What is the practical interpretation of the 160?
    (b) Why is it reasonable that the p term has a negative sign?
    (c) The number of people who attend a movie at ticket price $\$p$ is $175 - p$. If tickets are the same price, does the concert or the movie draw the larger audience?

(d) The number of people who attend a dance performance at ticket price $\$p$ is $160 - 2p$. If tickets are the same price, does the concert or the dance performance draw the larger audience?

Oil well number 1 produces r_1 barrels per day, and oil well number 2 produces r_2 barrels per day. Each expression in Problems 35–40 describes the production at another well. What does the expression tell you about the well?

35. $r_1 + r_2$ **36.** $2r_1$ **37.** $\dfrac{1}{2}r_2$

38. $r_1 - 80$ **39.** $\dfrac{r_1 + r_2}{2}$ **40.** $3(r_1 + r_2)$

41. The perimeter of a rectangle of length l and width w is $2l + 2w$. Write a brief explanation of where the constants 2 in this expression come from.

42. A rectangle with base b and height h has area bh. A triangle with the same base and height has area $(1/2)bh$. Write a brief explanation of where the $1/2$ in this expression comes from by comparing the area of the triangle to the area of the rectangle.

In Problems 43–45 assume a person originally owes $\$B$ and has made n payments of $\$p$ each. Assume no interest is charged.

43. What does the expression $B - np$ represent in terms of the person's debt?

44. Write an expression for the number of $\$p$ payments remaining before this person pays off the debt after he has made n payments.

45. Suppose $B = 5np$. How many payments are required in total to pay it off? How many more payments must be made after n payments have been made?

In Problems 46–49, say which of the given pair of expressions is larger.

46. $10 + t^2$ and $9 - t^2$

47. $\dfrac{6}{k^2 + 3}$ and $k^2 + 3$

48. $(s^2 + 2)(s^2 + 3)$ and $(s^2 + 1)(s^2 + 2)$

49. $\dfrac{12}{z^2 + 4}$ and $\dfrac{12}{z^2 + 3}$

50. A teacher calculates the course grade by adding the four semester grades g_1, g_2, g_3, and g_4, then adding twice the final grade, f, then dividing the total by 6. Write an expression for the course grade.

51. It costs a contractor $\$p$ to employ a plumber, $\$e$ to employ an electrician, and $\$c$ to employ a carpenter.

(a) Write an expression for the total cost to employ 4 plumbers, 3 electricians, and 9 carpenters.

(b) Write an expression for the fraction of the total cost in part (a) that is due to plumbers.

(c) Suppose the contractor hires P plumbers, E electricians, and C carpenters. Write expressions for the total cost for hiring these workers and the fraction of this cost that is due to plumbers.

In Problems 52–55, decide whether the expressions can be put in the form
$$\frac{ax}{a + x}.$$
For those that are of this form, identify a and x.

52. $\dfrac{3y}{y + 3}$ **53.** $\dfrac{b^2\theta^2}{b^2 + \theta^2}$

54. $\dfrac{8y}{4y + 2}$ **55.** $\dfrac{5(y^2 + 3)}{y^2 + 8}$

In Problems 56–58, you have p pennies, n nickels, d dimes, and q quarters.

56. Write an expression for the total number of coins.

57. Write an expression for the dollar value of these coins.

58. Write an expression for the total number of coins if you change your quarters into nickels and your dimes into pennies.

59. A car travels from Tucson to San Francisco, a distance of 870 miles. It has rest stops totaling 5 hours. While driving, it maintains a speed of v mph. Give an expression for the time it takes. What is the difference in time taken between a car that travels 65 mph and a car that travels 75 mph?

60. Write an expression that is the sum of two terms, the first being the quotient of x and z, the second being the product of $L + 1$ with the sum of y and $2k$.

61. To determine the number of tiles needed to cover A square feet of wall, a tile layer multiplies A by the number of tiles in a square foot and then adds 5% to the result to allow for breakage. If each tile is a square with side length 4 inches, write an expression for the number of tiles.

62. Five quizzes are taken and the average score is x. A sixth quiz is taken and the score on it is s. Write an expression for the average score on all six quizzes.

■ In Problems **63–64**, a farm planted with A acres of corn yields b bushels per acre. Each bushel brings $\$p$ at market, and it costs $\$c$ to plant and harvest an acre of corn.

63. What do the expressions cA and pb represent for the farm?

64. Write a simplified expression for the farm's overall profit (or loss) for planting, harvesting, and selling its corn.

■ In Problems **65–67**, each row of the table is obtained by performing the same operation on all the entries of the previous row. Fill in the blanks in the table.

65.

-2	-1	0	1	2	a
-1	0	1	2	3	
$-1/2$	0	$1/2$	1	$3/2$	
$1/2$	0	$-1/2$	-1	$-3/2$	
$3/2$	1	$1/2$	0	$-1/2$	

66.

-2	-1	0	1	2	a
-1	$-1/2$	0	$1/2$	1	
1	$1/2$	0	$-1/2$	-1	
$3/2$	1	$1/2$	0	$-1/2$	

67.

1	2	3	4	5	x	$x-1$
2		6	8	10		B
3	5	7		11		

68. The volume of a cone with radius r and height h is given by the expression $(1/3)\pi r^2 h$. Write an expression for the volume of a cone in terms of the height h if the radius is equal to half of the height.

1.2 EQUATIONS

An *equation* is a statement that two expressions are equal. Equations often arise when we want to find a value of a variable that makes two different expressions have the same value. A value of the variable that makes the expressions on either side of an equation equal to each other is called a *solution* of the equation. A solution makes the equation a true statement.

Example 1   You have a $\$10$ discount certificate for a pair of pants. When you go to the store you discover that there is a 25% off sale on all pants, but no further discounts apply. For what tag price do you end up paying the same amount with each discount method?

Solution   Let p be the tag price, in dollars, of a pair of pants. With the discount certificate you pay $p - 10$, and with the store discount you pay 75% of the price, or $0.75p$. You want to know what values of p make the following statement true:

$$p - 10 = 0.75p.$$

Table 1.1 shows that you pay the same price with either discount method when the tag price is $\$40$, but different prices when $p = 20, 30, 50$ or 60.

Table 1.1   *Comparison of two discount methods*

Tag price, p (dollars)	20	30	40	50	60
Discount certificate, $p - 10$	10	20	30	40	50
Store discount, $0.75p$	15	22.50	30	37.50	45

So $p = 40$ is a solution to the equation, but the other values of p are not.

In Example 1 we found one solution to the equation, but we did not show that there were not any other solutions (although it seems unlikely). Finding all the solutions to an equation is called *solving the equation*.

In general, an equation is true for some values of the variables and false for others. To test whether a number is a solution, we evaluate the expression on each side of the equation and see if we have an equality.

Example 2   For each of the following equations, which of the given values is a solution?
(a) $3 - 4t = 5 - (2 + t)$, for the values $t = -3, 0$.
(b) $3x^2 + 5 = 8$, for the values $x = -1, 0, 1$.

Solution   (a) We have:

$$\text{for } t = -3, \text{ the equation says} \quad 3 - 4(-3) = 5 - (2 + (-3)), \quad \text{that is,} \quad 15 = 6,$$
$$\text{for } t = 0, \text{ the equation says} \quad 3 - 4 \cdot 0 = 5 - (2 + 0), \quad \text{that is,} \quad 3 = 3.$$

So $t = -3$ is not a solution to the equation, and $t = 0$ is a solution.
(b) We have:

$$\text{for } x = -1, \text{ the equation says} \quad 3(-1)^2 + 5 = 8, \quad \text{that is,} \quad 8 = 8$$
$$\text{for } x = 0, \text{ the equations says} \quad 3 \cdot 0^2 + 5 = 8, \quad \text{that is,} \quad 5 = 8$$
$$\text{for } x = 1, \text{ the equations says} \quad 3 \cdot 1^2 + 5 = 8, \quad \text{that is,} \quad 8 = 8.$$

Both $x = -1$ and $x = 1$ are solutions, but $x = 0$ is not a solution. Notice that the equation has more than one solution.

The Difference between Equations and Expressions

An equation must have an equal sign separating the expressions on either side. An expression never has an equal sign. For example, $3(x - 5) + 6 - x$ is an expression, but $3(x - 5) + 6 - x = 0$ is an equation. Although they look similar, they mean quite different things. One way to avoid confusion is to interpret the meaning of an expression or an equation.

Example 3   You have $10.00 to spend on n bottles of soda, costing $1.50 each. Are the following expressions? Equations? Give an interpretation.
(a) $1.50n$
(b) $1.50n = 6.00$
(c) $10 - 1.50n$
(d) $10 - 1.50n = 2.50$

Solution   (a) This is an expression representing the cost of n bottles of soda.
(b) This is an equation, whose solution is the number of bottles that can be purchased for $6.00.
(c) This is an expression representing the amount of money left after buying n bottles of soda.
(d) This is an equation, whose solution is the number of bottles you bought if the change you received was $2.50.

Solving Equations

In Example 2 we saw how to look for solutions to equations by trial and error. However, we cannot be sure we have all the solutions by this method. But if an equation is simple enough, we can often reason out the solutions using arithmetic.

Example 4   Find all the solutions to the following equations.

(a) $x + 5 = 17$ (b) $20 = 4a$ (c) $s/3 = 22$ (d) $g^2 = 49$

Solution   (a) There is only one number that gives 17 when you add 5, so the only solution is $x = 12$.

(b) Here the product of 4 and a is 20, so $a = 5$ is the only solution.

(c) Here 22 must be one third of s, so s must be $3 \cdot 22 = 66$.

(d) The only numbers whose squares are 49 are 7 and -7, so $g = 7$ and $g = -7$ are the solutions.

In later chapters we see how to solve more complicated equations.

Finding Solutions by Looking for Structure in Equations

Even when it is not easy to find all solutions to an equation, it is sometimes possible to see one of the solutions by comparing the structure of the expressions on each side.

Example 5   Give a solution to each equation.

(a) $\dfrac{x+1}{5} = 1$

(b) $\dfrac{x-7}{3} = 0$

(c) $9 - z = z - 9$

(d) $t^3 - 5t^2 + 5t - 1 = 0$

Solution   (a) The only way a fraction can equal one is for the numerator to equal the denominator. Therefore, $x + 1$ must be 5, so $x = 4$.

(b) The only way a fraction can equal zero is for the numerator to be zero. Therefore $x = 7$.

(c) The left side is the negative of the right side, so the equation requires a number to equal its own negative. The only such number is zero. So $9 - z$ and $z - 9$ are both zero, and $z = 9$.

(d) The constants multiplying the powers of x on the left side add to zero. Therefore, if $t = 1$,

$$t^3 - 5t^2 + 5t - 1 = 1 - 5 + 5 - 1 = 0,$$

so $t = 1$ is a solution. There could be other solutions as well.

With some equations, it is possible to see from their structure that there is no solution.

Example 6   For each of the following equations, why is there no solution?

(a) $x^2 = -4$ (b) $t = t + 1$ (c) $\dfrac{3x+1}{3x+2} = 1$ (d) $\sqrt{w+4} = -3$

Solution   (a) Since the square of any number is positive, this equation has no solutions.

(b) No number can equal one more than itself, so there are no solutions.

(c) A fraction can equal one only when its numerator and denominator are equal. Since the denominator is one larger than the numerator, the numerator and denominator can never be equal.

(d) The square root of a number must be positive. Therefore, this equation has no solutions.

What Sort of Numbers Are We Working With?

If you know about complex numbers, then you know that Example 6(a) has two solutions that are complex numbers, $2i$ and $-2i$ (we review complex numbers in Section 9.5). Thus it has no real solutions, but has two complex solutions. Unless otherwise specified, in this book we will be concerned only with solutions that are real numbers.

Using Equations to Solve Problems

A problem given verbally can often be solved by translating it into symbols and solving an equation.

Example 7   You will receive a C in a course if your average on four tests is between 70 and 79. Suppose you scored 50, 78, and 84 on the first three tests. Write an equation that allows you to determine the lowest score you need on the fourth test to receive a C. What is the lowest score?

Solution   To find the average, we add the four test scores and divide by 4. If we let g represent the unknown fourth score, then we want

$$\frac{50 + 78 + 84 + g}{4} = \frac{212 + g}{4} = 70.$$

Since the numerator must be divided by 4 to obtain 70, the numerator has to be 280. So $g = 68$.

Example 8   A drought in Central America causes the price of coffee to rise 25% from last year's price. You have a $3 discount coupon and spend $17.00 for two pounds of coffee. What did coffee cost last year?

Solution   If we let c represent the price of one pound of coffee last year, then this year's price would be 25% more, which is

$$c + 0.25c = 1.25c.$$

Two pounds of coffee would cost $2(1.25c)$. When we subtract the discount coupon, the cost would be $2(1.25c) - 3$. Since we spent $17.00, our equation is

$$2(1.25c) - 3 = 17,$$

or

$$2.5c - 3 = 17.$$

Thus

$$2.5c = 17 + 3 = 20,$$

so

$$c = 20/2.5 = 8.$$

This means that the price of coffee last year was $8 per pound.

Exercises for Section 1.2

EXERCISES

In Exercises 1–4, write an equation representing the situation if p is the price of the dinner in dollars.

1. The cost for two dinners is $18.

2. The cost for three dinners plus a $5 tip is $32.

3. The cost of a dinner plus a 20% tip is $10.80.

4. The cost of two dinners plus a 20% tip is $21.60.

In Exercises 5–9, write in words the statement represented by the equation.

5. $2x = 16$

6. $x = x^2$

7. $s + 10 = 2s$

8. $0.5t = 250$

9. $y - 4 = 3y$

In Exercises 10–15, is the value of the variable a solution to the equation?

10. $t + 3 = t^2 + 9, t = 3$ 11. $x + 3 = x^2 - 9, x = -3$

12. $x + 3 = x^2 - 9, x = 4$ 13. $\dfrac{a+3}{a-3} = 1, a = 0$

14. $\dfrac{3+a}{3-a} = 1, a = 0$

15. $4(r - 3) = 4r - 3, r = 1$

Is $t = 0$ a solution to the equations in Exercises 16–19?

16. $20 - t = 20 + t$ 17. $t^3 + 7t + 5 = 5 - \dfrac{rt}{n}$

18. $t + 1 = \dfrac{1 - t}{t}$

19. $t(1 + t(1 + t(1 + t))) = 1$

20. Which of the numbers in (a)–(d) is a solution to the equation?

$$\frac{-3x^2 + 3x + 8}{2} = 3x(x + 1) + 1$$

(a) 1   (b) 0   (c) −1   (d) 2

21. Which of the following are equations?

(a) $3(x + 5) = 6 - 2(x - 5)$
(b) $ax^2 + bx + c = 0$
(c) $5(2x - 1) + (5 - x)(x + 3)$
(d) $t = 7(t + 2) - 1$

22. (a) Construct a table showing the values of the expression $1 + 5x$ for $x = 0, 1, 2, 3, 4$.
(b) For what value of x does $1 + 5x = 16$?

23. (a) Construct a table showing the values of the expression $3 - a^2$ for $a = 0, 1, 2, 3, 4$.
(b) For what value of a does $3 - a^2 = -6$?

Solve each equation in Exercises 24–36.

24. $x + 3 = 8$ 25. $2x = 10$

26. $x - 4 = 15$ 27. $\dfrac{x}{3} = 1$

28. $12 = x - 7$ 29. $24 = 6w$

30. $\dfrac{T}{3} = 10$ 31. $20 - x = 13$

32. $5x = 20$ 33. $t + 5 = 20$

34. $w/5 = 20$ 35. $y - 5 = 20$

36. $20 = 5 - x$

PROBLEMS

37. The value of a computer t years after it is purchased is:

Value of the computer $= 3500 - 700t$.

(a) What is value of the computer when it is purchased?
(b) Write an equation whose solution is the time when the computer is worth nothing.

38. Eric plans to spend $20 on ice cream cones at $1.25 each. Write an equation whose solution is the number of ice cream cones he can buy, and find the number.

39. Hannah has $100 in a bank account and deposits $75 more. Write an equation whose solution is the amount she needs to deposit for the balance to be $300, and find the amount.

40. Dennis is on a diet of 1800 Calories per day. His dinner is 1200 Calories and he splits the remaining Calories equally between breakfast and lunch. Write an equation whose solution is the number of Calories he can eat at breakfast, and find the number.

41. A town's total allocation for firemen's wage and benefits is $600,000. If wages are calculated at $40,000 per fireman and benefits at $20,000 per fireman, write an equation whose solution is the maximum number of firemen the town can employ, and solve the equation.

A ball thrown vertically upward at a speed of v ft/sec rises a distance d feet in t seconds, given by $d = 6 + vt - 16t^2$. In Problems 42–45, write an equation whose solution is the given value. Do not solve the equation.

42. The time it takes a ball thrown at a speed of 88 ft/sec to rise 20 feet.

43. The time it takes a ball thrown at a speed of 40 ft/sec to rise 15 feet.

44. The speed with which the ball must be thrown to rise 20 feet in 2 seconds.

45. The speed with which the ball must be thrown to rise 90 feet in 5 seconds.

46. Verify that $t = 1, 2, 3$ are solutions to the equation

$$t(t - 1)(t - 2)(t - 3)(t - 4) = 0.$$

Can you find any other solutions?

Table 1.2 shows values of three unspecified expressions in x for various different values of x. Give as many solutions of the equations in Problems 47–49 as you can find from the table.

Table 1.2

x	-1	0	1	2
Expression 1	1	2	-1	0
Expression 2	1	0	-1	0
Expression 3	0	2	-1	-1

47. Expression 1 = Expression 2
48. Expression 1 = Expression 3
49. Expression 2 = Expression 3

In Problems 50–56, does the equation have a solution? Explain how you know without solving it.

50. $2x - 3 = 7$
51. $x^2 + 3 = 7$
52. $4 = 5 + x^2$
53. $2 + 5x = 6 + 5x$
54. $\dfrac{x+3}{2x+5} = 1$
55. $\dfrac{x+3}{5+x} = 1$
56. $\dfrac{x+3}{2x+6} = 1$

In Problems 57–60, explain how you can tell from the form of the equation that it has no solution.

57. $1 + 3a = 3a + 2$
58. $\dfrac{3x^2}{3x^2 - 1} = 1$
59. $\dfrac{1}{4z^2} = -3$
60. $\dfrac{a+1}{2a} = \dfrac{1}{2}$

61. Given that $x = 4$ is a solution to
$$2jx + z = 3,$$
evaluate the expression $16j + 2z$.

62. The equation
$$-x\sqrt{7 + x} = 2$$
has two solutions. Are they positive, zero, or negative? Give an algebraic reason why this must be the case. You need not find the solutions.

63. The equation $x^2 + 1 = 2x + \sqrt{x}$ has two solutions. Are they positive, zero, or negative? Give an algebraic reason why this must be the case. You need not find the solutions.

1.3 EQUIVALENT EXPRESSIONS

Two expressions are equivalent if they have the same value for every value of the variables. For example, $\dfrac{a}{2}$ is equivalent to $\dfrac{1}{2}a$ because no matter what value of a we choose, we get the same value for both expressions; dividing by 2 is the same as multiplying $1/2$. If $a = 10$, for instance, then
$$\frac{a}{2} = \frac{10}{2} = 5 \quad \text{and} \quad \frac{1}{2}a = \frac{1}{2} \cdot 10 = 5.$$

Since the variables in an expression stand for numbers, the rules of arithmetic can be used to determine when two expressions are equivalent.

Example 1   Use the rules of arithmetic to explain why $\dfrac{a}{2}$ is equivalent to $\dfrac{1}{2}a$.

Solution   We have
$$2\left(\frac{1}{2}a\right) = \left(2 \cdot \frac{1}{2}\right)a = 1a = a.$$

Dividing both sides by 2, we get
$$\frac{1}{2}a = \frac{a}{2}.$$

Example 1 illustrates that we can rewrite division as multiplication. For example, we can rewrite $12/6$ as $12 \cdot \frac{1}{6}$. In general we have

$$\frac{a}{b} = a \cdot \frac{1}{b}. \quad \text{rewrite division as multiplication}$$

If $b \neq 0$, the number $1/b$ is called the *reciprocal* of b.

At the beginning of Example 1 we regrouped the multiplication to see that $2\left(\frac{1}{2}a\right) = \left(2 \cdot \frac{1}{2}\right)a$. We review rules for transforming expressions, such as regrouping and reordering multiplication and addition, in Section 2.1.

Using Evaluation to See When Expressions Are Not Equivalent

Many common errors in algebra result from thinking that expressions are equivalent when in fact they are not. It is usually easy to find out when two expressions are not equivalent by evaluating them. If you find a value of the variable that makes the expressions have different values, then they are not equivalent.

Example 2 A student accidentally replaced the expression $(x+y)^2$ with the expression $x^2 + y^2$. Choose values of x and y that show this is wrong.

Solution Since the parentheses change the order in which we perform the operations of squaring and adding, we do not expect these two expressions to be equivalent. If, for example, $x = 4$ and $y = 3$, the expressions have different values:

$$
\begin{aligned}
(x+y)^2: \quad & (4+3)^2 = \quad 7^2 \quad = 49 \quad \text{add first, then square}\\
x^2 + y^2: \quad & 4^2 + 3^2 = 16 + 9 = 25 \quad \text{square first, then add.}
\end{aligned}
$$

Mistakes often occur in working with fractions. Again, a quick check with actual values of the variables will usually catch a mistake.

Example 3 (a) Show $\dfrac{2}{x+y}$ is not equivalent to $\dfrac{1}{x} + \dfrac{1}{y}$.

(b) Show $\dfrac{2}{x+y}$ is not equivalent to $\dfrac{2}{x} + y$.

Solution (a) One way to see this is to let $x = 1$ and $y = 1$ in each expression. We obtain

$$\frac{1}{x} + \frac{1}{y} = \frac{1}{1} + \frac{1}{1} = 1 + 1 = 2,$$

$$\frac{2}{x+y} = \frac{2}{1+1} = \frac{2}{2} = 1.$$

Since $2 \neq 1$, the expression $1/x + 1/y$ is not equivalent to $2/(x+y)$.

(b) We do not expect the two expressions to be equivalent since they represent quite different calculations:

$$\frac{2}{x+y} \quad \text{means} \quad \text{divide 2 by the sum } (x+y)$$

$$\frac{2}{x} + y \quad \text{means} \quad \text{divide 2 by } x \text{ then add } y.$$

For example, if $x = 1$ and $y = 1$, the first calculation gives 1 and the second gives 3.

Warning: Checking Values Cannot Show That Two Expressions Are Equivalent

Even if two expressions are not equivalent, it is possible that they could have the same value every now and then. You cannot show two expressions are equivalent by checking values. Rather, you need to use some reasoning such as in Example 1.

Example 4 The expressions $2x^2 - 5x + 3$ and $x^2 - 2x + 1$ have the same value at $x = 1$ and $x = 2$. Are they equivalent?

Solution We have

$$2 \cdot 1^2 - 5 \cdot 1 + 3 = 0 \qquad 1^2 - 2 \cdot 1 + 1 = 0 \quad \text{equal values at } x = 1$$
$$2 \cdot 2^2 - 5 \cdot 2 + 3 = 1 \qquad 2^2 - 2 \cdot 2 + 1 = 1 \quad \text{equal values at } x = 2.$$

However, trying one more value, $x = 3$, we find

$$2 \cdot 3^2 - 5 \cdot 3 + 3 = 6 \qquad 3^2 - 2 \cdot 3 + 1 = 4 \quad \text{different values at } x = 3.$$

So the expressions are not equivalent.

Using Equivalent Expressions

Algebra is a powerful tool for analyzing calculations and reasoning with numbers. We can use equivalent expressions to decide when seemingly different calculations give the same answer.

Example 5 To convert from miles to kilometers, Abby doubles the number of miles, m, then decreases the result by 20%. Renato first divides the number of miles by 5, and then multiplies the result by 8.

(a) Write an algebraic expression for each method.
(b) Use your answer to part (a) to decide if the two methods give the same answer.

Solution (a) Abby's method starts by doubling m, giving $2m$. She then takes 20% of the result, giving $0.2(2m)$. Finally she subtracts this from $2m$, giving $2m - (0.2)2m$. Renato's method starts by dividing m by 5, giving $m/5$, and then multiplies the result by 8, giving $8(m/5)$.

(b) Abby's expression can be simplified as follows:

$$2m - (0.2)2m = 2m - 0.4m = (2 - 0.4)m = 1.6m.$$

(The step where we rewrite $2m - 0.4m$ as $(2 - 0.4)$ uses the distributive law, which we review in Section 2.2.) Renato's method gives

$$8 \cdot \frac{m}{5} = 8 \cdot \frac{1}{5} \cdot m = \frac{8}{5} \cdot m = 1.6m.$$

So the two methods give the same answer and the expressions are equivalent.

Equivalent Expressions, Equations, and Identities

When we say two expressions, such as $x + x$ and $2x$, are equivalent we are really saying: "For all numbers x, we have $x + x = 2x$." This statement looks like an equation. In order to distinguish this use of equations, we refer to

$$x + x = 2x$$

as an *identity*. An identity is really a special equation, one that is satisfied by all values of the variables. For simple identities, we can sometimes see that the two expressions on either side are equivalent by visualizing them on the number line.

Example 6   On Figure 1.1, indicate an interval of length $1 - x$, and then use this to indicate an interval of length $1 - (1 - x)$. What two expressions does this suggest are equivalent?

Figure 1.1: Subdividing an interval

Solution   The interval of length 1 is divided into two pieces, the first one being of length x. So the second one is what is left over after you take x from 1, so it must be of length $1 - x$. But now the original piece is what is left over when you take $1 - x$ away from 1, so it must be of length $1 - (1 - x)$. See Figure 1.2.

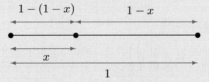

Figure 1.2: The first interval has length x
and also length $1 - (1 - x)$

Since the intervals x and $1 - (1 - x)$ are the same, this suggests that $1 - (1 - x)$ is equivalent to x. In fact

$$1 - (1 - x) = x \quad \text{for all values of } x.$$

Exercises for Section 1.3

EXERCISES

■ In Exercises 1–4, find a value of x to show that the two expressions are not equivalent.

1. $2x$ and $8x$

2. $2x + 10$ and $x + 5$

3. $x/5$ and $5/x$

4. $2x^2 - 5x$ and $25 - 3x^2$

■ In Exercises 5–14, are the expressions equivalent?

5. $a + (2 - d)$ and $(a + 2) - d$

6. $6 + r/2$ and $3 + 0.5r$

7. $3(z + w)$ and $3z + 3w$ 8. $(a - b)^2$ and $a^2 - b^2$

9. $\sqrt{a^2 + b^2}$ and $a + b$ 10. $-a + 2$ and $-(a + 2)$

11. $(3 - 4t)/2$ and $3 - 2t$ 12. $bc - cd$ and $c(b - d)$

13. $x^2 + 4x^2$, $5x^2$, and $4x^4$

14. $(x + 2)^2$, $x^2 + 4$, and $\sqrt{x^4 + 16}$

In Exercises **15–17**,

(a) Write an algebraic expression representing each of the given operations on a number b.

(b) Are the expressions equivalent? Explain what this tells you.

15. "Multiply by one fifth"

　　"Divide by five"

16. "Multiply by 0.4"

　　"Divide by five-halves"

17. "Multiply by eighty percent"

　　"Divide by eight-tenths"

18. The area of a triangle is often expressed as $A = (1/2)bh$. Is the expression $bh/2$ equivalent to the expression $(1/2)bh$?

19. Show that the following expressions have the same value at $x = -8$:

$$\frac{12 + 2x}{4 + x} \quad \text{and} \quad \frac{12}{4 - x}.$$

Does this mean these expressions are algebraically equivalent? Explain your reasoning.

Which of the equations in Exercises **20–23** are identities?

20. $x^2 + 2 = 3x$　　　　21. $2x^2 + 3x^2 = 5x^2$

22. $2u^2 + 3u^3 = 5u^5$

23. $t + 1/(t^2 + 1) = (t + 1)/(t^2 + 1)$

PROBLEMS

24. Explain why

$$\frac{5}{x - 1} + \frac{3}{1 - x} = \frac{2}{x - 1} \quad \text{for all } x \neq 1.$$

A person's monthly income is $\$I$, her monthly rent is $\$R$, and her monthly food expense is $\$F$. In Problems **25–28**, do the two expressions have the same value? If not, say which is larger, or that there is not enough information to decide. Explain your reasoning in terms of income and expenses.

25. $I - R - F$ and $I - (R + F)$

26. $12(R + F)$ and $12R + 12F$

27. $I - R - F + 100$ and $I - R - (F + 100)$

28. $\dfrac{R + F}{I}$ and $\dfrac{I - R - F}{I}$

29. To convert kilograms to pounds, Abby halves the number of kilograms, n, then subtracts 10% from the result of that calculation, whereas Renato subtracts 10% first and then halves the result.

(a) Write an algebraic expression for each method.

(b) Do the methods give the same answer?

30. Professor Priestley calculates your final grade by averaging the number of points, x, that you receive on the midterm with the number of points, y, that you receive on the final. Professor Alvorado takes half the points on each exam and adds them together. Are the two methods the same? Explain your answer algebraically.

31. Your older sister, who has more money than you, proposes that she give you half the difference between the amount of money, $\$q$, that she has and the amount of money, $\$p$, that you have. You propose instead that you give her half your money and she give you half hers. Is there any difference between the methods? Explain your answer algebraically.

32. Consider the following sequence of operations on a number n: "Add four, double the result, add the original number, subtract five, divide by three."

(a) Write an expression in n giving the result of the operations.

(b) Show that the result is always one more than the number you start with.

33. On Figure 1.3, indicate intervals of length

(a) $x + 1$　　　　(b) $2(x + 1)$

(c) $2x$　　　　　(d) $2x + 1$

What does your answer tell you about whether $2x + 1$ and $2(x + 1)$ are equivalent?

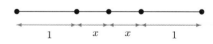

Figure 1.3

In Problems **34–36**, determine whether the sentence describes an identity.

34. Eight more than a number n is the same as two less than six times the number.

35. Twice the combined income of Carlos and Jesse equals the sum of double Carlos' income and double Jesse's income.

36. In a store, N bottles of one brand of bottled water, at $1.19 a bottle, plus twice that many bottles of another brand at $1.09 a bottle, cost $6.74.

In Problems **37–41**, complete the table. Which, if any, of the expressions in the left-hand column are equivalent to each other? Justify your answer algebraically.

37.

x	-11	-7	0	7	11
$2x$					
$3x$					
$2x + 3x$					
$5x$					

38.

t	-11	-7	0	7	11
$2t$					
$-3t$					
$2t - 3t$					
$-t$					

39.

m	-1	0	1
m^4			
$2m^2$			
$2m^2 + 2m^2$			
$4m^4$			
$4m^2$			

40.

I	-2	-1	0	1	2
$-I$					
$-(-I)$					

41.

x	-2	-1	0	1	2
$x + 3$					
$-(x + 3)$					
$-x$					
$-x + 3$					
$-x - 3$					

42. Write a sentence explaining what it means for two expressions to be equivalent.

1.4 EQUIVALENT EQUATIONS

In Example 5 of Section 1.2, the equations were simple enough to solve by direct reasoning about numbers. But how can we be sure of finding a solution to an equation, and how can we be sure that we have found all the solutions? For a more complicated equation, we try to find a simpler equation having the same solutions. A common way of doing this is to perform the same operation on both sides of the equation.

Example 1   What operation transforms the first equation into the second equation? Check to see that the solutions of the second equation are also solutions of the first equation.

(a) $2x - 10 = 12$
       $2x = 22$

(b) $5(w + 1) = 20$
       $w + 1 = 4$

(c) $3t + 58.5 = 94.5$
       $3t = 36$

(d) $a^2/1.6 = 40$
       $a^2 = 64$

Solution   (a) We add 10 to both sides of the equation to get

$$2x - 10 + 10 = 12 + 10$$
$$2x = 22.$$

The solution to the last equation is 11. Substituting $x = 11$ into the first equation gives $2(11) - 10 = 22 - 10 = 12$, so 11 is also a solution to the first equation.

(b) We divide both sides by 5 to get

$$\frac{5(w+1)}{5} = \frac{20}{5}$$
$$w + 1 = 4.$$

The solution to the last equation is 3. Substituting $w = 3$ into the first equation gives $5(3+1) = 5(4) = 20$, so 3 is also a solution to the first equation.

(c) We subtract 58.5 from both sides of the first equation to get

$$3t + 58.5 - 58.5 = 94.5 - 58.5$$
$$3t = 36.$$

The solution to the last equation is 12. Substituting $t = 12$ into the first equation gives $3(12) + 58.5 = 36 + 58.5 = 94.5$ so 12 is also a solution of the first equation.

(d) We multiply both sides of the first equation by 1.6 to get

$$1.6(a^2/1.6) = 40(1.6)$$
$$a^2 = 64.$$

There are two solutions to the second equation, 8 and -8. Substituting these values into the first equation gives $8^2/1.6 = 64/1.6 = 40$ and $(-8)^2/1.6 = 64/1.6 = 40$, so 8 and -8 are both solutions to the first equation as well.

In general,

Equivalent Equations

We say two equations are *equivalent* if they have exactly the same solutions.

How can we tell when two equations are equivalent? We can think of an equation like a scale on which things are weighed. When the two sides are equal, the scale balances, and when they are different it is unbalanced. If we change the weight on one side of the scale, we must change the other side in exactly the same way to be sure that the scale remains in the same state as before, balanced or unbalanced. Similarly, in order to transform an equation into an equivalent one, we must perform an operation on both sides of the equal sign that keeps the relationship between the two sides the same, either equal or unequal. In that way we can be sure that the new equation has the same solutions—the same values that make the scale balance—as the original one.

What Operations Can We Perform on an Equation?

In Example 1, we solved the first equation by adding 10 to both sides, and the second equation by dividing both sides by 5. In each case, we chose an operation that would make the equation simpler. Being able to anticipate the effect of an operation is an important skill in solving equations.

Example 2   Which operation should we use to solve each equation?

(a) $x + 5 = 20$ (b) $5x = 20$ (c) $x/5 = 20$

Solution   (a) In this equation, 5 is added to x, so we should subtract 5 from both sides of the equation.
(b) Because x is multiplied by 5, we should divide both sides of the equation by 5.
(c) Here x is divided by 5, so we should multiply both sides of the equation by 5.

In general,

> We can transform an equation into an equivalent equation using any operation that does not change the balance between the two sides. This includes:
> - Adding or subtracting the same number to both sides
> - Multiplying or dividing both sides by the same number, provided it is not zero
> - Replacing any expression in an equation by an equivalent expression.

These operations ensure that the original equation has the same solutions as the new equation, even though the expressions on each side might change. We explore other operations in later chapters.

Example 3   Without solving, explain why the equations in each pair have the same solution.

(a) $2.4(v - 2.1)^2 = 15$ (b) $y^3 + 4y = y^3 + 2y + 7$
        $(v - 2.1)^2 = 6.25$ $4y = 2y + 7$

Solution   (a) We divide both sides of the first equation by 2.4 to obtain the second equation.
(b) We subtract y^3 from both sides of the first equation to get the second.

Warning: Dividing Both Sides by an Expression That Might Be Zero

Not every operation that we can perform on both sides of an equation leads to an equivalent equation. If you divide both sides by an expression that could be equal to zero, then you might lose some solutions.

Example 4   What operation transforms the first equation into the second equation? Explain why this operation does not produce an equivalent equation.

$$x^2 = 3x$$
$$x = 3.$$

Solution   We divide both sides of $x^2 = 3x$ by x to obtain $x = 3$. This step does not produce an equivalent equation because x might be equal to zero. In fact, both $x = 3$ and $x = 0$ are solutions to the first equation, but only $x = 3$ is a solution of the second equation. When we divide by x, we lose one of the solutions of the original equation.

Since it is not helpful to lose solutions to an equation, dividing both sides of an equation by an expression that could take the value zero is not regarded as a valid step in solving equations.

Deciding Which Operations to Use

In Example 2 we chose an operation that produced an equation of the form $x =$ Number. We sometimes describe this as *isolating the variable* on one side of the equation. In general, we want to choose operations that head toward a form in which the variable is isolated.

When we evaluate the expression $2x + 5$, we first multiply the x by 2, then add the 5. When we solve the equation

$$2x + 5 = 13$$

we first subtract the 5 from both sides, then divide by 2. Notice that when solving the equation, we undo in reverse order the operations used to evaluate the expression.

Example 5   Solve each equation.

(a) $3x - 10 = 20$ (b) $2(x + 3) = 50$

Solution

(a) In the expression on the left side of the equal sign, we first multiply x by 3 and then subtract 10. To solve for x in the equation, we undo these operations in reverse order. We first add 10 and then divide by 3 on both sides of the equation to produce equivalent equations.

$$3x - 10 = 20$$
$$3x = 30 \quad \text{add 10 to both sides}$$
$$x = 10 \quad \text{divide both sides by 3.}$$

We check to see that 10 is a solution: $3(10) - 10 = 30 - 10 = 20$.

(b) In the expression on the left side of the equal sign, we first add 3 to x and then multiply by 2. To solve for x in the equation, we undo these operations in the reverse order. We first divide by 2 and then subtract 3 from both sides to produce equivalent equations.

$$2(x + 3) = 50$$
$$x + 3 = 25 \quad \text{divide both sides by 2}$$
$$x = 22 \quad \text{subtract 3 from both sides.}$$

We check to see that 22 is a solution: $2(22 + 3) = 2(25) = 50$.

Reasoning About Equations

In future chapters we will encounter more operations that we might want to perform on both sides of an equation to solve it. It is not always easy to decide whether operations might change the solutions, as in Example 4. Sometimes it is simpler just to think about the numbers involved in the equation.

Example 6   Solve the equation $\dfrac{1}{z} = 2.5$.

Solution

To undo the operation of dividing by z, we might think about multiplying both sides by z. Since z might be zero, this would not normally be allowed. However, in this case, z cannot be equal to zero in the original equation, since the left-hand side would be undefined. So if

$$\text{if} \quad \frac{1}{z} = 2.5 \quad \text{then} \quad 1 = z \cdot 2.5.$$

The second equation can be solved by dividing both sides by 2.5 to get

$$z = \frac{1}{2.5} = 0.4.$$

We check to see that 0.4 is a solution: $1/0.4 = 2.5$.

Equivalent Equations Versus Equivalent Expressions

Let us compare what we have learned about expressions and equations. To transform an equation into an equivalent equation, we can use any operation that does not change the equality of the two sides. However, when we solve an equation, the expressions on either side of the equal sign may not be equivalent to the previous expressions. Consider the equation

$$2x + 5 = 13,$$

whose solution is 4. If we subtract 5 from both sides of the equation, we get

$$2x = 8$$

whose solution is also 4. Although we can subtract 5 from both sides of the equation, we cannot subtract 5 from the expression $2x + 5$ alone without changing its value, because clearly $2x + 5$ does not equal $2x$. Thus, the operations we can use to create an equivalent equation include some that we cannot use to create an equivalent expression.

Example 7   (a) Is the equation $2x + 6 = 10$ equivalent to the equation $x + 3 = 5$?
(b) Is the expression $2x + 6$ equivalent to the expression $x + 3$?

Solution   (a) If we divide both sides of the first equation by 2, we have

$$\begin{aligned} 2x + 6 &= 10 \\ \frac{2x + 6}{2} &= \frac{10}{2} \quad \text{dividing both sides by 2} \\ x + 3 &= 5. \end{aligned}$$

Therefore, the two equations are equivalent. Dividing both sides of an equation by 2 produces an equivalent equation. You can verify that $x = 2$ is the solution to both equations.

(b) The expression $2x + 6$ is not equivalent to $x + 3$. This can be seen by substituting $x = 0$ into each expression. The first expression becomes

$$2x + 6 = 2(0) + 6 = 6,$$

but the second expression becomes

$$x + 3 = 0 + 3 = 3.$$

Since 6 is not equal to 3, the expressions are not equivalent. We cannot divide this expression by 2 without changing its value.

Exercises for Section 1.4

EXERCISES

In Exercises 1–8, what operation on both sides of the equation isolates the variable on one side? Give the solution of the equation.

1. $0.1 + t = -0.1$

2. $-10 = 3 + r$

3. $-t + 8 = 0$

4. $5y = 19$

5. $-x = -4$

6. $7x = 6x - 6$

7. $\dfrac{-x}{5} = 4$

8. $0.5x = 3$

Each of the equations in Exercises 9–12 can be solved by performing a single operation on both sides. State the operation and solve the equation.

9. $x + 7 = 10$

10. $3x = 12$

11. $\dfrac{x}{9} = 17$

12. $x^3 = 64$

Each of the equations in Exercises 13–16 can be solved by performing two operations on both sides. State the operations in order of use and solve the equation.

13. $2x + 3 = 13$

14. $2(x + 3) = 13$

15. $\dfrac{x}{3} + 5 = 20$

16. $\dfrac{x + 5}{3} = 20$

In Exercises 17–24, is the second equation the result of a valid operation on the first? If so, what is the operation?

17. $3 + 5x = 1 - 2x$
    $3 + 7x = 1$

18. $3 + 2x = 5$
    $3 = 2x + 5$

19. $1 - 2x^2 + x = 1$
    $x - 2x^2 = 0$

20. $\dfrac{x}{3} - \dfrac{3}{4} = 0$
    $4x - 9 = 0$

21. $9x - 3x^2 = 5x$
    $9 - 3x = 5$

22. $\dfrac{3}{4} - \dfrac{x}{3} = 2$
    $9 - 3x = 2$

23. $5x^2 - 20x = 90$
    $x^2 - 4x = 18$

24. $\dfrac{x + 2}{5} = 1 - 3x$
    $x + 2 = \dfrac{1 - 3x}{5}$

In Exercises 25–32, are the two equations equivalent? If they are, what operation transforms the first into the second?

25. $2(x + 3) = 10$
    $x + 3 = 10$

26. $2x + 5 = 22$
    $2x = 27$

27. $5 - 3x = 10$
    $5 = 10 + 3x$

28. $x^2 = 5x$
    $x = 5$

29. $3x - 6 = 10$
    $3x = 16$

30. $2x = 5x + 8$
    $3x = 8$

31. $x = a$
    $x^2 = xa$

32. $\frac{x}{3} = 12$
    $x = 36$

33. Which of the following equations have the same solution? Give reasons for your answer that do not depend on solving the equations.

I. $x + 3 = 5x - 4$

II. $x - 3 = 5x + 4$

III. $2x + 8 = 5x - 3$

IV. $10x + 6 = 2x - 8$

V. $10x - 8 = 2x + 6$

VI. $0.3 + \dfrac{x}{10} = \dfrac{1}{2}x - 0.4$

34. You can verify that $t = 2$ is a solution to the two equations $t^2 = 4$ and $t^3 = 8$. Are these equations equivalent?

Solve the equations in Exercises 35–46.

35. $5x + 12 = 90$

36. $10 - 2x = 60$

37. $3(x - 5) = 12$

38. $\dfrac{x + 2}{5} = 10$

39. $3x = 18$

40. $-2y = 14$

41. $3z = 22$

42. $x + 3 = 13$

43. $y - 7 = 21$

44. $w + 23 = -34$

45. $2x + 5 = 13$

46. $2x + 5 = 4x - 9$

PROBLEMS

■In Problems 47–54, which of the following operations on both sides transforms the equation into one whose solution is easiest to see?

(a) Add 5   (b) Add x (c) Collect like terms
(d) Multiply by 3 (e) Divide by 2

47. $x - 5 = 6$

48. $2x = 2$

49. $\dfrac{x}{3} - \dfrac{1}{3} = 0$

50. $5 - x = 0$

51. $2x - 7 - x = 3$

52. $2 - 2x = 2 - x$

53. $\dfrac{2x}{3} + \dfrac{x}{3} = 2$

54. $5 - 2x = 0$

■In Problems 55–58, the solution depends on the constant a. Assuming a is positive, what is the effect of increasing a on the solution? Does it increase, decrease, or remain unchanged? Give a reason for your answer that can be understood without solving the equation.

55. $x - a = 0$

56. $ax = 1$

57. $ax = a$

58. $\dfrac{x}{a} = 1$

59. (a) Does $x/3 + 1/2 = 4x$ have the same solution as $2x + 3 = 24x$?
    (b) Is $x/3 + 1/2$ equivalent to $2x + 3$?

60. (a) Does $8x - 4 = 12$ have the same solution as $2x - 1 = 3$?
    (b) Is $8x - 4$ equivalent to $2x - 1$?

61. Which of the equations in parts (a)–(d) are equivalent to
$$2x - (x + 3) = 4 + \frac{x - 3}{10}?$$
Give reasons for your answers.

(a) $2x - x + 3 = 4 + \dfrac{x - 3}{10}$

(b) $2x - (x + 3) = 4 + 0.1x - 0.3$
(c) $x - 3 = 3.7 + 0.1x$
(d) $20x - 10(x + 3) = 4 + x - 3$

62. The equation
$$x^2 - 5x + 6 = 0$$
has solutions $x = 2$ and $x = 3$. Is
$$x^3 - 5x^2 + 6x = 0$$
an equivalent equation? Explain your reasoning.

■Each of the equations in Problems 63–68 can be solved by performing a single operation on both sides. State the operation and solve the equation.

63. $13 = -2z$

64. $-11 = \dfrac{1}{5}s$

65. $\dfrac{3}{7}M = \dfrac{4}{3}$

66. $\sqrt{r} = 16$

67. $y^3 = -8$

68. $\dfrac{1}{B} = \dfrac{5}{14}$

69. Suppose $x = 5$ is a solution to the equation $b(x - r) = 3$. Square both sides of this equation and then add 4 to both sides. Give the resulting equation and find a solution.

70. Show that $x = 5$ is a solution to both the equations
$$2x - 10 = 0 \quad \text{and} \quad (x - 3)^2 - 4 = 0.$$
Does this mean these equations are equivalent?

71. Write a short explanation describing the difference between an expression and an equation.

REVIEW EXERCISES AND PROBLEMS FOR CHAPTER 1

EXERCISES

1. For m CDs costing $16.99 each and n DVDs costing $29.95 each, Norah's total cost is given by

$$\text{Total cost} = 16.99m + 29.95n.$$

Find her total cost if she buys 5 CDs and 3 DVDs.

2. (a) Kari has $20 and buys m muffins at $1.25 each. Write an expression for the amount of money she has left.
   (b) Find the amount remaining if she buys 9 muffins.

3. Write an expression for a person's body mass index, which is 704.5 times their weight in pounds, w, divided by the square of their height in inches, h.

4. The volume of a cone with radius r and height h is represented by the expression

$$\text{Volume of cone} = \frac{1}{3}\pi r^2 h.$$

Which has greater volume: a short fat cone with height 2 and radius 3 or a tall thin cone with height 5 and radius 1?

The surface area of a cylinder with radius r feet and height h feet is $2\pi r^2 + 2\pi rh$ square feet. In Exercises 5–8, find the surface area of the cylinder with the given radius and height.

5. Radius 5 feet and height 10 feet.

6. Radius 10 feet and height 5 feet.

7. Radius 6 feet and height half the radius.

8. Radius b feet and height half the radius.

For Exercises 9–12, write an expression that gives the result of performing the indicated operations on the number x.

9. Add 2, double the result, then subtract 4.

10. Divide by 5, subtract 2 from the result, then divide by 3.

11. Subtract 1, square the result, then add 1.

12. Divide by 2, subtract 3 from the result, then add the result to the result of first subtracting 3 from x and then dividing the result by 2.

In Exercises 13–16, evaluate the expressions given that $u = -2, v = 3, w = 2/3$.

13. $uv - vw$

14. $v + \dfrac{u}{w}$

15. $u^2 + v^2 - (u - v)^2$

16. $u^v + w^v$

In Exercises 17–20, find values of x and y to show that the two expressions are not equivalent.

17. $x + y$ and xy

18. $2x + y$ and $x + 2y$

19. $2(xy)$ and $(2x)(2y)$

20. $\sqrt{x + y}$ and $\sqrt{x} + \sqrt{y}$

In Exercises 21–26, are the expressions equivalent?

21. $2x + 6$ and $x + 3$

22. $\dfrac{1}{3}x$ and $\dfrac{1}{3x}$

23. $\dfrac{1}{5}x$ and $\dfrac{x}{5}$

24. $\dfrac{x}{3} + \dfrac{y}{3}$ and $\dfrac{x + y}{3}$

25. $\dfrac{x}{2} + \dfrac{y}{2}$ and $\dfrac{x + y}{4}$

26. $(x + y)^2$ and $x^2 + y^2$

In Exercises 27–34, for what values of t is the expression zero?

27. $t^2 - 4t$

28. $t^3 - 25t$

29. $t^3 + 16t$

30. $t(t - 3) + 2(t - 3)$

31. $4t^{1/2} - 4t$

32. $t^2(t + 5) + 9(t + 5)$

33. $t^2(t + 8) - 36(t + 8)$

34. $4t^2(2t - 3) - 36(2t - 3)$

Write an equation for each situation presented in Exercises 35–39, letting x stand for the unknown number.

35. Twice a number is 24.

36. A number increased by 2 is twice the number.

37. Six more than a number is the negative of the number.

38. Six less than a number is 30.

39. A number is doubled and then added to itself. The result is 99.

For Exercises 40–47, which of the given values of the variable are solutions?

40. $x^2 + 2 = 3x$ for $x = 0, 1, 2$

41. $2x^2 + 3x^3 = 5x^5$ for $x = -1, 0, 1$

42. $t + 1/(t^2 + 1) = (t + 1)/(t^2 + 1)$ for $t = -1, 0, 1$

43. $2(u-1)+3(u-2) = 7(u-3)$ for $u = 1, 2, 3$, and 6.5

44. $2(r - 6) = 5r + 12$, for $r = 8, -8$

45. $n^2 - 3n = 2n + 24$, for $n = 8, -3$

46. $n^2 - 3n = 2n - 24$, for $n = -8, 3$

47. $s^3 - 8 = -16$, for $s = -2, 2$

In Exercises 48–52, write in words the statement represented by the equation.

48. $1 + 3r = 15$

49. $s^3 = s^2$

50. $w + 15 = 3w$

51. $0.25t = 100$

52. $2a - 7 = 5a$

In Exercises 53–57, solve the equation.

53. $7r = 21$

54. $a + 7 = 21$

55. $J/8 = 4$

56. $b - 15 = 25$

57. $5 = 1 - d$

In Exercises 58–62, explain what operation can be used to transform the first equation into the second equation.

58. $2x + 5 = 13$ and $2x = 8$

59. $x - 11 = 26$ and $x = 37$

60. $x/2 = 40$ and $x = 80$

61. $(2y)/3 = 20$ and $y = 30$

62. $t/2 = (t + 1)/4$ and $2t = t + 1$

In Exercises 63–68, identify each description as either an expression or equation and write it algebraically using the variables given.

63. Twice n plus three more than n.

64. Twice n plus three more than n is 21.

65. The combined salary of Jason, J, and Steve, S.

66. Twice the combined salary of Jason, J, and Steve, S, is \$140,000.

67. 225 pounds is ten pounds more than Will's weight, w.

68. 50 pounds less than triple Bob's weight, w.

PROBLEMS

69. A company outsources the manufacturing of widgets to two companies, A and B, which supply a and b widgets respectively. However, 10% of a and 5% of b are defective widgets. What do the following expressions represent in practical terms?

(a) $a + b$

(b) $a/(a + b)$

(c) $10a + 5b$

(d) $0.1a$

(e) $0.1a + 0.05b$

(f) $(0.1a+0.05b)/(a+b)$

(g) $(0.9a+0.95b)/(a+b)$

70. A total of 5 bids are entered for a government contract. Let p_1 be the highest bid (in dollars), p_2 be the second highest, and so forth to p_5, the lowest bid. Explain what the following expression means in terms of the bids:

$$p_1 - \frac{p_1 + p_2 + p_3 + p_4 + p_5}{5}.$$

71. Weight Watchers© assigns points to various foods, and limits the number of points you can accumulate in a day. If c is the number of calories, g the grams of fat, and f the grams of dietary fiber, then the number of points for a piece of food is

$$\text{Number of points} = \frac{c}{50} + \frac{g}{12} - \frac{f}{4}.$$

How many grams of fiber can you trade for three grams of fat without changing the number of points?

72. If m is the number of males and f is the number of females in a population, which of the following expresses the fact that 47% of the population is male and 53% of the population is female?

(a) $P - 0.47m + 0.53f$

(b) $P = 0.53m + 0.47f$

(c) $m = 0.47P$ and $f = 0.53P$

(d) $P = 0.47m$ and $P = 0.53f$

In Problems 73–75 assume v tickets are sold for $\$p$ each and w tickets are sold for $\$q$ each.

73. What does the expression $vp + wq$ represent in terms of ticket sales?

74. Write an expression for the average amount spent per ticket.

75. Suppose $v = 2w$ and $q = 4p$. Rewrite the expression in Problem 73 in terms of w and q.

In Problems 76–82, both a and x are positive. What is the effect of increasing a on the value of the expression? Does the value increase? Decrease? Remain unchanged?

76. $ax + 1$

77. $x + a$

78. $x - a$

79. $\dfrac{x}{a} + 1$

80. $x + \dfrac{1}{a}$

81. $ax - \dfrac{1}{a}$

82. $a + x - (2 + a)$

Give possible values for A and B that make the expression

$$\frac{1 + A + B}{AB}$$

equal to the expressions in Problems 83–89.

83. $\dfrac{1 + r + s}{rs}$

84. $\dfrac{1 + 2r + 3s}{6rs}$

85. $\dfrac{1 + n + m + z^2}{(n + m)z^2}$

86. $\dfrac{1 + n + m + z^2}{n(m + z^2)}$

87. $\dfrac{1 + 2x}{x^2}$

88. $\dfrac{13}{35}$

89. 0

The expressions in Problems 90–93 can be put in the form

$$ax^2 + x.$$

For each expression, identify a and x.

90. $bz^2 + z$

91. $r(n + 1)^2 + n + 1$

92. $\dfrac{t^2}{7} + t$

93. $12d^2 + 2d$

94. Group expressions (a)–(f) together so that expressions in each group are equivalent. Note that some groups may contain only one expression.

(a) $\dfrac{2}{k} + \dfrac{3}{k}$ (b) $\dfrac{5}{2k}$ (c) $\dfrac{10}{2k}$

(d) $2.5k$ (e) $\dfrac{10k}{4}$ (f) $\dfrac{1}{2k/5}$

Verify the identities in Problems 95–96.

95. $\frac{1}{2}(a + b) + \frac{1}{2}(a - b) = a$

96. $\frac{1}{2}(a + b) - \frac{1}{2}(a - b) = b$

97. Find the expression for the volume of a rectangular solid in terms of its width w where the length l is twice the width and the height h is five more than the width.

98. Two wells produce r_1 and r_2 barrels of oil per day. Write an expression in terms of r_1 and r_2 describing a well that produces 100 fewer barrels per day than the first two wells combined.

For Problems 99–101, assume that movie tickets cost $\$p$ for adults and $\$q$ for children.

99. Write expressions for the total cost of tickets for:

(a) 2 adults and 3 kids   (b) No adults and 4 kids

(c) No kids and 5 adults   (d) A adults and C kids

100. Write expressions for the average cost per ticket for:

(a) 2 adults and 3 kids   (b) No adults and 4 kids

(c) No kids and 5 adults   (d) A adults and C kids

101. A family of two adults and three children has an entertainment budget equal to the cost of 10 adult tickets. An adult ticket costs twice as much as a child ticket. How much money will they have left after seeing two movies? Can they afford to see a third movie?

An airline has four different types of jet in its fleet:
- p 747s with a capacity of 400 seats each,
- q 757s with a capacity of 200 seats each,
- r DC9s with a capacity of 120 seats each, and
- s Saab 340s with a capacity of 30 seats each.

Answer Problems 102–103 given this information.

102. Write an expression representing the airline's total capacity (in seats).

103. To cut costs, the airline decides to maintain only two types of airplanes in its fleet. It replaces all 747s with 757s and all DC9s with Saab 340s. It keeps all the original 757s and Saab 340s. If it retains the same total capacity (in seats), write an expression representing the total number of airplanes it now has.

104. Here are the line-by-line instructions for calculating the deduction on your federal taxes for medical expenses.

1	Enter medical expenses	
2	Enter adjusted gross income	
3	Multiply line 2 by 7.5% (0.075)	
4	Subtract line 3 from line 1. If line 3 is more than line 1, enter 0.	

Write an expression for your medical deduction (line 4) in terms of your medical expenses, E, and your adjusted gross income, I.

105. You plan to drive 300 miles at 55 miles per hour, stopping for a two-hour rest. You want to know t, the number of hours the journey is going to take. Which of the following equations would you use?

(A) $55t = 190$ **(B)** $55 + 2t = 300$

(C) $55(t + 2) = 300$ **(D)** $55(t - 2) = 300$

106. A tank contains $20 - 2t$ cubic meters of water, where t is in days.

(a) Construct a table showing the number of m^3 of water at $t = 0, 2, 4, 6, 8, 10$.

(b) Use your table to determine when the tank

(i) Contains 12 m^3

(ii) Is empty.

107. A vending machine contains $40 - 8h$ bags of chips h hours after 9 am.

(a) Construct a table showing the number of bags of chips in the machine at $h = 0, 1, 2, 3, 4, 5, 6$.

(b) How many bags of chips are in the machine at 9 am?

(c) At what time does the machine run out of chips?

(d) Explain why the number of bags you found for $h = 6$ using $40 - 8h$ is not reasonable. How many bags are probably in the machine at $h = 6$?

108. (a) Does $\dfrac{x}{4} + \dfrac{1}{2} = 3x$ have the same solution as $x + 2 = 12x$?

(b) Is $\dfrac{x}{4} + \dfrac{1}{2}$ equivalent to $x + 2$?

109. (a) Does $\dfrac{5x}{3} + 2 = 1$ have the same solution as $5x + 6 = 1$?

(b) Is $\dfrac{5x}{3} + 2$ equivalent to $\dfrac{5x + 6}{3}$?

110. The equation

$$x^2 = 12 + x$$

has solutions $x = 4$ and $x = -3$. Is

$$x = \sqrt{12 + x}$$

an equivalent equation? Explain your reasoning.

111. Which of the following is *not* the result of a valid operation on the equation $3(x - 1) + 2x = -5 + x^2 - 2x$?

(a) $5x - 3 = x^2 - 2x - 5$

(b) $4x + 3(x - 1) = -5 + x^2$

(c) $3(x - 1) = -5 + x^2 - 2x + 2x$

(d) $3(x - 1) + 2x - 5 = -10 + x^2 - 2x$

112. Use the fact that $x = 2$ is a solution to the equation $x^2 - 5x + 6 = 0$ to find a solution to the equation

$$(2w - 10)^2 - 5(2w - 10) + 6 = 0.$$

■ A government buys x fighter planes at f dollars each, and y tons of wheat at w dollars each. It spends a total of B dollars, where $B = xf + yw$. In Problems **113–115**, write an equation whose solution is the given value.

113. The number of tons of wheat it can afford to buy if it spends a total of $100 million, wheat costs $300 per ton, and it must buy 5 fighter planes at $15 million each.

114. The price of fighter planes if it bought 3 of them, 10,000 tons of wheat at $500 a ton, and spent a total of $50,000,000.

115. The price of a ton of wheat if it buys 20 fighter planes and 15,000 tons of wheat for a total expenditure of $90,000,000, and a fighter plane costs 100,000 times a ton of wheat.

■ In Problems **116–118**, decide if the statement is true or false. Justify your answers using algebraic expressions.

116. The sum of three consecutive integers is a multiple of 3.

117. The sum of three consecutive integers is even.

118. The sum of three consecutive integers is three times the middle integer.

Chapter 2

Rules for Expressions and the Reasons for Them

CONTENTS

2.1   REORDERING AND REGROUPING

Since an expression represents a calculation with numbers, the rules about how we can manipulate expressions come from the rules of arithmetic, which tell us how we can rearrange calculations without changing the result. For example, since the rules of arithmetic say we can add two numbers in any order, we have

$$x + 1 = 1 + x \quad \text{for all values of } x.$$

Thus, we can replace $x + 1$ with $1 + x$ in any expression without changing its value. On the other hand, we cannot replace $x - 1$ by $1 - x$, because they usually have different values. For instance, $5 - 1 = 4$ but $1 - 5 = -4$.

We can reorder and regroup addition, and reorder and regroup multiplication, without changing the value of a numerical expression. For example, reordering allows us to write

$$3 \cdot 5 = 5 \cdot 3 \quad \text{and} \quad 3 + 5 = 5 + 3,$$

while regrouping allows us to write

$$2 \cdot (3 \cdot 5) = (2 \cdot 3) \cdot 5 \text{ and } 2 + (3 + 5) = (2 + 3) + 5.$$

In general, for reordering and regrouping respectively,

$$ab = ba \qquad \text{and} \qquad a + b = b + a, \qquad \text{for all values of } a \text{ and } b$$
$$a(bc) = (ab)c \quad \text{and} \quad a + (b + c) = (a + b) + c \quad \text{for all values of } a, b, \text{ and } c.$$

Example 1   In each of the following, an expression is changed into an equivalent expression by reordering addition, reordering multiplication, regrouping addition, regrouping multiplication, or a combination of these. Which principles are used where?

(a) $(x + 2)(3 + y) = (3 + y)(x + 2)$ (b) $(2x)x = 2x^2$
(c) $(2c)d = c(2d)$

Solution   (a) The product of the two factors on the left is reversed on the right, so multiplication is reordered.
(b) We have a product of two terms, $2x$ and x, on the left. On the right, the two xs are grouped together to make an x^2. The multiplication is regrouped.
(c) Here both regrouping and reordering of multiplication are used. First, $2c$ is rewritten as $c \cdot 2$, and then the product is regrouped. In symbols, we have

$$(2c)d = (c \cdot 2)d = c(2d).$$

Comparing Expressions Using Reordering and Regrouping

In addition to using reordering and regrouping to find equivalent expressions, we can also use them to see the difference between two expressions that are not equivalent.

Example 2   During a normal month, a bike shop sells q bicycles at p dollars each, so its gross revenue is qp dollars. During a sale month it halves the price but sells three times as many bicycles. Write an expression for the revenue during a sale month. Does the revenue change, and if so, how?

Solution

Since the bike shop sells q bicycles during a normal month, and since that number triples during a sale month, it sells $3q$ bicycles during a sale month. Similarly, the price is $(1/2)p$ during a sale month. Thus,

$$\text{Revenue during sale month} = (3q)\left(\frac{1}{2}p\right).$$

By regrouping and reordering the factors in this product we get

$$\text{Revenue during sale month} = (3q)\left(\frac{1}{2}p\right) = 3 \cdot \frac{1}{2}qp = \frac{3}{2}qp.$$

Since the original revenue is qp, the new revenue is $3/2$, or 1.5, times the original revenue.

Example 3

In a population of 100 prairie dogs there are b births and d deaths over a one-year period. Is each pair of expressions equivalent? Explain your answer in terms of population.

(a) $b + (100 - d)$ and $100 + (b - d)$ (b) $100 - (d - b)$ and $100 - d - b$

(c) $b - d/2$ and $0.5b - 0.5d$

Solution

(a) Yes. Each expression represents the new population. The first expresses this by adding to the births what is left after subtracting the deaths from the previous population of 100. The second expresses it by adding the net growth, births minus deaths, to the previous population of 100.

(b) No. The first subtracts the net loss, deaths minus births, from the population of 100, and thus represents the new population. The second subtracts both the births and the deaths from 100, giving the number of prairie dogs who survive the whole year.

(c) No. The first represents the net growth of the population if the number of deaths is cut in half, whereas the second represents the net growth if both the number of births and the number of deaths are cut in half.

Reordering and Regrouping with Subtraction

To reorder and regroup correctly with subtraction, we need to remember that subtraction can be rewritten as addition. For example, we can rewrite $3 - 5$ as $3 + (-5)$. In general:

> To rewrite subtraction as addition,
> $$a - b = a + (-b)$$

Example 4

If $x + y + z = 1$ find the value of $(x + 10) + (y - 8) + (z + 3)$.

Solution

Writing $y - 8$ as $y + (-8)$, we can reorder the terms in the expression to get

$$(x + 10) + (y - 8) + (z + 3) = x + 10 + y + (-8) + z + 3 = x + y + z + 10 + (-8) + 3.$$

Then, grouping the first and last three terms together, we find

$$(x + y + z) + (10 - 8 + 3) = (x + y + z) + 5.$$

Since $x + y + z = 1$, it follows that $(x + 10) + (y - 8) + (z + 3) = 1 + 5 = 6$.

Combining Like Terms

One purpose for reordering terms in an expression is to put like terms next to each other so that they can be combined. For example,

$$2x^2 + 3 + 5x^2 = 2x^2 + 5x^2 + 3 = (2 + 5)x^2 + 3 = 7x^2 + 3.$$

We say $2x^2$ and $5x^2$ are *like terms* because they have the same variables raised to the same powers. In contrast, the terms in the expression $3x + 4x^2$ may not be combined because the variables are raised to different powers.

Example 5   Combine like terms in each of the following expressions:

    (a) $3x^2 - 0.5x + 9x - x^2$ (b) $-z^3 + 5z^3 - 3z.$

Solution   (a) We begin by reordering:

$$3x^2 - 0.5x + 9x - x^2 = 3x^2 - x^2 - 0.5x + 9x.$$

We now regroup and combine like terms:

$$3x^2 - x^2 - 0.5x + 9x = (3x^2 - x^2) + (-0.5x + 9x) = (3 - 1)x^2 + (-0.5 + 9)x = 2x^2 + 8.5x.$$

(b) We can combine only the z^3 terms:

$$-z^3 + 5z^3 - 3z = (-1 + 5)z^3 - 3z = 4z^3 - 3z.$$

Example 6   Simplify $3(x + 3) - x - 14$.

Solution   In order to combine like terms we first have to rewrite $3(x + 3)$ as $3x + 3 \cdot 3 = 3x + 9$. Then, reordering and combining like terms, we obtain

$$3(x + 3) - x - 14 = 3x + 9 - x - 14 = \overbrace{3x - x}^{\text{Combine}} + \overbrace{9 - 14}^{\text{Combine}} = 2x - 5.$$

The rule of arithmetic that allows us to rewrite $3(x + 3)$ as $3x + 9$ is called the *distributive law*, which is the subject of the next section. In fact, combining like terms is really a special case of the distributive law.

Exercises and Problems for Section 2.1

EXERCISES

In Exercises **1–6**, are the two expressions equivalent?

  1. $x(3x)$ and $4x$

  2. $2c + d$ and $c + 2d$

  3. $5 - x$ and $-x + 5$

  4. $2xy$ and $(2x)(2y)$

  5. $(3x)(4y)(2x)$ and $24x^2y$

  6. $(x + 3)(4 + x)$ and $(x + 3)(x + 4)$

For Exercises **7–10**, is the attempt to combine like terms correct?

  7. $2x^2 + 3x^3 = 5x^5$

  8. $2AB^2 + 3A^2B = 5A^3B^3$

  9. $3h^2 + 2h^2 = 5h^2$

  10. $3b + 2b^2 = 5b^3$

In Exercises **11–18**, write the expression in a simpler form, if possible.

11. $(2x + 1) + (5x + 8)$

12. $(4 - 2x) + (5x - 9)$

13. $(x + 1) + (x + 2) + (x + 3)$

14. $(7x + 1) + (5 - 3x) + (2x - 4)$

15. $5x^2 + 5x + 3x^2$

16. $3a + 5ab + 2b$

17. $(2x)(3y) + 4x + 5y + (6x)(3y)$

18. $(2x)(5x) + (3x)(2x) + 5(3x) + x(3x)$

PROBLEMS

19. If $a + b + c = 12$, find the value of

$$(a + 5) + (b - 3) + (c + 8).$$

20. If $x + y + z = 25$, find the value of

$$(y - 10) + (z + 8) + (x - 5).$$

21. If $xyz = 100$, find the value of $(3x)(2y)(5z)$.

22. If $xyz = 20$, find the value of $(2z)(\frac{x}{4})(6y)$.

23. Rewrite the expression $a + 2(b - a) - 3(c + b)$ without using parentheses. Simplify your answer.

24. A car travels 200 miles in t hours at a speed of r mph. If the car travels half as fast but three times as long, how far does it travel? (Use the fact that distance equals speed times time.)

25. The area of a rectangle is 50 square meters. If the length is increased by 25% and the width is increased by 10%, what is the new area?

26. Quabbin Reservoir in Massachusetts provides much of Boston's water. At the start of 2009 the reservoir contained 412 billion gallons of water.

(a) If in January 2009 the amount (in billions of gallons) flowing into the reservoir is A and the amount flowing out is B, write an expression for the amount of water in the reservoir at the end of January 2009.

(b) If in February 2009 the amount flowing in is 20 billion gallons less than the previous month and the amount flowing out is 12 billion gallons more than the previous month, write an expression for the amount of water in the reservoir at the end of February 2009.

(c) Combine like terms to simplify your answer to part (b).

2.2 THE DISTRIBUTIVE LAW

In Examples 5 and 6 on page 32, we use the distributive law. The distributive law gives two different ways of calculating the same number. For example, if Abby's bill at a restaurant is B dollars and Renato's is b dollars, and they each want to pay a 20% tip, there are two ways to calculate the total tip

- The total bill is $B + b$, so the total tip is 20% of $B + b$, or $0.2(B + b)$.

- Alternatively, Abby's tip is $0.2B$ and Renato's tip is $0.2b$, so the total tip is $0.2B + 0.2b$.

The distributive law tells us that the result has to be the same either way:

$$0.2(B + b) = 0.2B + 0.2b$$

In general, we have:

The Distributive Law

For any numbers a, b, and c,
$$a(b + c) = ab + ac.$$

We say that multiplication by the number a *distributes over* the sum $b + c$.

Example 1	Use the distributive law to write two equivalent expressions for the perimeter of a rectangle, and interpret each expression in words.
Solution	Let l be the length and w the width of the rectangle. The perimeter is the sum of the lengths of the 4 sides. Since there are 2 sides of length l and 2 of length w, the perimeter is given by the expression $2l + 2w$. By the distributive law, this is equivalent to $2(l + w)$. The expression $2l + 2w$ says "double the length, double the width, and add." The expression $2(l + w)$ says "add the length to the width and double."

Distributing Over More Than Two Terms

Applying the distributive law repeatedly, we can see that multiplication distributes over a sum with any number of terms. For example,

$$a(b + c + d) = a(b + c) + ad = ab + ac + ad.$$

Example 2	Use the distributive law to rewrite each expression. (a) $6(s + 3w)$ (b) $4(s - 3t + 4w)$ (c) $a(b + c - d)$
Solution	(a) Here we have $6(s + 3w) = 6s + 18w$. (b) Distributing over the three terms in the sum, we have $4(s - 3t + 4w) = 4s - 12t + 16w$. (c) Again, distributing over three terms, we get $a(b + c - d) = ab + ac - ad$.

Distributing a Negative Sign

Since $-x = (-1)x$, the distributive law tells us how to distribute a negative sign:

$$-(b + c) = \underbrace{(-1)}_{a}(b + c)$$
$$= \underbrace{(-1)b}_{ab} + \underbrace{(-1)c}_{ac} \quad \text{by the distributive law}$$
$$= -b - c.$$

Example 3	Express $-(c - d) - (d + c)$ in a simpler way.
Solution	Distributing the negative signs, we get $$-(c - d) - (d + c) = -c + d - d - c = -2c.$$

Example 4	Let n and k be positive integers. Show that $(n - (n - k))^3$ does not depend on the value of n, but only on the value of k.

Solution   Simplifying inside the cube gives

$$n - (n - k) = n - n + k = k.$$

Thus our expression is

$$(n - (n - k))^3 = (k)^3 = k^3,$$

which depends only on the value of k, not n.

Distributing Division

Since division is multiplication by the reciprocal, the distributive law allows us to distribute division over a sum, which can be used to split apart fractions with sums in the numerator.

Example 5   Show that $\dfrac{-x + y}{-3}$ is equivalent to $\dfrac{x}{3} - \dfrac{y}{3}$.

Solution   Since dividing $(-x + y)$ by -3 is the same as multiplying it by $-1/3$, we have

$$\begin{aligned}
\frac{-x + y}{-3} &= -\frac{1}{3}(-x + y) \\
&= -\frac{1}{3} \cdot (-x + y) \\
&= -\frac{1}{3} \cdot (-x) + \left(-\frac{1}{3}\right) \cdot y \quad \text{distributing } -1/3 \\
&= \frac{1}{3} \cdot x - \frac{1}{3} \cdot y \quad \text{rewriting addition as subtraction} \\
&= \frac{x}{3} - \frac{y}{3} \quad \text{rewriting multiplication as division.}
\end{aligned}$$

Warning: You Cannot Distribute Every Operation

The distributive law allows us to distribute multiplication over a sum. A common mistake is believing that the distributive law applies to all types of expressions that involve parentheses. For example, students sometimes mistakenly distribute powers over sums, and replace $(x + y)^2$ with $x^2 + y^2$, whereas we saw in Example 2 on page 14 that these expressions are not equivalent.

Example 6   Are the two expressions equivalent?

(a) $10(x + y)$ and $10x + 10y$
(b) $2(xy)$ and $2x2y$.
(c) $\sqrt{s + t}$ and $\sqrt{s} + \sqrt{t}$

Solution   (a) These two expressions are equivalent by the distributive law.
(b) Since the expression inside parentheses is a product, not a sum, the distributive law does not apply. If $x = 5$ and $y = 3$, then $2(xy) = 2(5 \cdot 3) = 30$, while $2 \cdot 5 \cdot 2 \cdot 3 = 60$. In general, multiplication does not distribute over products.
(c) Let $s = 9$ and $t = 16$. Then $\sqrt{s + t} = \sqrt{9 + 16} = \sqrt{25} = 5$, while $\sqrt{s} + \sqrt{t} = 3 + 4 = 7$. The expressions are not equivalent; taking square roots does not distribute over sums.

Using the Distributive Law to Interpret Expressions

By giving us two different ways of calculating the same result, the distributive law allows us to interpret equivalent expressions.

Example 7   In the expression for a student's grade from Example 10 on page 5, the term

$$0.6\left(\frac{t_1 + t_2 + t_3}{3}\right),$$

represents the contribution from the three test scores t_1, t_2, and t_3. Show that this expression is equivalent to

$$0.2t_1 + 0.2t_2 + 0.2t_3,$$

and explain what this means in terms of grades.

Solution   Using the distributive law, we have

$$0.6\left(\frac{t_1 + t_2 + t_3}{3}\right) = 0.6\left(\frac{1}{3}\right)(t_1 + t_2 + t_3) = 0.2(t_1 + t_2 + t_3) = 0.2t_1 + 0.2t_2 + 0.2t_3.$$

Because

$$0.60\left(\frac{t_1 + t_2 + t_3}{3}\right) \quad \text{and} \quad 0.2t_1 + 0.2t_2 + 0.2t_3$$

have the same value, they are equivalent expressions. This says that if the average of three tests counts for 60%, then each test alone counts for 20%.

Whether you choose the form where the multiplication has been distributed or not depends on how you want to use an expression.

Example 8   Which of the two expressions in Example 7 would you use if you

(a) knew all of the test scores and wanted to compute their contribution to your final grade?
(b) wanted to know the effect on your final grade of getting 10 more points on the second test?

Solution   (a) In this case it would take fewer steps to average all the grades first and then multiply by 0.60, so you would use the first expression,

$$0.60\left(\frac{t_1 + t_2 + t_3}{3}\right).$$

(b) In this case the second expression is useful, because it tells us that the contribution of the second test to the course grade is $0.2t_2$. So if we add 10 to t_2 then that contribution becomes

$$0.2(t_2 + 10) = 0.2t_2 + 0.2 \cdot 10 = 0.2t_2 + 2.$$

The coefficient of 0.2 next to the t_2 tells us that increasing the score on the second exam by 10 increases the course grade by 2 points.

Taking Out a Common Factor: Using the Distributive Law in Both Directions

Sometimes we use the distributive law from left to right:

$$a(b + c) \longrightarrow ab + ac.$$

We call this distributing a over $b + c$. Other times, we use the distributive law from right to left:

$$a(b + c) \longleftarrow ab + ac.$$

We call this *taking out a common factor* of a from $ab + ac$.

Example 9   Take out a common factor in   (a) $2x + xy$ (b) $-wx - 2w$ (c) $12lm^2 - m^2$.

Solution
(a) We factor out an x, giving $2x + xy = x(2 + y)$.
(b) We factor out the $-w$, giving $-w(x + 2)$.
(c) Here there is a common factor of m^2. Factoring it out gives $m^2(12l - 1)$.

In the expression $a(b + c)$, the quantities $a, b,$ and c often stand for more complicated terms than single letters or numbers. Nevertheless, the the distributive law $a(b + c) = ab + ac$ still can be applied.

Example 10   Use the distributive law to rewrite each expression.
(a) $6xy - 2xz$ (b) $12pq + 3p$ (c) $-4xyz - 8xy$
(d) $-ab + a^2b - ab^2$ (e) $t(p + r) - 7(p + r)$ (f) $w^2(a+2)+w(a+2)-(a+2)$

Solution
(a) Writing $6xy$ as $(2x)(3y)$, we see that both terms have a common factor of $2x$, so

$$6xy - 2xz = 2x(3y - z).$$

(b) Both terms have a factor of $3p$, so

$$12pq + 3p = 3p(4q + 1).$$

(c) Taking out a common factor of $-4xy$ we have

$$-4xyz - 8xy = -4xy(z + 2).$$

(d) Each term has a factor ab, so we have

$$-ab + a^2b - ab^2 = ab(-1 + a - b).$$

We can instead take out a common factor of $-ab$

$$-ab + a^2b - ab^2 = -ab(1 - a + b).$$

(e) Notice that each term has a factor $(p + r)$, so

$$t(p + r) - 7(p + r) = (p + r)(t - 7).$$

(f) Taking out a common factor of $a + 2$ we get

$$w^2(a + 2) + w(a + 2) - (a + 2) = (a + 2)(w^2 + w - 1).$$

Exercises and Problems for Section 2.2

EXERCISES

In Exercises 1–8, use the distributive law to rewrite each expression as an equivalent expression with no parentheses.

1. $2(x + 3y)$

2. $3x(x + 4)$

3. $-5(2x - 3)$

4. $3ab(2a - 5b)$

5. $2x(x^2 - 3x + 4)$

6. $-3x(5 - 3x - 2x^2)$

7. $2(5x - 3y) + 5$

8. $3(C - D)D$

In Exercises 9–19, rewrite the expression by taking out the common factors.

9. $2ax - 3bx$

10. $5x + 100$

11. $\dfrac{x}{5} + \dfrac{y}{5}$

12. $2x^2 - 6x$

13. $-m^2n - 3mn^2$

14. $9x^2 + 18x + 3$

15. $-4a^2b - 6ab^2 - 2ab$

16. $5x(x + 1) + 7(x + 1)$

17. $b(b + 3) - 6(b + 3)$

18. $6r(s - 2) - 12(s - 2)$

19. $4ax(x+4) - 2x(x+4)$

In Exercises 20–27, are the two expressions equivalent?

20. $-2(x - 4)$ and $-2x - 8$

21. $3x^2 + 6x + 3$ and $3(x^2 + 2x)$

22. $ab(a + b + 1)$ and $a^2b + ab^2 + ab$

23. $5x + 100$ and $5(x + 500)$

24. $(x + 2)(x - 3)$ and $x(x - 3) + 2(x - 3)$

25. $2(3x \cdot 4y)$ and $6x \cdot 8y$

26. $(x + y)^3$ and $x^3 + y^3$

27. $\dfrac{a}{b + c}$ and $\dfrac{a}{b} + \dfrac{a}{c}$

PROBLEMS

28. If $pqr = 17$, what is $p(2qr + 3r) + 3r(pq - p)$?

29. If $a - b + c = 17$, what is $2(a+1) - (b+3) + (2c-b)$?

30. Which of the following expressions is equivalent to $3(x^2 + 2) - 3x(1 - x)$?

    (i) $6 + 3x$
    (ii) $-3x + 6x^2 + 6$
    (iii) $3x^2 + 6 - 3x - 3x^2$
    (iv) $3x^2 + 6 - 3x$

31. An order is placed for n items each costing p dollars and twice as many items each costing $1 more. Write a simplified expression for the total cost of the order.

32. In January, a company's three factories produce q units, r units, and s units of a product. In February, the company doubles its output of the product.

    (a) Write the expression for February's output if we take the total output in January and double it.
    (b) Write the expression for February output if we double the output at each factory and add them up.
    (c) Are the expressions in parts (a) and (b) equivalent? Explain.

In Problems 33–38, the expression is equivalent to an expression in the form $k(x + A)$. Write the expression in this form, and give the values of k and A.

33. $2x + 50$

34. $3x - 18$

35. $15 - 5x$

36. $0.05x + 100$

37. $0.2x - 60$

38. $50 - 0.1x$

39. Is the fraction $\dfrac{x + 3}{x}$ equivalent to $1 + \dfrac{3}{x}$?

40. Is the fraction $\dfrac{x}{x + 3}$ equivalent to $1 + \dfrac{x}{3}$?

41. Complete the table. Are the two expressions in the left column equivalent? Justify your answer.

a	-2	-1	0	1	2
$-(1/2)(a + 1) + 1$					
$-(1/2)a + (1/2)$					

In Problems 42–43, explain how the distributive law $a(b + c) = ab + ac$ has been used in the identity.

42. $(2x + 3)^3 = (2x + 3)^2 \cdot 2x + (2x + 3)^2 \cdot 3$

43. $x^2(x + r + 3) = x^2(x + r) + 3x^2$

44. A contractor is managing three different job sites. It costs her $c to employ a carpenter, $p to employ a plumber, and $e to employ an electrician. The total cost to employ carpenters, plumbers, and electricians at each site is

$$\text{Cost at site } 1 = 12c + 2p + 4e$$
$$\text{Cost at site } 2 = 14c + 5p + 3e$$
$$\text{Cost at site } 3 = 17c + p + 5e.$$

Write expressions in terms of c, p, and e for:

(a) The total employment cost for all three sites.

(b) The difference between the employment cost at site 1 and site 3.

(c) The amount remaining in the contractor's budget after accounting for the employment cost at all three sites, given that originally the budget is $50c + 10p + 20e$.

2.3 EXPANDING AND FACTORING

In Section 2.2 we use the distributive law to expand products like $a(b + c)$ and to take common factors out of expressions like $ab + ac$. Now we consider products in which both factors are sums of more than one term. Such products can be multiplied out or *expanded* by repeated use of the distributive law.

Example 1 Expand the product $(x - 4)(x + 6)$.

Solution When we expand, we can combine the like terms, $-4x$ and $6x$, giving

$$(x - 4)(x + 6) = x^2 - 4x + 6x - 24 = x^2 + 2x - 24.$$

In general, we have:

Expanding Using the Distributive Law

$$(x + r)(y + s) = (x + r)y + (x + r)s \quad \text{distribute } x + r \text{ over } y + s$$
$$= xy + ry + xs + rs \quad \text{use the distributive law twice more}$$

Notice that in the final result, each term in the first factor, $x + r$, has been multiplied by each term in the second factor, $y + s$.

Quadratic Expressions

An expression of the form

$$ax^2 + bx + c, \quad \text{where } a, b, \text{ and } c \text{ are constants,}$$

is called a *quadratic expression* in x. The constants a, b, and c are called the *coefficients* in the expression. Expanding a product of the form $(x + r)(x + s)$ always gives a quadratic expression:

$$(x + r)(x + s) = (x + r)x + (x + r)s$$
$$= x^2 + rx + sx + rs$$
$$= x^2 + (r + s)x + rs.$$

Notice that the sum $r + s$ is the coefficient of the x term, and the product rs is the constant term. This gives a quick way to expand $(x + r)(x + s)$. For instance, in Example 1, we have $-4 + 6 = 2$, giving a term $2x$, and $-4 \cdot 6 = -24$, giving a constant term -24.

Going in the Other Direction: Factoring Quadratic Expressions

What if we are given a quadratic expression and want to express it as the product of two linear expressions? One approach to writing a quadratic expression in the form $(x + r)(x + s)$ is to find r and s by trying various values and seeing which ones fit, if any.

Example 2   If possible, factor into $(x + r)(x + s)$, where r and s are integers:

(a) $x^2 - 9x + 18$ (b) $x^2 + 3x + 4$.

Solution   (a) If $x^2 - 9x + 18 = (x + r)(x + s)$, then we must have $r + s = -9$ and $rs = 18$, so we look for numbers satisfying these conditions. The numbers $r = -3$ and $s = -6$ satisfy both conditions, so $x^2 - 9x + 18 = (x - 3)(x - 6)$.

(b) If $x^2 + 3x + 4 = (x + r)(x + s)$, then we must have $r + s = 3$ and $rs = 4$. There are no integers that satisfy both conditions. Therefore, $x^2 + 3x + 4$ cannot be factored in this way.

Example 3   Factor $x^2 + 10xy + 24y^2$.

Solution   If the expression $x^2 + 10xy + 24y^2$ can be factored, it can be written as $(x + ry)(x + sy)$, where $r + s = 10$ and $rs = 24$. Since 4 and 6 satisfy these conditions, we have

$$x^2 + 10xy + 24y^2 = (x + 4y)(x + 6y).$$

What If the Coefficient of x^2 Is Not 1?

Expanding products of the form $(px + r)(qx + s)$ gives a quadratic expression in which the coefficient of x^2 is not necessarily 1.

Example 4   Expand $(2x + 1)(x + 3)$.

Solution   Expanding using the distributive law gives

$$\begin{aligned}
(2x + 1)(x + 3) &= x(2x + 1) + 3(2x + 1) \\
&= 2x^2 + x + 6x + 3 \\
&= 2x^2 + 7x + 3.
\end{aligned}$$

To factor the expression $2x^2 + 7x + 3$, we can try to reverse the steps in Example 4.

Example 5   Factor $2x^2 + 7x + 3$.

Solution   Following Example 4, we break $7x$ into $x + 6x$, so

$$2x^2 + 7x + 3 = 2x^2 + x + 6x + 3.$$

Then we group the terms into pairs and pull out a common factor from each pair:

$$2x^2 + x + 6x + 3 = (2x^2 + x) + (6x + 3)$$
$$= x(2x + 1) + 3(2x + 1)$$
$$= (2x + 1)(x + 3).$$

We call this method *factoring by grouping*. Notice that this works because we get a common factor $(2x + 1)$ at the last stage. If we had split the $7x$ up differently, this might not have happened.

How would we know to write $7x = x + 6x$ in Example 5, without Example 4 to guide us? There is a method that always leads to the right way of breaking up the x term, if it can be done at all. In the expression $ax^2 + bx + c$ that we want to factor, we form the product of the constant term and the coefficient of x^2. In Example 5, this is $2 \cdot 3 = 6$. Then we try to write the x term as a sum of two terms whose coefficients multiply to this product. In Example 5, we find $7x = x + 6x$ works, since $1 \cdot 6 = 6$.

Example 6   Factor $8x^2 + 14x - 15$.

Solution   We multiply the coefficient of the x^2 term by the constant term: $8 \cdot (-15) = -120$. Now we try to write $14x$ as a sum of two terms whose coefficients multiply to -120. Writing $14x = -6x + 20x$ works, since $-6 \cdot 20 = -120$. Breaking up the $14x$ in this way enables us to factor by grouping:

$$8x^2 + 14x - 15 = 8x^2 - 6x + 20x - 15$$
$$= (8x^2 - 6x) + (20x - 15)$$
$$= 2x(4x - 3) + 5(4x - 3)$$
$$= (4x - 3)(2x + 5).$$

Here is the general process:

Factoring Quadratic Expressions

If a quadratic expression is factorable, the following steps work:
- Factor out all common constant factors, giving $k(ax^2 + bx + c)$.
- In the remaining expression, multiply the coefficient of the x^2 term by the constant term, giving ac.
- Find two numbers that multiply to ac and sum to b, the coefficient of the x term.
- Break the middle term, bx, into two terms using the result of the previous step.
- Factor the four terms by grouping.

Example 7   Factor $12x^2 - 44x + 24$.

Solution   First, take out the common factor of 4. This gives

$$12x^2 - 44x + 24 = 4(3x^2 - 11x + 6).$$

To factor $3x^2 - 11x + 6$, we multiply 3 and 6 to get 18, and then we look for two numbers that multiply to 18 and sum to -11. We find that -9 and -2 work, so we write $-11x = -9x - 2x$. This gives

$$\begin{aligned}
3x^2 - 11x + 6 &= 3x^2 - 9x - 2x + 6 \\
&= (3x^2 - 9x) + (-2x + 6) \\
&= 3x(x - 3) - 2(x - 3) \\
&= (x - 3)(3x - 2).
\end{aligned}$$

Therefore,

$$12x^2 - 44x + 24 = 4(x - 3)(3x - 2).$$

In Example 7, how can we find the numbers -9 and -2 if they don't jump out at us? A systematic method is to list all the ways of factoring 18 into two integer factors and to pick out the pair with the correct sum. The factorizations are

$$18 = 18 \times 1 = 6 \times 3 = 9 \times 2 = -18 \times -1 = -6 \times -3 = -9 \times -2.$$

Only the last pair, -9 and -2, sums to -11.

All the methods of factorization in this section are aimed at finding factors whose coefficients are integers. If we allow square roots, further factorizations might be possible.

Perfect Squares

If the two factors in a quadratic expression are the same, then the expanded form has a particular shape. For example,

$$\begin{aligned}
(2z + 5)^2 &= (2z + 5)(2z + 5) \\
&= 2z(2z + 5) + 5(2z + 5) \\
&= 4z^2 + 10z + 10z + 25 \\
&= 4z^2 + 20z + 25.
\end{aligned}$$

Notice that the first and last term of the final expression are squares of the first and last term of the expression $2z + 5$. The middle term is twice the product of the first and last term. This pattern works in general:

Perfect Squares

$$x^2 + 2rx + r^2 = (x + r)^2$$

$$x^2 - 2rx + r^2 = (x - r)^2$$

We can sometimes recognize an expression as a perfect square and thus factor it. The clue is to recognize that two terms of a perfect square are squares of other terms and that the third term is twice the product of those two other terms.

Example 8

If possible, factor as perfect squares.

(a) $4r^2 + 10r + 25$ 　　　　(b) $9p^2 + 60p + 100q^2$ 　　　　(c) $25y^2 - 30yz + 9z^2$

Solution

(a) In the expression $4r^2 + 10r + 25$ we see that the first and last terms are squares:

$$4r^2 = (2r)^2 \quad \text{and} \quad 25 = 5^2.$$

Twice the product of $2r$ and 5 is $2(2r)(5) = 20r$. Since $20r$ is not the middle term of the expression to be factored, $4r^2 + 10r + 25$ does not appear to be a perfect square.

(b) We have

$$9p^2 = (3p)^2 \quad \text{and} \quad 100q^2 = (10q)^2 \quad \text{and} \quad 2(3p)(10q) = 60pq.$$

Since the middle term of does not contain a q, it appears that $9p^2 + 60p + 100q^2$ is not a perfect square.

(c) In the expression $25y^2 + 30yz + 9z^2$ we see that the first and last terms are squares:

$$25y^2 = (5y)^2 \quad \text{and} \quad 9z^2 = (3z)^2.$$

Twice the product of $5y$ and $3z$ is $2(5y)(3z) = 30yz$, which is the opposite of the middle term of the expression to be factored. However, we can also write

$$9z^2 = (-3z)^2,$$

and then twice the product of $5y$ and $-3z$ is $2(5y)(-3z) = -30yz$. Thus, $25y^2 - 30yz + 9z^2$ is a perfect square, and $25y^2 - 30yz + 9z^2 = (5y - 3z)^2$.

In summary:

Determining if an Expression is a Perfect Square

An expression with three terms is a perfect square if
- Two terms are squares, and
- The third term is twice the product of the expressions whose squares are the other terms.

Difference of Squares

Another special form for a quadratic expression is the difference of two squares:

Difference of Squares

If an expression is in the form $x^2 - r^2$, it can be factored as

$$x^2 - r^2 = (x - r)(x + r).$$

Notice that the expanded form of the expression has no x term, because the terms rx and $-rx$ in the expansion cancel each other out. If we see the special from $x^2 - r^2$, we know it can be factored as $(x - r)(x + r)$.

Example 9   If possible, factor the following expressions as the difference of squares.

(a) $x^2 - 100$
(c) $49y^2 + 25$
(b) $8a^2 - 2b^2$
(d) $18(t + 1)^2 - 32$

Solution   (a) We see that $100 = 10^2$, so $x^2 - 100$ is a difference of squares and can be factored as

$$x^2 - 100 = (x - 10)(x + 10).$$

(b) First we take out a common factor of 2:

$$8a^2 - 2b^2 = 2(4a^2 - b^2).$$

Since $4a^2$ is the square of $2a$, we see that $4a^2 - b^2$ is a difference of squares. Therefore,

$$8a^2 - 2b^2 = 2(2a + b)(2a - b).$$

(c) We have $49y^2 = (7y)^2$ and $25 = 5^2$. However, $49y^2 + 25$ is a sum of squares, not a difference. Therefore it cannot be factored as the difference of squares.

(d) We take out a common factor of 2:

$$2(9(t + 1)^2 - 16).$$

Since $3(t+1)$ squared is $9(t+1)^2$ and 4 squared is 16, we see that $9(t+1)^2 - 16$ is a difference of squares. Hence,

$$18(t + 1)^2 - 32 = 2(3(t + 1) - 4)(3(t + 1) + 4).$$

Using the distributive property and collecting like terms, we have

$$18(t + 1)^2 - 32 = 2(3(t + 1) - 4)(3(t + 1) + 4) = 2(3t - 1)(3t + 7).$$

Example 10   Factor $4a^2 + 4ab + b^2 - 4$.

Solution   Since there are four terms with no common factors, we might try to group any two terms that have a common factor and hope that a common expression will be in each term. One possibility is

$$4a^2 + 4ab + b^2 - 4 = (4a^2 + 4ab) + (b^2 - 4) = 4a(a + b) + (b^2 - 4).$$

Another possibility is

$$4a^2 + 4ab + b^2 - 4 = (4a^2 - 4) + (4ab + b^2) - 4(a^2 - 1) + b(4a + b).$$

The third possibility is

$$4a^2 + 4ab + b^2 - 4 = (4a^2 + b^2) + (4ab - 4) = (4a^2 + b^2) + 4(ab - 1).$$

Unfortunately, none of these groupings leads to a common factor. Another approach is to notice that the first three terms form a perfect square. Therefore,

$$
\begin{aligned}
4a^2 + 4ab + b^2 - 4 &= (4a^2 + 4ab + b^2) - 4 && \text{grouping the first three terms} \\
&= (2a + b)^2 - 4 && \text{factoring the perfect square} \\
&= ((2a + b) - 2)((2a + b) + 2) && \text{factoring the difference of squares} \\
&= (2a + b - 2)(2a + b + 2).
\end{aligned}
$$

Factoring Expressions That Are Not Quadratic

Sometimes, taking out a common factor from an expression leaves a quadratic expression that can be factored.

Example 11   Factor (a) $x^3 - 6x^2 - 16x$ (b) $p^4 + p^2 - 2$.

Solution   (a) We first take out a common factor of x, giving

$$
x^3 - 6x^2 - 16x = x(x^2 - 6x - 16).
$$

The quadratic expression on the right factors as $x^2 - 6x - 16 = (x + 2)(x - 8)$, so

$$
x^3 - 6x^2 - 16x = x(x - 8)(x + 2).
$$

(b) By recognizing p^4 as $(p^2)^2$, we can see this as a quadratic expression in p^2:

$$
p^4 + p^2 - 2 = (p^2)^2 + p^2 - 2.
$$

If we think of p^2 as x, then this has the form $x^2 + x - 2$, which factors into $(x - 1)(x + 2)$. So

$$
p^4 + p^2 - 2 = (p^2 - 1)(p^2 + 2).
$$

The first factor is a difference of two squares, and factors even further, giving

$$
p^4 + p^2 - 2 = (p - 1)(p + 1)(p^2 + 2)
$$

Exercises and Problems for Section 2.3

EXERCISES

In Exercises 1–25, expand and combine like terms.

1. $(x + 5)(x + 2)$

2. $(y + 3)(y - 1)$

3. $(z - 5)(z - 6)$

4. $(2a + 3)(3a - 2)$

5. $(3b + c)(b + 2c)$

6. $(a + b + c)(a - b - c)$

7. $3(x - 4)^2 + 8x - 48$

8. $(x + 6)^2$

9. $(x - 8)^2$

10. $(x + 11)^2$

11. $(x - 13)^2$

12. $(x + 7)(x - 7)$

13. $(x + y)^2$

14. $(2a + 3b)^2$

15. $(5p^2 - q)^2$

16. $(x - y)^3$

17. $(2a - 3b)^3$

18. $(x + 9)(x - 9)$

19. $(x - 8)(x + 8)$

20. $(x - 12)(x + 12)$

21. $(x - 11)(3x + 2)$

22. $x(4x - 7)(2x + 2)$

23. $x(3x + 7)(5x - 8)$

24. $2x(5x + 8)(7x + 2)$

25. $(s - 3)(s + 5) + s(s - 2)$.

■ Factor the expressions in Exercises 26–57.

26. $x^2 + 5x + 6$

27. $y^2 - 5y - 6$

28. $n^2 - n - 30$

29. $g^2 - 12g + 20$

30. $t^2 - 27t + 50$

31. $q^2 + 15q + 50$

32. $b^2 + 2b - 24$

33. $x^2 + 11xy + 24y^2$

34. $2z^2 + 12z - 14$

35. $4z^2 + 19z + 12$

36. $y^2 - 6y + 7$

37. $3w^2 + 12w - 36$

38. $2n^2 - 12n - 54$

39. $a^2 - a - 16$

40. $x^2 - 16$

41. $s^2 - 12st + 36t^2$

42. $x^2 + 7x$

43. $x^2 - 36$

44. $x^2 - 169$

45. $x^2 + 10x + 25$

46. $x^2 + 26x + 169$

47. $x^2 + 15x + 56$

48. $x^2 - 19x + 90$

49. $5x^2 - 37x - 72$

50. $x^2 - 5x + dx - 5d$

51. $8x^2 - 4xy - 6x + 3y$

52. $2qx^2 + pqx - 14x - 7p$

53. $x^3 - 16x^2 + 64x$

54. $x^6 - 2x^3 - 63$

55. $r^3 - 14r^2s^3 + 49rs^6$

56. $8a^3 + 50ab^2$

57. $18x^7 + 48x^4z^2 + 32xz^4$

PROBLEMS

■ In Problems 58–68, factor each expression completely.

58. $ay - a^3y^3$

59. $18x^2 - 2x^4z^6$

60. $(a + b)^2 - 100$

61. $12 - 27(t + 1)^2$

62. $x^2 + 8x + 16 - y^2$

63. $q^4 - q^8$

64. $(t + 1)^3 - 25(t + 1)$

65. $2w^3 - 16w^2 + 32w$

66. $12a^2 + 60a + 75$

67. $16s^3t - 24s^2t^2 + 9st^3$

68. $(r + 1)^2 + 12t(r + 1) + 36t^2$

■ For Problems 69–71, expand and combine like terms.

69. $((x + h) + 1)((x + h) - 1)$

70. $(2 + 3(a + b))^2$

71. $8y \left(\dfrac{y^3}{2} - \dfrac{1}{4} \right) \left(\dfrac{y^3}{2} + \dfrac{1}{4} \right)$

2.4 ALGEBRAIC FRACTIONS

An *algebraic fraction* is an expression in which one expression is divided by another.

Cancelation

Sometimes we can simplify the form of an algebraic fraction by dividing a common factor from the numerator and the denominator. We call this canceling the common factor.

Example 1   Simplify $\dfrac{6x - 6}{3}$.

Solution   The numerator has a common factor of 6:

$$6x - 6 = 6(x - 1).$$

Now, dividing both the numerator and denominator by 3, we obtain

$$\frac{6x - 6}{3} = \frac{{}^2\cancel{6}(x - 1)}{{}^1\cancel{3}} = 2(x - 1).$$

We can cancel the 3 because it amounts to expressing the fraction as a product with the number 1:

$$\frac{6(x - 1)}{3} = \frac{3 \cdot 2(x - 1)}{3} = \overset{1}{\overbrace{\frac{3}{3}}} \cdot \frac{2(x - 1)}{1} = 2(x - 1).$$

Example 2   Simplify $\dfrac{10m + 20n}{-15m - 5n}$.

Solution   The numerator contains a common factor of 10 and the denominator has a common factor of -5. Rewriting, we have

$$\frac{10m + 20n}{-15m - 5n} = \frac{10(m + 2n)}{-5(3m + n)} = -\frac{2(m + 2n)}{3m + n}.$$

Canceling Expressions

We can cancel expressions from the numerator and denominator of a fraction, but it is important to remember that cancelation is valid only when the factor being canceled is not zero.

Example 3   Simplify $\dfrac{5a^3 + 10a}{10a^2 + 20}$.

Solution   We factor $5a$ out of the numerator, giving

$$5a^3 + 10a = 5a(a^2 + 2),$$

and we factor 10 out of the denominator, giving

$$10a^2 + 20 = 10(a^2 + 2).$$

Thus, our original fraction becomes

$$\frac{5a^3 + 10a}{10a^2 + 20} = \frac{5a(a^2 + 2)}{10(a^2 + 2)}.$$

We simplify by dividing both the numerator and denominator by 5 and $(a^2 + 2)$.

$$\frac{5a^3 + 10a}{10a^2 + 20} = \frac{{}^1\cancel{5}a\cancel{(a^2 + 2)}}{{}^2\cancel{10}\cancel{(a^2 + 2)}} = \frac{a}{2}.$$

Since the quantity $(a^2 + 2)$ is never 0, the quantity

$$\frac{a^2 + 2}{a^2 + 2}$$

is equal to 1 for all real values of a, so $a/2$ is equivalent to the original expression.

If the expression being factored out is zero for some values of the variables, then the simplification is still valid if we avoid those values.

Example 4   Factor the numerator and denominator to simplify $\dfrac{3 - x}{2x - 6}$.

Solution   Notice that we can factor out -1 from the numerator:

$$3 - x = -1(-3 + x) = -1(x - 3).$$

We can also factor a 2 from the denominator:

$$2x - 6 = 2(x - 3).$$

Both numerator and denominator have $x - 3$ as a factor. So

$$\frac{3 - x}{2x - 6} = \frac{-1\cancel{(x - 3)}}{2\cancel{(x - 3)}} = -\frac{1}{2}, \quad \text{provided } x \neq 3.$$

Example 5   (a) If $w \neq r$, show that $\dfrac{-(w - r)}{r - w} = 1$.

(b) If $m \neq n$, show that $\dfrac{m - n}{n - m} = -1$.

Solution   (a) Since $\dfrac{-(w - r)}{r - w} = \dfrac{-1(w - r)}{r - w}$, we can distribute the -1 to get $\dfrac{-(w - r)}{r - w} = \dfrac{-w + r}{r - w}$. Then, reordering the numerator, we have

$$\frac{-(w - r)}{r - w} = \frac{-w + r}{r - w} = \frac{r - w}{r - w} = 1.$$

(b) If we factor out a -1 in the numerator and then reorder, we have

$$\frac{m - n}{n - m} = \frac{-1(-m + n)}{n - m} = \frac{-1(n - m)}{n - m} = -1.$$

The techniques for factoring quadratic expressions from Section 2.3 often come in handy for simplifying algebraic fractions.

Example 6   Simplify $\dfrac{2k^2 - 8}{2 - k}$.

Solution   After factoring out a common factor of 2 in the numerator we are left with an expression that is the difference of squares. Therefore,

$$\frac{2k^2 - 8}{2 - k} = \frac{2(k^2 - 4)}{2 - k} = \frac{2(k + 2)(k - 2)}{2 - k}.$$

If we factor a -1 from the denominator, we have

$$\frac{2k^2 - 8}{2 - k} = \frac{2(k^2 - 4)}{2 - k} = \frac{2(k + 2)\cancel{(k - 2)}}{-1\cancel{(k - 2)}}.$$

Therefore,

$$\frac{2k^2 - 8}{2 - k} = -2(k + 2) \quad \text{provided } k \neq -2.$$

Adding Algebraic Fractions

Another situation where the distributive law is useful is in adding algebraic fractions.

Example 7   Express $\dfrac{j}{7} + \dfrac{2j}{7}$ as a single fraction.

Solution   We have

$$\begin{aligned}
\frac{j}{7} + \frac{2j}{7} &= \frac{1}{7} \cdot j + \frac{1}{7} \cdot 2j && \text{rewriting division as multiplication} \\
&= \frac{1}{7} \cdot (j + 2j) && \text{factoring out 1/7} \\
&= \frac{1}{7}(3j) && \text{collecting like terms} \\
&= \frac{3j}{7} && \text{rewriting multiplication as division.}
\end{aligned}$$

This is the same answer we get by adding numerators and dividing by their common denominator.

We can also use the distributive law to reverse the process in Example 7 and express a fraction as a sum of two other fractions.

Example 8   Express $\dfrac{5x + y}{6}$ as the sum of two algebraic fractions.

Solution   We have

$$\frac{5x + y}{6} = \frac{5x}{6} + \frac{y}{6}.$$

If two fractions do not have a common denominator, we replace them with equivalent fractions that do have a common denominator before adding them.

Example 9   Simplify the expression

$$\frac{c}{25} + \frac{h}{30}.$$

by writing it as a single fraction.

Solution   We have

$$\frac{c}{25} + \frac{h}{30} = \frac{30c}{25 \cdot 30} + \frac{25h}{25 \cdot 30} = \frac{30c + 25h}{25 \cdot 30} = \frac{30c + 25h}{750} = \frac{6c + 5h}{150}.$$

An Application of Adding Fractions: Fuel Efficiency

Suppose a car's fuel efficiency is 25 mpg in the city and 30 mpg on the highway. We want to know the average fuel efficiency if the car drives 150 miles in the city and 300 miles on the highway. It uses $150/25 = 6$ gallons in the city, and $300/30 = 10$ gallons on the highway. So

$$\text{Average fuel efficiency} = \frac{\text{Total miles driven}}{\text{Total gallons gas}} = \frac{\text{City miles} + \text{Highway miles}}{\text{City gallons} + \text{Highway gallons}}$$

$$= \frac{150 + 300}{6 + 10} = \frac{450}{16} = 28.125 \text{ mpg}.$$

Example 10   Find an algebraic expression for the fuel efficiency if the car drives c miles in the city and h miles on the highway.

Solution   We have

$$\text{Average fuel efficiency} = \frac{\overbrace{\text{City miles}}^{c} + \overbrace{\text{Highway miles}}^{h}}{\underbrace{\text{City gallons}}_{c/25} + \underbrace{\text{Highway gallons}}_{h/30}} = \frac{c + h}{\dfrac{c}{25} + \dfrac{h}{30}}.$$

Using Example 9, we can write this as

$$\frac{c + h}{\dfrac{c}{25} + \dfrac{h}{30}} = \frac{c + h}{\dfrac{6c + 5h}{150}} = \frac{150(c + h)}{6c + 5h}.$$

Rules for Operations on Algebraic Fractions

Example 9 illustrates some of the general rules for adding and multiplying algebraic fractions, which are the same as the rules for ordinary fractions.

Addition and Subtraction

$$\frac{a}{b} + \frac{c}{d} = \frac{ad + cb}{bd} \quad \text{and} \quad \frac{a}{b} - \frac{c}{d} = \frac{ad - cb}{bd}, \quad \text{provided } b \text{ and } d \text{ are not zero.}$$

We call bd a common denominator of the two fractions.

Multiplication

$$\frac{a}{b} \cdot \frac{c}{d} = \frac{ac}{bd}, \quad \text{provided } b \text{ and } d \text{ are not zero.}$$

Division

$$\frac{a/b}{c/d} = \frac{a}{b} \cdot \frac{d}{c} = \frac{ad}{bc}, \quad \text{provided } b, c, \text{ and } d \text{ are not zero.}$$

Example 11   Express as a single fraction   (a) $\dfrac{3}{p} + \dfrac{2}{p+2}$ (b) $\dfrac{t}{s+t} - \dfrac{s}{s-t}$

Solution   (a) Using the rule $\dfrac{a}{b} + \dfrac{c}{d} = \dfrac{ad + bc}{bd}$, we have

$$\frac{3}{p} + \frac{2}{p+2} = \frac{3(p+2) + 2p}{p(p+2)}.$$

Now, using the distributive law and collecting like terms, we get

$$\frac{3}{p} + \frac{2}{p+2} = \frac{3p + 6 + 2p}{p(p+2)}$$
$$= \frac{5p + 6}{p(p+2)}.$$

(b) Using the rule $\dfrac{a}{b} + \dfrac{c}{d} = \dfrac{ad + bc}{bd}$, we have

$$\frac{t}{s+t} - \frac{s}{s-t} = \frac{t(s-t) - s(s+t)}{(s+t)(s-t)} = \frac{st - t^2 - s^2 - st}{s^2 - t^2}$$

Factoring out a -1 from the numerator and reordering, we get

$$\frac{t}{s+t} - \frac{s}{s-t} = -\frac{s^2 + t^2}{s^2 - t^2}$$

Example 12   Express as a single fraction   (a) $\dfrac{3}{v+w} \cdot \dfrac{vw^2}{7}$ (b) $\dfrac{3/(v+w)}{vw^2/7}$

Solution   (a) Using the rule $\dfrac{a}{b} \cdot \dfrac{c}{d} = \dfrac{ac}{bd}$, we have

$$\frac{3}{v+w} \cdot \frac{vw^2}{7} = \underbrace{\overbrace{3}^{a}}_{b}\underbrace{\overbrace{vw^2}^{c}}_{d} = \underbrace{\frac{3 \cdot vw^2}{(v+w) \cdot 7}}_{bd} = \frac{3vw^2}{7(v+w)}.$$

(b) Using the rule $\dfrac{a/b}{c/d} = \dfrac{a}{b} \cdot \dfrac{d}{c}$, we have

$$\frac{3/(v+w)}{vw^2/7} = \frac{\overbrace{3}^{a} / \overbrace{(v+w)}^{b}}{\underbrace{vw^2}_{c} / \underbrace{7}_{d}} = \frac{3}{v+w} \cdot \frac{7}{vw^2}$$

$$= \frac{21}{(v+w)vw^2}.$$

Example 13   Express as a single fraction (a) $\dfrac{z^2 + z - 6}{z^2 - 1} \cdot \dfrac{z-1}{z-2}$ (b) $\dfrac{(r^2 - 25)/5r}{(r^2 - 10r + 25)/(5r - 25)}$

Solution   (a) Expressing the fractions in factored form gives

$$\frac{z^2 + z - 6}{z^2 - 1} \cdot \frac{z-1}{z-2} = \frac{(z+3)(z-2)}{(z+1)(z-1)} \cdot \frac{z-1}{z-2}$$

Canceling out like terms gives

$$\frac{z^2 + z - 6}{z^2 - 1} \cdot \frac{z-1}{z-2} = \frac{(z+3)\cancel{(z-2)}}{(z+1)\cancel{(z-1)}} \cdot \frac{\cancel{(z-1)}}{\cancel{(z-2)}} = \frac{z+3}{z+1}.$$

Be sure you understand why z cannot be equal to $-1, 1$, or 2.

(b) Using the rule for division of fractions, factoring, and canceling common factors (provided $r \neq 5$), we obtain

$$\frac{(r^2 - 25)/5r}{(r^2 - 10r + 25)/(5r - 25)} = \frac{(r+5)(r-5)}{5r} \cdot \frac{5(r-5)}{(r-5)(r-5)} = \frac{r+5}{r}.$$

Exercises and Problems for Section 2.4

EXERCISES

Write each of the expressions in Exercises 1–12 as a single fraction.

1. $\dfrac{m}{2} + \dfrac{m}{3}$

2. $2 + \dfrac{3}{x}$

3. $\dfrac{1}{x-2} - \dfrac{1}{x-3}$

4. $\dfrac{3}{x} + \dfrac{4}{x-1}$

5. $\dfrac{-1}{x} - \dfrac{1}{-x} + \dfrac{-1}{-x}$

6. $z + \dfrac{z}{2} + \dfrac{2}{z}$

7. $\frac{1}{4}(e/2)$

8. $\dfrac{1}{a} + \dfrac{1}{b}$

9. $\dfrac{1}{a-b} + \dfrac{1}{a+b}$

10. $\dfrac{1}{x-a} - \dfrac{1}{x-b}$

11. $\dfrac{\frac{1}{3}r + r/4}{2r/5 - \frac{1}{11}(3r)}$

12. $1 + \dfrac{1}{1+\frac{1}{x}}$

In Exercises 13–24, multiply and simplify. Assume any factors you cancel are not zero.

13. $\dfrac{5p}{6q^2} \cdot \dfrac{3pq}{5p}$

14. $\dfrac{3xy^2}{4x^2z} \cdot \dfrac{8xy^3z}{6xy^5}$

15. $\dfrac{2ab}{5b} \cdot \dfrac{10a^2b^2}{6a}$

16. $\dfrac{4}{6x+12y} \cdot \dfrac{3x+6y}{10}$

17. $\dfrac{2r+3s}{4s} \cdot \dfrac{6r}{6r+9s}$

18. $\dfrac{x+3}{x+4} \cdot \dfrac{2x+8}{4x+12}$

19. $\dfrac{1}{ab+abc} \cdot (c(a+b) + (a+b))$

20. $\dfrac{p^2+4p}{p^2-2p} \cdot \dfrac{3p-6}{3p+12}$

21. $\dfrac{w^2r+4wr}{2r^2w+2wr} \cdot \dfrac{r+r^2}{4w+16}$

22. $\dfrac{cd+c}{cd-8d} \cdot \dfrac{16-2c}{4+4d}$

23. $\dfrac{5v^2+15v}{vw-v} \cdot \dfrac{3w+3}{5v+15}$

24. $\dfrac{ab+b}{2b^2+6b} \cdot \dfrac{3a^2+6a}{a+a^2}$

In Exercises 25–43, write each expression as a sum, difference, or product of two or more algebraic fractions. There is more than one correct answer. Assume all variables are positive.

25. $\dfrac{z+1}{2}$

26. $\dfrac{w}{10}$

27. $\dfrac{4}{c+2}$

28. $\dfrac{5}{6p}$

29. $\dfrac{6p-3}{6}$

30. $\dfrac{1}{xyz}$

31. $\dfrac{8-5x}{-2}$

32. $\dfrac{-3t}{9s}$

33. $\dfrac{2xh+h^2}{h}$

34. $\dfrac{c}{ab}$

35. $\dfrac{c}{a+b}$

36. $\dfrac{h(B+b)}{2}$

37. $\dfrac{3}{t(r+s)}$

38. $\dfrac{4}{y+x}$

39. $\dfrac{1+2a+3b}{4}$

40. $\dfrac{p+prt}{p}$

41. $\dfrac{(x+1)^2-y}{xy}$

42. $\dfrac{1}{(p+2)^2+b}$

43. $\dfrac{x}{x-1}$

PROBLEMS

44. In electronics, when two resistors, with resistances A and B, are connected in a parallel circuit, the total resistance is
$$\dfrac{1}{1/A + 1/B}.$$
Rewrite this so there are no fractions in the numerator or denominator.

In Problems 45–50, find an expression equivalent to one of parts (a)–(f), if possible.

(a) $2x$

(b) $\dfrac{1}{2x}$

(c) $\dfrac{2}{x}$

(d) $\dfrac{4}{x+2}$

(e) $\dfrac{1}{1-x}$

(f) $\dfrac{1}{x+1}$

45. $\dfrac{1}{x} + \dfrac{1}{x}$

46. $\dfrac{x}{0.5}$

47. $\dfrac{1/x}{2}$

48. $-\dfrac{1}{-1-x}$

49. $\dfrac{3}{x+1}$

50. $-\dfrac{1}{x-1}$

■ For Exercises **51–64**, simplify each expression. Assume any factors you cancel are not zero.

51. $\dfrac{\dfrac{4ab^3}{3}}{\dfrac{2b}{a^2}}$

52. $\dfrac{\dfrac{1}{s}+\dfrac{1}{t}}{st}$

53. $\dfrac{p+q}{\dfrac{p}{12}+\dfrac{q}{18}}$

54. $\dfrac{\dfrac{m+2}{3}}{\dfrac{m^2-4}{6m}}$

55. $\dfrac{\dfrac{2}{x}-3}{\dfrac{2-3x}{2}}$

56. $\dfrac{\dfrac{t}{t-3}}{\dfrac{4}{4t-12}}$

57. $\dfrac{\dfrac{1}{25}-\dfrac{1}{x^2}}{\dfrac{1}{x}-\dfrac{1}{5}}$

58. $\dfrac{\dfrac{x}{y}-\dfrac{y}{x}}{\dfrac{1}{y}+\dfrac{1}{x}}$

59. $\dfrac{\dfrac{1}{c^2}-\dfrac{1}{d^2}}{\dfrac{d-c}{c^2d}}$

60. $\dfrac{\dfrac{1}{k+1}-1}{\dfrac{1}{k+1}+1}$

61. $\dfrac{\dfrac{1}{m-1}+\dfrac{2}{m+2}}{\dfrac{3}{m+2}}$

62. $2-\dfrac{2}{2+\frac{2}{2+2}}$

63. $\dfrac{1+\dfrac{2}{w}-\dfrac{24}{w^2}}{1-\dfrac{1}{w}-\dfrac{12}{w^2}}$

64. $\dfrac{12+\dfrac{12}{a}+\dfrac{3}{a^2}}{12+\dfrac{6}{a}}$

REVIEW EXERCISES AND PROBLEMS FOR CHAPTER 2

EXERCISES

■ In Exercises **1–4**, are the two expressions equivalent?

1. $x(5x)$ and $5x^2$

2. $2a+b$ and $b+2a$

3. $(2x)(5y)$ and $7xy$

4. $(a+3)+(b+2)$ and $(a+b)+5$

■ Given the values of r and z in Exercises **5–6**, evaluate and simplify the expression

$$3(z+r)^2+6rz+\frac{r-z-2}{r+z}.$$

Your answers may involve other letters.

5. $r=-5, z=2$ **6.** $r=5k, z=-3k$

■ In Exercises **7–12**, write the expression in a simpler form, if possible.

7. $3p^2-2q^2+6pq-p^2$

8. $y^3+2xy-4y^3+x-2xy$

9. $(1/2)A+(1/4)A-(1/3)A$

10. $(a+4)/3+(2a-4)/3$

11. $3(2t-4)-t(3t-2)+16$

12. $5z^4+5z^3-3z^4$

■ In Exercises **13–22**, simplify by using the distributive law and combining like terms.

13. $2(x+5)+3(x-4)$

14. $7(x-2)-5(2x-5)$

15. $3(2x-5)-2(5x+4)+5(10-3x)$

16. $6(2x+1)-5(3x-4)+6x-10$

17. $2x(3x+4)+3(x^2-5x+6)$

18. $3x(x+5)-4x(3x-1)+5(6x-3)$

19. $2a(a+b)-5b(a+b)$

20. $5a(a+2b)-3b(2a-5b)+ab(2a-1)$

21. $mn(m+2n)+3mn(2m+n)+5m^2n$

22. $5(x^2-3x-5)-2x(x^2-5x-7)$

■ Factor the expressions in Exercises **23–52**.

23. z^2-5z+6 **24.** $a^2+8a-48$

25. $v^2-4v-32$ **26.** $b^2-23b-50$

27. $w^2+2w+24$ **28.** x^2+x-72

29. $6x^2+5x-6$ **30.** $30t^2+26t+4$

31. $q^2-4qz+3z^2$ **32.** $12s^2+17s-5$

33. $12w^2 - 10w - 8$

34. $10z^2 - 21z - 10$

35. $r^2 + 4$

36. $(c+3)^2 - d^4$

37. $x^3 + 4x$

38. $x^2 - 81$

39. $x^2 - 144$

40. $x^2 - 14x + 49$

41. $x^2 - 22x + 121$

42. $x^2 - 3x - 54$

43. $3x^2 + 22x + 35$

44. $x^2 + x + 3x + 3$

45. $ax + bx - ay - by$

46. $12a^2 + 2b + 24a + ab$

47. $x^3 + 23x^2 + 132x$

48. $x^4 - 18x^3 + 81x^2$

49. $y^3 + 7y^2 - 18y$

50. $7y^5 - 28yz^6$

51. $-3t^7 + 24t^5 v^2 - 48t^3 v^4$

52. $n^2 + 10n + 25q^2$

For Exercises **53–67**, simplify each expression. Assume any factors you cancel are not zero.

53. $\dfrac{4a - 8}{16}$

54. $\dfrac{10y^3 - 5y^2}{15y}$

55. $\dfrac{12w - 36w^3}{24w^2}$

56. $\dfrac{4x - 8}{10x - 20}$

57. $\dfrac{3t^3 + 12t}{4t^2 + 16}$

58. $\dfrac{3s^3 - 12s}{s - 2}$

59. $\dfrac{(x - y)^2}{x^2 - y^2}$

60. $\dfrac{2x^2 - 32}{x^2 - 2x - 8}$

61. $\dfrac{p^2 q - pq^2}{(p - q)^2}$

62. $\dfrac{2(y - 4)}{4(4 - y)}$

63. $\dfrac{9z - 3z^2}{z^2 - 9}$

64. $\dfrac{6a - 2a^2}{a^2 - a - 6}$

65. $\dfrac{r^4 - 1}{r^3 p - rp}$

66. $\dfrac{9k^2 + 12k + 4}{12k + 8}$

67. $\dfrac{r(r + s) - s(r + s)}{r^2 s + rs^2}$

PROBLEMS

68. A store has a 10% off sale, and you plan to buy four items costing a, b, c, and d dollars, respectively.

   (a) Write an expression for the total cost of the four items if the 10% is deducted off the total amount.
   (b) Write an expression for the total cost of the four items if the price of each item is reduced by 10% separately before totaling up.
   (c) Are the expressions in (a) and (b) equivalent?
   (d) Does it matter whether the 10% is taken off each item separately or is taken off the total at the end?

69. If $a/b = 2/3$, what is $3a/2b$?

70. If $m + n = s + 2$, what is $2(m + n) - (s - 2) + (1 - s)$?

71. The area of a circle is 30 square meters. What is the area of a circle whose radius is three times as long? (The area of a circle of radius r is πr^2.)

72. The volume of a cylinder is 100 cubic centimeters. If the radius of the cylinder is doubled and the height of the cylinder is tripled, what is the volume of the new cylinder? (The volume of a cylinder of radius r and height h is $\pi r^2 h$.)

In Problems **73–75**, put the expression $6x^2 + 12$ in the form given. Identify the values of r, v, and w, as applicable.

73. $2r$

74. $3v + 3$

75. $rv + (r + 1)w$

In Problems **76–77**, write the expression in the form $k - \dfrac{mx}{x - n}$, and identify the values of the constants k, m, and n.

76. $\dfrac{4x}{2x - 6} - 5$

77. $\dfrac{3(x + 5) - 7x}{x + 5}$

78. Given $w = -2$ and $v = k - 1$, evaluate and simplify:

$$2w^3 + 3v^2 + 3w^2v - 3(v - w)^2.$$

79. An employee holds two different part-time positions with the same company, earning A a week in the first position and B a week in the second position.

    (a) How much does the employee earn a week from this company?

    (b) If the employee's total earnings from the company increase by 5%, how much does she earn a week?

    (c) If the first position earnings increase by 5% and the second position earnings stay the same, how much does she earn a week?

    (d) If the first position earnings stay the same and the second position earnings increase by 5%, how much does she earn a week?

    (e) If the pay for each of the two positions increases by 5%, how much does she earn a week?

    (f) Which, if any, of the expressions given in (a), (b), (c), (d), (e) are equivalent?

■ In Problems **80–81**, explain how the distributive law $a(b + c) = ab + ac$ has been used in the identity.

80. $\left(x^2 + x + 6\right)\left(3x^3 + 6x^2\right) = x^2\left(3x^3 + 6x^2\right) + (x + 6)\left(3x^3 + 6x^2\right)$

81. $\left(x^2 + x + 6\right)\left(3x^3 + 6x^2\right) = 3x^2\left(x^2 + x + 6\right)(x + 2)$

Chapter 3

Rules for Equations and the Reasons for Them

CONTENTS

3.1 SOLVING EQUATIONS

In Chapter 1 we saw that the basic principle for solving equations is to replace the equation with a simpler equivalent equation. In this section we look at the practical techniques for doing this.

Using the Operations of Arithmetic to Solve Equations

Often, mathematics problems that arise in everyday life can be solved by basic arithmetic reasoning. For example, suppose you want to know the price p you can pay for a car if the tax is 5% and you have a total of $4200 to spend. Since 5% tax on p dollars is $0.05p$, the total amount you will spend on a car with a sticker price p is $p + 0.05p$. So you want to know when

$$p + 0.05p = 4200.$$

Using the distributive law we get

$$\begin{array}{ll} (1 + 0.05)p = 4200 & \text{distributive law} \\ 1.05p = 4200 & \\ p = \dfrac{4200}{1.05} = 4000 & \text{divide both sides by 1.05.} \end{array}$$

So if you have $4200 to spend, you can afford a car whose sticker price is $4000.

You can go quite a long way in solving equations with basic arithmetic operations for producing an equivalent equation:

- Adding or subtracting the same number to both sides

- Multiplying or dividing both sides by the same number, provided it is not zero

- Replacing any expression in an equation by an equivalent expression.

Example 1   Solve (a) $7 + 3p = 1 + 5p$ (b) $1 - 2(3 - x) = 10 + 5x.$

Solution   (a) For this equation we collect the p terms on the left and the constant terms on the right.

$$\begin{array}{ll} 7 + 3p = 1 + 5p & \\ 3p = 5p - 6 & \text{subtract 7 from both sides} \\ -2p = -6 & \text{subtract } 5p \text{ from both sides} \\ p = \dfrac{-6}{-2} = 3 & \text{divide both sides by } -2. \end{array}$$

We verify that $p = 3$ is the solution by putting it into the original equation:

$$7 + 3 \cdot 3 = 1 + 5 \cdot 3 \quad \longrightarrow \quad 16 = 16.$$

(b) First we distribute the -2 and collect like terms:

$$\begin{array}{ll} 1 - 2(3 - x) = 10 + 5x & \\ 1 - 6 + 2x = 10 + 5x & \text{distribute } -2 \\ -5 + 2x = 10 + 5x & \text{simplify left side} \\ -5 - 3x = 10 & \text{subtract } 5x \text{ from both sides} \end{array}$$

$$-3x = 15 \qquad \text{add 5 to both sides}$$
$$x = \frac{15}{-3} = -5 \quad \text{divide both sides by } -3.$$

Checking $x = -5$:

$$1 - 2(3 - (-5)) = 10 + 5(-5) \quad \longrightarrow \quad -15 = -15.$$

Solving Equations with Constants Represented by Letters

In the previous examples the coefficients and the constant terms were specific numbers. Often we have to solve an equation where there are unspecified constants represented by letters. The general method for solving such equations is the same.

Example 2 Solve each of the equations $2x + 12 = 20$ and $2x + z = N$ for the variable x.

Solution

$$2x + 12 = 20 \qquad\qquad\qquad 2x + z = N$$
$$2x = 20 - 12 \quad \text{subtract 12} \qquad 2x = N - z \quad \text{subtract } z$$
$$x = \frac{20 - 12}{2} \quad \text{divide by 2.} \qquad x = \frac{N - z}{2} \quad \text{divide by 2.}$$

We use the same operations to solve both equations: subtraction in the first step and division in the second. We can further simplify the solution on the left by writing $(20 - 12)/2$ as 4, but there is no corresponding simplification of the solution for the equation on the right.

Example 3 Solve $Aw + 3w - 5 = 12$ for w. Assume that A is a positive constant.

Solution We collect terms involving w on the left side of the equal sign and the remaining terms on the right.

$$Aw + 3w - 5 = 12$$
$$Aw + 3w = 17 \qquad \text{add 5 to both sides}$$
$$w(A + 3) = 17 \qquad \text{factor out } w \text{ on the left side}$$
$$w = \frac{17}{A + 3} \qquad \text{divide both sides by } A + 3.$$

Dividing both sides by $A + 3$ is not a legitimate operation unless we know that $A + 3$ is not zero. Since A is positive, $A + 3$ cannot be zero, and the operation is allowed.

Example 4 Solve the equation $2q^2 p + 5p + 10 = 0$ for p.

Solution   We collect terms involving p on the left of the equal sign and the remaining terms on the right.

$$2q^2p + 5p + 10 = 0$$
$$2q^2p + 5p = -10 \qquad \text{subtract 10 from both sides}$$
$$p(2q^2 + 5) = -10 \qquad \text{factor out } p$$
$$p = \frac{-10}{2q^2 + 5} \qquad \text{divide by } 2q^2 + 5.$$

Since $2q^2 + 5$ cannot be zero, the last step is legitimate.

What Other Operations Can We Use?

As we learn about more and more complicated equations, we need more and more operations to solve them, in addition to the basic arithmetic ones we have discussed here. To decide whether an operation is allowable, we look at whether or not it changes the equality between two quantities. If in doubt, we check the solutions we have obtained back in the original equation.

Example 5   Solve the equation $\sqrt[3]{z} = 8$.

Solution   To undo the operation of taking the cube root, we cube both sides of the equation. Since cubing the two sides does not change whether they are equal or not, this produces an equivalent equation.

$$\sqrt[3]{z} = 8$$
$$z^{1/3} = 8 \qquad \text{rewriting in exponential form}$$
$$(z^{1/3})^3 = 8^3 \qquad \text{cube both sides}$$
$$z = 512.$$

We check to see that 512 is a solution: $\sqrt[3]{512} = 8$.

In Example 5 we performed a new operation on both sides of an equation: cubing. Cubing two numbers does not change whether they are equal or not: different numbers have different cubes. So this operation produces an equivalent equation.

Example 6   Solve the equation $\sqrt{x} = -2$.

Solution   To undo the operation of taking the square root, we try squaring both sides of the equation.

$$\sqrt{x} = -2$$
$$(\sqrt{x})^2 = (-2)^2 \qquad \text{square both sides}$$
$$x = 4.$$

If we check the result, we see that
$$\sqrt{4} = 2 \neq -2,$$

so $x = 4$ is *not* a solution to the equation. Since square roots are always positive, we recognize that $\sqrt{x}$ cannot equal -2 and the equation has no solutions.

Since different numbers do not necessarily have different squares, the operation of squaring both sides of an equation (or taking the square root of both sides of an equation) does not give an equivalent equation. Just as we need to be careful about multiplying or dividing by zero, we also need to be careful about raising both sides of an equation to an even power.

Equations with Fractional Expressions

An equation involving a fractional expression can sometimes be solved by first transforming it into a simpler equation. To do this we clear the equation of fractions. One way to do this is to multiply by the product of the denominators. The resulting equation is equivalent to the original, except at values of the variable that make one of the denominators zero.

Example 7 Solve for x (a) $\dfrac{2}{3+x} = \dfrac{5}{6-x}$ (b) $s = \dfrac{a}{1-x}$, if $s \neq 0$ and $a \neq 0$

Solution (a) Multiply both sides of the equation by $(3+x)(6-x)$:

$$\frac{2}{3+x} \cdot (3+x)(6-x) = \frac{5}{6-x} \cdot (3+x)(6-x).$$

Canceling $(3+x)$ on the left and $(6-x)$ on the right, we have

$$2(6-x) = 5(3+x).$$

This is now a simpler equation, with solution given by

$$12 - 2x = 15 + 5x$$
$$-7x = 3$$
$$x = -\frac{3}{7}.$$

Since we cleared our fractions by multiplying by a quantity that can be zero, we need to check the solution by substitution in the original equation.

$$\frac{2}{3+x} = \frac{2}{3 - 3/7} = \frac{2}{(21-3)/7} = 2 \cdot \frac{7}{18} = \frac{7}{9}$$
$$\frac{5}{6-x} = \frac{5}{6 + 3/7} = \frac{5}{(42+3)/7} = 5 \cdot \frac{7}{45} = \frac{7}{9}.$$

Thus the solution is valid.

(b) Multiplying both sides by $1 - x$ gives

$$s(1-x) = a$$
$$s - sx = a$$
$$s - a = sx$$
$$x = \frac{s-a}{s}.$$

We can divide by s because it cannot be zero. Again, we need to check our solution. Substituting it into the right side of the equation, we get

$$\text{Right side} = \frac{a}{1 - \frac{s-a}{s}} = \frac{a}{\frac{s-(s-a)}{s}} = \frac{sa}{a} = s = \text{Left side}.$$

Exercises and Problems for Section 3.1

EXERCISES

■ Solve the equations in Exercises 1–14.

1. $0.5x - 3 = 11$

2. $\dfrac{5}{3}(y + 4) = \dfrac{1}{2} - y$

3. $2(a + 3) = 10$

4. $-9 + 10r = -3r$

5. $4p - 1.3 = -6p - 16.7$

6. $6n - 3 = -2n + 37$

7. $\dfrac{1}{2}r - 2 = 3r + 5$

8. $0.2(g - 6) = 0.6(g - 4)$

9. $-4(2m - 5) = 5$

10. $5 = \dfrac{1}{3}(t - 6)$

11. $\dfrac{2}{3}(3n - 12) = \dfrac{3}{4}(4n - 3)$

12. $3d - \dfrac{1}{2}(2d - 4) = -\dfrac{5}{4}(d + 4)$

13. $B - 4(B - 3(1 - B)) = 57$

14. $1.06s - 0.01(240 - s) = 22.67s$

■ Solve the equations in Exercises 15–25.

15. $\dfrac{3}{z - 2} = \dfrac{2}{z - 3}$

16. $\dfrac{2}{2 - x} - \dfrac{3}{x - 5} = 0$

17. $\dfrac{3}{2x - 1} + \dfrac{5}{3 - 2x} = 0$

18. $\dfrac{-3}{x - 2} - \dfrac{2}{x - 3} = 0.$

19. $\dfrac{1}{1 + \dfrac{1}{2 - x}} = \dfrac{2}{3 + \dfrac{1}{2 - x}}.$

20. $\dfrac{x}{x - 4} = \dfrac{2}{3}$

21. $\dfrac{x + 5}{5} - \dfrac{2x + 3}{4} = 0$

22. $\dfrac{17}{3n - 4} = \dfrac{7}{n}$

23. $\dfrac{3a}{a + 2} = 5$

24. $\dfrac{6h - 1}{4h + 1} = \dfrac{3}{5}$

25. $\dfrac{y - 3}{y + 3} = \dfrac{5}{7}$

PROBLEMS

■ In Exercises 26–37, solve for the indicated variable. Assume all constants are non-zero.

26. $A = l \cdot w$, for w.

27. $y = 3\pi t$, for t.

28. $t = t_0 + \dfrac{k}{2}w$, for w.

29. $s = v_0 t + \dfrac{1}{2}at^2$, for a.

30. $ab + aw = c - aw$, for w.

31. $3xt + 1 = 2t - 5x$, for t, if $x \neq 2/3$.

32. $u(m + 2) + w(m - 3) = z(m - 1)$, for m, if $u + w - z \neq 0$.

33. $x + y = z$, for y.

34. $ab = c$, for b.

35. $2r - t = r + 2t$, for r.

36. $6w - 4x = 3w + 5x$, for w.

37. $3(3g - h) = 6(g - 2h)$, for g.

38. Solve for L:
$$\dfrac{3kL - 8}{rkL - 7} = 5.$$
Your answer may involve other letters.

39. Suppose $x = 3$ is a solution to the equation $2zx + 1 = j$, where z and j are constants. Find a solution to the equation
$$4zx + 5 = 2j + 3.$$

3.2 SOLVING INEQUALITIES

Inequality Notation

Whereas an equation is a statement that two quantities are equal, an inequality is a statement that one quantity is greater than (or less than) another. We use four symbols for expressing inequalities:

$$x < y \quad \text{means} \quad x \text{ is less than } y$$

$$x > y \quad \text{means} \quad x \text{ is greater than } y$$
$$x \leq y \quad \text{means} \quad x \text{ is less than or equal to } y$$
$$x \geq y \quad \text{means} \quad x \text{ is greater than or equal to } y.$$

If z is between x and y, we write $x < z < y$.

Example 1 Under normal conditions, water freezes at $0°C$ and boils at $100°C$. If H stands for the temperature in $°C$ of a quantity of water, write an inequality describing H given that

(a) The water has frozen solid.   (b) The water has boiled into vapor.
(c) The water is liquid and is not boiling.

Solution (a) The value of H must be less than or equal to $0°C$, so $H \leq 0$.
(b) The value of H must be greater than or equal to $100°C$, so $H \geq 100$.
(c) The value of H must be greater than $0°C$ and less than $100°C$, so $0 < H < 100$.

Solutions to Inequalities

A value of a variable that makes an inequality a true statement is a solution to the inequality.

Example 2 Which of the given values is a solution to the inequality?

(a) $20 - t > 12$, for $t = -10, 10$
(b) $16 - t^2 < 0$, for $t = -5, 0, 5$.

Solution (a) We have:

For $t = -10$: $\underbrace{20 - (-10)}_{30} > 12$, a true statement.

For $t = 10$: $\underbrace{20 - 10}_{10} > 12$, a false statement.

So $t = -10$ is a solution but $t = 10$ is not.
(b) We have:

For $t = -5$: $\underbrace{16 - (-5)^2}_{-9} < 0$, a true statement.

For $t = 0$: $\underbrace{16 - 0^2}_{16} < 0$, a false statement.

For $t = 5$: $\underbrace{16 - 5^2}_{-9} < 0$, a true statement.

So $t = -5$ and $t = 5$ are solutions but $t = 0$ is not.

Solving Inequalities

We solve inequalities the same way we solve equations, by deducing simpler inequalities having the same solutions.

Example 3   Solve the equation $3x - 12 = 0$ and the inequality $3x - 12 < 0$.

Solution   In this case the steps for solving the equation also work for solving the inequality:

$$
\begin{array}{l|ll}
3x - 12 = 0 & 3x - 12 < 0 & \\
\quad\ 3x = 12 & \quad\ 3x < 12 & \text{add 12 to both sides} \\
\quad\ \ x = 4 & \quad\ \ x < 4 & \text{multiply both sides by 1/3.}
\end{array}
$$

In Example 3, adding 12 to both sides of the inequality, and multiplying both sides by 1/3, did not change the truth of the inequality. But we have to be careful when multiplying or dividing both sides of an inequality by a negative number. For example, although $2 > 1$, this does not mean that $(-3) \cdot 2 > (-3) \cdot 1$. In fact, it is the other way around: $-6 < -3$. When we multiply or divide both sides of an inequality by a negative number, we must reverse the direction of the inequality to keep the statement true.

Example 4   Solve the equation $30 - x = 50$ and the inequality $30 - x > 50$.

Solution   Solving the equation gives

$$
\begin{aligned}
30 - x &= 50 & \\
-x &= 20 & \text{subtract 30 from both sides} \\
x &= -20 & \text{multiply both sides by } -1.
\end{aligned}
$$

To solve the inequality, we start out the same way, since subtracting 30 from both sides of an inequality does not change whether it is true or not:

$$
\begin{aligned}
30 - x &> 50 \\
-x &> 20
\end{aligned}
$$

The last inequality states that, when x is multiplied by -1, the result must be greater than 20. This is true for numbers *less than* -20. We conclude that the correct solution is

$$
x < -20.
$$

Notice that multiplying both sides by -1 without thinking could have led to the incorrect solution $x > -20$.

Another way to solve Example 4 is to write

$$
\begin{aligned}
30 - x &> 50 & \\
30 &> 50 + x & \text{add } x \text{ to both sides} \\
30 - 50 &> x & \text{subtract 50 from both sides} \\
-20 &> x.
\end{aligned}
$$

The statement $-20 > x$ is equivalent to the statement $x < -20$.

Example 5   Solve the inequality $4 - (3x + 2) \geq 6 + x$.

Solution   We have

$$4 - (3x + 2) \geq 6 + x$$
$$4 - 3x - 2 \geq 6 + x$$
$$2 - 3x \geq 6 + x$$
$$-4x \geq 4$$
$$x \leq -1.$$

Notice that in the last step of the solution, we had to reverse the direction of the inequality since we multiplied both sides by the negative number $-1/4$.

Example 6   Michele will get a final course grade of $B+$ if the average on four exams is greater than or equal to 85 but less than 90. Her first three exam grades were $98, 74$ and 89. What fourth exam grade will result in a $B+$ for the course?

Solution   For Michele's average to be greater than or equal to 85 but less than 90 we write

$$85 \leq \frac{98 + 74 + 89 + g}{4} < 90.$$

Collecting like terms, we have

$$85 \leq \frac{261 + g}{4} < 90.$$

Multiplying the inequality by four gives

$$340 \leq 261 + g < 360.$$

Subtracting 261 produces

$$79 \leq g < 99.$$

Therefore, anything from a 79 up to but not including a 99 will give Michele a B^+ in the course.

Determining the Sign of an Expression from Its Factors

We are often interested in determining whether a quantity N is positive, negative, or zero, which can be expressed in terms of inequalities:

N is positive: $N > 0$

N is negative: $N < 0$

N is zero: $N = 0.$

In practice, determining the sign of an expression often amounts to inspecting its factors and using the following facts:
- A negative number times a negative number is positive.
- A negative number times a positive number is negative.
- A positive number times a positive number is positive.

Example 7

For what values of x is the quantity $5x - 105$ positive?

Solution

Factoring $5x - 105$, we obtain $5(x - 21)$. This shows that if x is greater than 21, the quantity $5x - 105$ is positive.

Example 8

What determines the sign of $a^2x - a^2y$ if $a \neq 0$?

Solution

Factoring, we obtain

$$a^2x - a^2y = a^2(x - y).$$

Since a^2 is always positive,

- If $x > y$, then $a^2(x - y)$ is positive.
- If $x = y$, then $a^2(x - y)$ is zero.
- If $x < y$, then $a^2(x - y)$ is negative.

Example 9

The quantity H takes on values between 0 and 1, inclusive, that is, $0 \leq H \leq 1$. For positive A, what are the possible values of

(a) $1 + H$ (b) $1 - H$ (c) $A + AH$ (d) $3A - 2HA$?

Solution

(a) Since H takes values between 0 and 1, the value of $1 + H$ varies between a low of $1 + 0 = 1$ and a high of $1 + 1 = 2$. So

$$1 \leq 1 + H \leq 2.$$

(b) The value of $1 - H$ varies between a low of $1 - 1 = 0$ and a high of $1 - 0 = 1$. So

$$0 \leq 1 - H \leq 1.$$

(c) The value of $A + AH$ can be investigated directly, or by factoring out an A and using the results of part (a).

$$A + AH = A(1 + H).$$

Since A is positive,

$$A \cdot 1 \leq A(1 + H) \leq A \cdot 2.$$

So

$$A \leq A + AH \leq 2A.$$

(d) Factoring out an A gives

$$3A - 2HA = A(3 - 2H).$$

Since H varies between 0 and 1, the value of $3 - 2H$ varies between a low of $3 - 2 \cdot 1 = 1$ and a high of $3 - 2 \cdot 0 = 3$. Thus

$$A \cdot 1 \leq A(3 - 2H) \leq A \cdot 3$$
$$A \leq 3A - 2AH \leq 3A.$$

Exercises and Problems for Section 3.2

EXERCISES

In Exercises **1–6**, write an inequality describing the given quantity. *Example*: An MP3 player can hold up to 120 songs. *Solution*: The number of songs is n where $0 \leq n \leq 120$, n an integer.

1. Chain can be purchased in one-inch lengths from one inch to twenty feet.

2. Water is a liquid above $32°$F and below $212°$F.

3. A 200-gallon holding tank fills automatically when its level drops to 30 gallons.

4. Normal resting heart rate ranges from 40 to 100 beats per minute.

5. Minimum class size at a certain school is 16 students, and state law requires fewer than 24 students per class.

6. An insurance policy covers losses of more than $1000 but not more than $20,000.

Each of the inequalities in Exercises **7–12** can be solved by performing a single operation on both sides. State the operation, indicating whether or not the inequality changes direction. Solve the inequality.

7. $12x \geq 60$

8. $-5t < 17.5$

9. $-4.1 + c \leq 2.3$

10. $-15.03 > s + 11.4$

11. $\dfrac{-3P}{7} < \dfrac{6}{14}$

12. $27 > -m$

Each of the inequalities in Exercises **13–18** can be solved by performing two operations on both sides. State the operations in order of use and solve the inequality.

13. $5y + 7 \leq 22$

14. $-2(n - 3) > 12$

15. $13 \leq 25 - 3a$

16. $3.7 - v \leq 5.3$

17. $\dfrac{5}{2}r - r < 6$

18. $\dfrac{4}{3}x \geq 2x - 3$

PROBLEMS

19. Table 3.1 shows values of z and the expression $4 - 2z$. For which values of z in the table is

    (a) $4 - 2z < 2$?   (b) $4 - 2z > 2$?
    (c) $4 - 2z = 2$?

    Table 3.1

z	0	1	2	3	4
$4 - 2z$	4	2	0	-2	-4

20. Table 3.2 shows values of a and the expressions $2 + a^2$ and $10 - 2a$. For which values of a in the table is

    (a) $2 + a^2 < 10 - 2a$?
    (b) $2 + a^2 > 10 - 2a$?
    (c) $2 + a^2 = 10 - 2a$?

    Table 3.2

a	0	1	2	3	4	5	6
$2 + a^2$	2	3	6	11	18	27	38
$10 - 2a$	10	8	6	4	2	0	-2

21. In 2008, the euro varied in value from $1.24/euro to $1.60/euro. Express the variation in the dollar cost, c, of a 140 euro hotel room during 2008 as an inequality.

22. An oven's heating element turns on when the the internal temperature falls to 20 degrees below the set temperature and turns off when the temperature reaches 10 degrees above the set temperature. You preheat the oven to 350 degrees before putting a cake in and baking it an hour at 350. Write an inequality to describe the temperature in the oven while the cake is baking.

23. A company uses two different-sized trucks to deliver sand. The first truck can transport x cubic yards, and the second y cubic yards. The first truck makes S trips to a job site, while the second makes T trips.

    (a) What do the following expressions represent in practical terms?

        (a) $S + T$ (b) $x + y$
        (c) $xS + yT$ (d) $(xS + yT)/(S + T)$
        (e) $xS > yT$ (f) $y > x$

    (b) If $xS > yT$ and $y > x$ what does this suggest about S and T? Verify your answer using algebra.

■ For what values of x are the quantities in Problems 24–27 negative?

26. $(x-3)(x+2)$ **27.** $x^2 + x - 2$

24. $x - 7$ **25.** $3x + 18$

3.3 ABSOLUTE VALUE EQUATIONS AND INEQUALITIES

The difference between two numbers, say 5 and 3, depends on the order in which we subtract them. Subtracting the smaller from the larger gives a positive number, $5 - 3 = 2$, but subtracting the larger from the smaller gives a negative number, $3 - 5 = -2$. Sometimes we are more interested in the *distance* between the two numbers than the difference. Absolute value measures the magnitude of a number, without distinguishing whether it is positive or negative. Thus, for example, the absolute value of $5 - 3$ is the same as the absolute value of $3 - 5$, namely 2.

Absolute Value

On the number line the *absolute value* of a number x, written $|x|$, is the distance between 0 and x. For example, $|5|$ is the distance between 0 and $+5$, so $|+5|$ equals 5. See Figure 3.1. Likewise, $|-5|$ is the distance between 0 and -5, so $|-5|$ also equals 5. See Figure 3.2. Since distance is always positive (or zero), $|x| \geq 0$ for every value of x.

Figure 3.1: Distance on the number line

Figure 3.2: Distance on the number line

Example 1   Evaluate the following expressions.

(a) $|100|$ (b) $|-3|$ (c) $|0|$

Solution   (a) The absolute value of 100 is the distance between 0 and 100, so $|100| = 100$.
(b) The absolute value of -3 is the distance between 0 and -3, so $|-3| = 3$.
(c) The absolute value of 0 is the distance between 0 and 0, so $|0| = 0$.

Measuring Distance

To find the distance between 2 and 5 on the number line, we subtract 2 from 5 to get 3. See Figure 3.3.

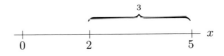

Figure 3.3: Distance on the number line

Notice that the distance between 2 to 5 is the same as the distance between 5 and 2. This means we could equally well subtract 5 from 2 as subtract 2 from 5, provided we take the absolute value:

$$\text{Distance from 2 to 5} = |5 - 2| = |+3| = 3$$
$$\text{Distance from 5 to 2} = |2 - 5| = |-3| = 3.$$

In general:

> The expression $|a - b|$ gives the distance between a and b on the number line.

Notice that $|a - b|$ is equivalent to $|b - a|$, because the distance between a and b is the same as the distance between b and a.

Example 2   Interpret each expression as a distance between two points and evaluate it.

  (a) $|5 - 3|$ (b) $|3 - 5|$ (c) $|5 + 3|$

Solution   (a) The expression $|5 - 3|$ gives the distance between 5 and 3, which is 2. Confirming this, we see that $|5 - 3| = |2| = 2$.

  (b) The expression $|3 - 5|$ gives the distance between 3 and 5, which is 2. Confirming this, we see that $|3 - 5| = |-2| = 2$, the same as the answer to part (a).

  (c) Writing $|5 + 3|$ in the form $|a - b|$, we have $|5 + 3| = |5 - (-3)|$. This expression gives the distance between 5 and -3, which is 8. Confirming this, we see that $|5 + 3| = |8| = 8$.

An Algebraic Definition for Absolute Value

When we define $|x|$ as the distance between 0 and x, we are giving a geometric definition, because distance is a geometric concept. We now give an *algebraic* definition of $|x|$. Notice that if x is a positive number (or zero), then $|x|$ is the same as x. However, if x is a negative number, then $|x|$ has the opposite sign as x, which means $|x|$ equals x multiplied by -1:

> ### Definition of Absolute Value
>
> If x is any real number then:
> - $|x| = x$ if $x \geq 0$
> - $|x| = -x$ if $x < 0$.

We can see how this rule works by looking at a few specific cases:

$$\text{For } x \geq 0: \quad \text{If } x = +3, \text{ then } |x| = x = +3 = 3$$
$$\text{If } x = +4, \text{ then } |x| = x = +4 = 4$$
$$\vdots \qquad \vdots \quad \vdots \qquad \vdots$$
$$\text{For } x < 0: \quad \text{If } x = -3, \text{ then } |x| = -x = -(-3) = 3$$
$$\text{If } x = -4, \text{ then } |x| = -x = -(-4) = 4$$
$$\vdots \qquad \vdots \quad \vdots \qquad \vdots$$

and so on.

Example 3 Evaluate the following expressions using the algebraic definition of absolute value.

(a) $|100|$ 　　　　(b) $|-3|$ 　　　　(c) $|0|$ 　　　　(d) $\dfrac{a}{|a|}, \quad a < 0.$

Solution

(a) Since 100 is positive, its absolute value is the same as itself, so $|100| = 100$.
(b) Since -3 is negative, its absolute value has the opposite sign, so $|-3| = -(-3) = 3$.
(c) The absolute value of 0 is the same as itself, so $|0| = 0$.
(d) Since a is negative, its absolute value has the opposite sign, so $|a| = -a$. Thus,

$$\frac{a}{|a|} = \frac{a}{-a} = -1.$$

Absolute Value Equations

We can use the geometric interpretation of $|x|$ to solve equations involving absolute values.

Example 4 Solve for x:

(a) $|x - 2| = 4$ 　　　　(b) $|x + 6| = 15$ 　　　　(c) $|x - 2| = -3$.

Solution

(a) The statement
$$|x - 2| = 4$$

means that the distance between x and 2 equals 4. We know that $x = 6$ is 4 units to the right of 2, and $x = -2$ is 4 units to the left of 2, so the solutions are $x = -2$ and $x = 6$. Substituting these values into the original equation, we have:

$$|6 - 2| = |4| \quad = 4 \quad \text{as required}$$
$$|(-2) - 2| = |-4| = 4 \quad \text{as required.}$$

(b) We can rewrite this equation as
$$|x - (-6)| = 15.$$

This statement means that the distance between x and -6 equals 15. We know that $x = 9$ is 15 units to the right of -6, and $x = -21$ is 15 units to the left of -6, so the solutions are $x = -21$ and $x = 9$. Substituting these values into the original equation, we have:

$$|-21 + 6| = |-15| = 15 \quad \text{as required}$$
$$|9 + 6| = |15| \quad = 15 \quad \text{as required.}$$

(c) The statement
$$|x - 2| = -3$$

means that the distance between x and 2 equals -3. We conclude there is no solution, because a distance cannot be negative.

Solving Absolute Value Equations Algebraically

In Example 4, we solved the equation $|x - 2| = 4$ by thinking geometrically. We can also solve it algebraically. The key is to notice that there are two possibilities:

$$\text{Since} \quad |x - 2| = 4,$$
$$\text{then either} \quad x - 2 = 4 \quad \longrightarrow x = 6$$
$$\text{or} \quad x - 2 = -4 \longrightarrow x = -2.$$

These solutions, $x = -2$ and $x = 6$, are the same as we got in part (a) of Example 4. Notice that we rewrote the equation $|x - 2| = 4$ as two separate equations: $x - 2 = 4$ and $x - 2 = -4$. This is similar to rewriting the equation $(x - 2)^2 = 4$ as two separate equations: $x - 2 = 2$ and $x - 2 = -2$.

Example 5   Solve:

(a) $|3t - 4| = 11$ (b) $3|2x - 5| - 7 = -1$

Solution   (a) We first rewrite the equation $|3t - 4| = 11$ as two separate equations:

$$
\begin{array}{c|c}
3t - 4 = +11 & 3t - 4 = -11 \\
3t = 15 & 3t = -7 \\
t = 5 & t = -\dfrac{7}{3}.
\end{array}
$$

Checking our answer, we see that:

$$|3 \cdot 5 - 4| = |15 - 4| = |+11| = 11 \quad \text{as required}$$
$$|3 \cdot (-7/3) - 4| = |-7 - 4| = |-11| = 11 \quad \text{as required.}$$

(b) First, we isolate the absolute value:

$$3|2x - 5| - 7 = -1$$
$$3|2x - 5| = 6$$
$$|2x - 5| = 2.$$

Next, we rewrite this as two separate equations:

$$
\begin{array}{c|c}
2x - 5 = +2 & 2x - 5 = -2 \\
2x = 7 & 2x = 3 \\
x = 3.5 & x = 1.5.
\end{array}
$$

Checking our answer, we see that:

$$3|2(3.5) - 5| - 7 = 3|7 - 5| - 7 = 3|+2| - 7 = 3 \cdot 2 - 7 = -1 \quad \text{as required}$$
$$3|2(1.5) - 5| - 7 = 3|3 - 5| - 7 = 3|-2| - 7 = 3 \cdot 2 - 7 = -1 \quad \text{as required.}$$

Absolute Value Inequalities

Since $|x|$ gives the distance between 0 and x, the inequality $|x| < 5$ tells us that the distance between 0 and x must be less than 5. Looking at the number line, we see in Figure 3.4 that this means that x must be between -5 and $+5$.

Figure 3.4: Solution of $|x| < 5$

We can write our solution to this inequality as $x < 5$ and $x > -5$. We say "and" because both conditions must be true. We can write this solution more compactly as

$$-5 < x < 5.$$

Example 6 Solve $|x| > 5$.

Solution The statement

$$|x| > 5$$

means that the distance between 0 and x is greater than 5. Looking at the number line, we conclude that x must *not* be between -5 and 5, not can it equal 5 or -5. See Figure 3.5.

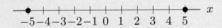

Figure 3.5: Solution of $|x| > 5$

We can write the solution to Example 6 as $x < -5$ or $x > 5$. Here, we say "or" instead of "and" because either condition (but not both) must apply. Note that there is no compact way of writing

$$x < -5 \text{ or } x > 5.$$

In particular, writing something like $5 < x > -5$ is incorrect!

Example 7 Solve each absolute value inequality.

(a) $|2x - 3| < 7$ (b) $|1 - 4x| \geq 10$

Solution (a) In order for $|2x - 3| < 7$, two things must be true:

$$2x - 3 > -7 \quad \text{and} \quad 2x - 3 < 7.$$

In other words, $2x - 3$ must be between -7 and 7. We have:

$$
\begin{array}{c|c}
2x - 3 > -7 & 2x - 3 < 7 \\
2x > -4 & 2x < 10
\end{array}
$$

$$x > -2 \quad | \quad x < 5.$$

Since both these statements must be true, we see that x must be between -2 and 5, and we can write $-2 < x < 5$.

(b) In order for $|1 - 4x| \geq 10$, one of two things must be true:

$$1 - 4x \leq -10 \quad \text{or} \quad 1 - 4x \geq 10.$$

In other words, $1 - 4x$ must *not* be between -10 and 10. We have:

$$
\begin{array}{c|c}
\begin{aligned}
1 - 4x &\leq -10 \\
1 &\leq -10 + 4x \\
11 &\leq 4x \\
2.75 &\leq x
\end{aligned}
&
\begin{aligned}
1 - 4x &\geq 10 \\
1 &\geq 10 + 4x \\
-9 &\geq 4x \\
-2.25 &\geq x.
\end{aligned}
\end{array}
$$

Thus, either $x \leq -2.25$ or $x \geq 2.75$.

Exercises and Problems for Section 3.3

EXERCISES

In Exercises **1–7**, are there values of x which satisfy the statements? Explain how you can tell without finding, or attempting to find, the values.

1. $|x| - 3 = 10$

2. $|x - 3| = 10$

3. $|x + 7| = -3$

4. $|2x + 5| = |-7 - 10|$

5. $|x - 1| > 2$

6. $|3 - x| - 1 < 0$

7. $|5 + x| + 7 < 0$

8. Interpret each of the following absolute values as a distance on the number line. Evaluate when possible.

    (a) $|3.5|$ **(b)** $|-14|$ **(c)** $|7 - 2|$

    (d) $|-7 - 2|$ **(e)** $|x - 4|$ **(f)** $|x + 4|$

9. Classify each statement as true or false.

    (a) $|-25| < 0$ **(b)** $-|-11| = 11$

    (c) $|5 - 7| = |5| - |7|$ **(d)** $|12 - 11| = |11 - 12|$

    (e) If $x < y$, $|x| < |y|$ **(f)** $|x| = |-x|$

10. Write an absolute value equation or inequality to describe each of the following situations.

    (a) The distance between x and zero is exactly 7.

    (b) The distance between x and 2 is exactly 6.

    (c) The distance between t and -2 is exactly 1.

    (d) The distance between x and zero is less than 4.

    (e) The distance between z and zero is greater than or equal to 9.

    (f) The distance between w and -5 is greater than 7.

In Exercises **11–15**, solve the absolute value equation by writing it as two separate equations.

11. $|x - 1| = 6$

12. $5 = |2x| - 3$

13. $|2t - 1| = 3$

14. $\left| 2 - \dfrac{r}{3} \right| = 7$

15. $2 = \dfrac{|p + 1|}{4}$

In Exercises **16–19**, solve the inequality.

16. $|5 + 2w| < 7$

17. $\left| \dfrac{x}{3} + 7 \right| \geq 2$

18. $|8 - 2x| < 6$

19. $|3z - 9| > 4$

PROBLEMS

In Problems **20–25**, write an absolute value equation or inequality whose solution matches the graph.

20.

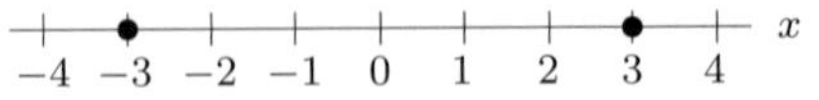

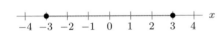

Figure 3.6

21.

Figure 3.7

22.

Figure 3.8

23.

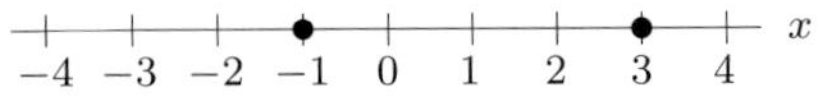

Figure 3.9

24.

Figure 3.10

25.

Figure 3.11

REVIEW EXERCISES AND PROBLEMS FOR CHAPTER 3

EXERCISES

Solve the equations in Exercises **1–4**.

1. $0.5x - 5 = 9$

2. $-7 + 9t = -4t$

3. $-6(2k - 1) = 5(3 - 2k)$

4. $4 - (r - 3) = 6(1 - r)$

In Exercises **5–7**, solve for the indicated variable. Assume all constants are non-zero.

5. $bx - d = ax + c$, for x, if $a - b \neq 0$.

6. $S = \dfrac{tL - a}{t - 1}$, for t, if $S \neq L$.

7. $2m + n = p$, for m.

Solve the equations in Exercises **8–11**.

8. $\dfrac{8x}{5} = x + 6$

9. $\dfrac{j + 3}{j - 4} + \dfrac{3}{7} = \dfrac{5}{7}$

10. $\dfrac{7}{8} + \dfrac{a + 4}{a - 3} = \dfrac{5}{8}$

11. $\dfrac{5}{6} + \dfrac{2r + 1}{r - 2} = \dfrac{1}{6}$

Each of the inequalities in Exercises **12–15** can be solved by performing a single operation on both sides. State the operation, indicating whether or not the inequality changes direction. Solve the inequality.

12. $15x \geq 45$

13. $14 \leq -q$

14. $10.7 + x > -8$

15. $\dfrac{-2r}{5} > \dfrac{4}{7}$

Each of the inequalities in Exercises **16–25** can be solved by performing two operations on both sides. State the operations in order of use and solve the inequality.

16. $6 + 3x < 13$

17. $7 - 4n \leq 5 - 5n$

18. $-3(t - 2) \geq -15$

19. $2k > 12 - 7k$

20. $8 - 4b \geq 22b$

21. $-13 - 12s < -17$

22. $3q + 2 < 50$

23. $-8 < -4(x - 5)$

24. $4.4 - s \leq 6.4$

25. $\dfrac{7}{3}r - r > 8$

In Exercises 26–29, are there values of x which satisfy the statements? Explain how you can tell without finding the values.

26. $|x| + 8 = 15$

27. $|x| + 8 = 7$

28. $|7x - 70| = |14 - 100|$

29. $|-7 - x| + 1 < 0$

In Exercise 30–35, solve the absolute value equation by writing it as two separate equations.

30. $|z - 3| = 7$

31. $|4 - 3t| = 5$

32. $\left| 5 - \dfrac{k}{7} \right| = 4$

33. $|3 - 4w| - 2 = 6$

34. $\dfrac{|3p + 7|}{4} + 2 = 5$

35. $3\,|x - 2| = 8 - |x - 2|$

In Exercises 36–39, solve the inequality.

36. $|2h + 7| < 5$

37. $\left| 4 - \dfrac{j}{3} \right| \geq 7$

38. $2 - \dfrac{|d - 3|}{4} \geq 5$

39. $14 - |3c - 4| \leq 6$

PROBLEMS

Solve the equations in Problems 40–43 for the indicated variable in terms of the other letters.

40. $p\sqrt{q} + np = k$, for q 41. $p\sqrt{q} + np = k$, for p

42. $\dfrac{1 + r\sqrt{s - 1}}{r + \sqrt{s - 1}} = z$, for r

43. $\dfrac{1 + r\sqrt{s - 1}}{r + \sqrt{s - 1}} = z$, for s

For what values of x are the quantities in Problems 44–47 positive?

44. $x + 5$ 45. $4x - 28$

46. $-7x - 35$ 47. $(x + 3)(x - 12)$

48. In 2008, shares of Total varied on the Paris bourse at prices from 33.18 euro to 58.90 euro.[1] Also during the year, the euro varied in value from $1.24 to $1.60. How many shares, s, of Total might you have been able to buy on any given day in 2008 with $10,000? Your answer should be an inequality.

49. A company's one-pound boxes of pasta actually weigh 1 lb plus or minus 1/32 lbs. You serve 4 friends with 2 boxes. Write and solve an inequality to express the amount of pasta per person. Do you think your friends would notice the difference between the heaviest and lightest packages?

50. A scale for weighing agricultural produce is accurate within plus or minus 2%. If the price of corn is $40/ton, write and solve an inequality describing how much money a farmer gets for 10 tons of corn.

---

[1] www.boursorama.com

Chapter 4

Functions, Expressions, and Equations

CONTENTS

4.1 WHAT IS A FUNCTION?

In everyday language, *function* expresses the notion of one thing depending on another. We might say that election results are a function of campaign spending, or that ice cream sales are a function of the weather. In mathematics, the meaning of *function* is more precise, but the idea is the same. If the value of one quantity determines the value of another, we say the second quantity is a function of the first.

A **function** is a rule that takes numbers as inputs and assigns to each input exactly one number as output. The output is a function of the input.

Example 1   A 20% tip on a meal is a function of the cost, in dollars, of the meal. What is the input and what is the output to this function?

Solution   The input is the amount of the bill in dollars, and the output is the amount of the tip in dollars.

Function Notation

We use *function notation* to represent the output of a function in terms of its input. The expression $f(20)$, for example, stands for the output of the function f when the input is 20. Here the letter f stands for the function itself, not for a number. If f is the function in Example 1, we have

$$f(20) = 4,$$

since the tip for a $20 meal is $(0.2)(20) = \$4$. In words, we say "$f$ of 20 equals 4." In general,

$$f(\text{Amount of bill}) = \text{Tip}.$$

Example 2   For the function in Example 1, let B stand for the bill in dollars and T stand for the tip in dollars.
(a) Express the function in terms of B and T.
(b) Evaluate $f(8.95)$ and $f(23.70)$ and interpret your answer in practical terms.

Solution   (a) Since the tip is 20% of B, or $0.2B$, we have

$$T = f(B) = 0.2B.$$

(b) We have

$$f(8.95) = 0.2(8.95) = 1.79$$

and

$$f(23.70) = 0.2(23.70) = 4.74.$$

This tells us the tip on a $8.95 meal is $1.79, and the tip on a $23.70 meal is $4.74.

Example 3   If you buy first-class stamps, the total cost in dollars is a function of the number of stamps bought.
(a) What is the input and what is the output of this function?
(b) Use function notation to express the fact that the cost of 14 stamps is $6.16.
(c) If a first-class stamp costs 44 cents, find a formula for the cost, C, in dollars, in terms of the number of stamps, n.

Solution   (a) The input is the number of stamps, and the output is the total cost:

$$\text{Total cost} = f(\text{Number of stamps}).$$

(b) If the total cost is $6.16 for 14 stamps, then

$$f(14) = 6.16.$$

(c) Since 44 cents is 0.44 dollars, n stamps cost $0.44n$ dollars, so

$$C = f(n) = 0.44n.$$

Dependent and Independent Variables

A variable used to stand for the input, such as B in Example 2 or n in Example 3, is also called the *independent variable*, and a variable used to stand for the output, such as T in Example 2 or C in Example 3, is also called the *dependent variable*. Symbolically,

$$\text{Output} = f(\text{Input})$$
$$\text{Dependent} = f(\text{Independent}).$$

Example 4   The cost in dollars of tuition, T, at most colleges is a function of the number of credits taken, c. Express the relationship in function notation and identify the independent and dependent variables.

Solution   We have
$$\text{Tuition cost} = f(\text{Number of credits}), \quad \text{or} \quad T = f(c).$$
The independent variable is c, and the dependent variable is T.

Example 5   The area of a circle of radius r is given by $A = \pi r^2$. What is the independent variable? What is the dependent variable?

Solution   We use this function to compute the area when we know the radius. Thus, the independent variable is r, the radius, and the dependent variable is A, the area.

In a situation like Example 5, where we have a formula like

$$A = \pi r^2$$

for the function, we often do not bother with function notation.

Evaluating Functions

If $f(x)$ is given by an algebraic expression in x, then finding the value of $f(5)$, for instance, is the same as evaluating the expression at $x = 5$.

Example 6   If $f(x) = 5 - \sqrt{x}$, evaluate each of the following:

   (a) $f(0)$ (b) $f(16)$ (c) $f(12^2)$.

Solution   (a) We have $f(0) = 5 - \sqrt{0} = 5$.
   (b) We have $f(16) = 5 - \sqrt{16} = 5 - 4 = 1$.
   (c) We have $f(12^2) = 5 - \sqrt{12^2} = 5 - 12 = -7$.

Example 6 (c) illustrates that the input to a function can be more complicated than a simple number. Often we want to consider inputs to functions that are numerical or algebraic expressions.

Example 7   Let $h(x) = x^2 - 3x + 5$. Evaluate the following:

   (a) $h(a) - 2$ (b) $h(a - 2)$ (c) $h(a - 2) - h(a)$.

Solution   To evaluate a function at an expression, such as $a - 2$, it is helpful to remember that $a - 2$ is the input, and to rewrite the formula for f as

$$\text{Output} = h(\text{Input}) = (\text{Input})^2 - 3 \cdot (\text{Input}) + 5.$$

(a) First input a, then subtract 2:

$$h(a) - 2 = \underbrace{(a)^2 - 3(a) + 5}_{h(a)} - 2$$
$$= a^2 - 3a + 3.$$

(b) In this case, Input $= a - 2$. We substitute and multiply out:

$$h(a - 2) = (a - 2)^2 - 3(a - 2) + 5$$
$$= a^2 - 4a + 4 - 3a + 6 + 5$$
$$= a^2 - 7a + 15.$$

(c) We must evaluate h at two different inputs, $a - 2$ and a. We have

$$h(a - 2) - h(a) = \underbrace{(a - 2)^2 - 3(a - 2) + 5}_{h(a-2)} - \underbrace{\left((a)^2 - 3(a) + 5\right)}_{h(a)}$$
$$= \underbrace{a^2 - 4a + 4 - 3a + 6 + 5}_{h(a-2)} - \underbrace{(a^2 - 3a + 5)}_{h(a)}$$
$$= -4a + 10.$$

Example 8   Let $n = f(p)$ be the average number of days a house in a particular community stays on the market before being sold for price p (in \$1000s), and let p_0 be the average sale price of houses in the community. What do the following expressions mean in terms of the housing market?

(a) $f(250)$ (b) $f(p_0 + 10)$ (c) $f(0.9p_0)$

Solution   (a) This is the average number of days a house stays on the market before being sold for \$250,000.
(b) Since $p_0 + 10$ is 10 more than p_0 (in thousands of dollars), $f(p_0 + 10)$ is the average number of days a house stays on the market before being sold for \$10,000 above the average sale price.
(c) Since $0.9p_0$ is 90% of p_0, or 10% less than p_0, $f(0.9p_0)$ is the average number of days a house stays on the market before being sold at 10% below the average sale price.

When interpreting the meaning of a statement about functions, it is often useful to think about the units of measurement for the independent and dependent variables.

Example 9   Let $f(A)$ be the number of gallons needed to paint a house with walls of area A ft^2.

(a) In the statement $f(3500) = 10$, what are the units of the 3500 and the 10? What does the statement tell you about painting the house?
(b) Suppose $f(A) = A/350$. How many gallons do you need to paint a house whose walls measure 5000 ft^2? 10,000 ft^2?
(c) Explain in words the relationship between the number of gallons and A. What is the practical interpretation of the 350 in the expression for the function?

Solution   (a) The input, 3500, is the area of the walls, so its units are ft^2. The output, 10, is the number of gallons of paint. The statement $f(3500) = 10$ tells us that we need 10 gallons to paint 3500 ft^2 of wall.
(b) For walls measuring 5000 ft^2 we need $f(5000) = 5000/350 = 14.3$ gallons. For walls measuring 10,000 ft^2 we need $f(10,000) = 10,000/350 = 28.6$ gallons.
(c) The expression tells us to divide the area by 350 to compute the number of gallons. This means that 350 ft^2 is the area covered by one gallon.

Tables and Graphs

We can often see key features of a function by making a table of output values and by drawing a graph of the function. For example, Table 4.1 and Figure 4.1 show values of $g(a) = a^2$ for $a = -2, -1, 0, 1$ and 2. We can see that the output values never seem to be negative, which is confirmed by the fact that a square is always positive or zero.

Table 4.1   *Values of* $g(a) = a^2$

a (input)	$g(a) = a^2$ (output)
-2	4
-1	1
0	0
1	1
2	4

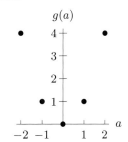

Figure 4.1: Graph of values of $g(a) = a^2$

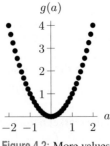

Figure 4.2: More values
of $g(a) = a^2$

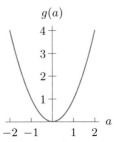

Figure 4.3: Graph of
$g(a) = a^2$

If we plot more values, we get Figure 4.2. Notice how the points appear to be joined by a smooth curve. If we could plot all the values we would get the curve in Figure 4.3. This is the *graph* of the function. A graphing calculator or computer shows a good approximation of the graph by plotting many points on the screen.

Example 10   Explain how Table 4.1 and Figure 4.3 illustrate that $g(a) = g(-a)$ for any number a.

Solution   Notice the pattern in the right column of Table 4.1: the values go from 4 to 1 to 0, then back to 1 and 4 again. This is because $(-2)^2$ and 2^2 both have the same value, namely 4, so

$$g(-2) = g(2) = 4.$$

Similarly,

$$g(-1) = g(1) = 1.$$

We can see the same thing in the symmetrical arrangement of the points in Figures 4.1 and 4.2 about the vertical axis. Since both a^2 and $(-a)^2$ have the same value, the point above a on the horizontal axis has the same height as the point above $-a$.

Example 11   For the function graphed in Figure 4.4, find $f(x)$ for $x = -3, -2, -1, 0, 1, 2, 3$.

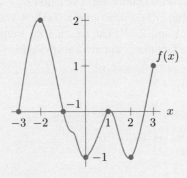

Figure 4.4

Solution The coordinates of a point on the graph of f are $(a, f(a))$ for some number a. So, since the point $(-3, 0)$ is on the graph, we must have $f(-3) = 0$. Similarly, since the point $(-2, 2)$ is on the graph, we must have $f(-2) = 2$. Using the other marked points, we get the values in Table 4.2.

Table 4.2

x	-3	-2	-1	0	1	2	3
$f(x)$	0	2	0	-1	0	-1	1

Exercises and Problems for Section 4.1

EXERCISES

In Exercises 1–2, write the relationship using function notation (that is, y is a function of x is written $y = f(x)$).

1. Weight, w, is a function of caloric intake, c.

2. Number of molecules, m, in a gas, is a function of the volume of the gas, v.

3. The number, N, of napkins used in a restaurant is $N = f(C) = 2C$, where C is the number of customers. What is the dependent variable? The independent variable?

4. A silver mine's profit, P, is $P = g(s) = -100{,}000 + 50{,}000s$ dollars, where s is the price per ounce of silver. What is the dependent variable? The independent variable?

In Exercises 5–6, evaluate the function for $x = -7$.

5. $f(x) = x/2 - 1$

6. $f(x) = x^2 - 3$

7. Let $g(x) = (12 - x)^2 - (x - 1)^3$. Evaluate

  (a) $g(2)$ **(b)** $g(5)$
  (c) $g(0)$ **(d)** $g(-1)$

8. Let $f(x) = 2x^2 + 7x + 5$. Evaluate

  (a) $f(3)$ **(b)** $f(a)$
  (c) $f(2a)$ **(d)** $f(-2)$

In Exercises 9–14, evaluate the expressions given that

$$f(x) = \frac{2x + 1}{3 - 5x} \qquad g(y) = \frac{1}{\sqrt{y^2 + 1}}.$$

9. $f(0)$ 10. $g(0)$ 11. $g(-1)$

12. $f(10)$ 13. $f(1/2)$ 14. $g\left(\sqrt{8}\right)$

In Exercises 15–20, evaluate the expressions given that

$$h(t) = 10 - 3t.$$

15. $h(r)$ 16. $h(2u)$ 17. $h(k - 3)$

18. $h(4 - n)$ 19. $h(5t^2)$ 20. $h(4 - t^3)$

21. If $f(x) = 1 - x^2 - x$, evaluate and simplify $f(1 - x)$.

22. The sales tax on an item is 6%. Express the total cost, C, in terms of the price of the item, P.

23. Let $f(T)$ be the volume in liters of a balloon at temperature $T°$C. If $f(40) = 3$,

  (a) What are the units of the 40 and the 3?
  (b) What is the volume of the balloon at $40°$C?

In Exercises 24–26 use the table to fill in the missing values. (There may be more than one answer.)

24. **(a)** $g(0) =?$ **(b)** $g(?) = 0$
   (c) $g(-5) =?$ **(d)** $g(?) = -5$

y	-10	-5	0	5	10
$g(y)$	-5	0	5	10	-10

25. **(a)** $h(0) =?$ **(b)** $h(?) = 0$
   (c) $h(-2) =?$ **(d)** $h(?) = -2$

t	-3	-2	-1	0	1	2	3
$h(t)$	-1	0	-3	-2	-1	-2	0

26. **(a)** $h(?) = 2h(0)$ **(b)** $h(?) = 2h(-3)+h(2)$

(c) $h(?) = h(-2)$ **(d)** $h(?) = h(1) + h(2)$

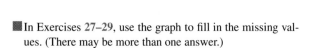

t	-3	-2	-1	0	1	2	3
$h(t)$	-1	0	-4	-2	-1	-2	0

28. **(a)** $h(0) =?$ **(b)** $h(?) = 0$

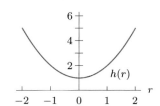

In Exercises **27–29**, use the graph to fill in the missing values. (There may be more than one answer.)

27. **(a)** $f(0) =?$ **(b)** $f(?) = 0$

29. **(a)** $g(0) =?$ **(b)** $g(?) = 0$

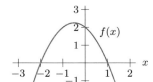

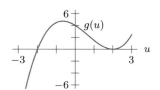

PROBLEMS

30. Table 4.3 shows the 5 top winning teams in the NBA playoffs between 2000 and May 20, 2007 and the number of games each team has won.

Table 4.3

Team	Playoff games won
Lakers	66
Spurs	66
Pistons	61
Nets	43
Mavericks	41

(a) Is the number of games a team won a function of the team? Why or why not?

(b) Is the NBA team a function of the number of games won? Why or why not?

31. Let $r(p)$ be the revenue in dollars that a company receives when it charges p dollars for a product. Explain the meaning of the following statements.

(a) $r(15) = 112{,}500$ **(b)** $r(a) = 0$

(c) $r(1) = b$ **(d)** $c = r(p)$

32. Let $d(v)$ be the braking distance in feet of a car traveling at v miles per hour. Explain the meaning of the following statements.

(a) $d(30) = 111$ **(b)** $d(a) = 10$

(c) $d(10) = b$ **(d)** $s = d(v)$

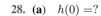The lower the price per song, the more songs are downloaded from an online music store. Let $n = r(p)$ give the average number of daily downloads as a function of the price (in cents). Let p_0 be the price currently being charged (in cents). What do the expressions in Problems **33–40** tell you about downloads from the store?

33. $r(99)$ **34.** $r(p_0)$

35. $r(p_0 - 10)$ **36.** $r(p_0 - 10) - r(p_0)$

37. $365r(p_0)$ **38.** $r(0.80p_0)$

39. $\dfrac{r(p_0)}{24}$ **40.** $p_0 \cdot r(p_0)$

Evaluate the expressions in Problems **41–42** given that

$$f(n) = \frac{1}{2}n(n + 1).$$

41. $f(100)$ **42.** $f(n + 1) - f(n)$

■Different strains of a virus survive in the air for different time periods. For a strain that survives t minutes, let $h(t)$ be the number of people infected (in thousands). The most common strain survives for t_0 minutes. What do the expressions in Problems 43–44 tell you about the number of people infected?

43. $h(t_0 + 3)$

44. $\dfrac{h(2t_0)}{h(t_0)}$

■A car with tire pressure P lbs/in^2 gets gas mileage $g(P)$ (in mpg). The recommended tire pressure is P_0. What do the expressions in Problems 45–46 tell you about the car's tire pressure and gas mileage?

45. $g(0.9P_0)$

46. $g(P_0) - g(P_0 - 5)$

4.2 FUNCTIONS AND EXPRESSIONS

A function is often defined by an expression. We find the output by evaluating the expression at the input. By paying attention to the form of the expression, we can learn properties of the function.

Using Expressions to Interpret Functions

Example 1

Bernardo plans to travel 400 miles over spring break to visit his family. He can choose to fly, drive, take the train, or make the journey as a bicycle road trip. If his average speed is r miles per hour, then the time taken is a function of r and is given by

$$\text{Time taken at } r \text{ miles per hour} = T(r) = \frac{400}{r}.$$

(a) Find $T(200)$ and $T(80)$ and give a practical interpretation of the answers.
(b) Use the algebraic form of the expression for $T(r)$ to explain why $T(200) < T(80)$, and explain why the inequality makes sense in practical terms.

Solution

(a) We have
$$T(200) = \frac{400}{200} = 2 \quad \text{and} \quad T(80) = \frac{400}{80} = 5.$$

This means that it takes Bernardo 2 hours traveling at 200 miles per hour and 5 hours traveling at 80 miles per hour.
(b) In the fraction $400/r$, the variable r occurs in the denominator. Since dividing by a larger number gives a smaller number, making r larger makes $T(r)$ smaller. This makes sense in practical terms, because if you travel at a faster speed you finish the journey in less time.

It is also useful to interpret the algebraic form graphically.

Example 2

For the function $T(r)$ in Example 1,

(a) Make a table of values for $r = 25, 80, 100$, and 200, and graph the function.
(b) Explain the shape of the graph using the form of the expression $400/r$.

Solution   (a) Table 4.4 shows values of the function, and Figure 4.5 shows the graph.

Table 4.4   *Values of* $T(r) = 400/r$

r	$T(r)$
25	16
80	5
100	4
200	2

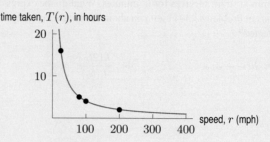

Figure 4.5: Graph of $400/r$

(b) The graph slopes downward as you move from left to right. This is because the points to the right on the graph correspond to larger values of the input, and so to smaller values of the output. Since the output is read on the vertical axis, as one moves to the right on the horizontal axis, output values become smaller on the vertical axis.

Constants and Variables

Sometimes an expression for a function contains letters in addition to the independent variable. We call these letters *constants* because for a given function, their value does not change.

Example 3   Einstein's famous equation $E = mc^2$ expresses energy E as a function of mass m. Which letters in this equation represent variables and which represent constants?

Solution   We are given that E is a function of m, so E is the dependent variable and m is the independent variable. The symbol c is a constant (which stands for the speed of light).

Which letters in an expression are constants and which are variables depends on the context in which the expression is being used.

Example 4   A tip of r percent on a bill of B dollars is given in dollars by

$$\text{Tip} = \frac{r}{100}B.$$

Which letters in the expression $(r/100)B$ would you call variables and which would you call constants if you were considering

(a)   The tip as a function of the bill amount?   (b)   The tip as a function of the rate?

Solution   (a) In this situation, we regard the tip as a function of the variable B and regard r as a constant. If we call this function f, then we can write

$$\text{Tip} = f(B) = \frac{r}{100}B.$$

(b) Here we regard the tip as a function of the variable r and regard B as a constant. If we call this function g, then

$$\text{Tip} = g(r) = \frac{r}{100}B.$$

Functions and Equivalent Expressions

A number can be expressed in many different ways. For example, $1/4$ and 0.25 are two different ways of expressing the same number. Similarly, we can have more than one expression for the same function.

Example 5   In Example 2 on page 78 we found the expression $0.2B$ for a 20% tip on a bill. Pares says she has an easy way to figure out the tip: she moves the decimal point in the bill one place to the left, then doubles the answer.

(a) Check that Pares' method gives the same answer on bill amounts of $8.95 and $23.70 as Example 2 on page 78.
(b) Does Pares' method always work? Explain your answer using algebraic expressions.

Solution   (a) For $8.95, first we move the decimal point to the left to get 0.895, then double to get a tip of $1.79. For $23.70, we move the decimal point to the left to get 2.370, then double to get a tip of $4.74. Both answers are the same as in Example 2 on page 78.

(b) Moving the decimal point to the left is the same as multiplying by 0.1. So first Pares multiplies the bill by 0.1, then multiplies the result by 2. Her calculation of the tip is

$$2(0.1B).$$

We can simplify this expression by regrouping the multiplications:

$$2(0.1B) = (2 \cdot 0.1)B = 0.2B.$$

This last expression for the tip is the same one we found in Example 2 on page 78.

For each value of B the two expressions, $2(0.1B)$ and $0.2B$, give equal values for the tip. Although the expressions appear different, they are equivalent, and therefore define the same function.

Example 6   Let $g(a) = a^2 - a$. Which of the following pairs of expressions are equivalent?

(a) $g(2a)$ and $2g(a)$
(b) $(1/2)g(a)$ and $g(a)/2$
(c) $g(a + 1)$ and $g(a) + 1$

Solution   (a) In order to find the output for g, we square the input value, a and subtract a from the square. This means that $g(2a) = (2a)^2 - (2a) = 4a^2 - 2a$. The value of $2g(a)$ is two times the value of $g(a)$. So $2g(a) = 2(a^2 - a) = 2a^2 - 2a$. The two expressions are not equivalent.

(b) Since multiplying by $1/2$ is the same as dividing by 2, the two expressions are equivalent:

$$\left(\frac{1}{2}\right)g(a) = \frac{1}{2}(a^2 - a) = \frac{a^2 - a}{2} = \frac{g(a)}{2}.$$

(c) To evaluate $g(a + 1)$ we have $g(a + 1) = (a + 1)^2 - (a + 1) = a^2 + 2a + 1 - a - 1 = a^2 + a$. To evaluate $g(a) + 1$ we add one to $g(a)$, so $g(a) + 1 = a^2 - a + 1$. The two expressions are not equivalent.

Often we need to express a function in a standard form to recognize what type of function it is. For example, in Section 4.5 we study functions of the form $Q = kt$, where k is a constant.

Example 7   Express each of the following functions in the form $Q = kt$ and give the value of k.

(a) $Q = 5t + rt$ (b) $Q = \dfrac{-t}{10}$ (c) $Q = t(t+1) - t(t-1)$

Solution   (a) We have

$$Q = 5t + rt = (5 + r)t,$$

which is the form $Q = kt$, with $k = 5 + r$.

(b) Rewriting the fraction as

$$Q = \frac{-t}{10} = \left(\frac{-1}{10}\right) t = -0.1t,$$

we see that $k = -0.1$.

(c) We have

$$Q = t(t+1) - t(t-1) = t^2 + t - t^2 + t = 2t,$$

so $k = 2$.

Exercises and Problems for Section 4.2

EXERCISES

■In Exercises 1–4

(a) Evaluate the function at the given input values. Which gives the greater output value?

(b) Explain the answer to part (a) in terms of the algebraic expression for the function.

1. $f(x) = 9 - x$, $x = 1, 3$

2. $g(a) = a - 2$, $a = -5, -2$

3. $C(p) = \dfrac{-p}{5}$, $p = 100, 200$

4. $h(t) = \dfrac{t}{5}$, $t = 4, 6$

■In Exercises 5–8, $f(t) = t/2 + 7$. Determine whether the two expressions are equivalent.

5. $\dfrac{f(t)}{3}, \dfrac{1}{3}f(t)$ 6. $f(t^2), (f(t))^2$

7. $2f(t), f(2t)$ 8. $f(4t^2), f((2t)^2)$

■In Exercises 9–16, are the two functions the same function?

9. $f(x) = x^2 - 4x + 5$ and $g(x) = (x-2)^2 + 1$

10. $f(x) = 2(x+1)(x-3)$ and $g(x) = x^2 - 2x - 3$.

11. $f(t) = 450 + 30t$, and $g(p) = 450 + 30p$

12. $A(n) = (n-1)/2$ and $B(n) = 0.5n - 0.5$

13. $r(x) = 5(x-2) + 3$ and $s(x) = 5x + 7$

14. $h(t) = t^2 - t(t-1)$ and $g(t) = t$

15. $Q(t) = \dfrac{t}{2} - \dfrac{3}{2}$ and $P(t) = t - 3$

16. $B(v) = 30 - \dfrac{480}{v}$ and $C(v) = 30\left(\dfrac{v-16}{v}\right)$.

■In Exercises 17–20, which letters stand for variables and which for constants?

17. $V(r) = (4/3)\pi r^2$ 18. $f(x) = b + mx$

19. $P(t) = A(1 - rt)$ 20. $B(r) = A(1 - rt)$

PROBLEMS

21. The number of gallons left in a gas tank after driving d miles is given by $G(d) = 17 - 0.05d$.

   (a) Which is larger, $G(50)$ or $G(100)$?
   (b) Explain your answer in terms of the expression for $G(d)$, and give a practical interpretation.

22. If you drive to work at v miles per hour, the time available for breakfast is $B(v) = 30 - 480/v$ minutes.

   (a) Which is greater, $B(35)$ or $B(45)$?
   (b) Explain your answer in terms of the expression for $B(v)$ and give a practical interpretation.

23. Abby and Leah go on a 5 hour drive for 325 miles at 65 mph. After t hours, Abby calculates the distance remaining by subtracting $65t$ from 325, whereas Leah subtracts t from 5 then multiplies by 65.

  (a) Write expressions for each calculation.
  (b) Do the expressions in (a) define the same function?

24. To calculate the balance after investing P dollars for two years at 5% interest, Sharif adds 5% of P to P, and then adds 5% of the result of this calculation to itself. Donald multiplies P by 1.05, and then multiplies the result of this by 1.05 again.

  (a) Write expressions for each calculation.
  (b) Do the expressions in (a) define the same function?

In Problems 25–30, put the functions in the form $Q = kt$ and state the value of k.

25. $Q = \dfrac{t}{4}$ **26.** $Q = t(n+1)$ **27.** $Q = bt + rt$

28. $Q = \dfrac{1}{2}t\sqrt{3}$ **29.** $Q = \dfrac{\alpha t - \beta t}{\gamma}$

30. $Q = (t-3)(t+3) - (t+9)(t-1)$

31. The price of apartments near a subway is given by

$$\text{Price} = \frac{1000 \cdot A}{10d} \text{ dollars},$$

where A is the area of the apartment in square feet and d is the distance in miles from the subway. Which letters are constants and which are variables if

  (a) You want an apartment of 1000 square feet?
  (b) You want an apartment 1 mile from the subway?
  (c) You want an apartment that costs $200,000?

4.3 FUNCTIONS AND EQUATIONS

Using Equations to Find Inputs from Outputs

In the last section we saw how to evaluate an expression to find the output of a function given an input. Sometimes we want to find the inputs that give a certain output. To do that, we must solve an equation. For example, if $f(x) = x^2$, and we want to know what inputs to f give the output $f(x) = 16$, we must solve the equation $x^2 = 16$. The solutions $x = 4$ and $x = -4$ are the inputs to f that give 16 as an output.

Example 1   In Example 1 on page 85 we saw that the function

$$T(r) = \frac{400}{r}$$

gives the time taken for Bernardo to travel 400 miles at an average speed of r miles per hour.

  (a) Write an equation whose solution is the average speed Bernardo would have to maintain to make the trip in 10 hours.
  (b) Solve the equation and represent your solution on a graph.

Solution   (a) We want the time taken to be 10 hours, so we want $T(r) = 10$. Since $T(r) = 400/r$, we want to solve the equation

$$10 = \frac{400}{r}.$$

  (b) Multiplying both sides by r we get

$$10r = 400,$$
$$r = \frac{400}{10} = 40,$$

so we see that the solution is $r = 40$. In function notation,

$$T(40) = 10.$$

See Figure 4.6. We also know the shape of the graph from Figure 4.5 on page 86.

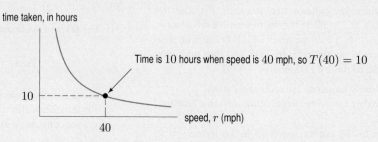

Figure 4.6: Solution to $400/r = 10$ is $r = 40$

Reading Solutions from the Graph

In Example 1 we visualized the solution on a graph. Sometimes it is possible to see the solution directly from a table or a graph.

Example 2   A town's population t years after it was incorporated is given by the function $f(t) = 30{,}000 + 2000t$.

(a) Make a table of values for the population at five-year intervals over a 20-year period starting at $t = 0$. Plot the results on a graph.

(b) Using the table, find the solution to the equation

$$f(t) = 50{,}000$$

and indicate the solution on your plot.

Solution   (a) The initial population in year $t = 0$ is given by

$$f(0) = 30{,}000 + 2000(0) = 30{,}000.$$

In year $t = 5$ the population is given by

$$f(5) = 30{,}000 + 2000(5) = 30{,}000 + 10{,}000 = 40{,}000.$$

In year $t = 10$ the population is given by

$$f(10) = 30{,}000 + 2000(10) = 30{,}000 + 20{,}000 = 50{,}000.$$

Similar calculations for year $t = 15$ and year $t = 20$ give the values in Table 4.5 and Figure 4.7.

Table 4.5   *Population over 20 years*

t, years	Population
0	30,000
5	40,000
10	50,000
15	60,000
20	70,000

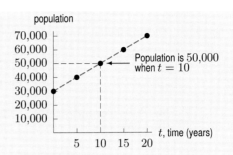

Figure 4.7: Town's population over 20 years

(b) Looking down the right-hand column of the table, we see that the population reaches 50,000 when $t = 10$, so the solution to the equation

$$f(t) = 50,000$$

is $t = 10$. The practical interpretation of the solution $t = 10$ is that the population reaches 50,000 in 10 years. See Figure 4.7.

It is important not to confuse the input and the output. Finding the output from the input is a matter of evaluating the function, whereas finding the input (or inputs) from the output is a matter of solving an equation.

Example 3   For the function graphed in Figure 4.8, give

(a) $f(0)$ (b) The value of x such that $f(x) = 0$.

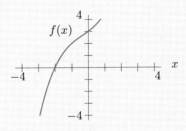

Figure 4.8

Solution   (a) Since the graph crosses the y-axis at the point $(0, 3)$, we have $f(0) = 3$.
(b) Since the graph crosses the x-axis at the point $(-2, 0)$, we have $f(-2) = 0$, so $x = -2$.

The value where the graph crosses the vertical axis is called the *vertical intercept* or *y-intercept*, and the values where it crosses the horizontal axis are called the *horizontal intercepts*, or *x-intercepts*. In Example 3, the vertical intercept is $y = 3$, and the horizontal intercept is $x = -2$.

How Do We Know When Two Functions Are Equal?

Often, we want to know when two functions are equal to each other. That is, we want to find the input value that produces the same output value for both functions. To do this, we set the two outputs equal to each other and solve for the input value.

Example 4   The populations, in year t, of two towns are given by the functions

$$\text{Town A :} \quad P(t) = 600 + 100(t - 2000)$$
$$\text{Town B :} \quad Q(t) = 200 + 300(t - 2000).$$

(a) Write an equation whose solution is the year in which the two towns have the same population.
(b) Make a table of values of the populations for the years 2000–2004 and find the solution to the equation in part (a).

Solution   (a) We want to find the value of t that makes $P(t) = Q(t)$, so we must solve the equation

$$600 + 100(t - 2000) = 200 + 300(t - 2000).$$

(b) From Table 4.6, we see that the two populations are equal in the year $t = 2002$.

Table 4.6

t	2000	2001	2002	2003	2004
$P(t)$	600	700	800	900	1000
$Q(t)$	200	500	800	1100	1400

Checking the populations in that year, we see

$$P(2002) = 600 + 100(2002 - 2000) = 800$$
$$Q(2002) = 200 + 300(2002 - 2000) = 800.$$

Functions and Inequalities

Sometimes, rather than wanting to know where two functions are equal, we want to know where one is bigger than the other.

Example 5   (a) Write an inequality whose solution is the years for which the population of Town A is greater than the population of Town B in Example 4.
(b) Solve the inequality by graphing the populations.

Solution   (a) We want to find the values of t that make $P(t) > Q(t)$, so we must solve the inequality

$$600 + 100(t - 2000) > 200 + 300(t - 2000).$$

(b) In Figure 4.9, the point where the two graphs intersect is $(2002, 800)$, because the populations are both equal to 800 in the year 2002. To the left of this point, the graph of P (the population of town A) is higher than the graph of Q (the population of town B), so town A has the larger population when $t < 2002$.

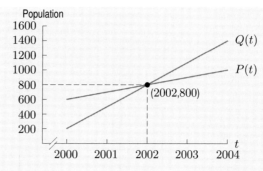

Figure 4.9: When is the population of Town A greater than Town B?

Exercises and Problems for Section 4.3

EXERCISES

In Exercises 1–6, solve $f(x) = 0$ for x.

1. $f(x) = \sqrt{x-2} - 4$ 2. $f(x) = 6 - 3x$

3. $f(x) = 4x - 9$ 4. $f(x) = 2x^2 - 18$

5. $f(x) = 2\sqrt{x} - 10$ 6. $f(x) = 2(2x - 3) + 2$

In Exercises 7–11, solve the equation $g(t) = a$ given that:

7. $g(t) = 6 - t$ and $a = 1$

8. $g(t) = (2/3)t + 6$ and $a = 10$

9. $g(t) = 3(2t - 1)$ and $a = -3$

10. $g(t) = \dfrac{t - 1}{3}$ and $a = 1$

11. $g(t) = 2(t - 1) + 4(2t + 3)$ and $a = 0$

Answer Exercises 12–13 based on the graph of $y = v(x)$ in Figure 4.10.

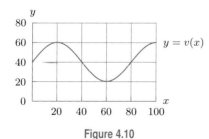

Figure 4.10

12. Solve $v(x) = 60$.   13. Evaluate $v(60)$.

In Exercises 14–16, give

(a) $f(0)$ (b) The values of x such that $f(x) = 0$.
Your answers may be approximate.

14. 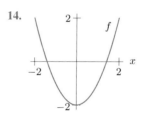   15.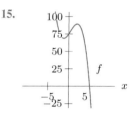

16.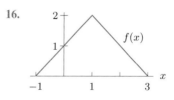

Figure 4.11

17. Chicago's average monthly rainfall, $R = f(t)$ inches, is given as a function of month, t, in Table 4.7. (January is $t - 1$.) Solve and interpret:

(a) $f(t) = 3.7$ (b) $f(t) = f(2)$

Table 4.7

t	1	2	3	4	5	6	7	8
R	1.8	1.8	2.7	3.1	3.5	3.7	3.5	3.4

PROBLEMS

18. The balance in a checking account set up to pay rent, m months after its establishment, is given by $\$4800 - 400m$.

(a) Write an equation whose solution is the number of months it takes for the account balance to reach $\$2000$.

(b) Make a plot of the balance for $m = 1, 3, 5, 7, 9, 11$ and indicate the solution $m = 7$ to the equation in part (a).

19. The number of gallons of gas in a car's tank, d miles after stopping for gas, is given by $15 - d/20$.

(a) Write an equation whose solution is the number of miles it takes for the amount of gas in the tank to reach 10 gallons.

(b) Make a plot of the balance for $d = 40, 60, 80, 100, 120, 140$, and indicate the solution $m = 100$ to the equation in part (a).

20. A car's distance (in miles) from home after t hours is given by $s(t) = 11t^2 + t + 40$.

(a) How far from home is the car at $t = 0$?

(b) Use function notation to express the car's position after 1 hour and then find its position.

(c) Use function notation to express the statement "For what value of t is the car 142 miles from home?"

(d) Write an equation whose solution is the time when the car is 142 miles from home.

(e) Use trial and error for a few values of t to determine when the car is 142 miles from home.

21. If $f(x) = \dfrac{x}{2 - 3x}$, solve $f(b) = 20$.

22. If $h(x) = 3 - 2/x$, solve $3h(x) + 1 = 7$.

23. If $w(t) = 3t + 5$, solve $w(t - 1) = w(2t)$.

24. If $w(x) = 0.5 - 0.25x$, solve

$$w(0.2x + 1) = 0.2w(1 - x).$$

25. In Table 4.8, for which values of x is

(a) $f(x) > g(x)$?   (b) $f(x) = g(x)$?

(c) $f(x) = 0$?   (d) $g(x) = 0$?

Table 4.8

x	-2	-1	0	1	2	3	4	5
$f(x)$	4	1	0	1	4	9	16	25
$g(x)$	1/4	1/2	1	2	4	8	16	32

26. Table 4.9 shows values of x and the expression $3x + 2$. For which values of x in the table is

(a) $3x + 2 < 8$?   (b) $3x + 2 > 8$?

(c) $3x + 2 = 8$?

Table 4.9

x	0	1	2	3	4
$3x + 2$	2	5	8	11	14

27. The height (in meters) of a diver s seconds after beginning his dive is given by the expression $10 + 2s - 9.8s^2$. For which values of s in Table 4.10 is

(a) $10 + 2s - 9.8s^2 < 9.89$?

(b) His height greater than 9.89 meters?

(c) Height $= 9.89$ meters?

Table 4.10

s	0	0.25	0.5	0.75	1
$10 + 2s - 9.8s^2$	10	9.89	8.55	5.99	2.2

28. Table 4.11 shows values of v and the expressions $12 - 3v$ and $-3 + 2v$. For which values of v in the table is

(a) $12 - 3v < -3 + 2v$?

(b) $12 - 3v > -3 + 2v$?

(c) $12 - 3v = -3 + 2v$?

Table 4.11

v	0	1	2	3	4	5	6
$12 - 3v$	12	9	6	3	0	-3	-6
$-3 + 2v$	-3	-1	1	3	5	7	9

29. Table 4.12 shows monthly life insurance rates, in dollars, for men and women. Let $m = f(a)$ be the rate for men at age a, and $w = g(a)$ be the rate for women at age a.

(a) Find $f(65)$.

(b) Find $g(50)$.

(c) Solve and interpret $f(a) = 102$.

(d) Solve and interpret $g(a) = 57$.

(e) For what values of a is $f(a) = g(a)$?

(f) For what values of a is $g(a) < f(a)$?

Table 4.12

Age	35	40	45	50	55	60	65	70
$f(x)$ (in dollars)	13	17	27	40	65	102	218	381
$g(x)$ (in dollars)	13	17	27	39	57	82	133	211

30. If a company sells p software packages, its profit per package is given by $\$10{,}000 - \dfrac{100{,}000}{p}$, as shown in Figure 4.12.

(a) From the graph, estimate the number of packages sold when profits per package are $8000.

(b) Check your answer to part (a) by substituting it into the equation

$$10{,}000 - \frac{100{,}000}{p} = 8000.$$

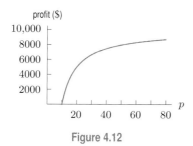

Figure 4.12

31. The tuition for a semester at a small public university t years from now is given by $\$3000 + 100t$, as shown in Figure 4.13.

(a) From the graph, estimate how many years it will take for tuition to reach $3700.

(b) Check your answer to part (a) by substituting it into the equation

$$3000 + 100t = 3700.$$

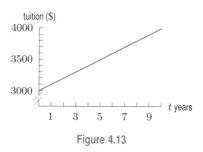

Figure 4.13

32. Antonio and Lucia are both driving through the desert from Tucson to San Diego, which takes each of them 7 hours of driving time. Antonio's car starts out full with 14 gallons of gas and uses 2 gallons per hour. Lucia's SUV starts out full with 30 gallons of gas and uses 6 gallons per hour.

(a) Construct a table showing how much gas is in each of their tanks at the end of each hour into the trip. Assume each stops for gas just as the tank is empty, and then the tank is filled instantaneously.

(b) Use your table to determine when they have the same amount of gas.

(c) If they drive at the same speed while driving and only stop for gas, which of them gets to San Diego first? (Assume filling up takes time.)

(d) Now suppose that between 1 hour and 6.5 hours outside of Tucson, all of the gas stations are closed unexpectedly. Does Antonio arrive in San Diego? Does Lucia?

(e) The amount of gas in Antonio's tank after t hours is $14 - 2t$ gallons, and the amount in Lucia's tank is $30 - 6t$ gallons. When does

   (i) $14 - 2t = 30 - 6t$?

   (ii) $14 - 2t = 0$?

   (iii) $30 - 6t = 0$?

4.4 FUNCTIONS AND CHANGE

Functions describe how quantities change, and by comparing values of the function for different inputs we can see how fast the output changes. For example, Table 4.13 shows the population $P = f(t)$ of a colony of termites t months after it was started. Using the Greek letter Δ (pronounced "delta") to indicate change, we have

Change in first 6 months $= \Delta P = f(6) - f(0) = 4000 - 1000 = 3000$ termites

Change in first 3 months $= \Delta P = f(3) - f(0) = 2500 - 1000 = 1500$ termites.

Although the colony grows by twice as much during the first 6 months as during the first 3 months (3000 as compared with 1500), it also had twice as long in which to grow. Sometimes it is

Table 4.13   *Population of a colony of termites*

t (months)	0	3	6	9	12
$P = f(t)$	1000	2500	4000	7000	2800

more useful to measure not the change in the output, but rather the change in the output divided by the change in the input, or *average rate of change*:

$$\text{Average rate of change in first 6 months} = \frac{\Delta P}{\Delta t} = = \frac{3000 \text{ termites}}{6 \text{ months}} = 500 \text{ termites/month}$$

$$\text{Average rate of change in first 3 months} = \frac{\Delta P}{\Delta t} = = \frac{1500 \text{ termites}}{3 \text{ months}} = 500 \text{ termites/month}.$$

Thus, on average, the colony adds 500 termites per month during the first 6 months, and also during first 3 months. Even though the population change over the first 3 months is less than over the first 6 months (1500 versus 3000), the average rate of change is the same (500 termites/month).

An Expression for the Average Rate of Change

If the input of the function $y = f(x)$ changes from $x = a$ to $x = b$, we write

$$\text{Change in input} = \text{New } x\text{-value} - \text{Old } x\text{-value} = b - a$$

$$\text{Change in output} = \text{New } y\text{-value} - \text{Old } y\text{-value} = f(b) - f(a).$$

The *average rate of change of $y = f(x)$ between $x = a$ and $x = b$* is given by

$$\text{Average rate of change} = \frac{\text{Change in output}}{\text{Change in input}} = \frac{\Delta y}{\Delta x} = \frac{f(b) - f(a)}{b - a}.$$

Example 1   Find the average rate of change of $g(x) = (x - 2)^2 + 3$ on the following intervals:

(a)   Between 0 and 3   (b)   Between -1 and 4.

Solution   (a) We have

$$g(0) = (0 - 2)^2 + 3 = 7$$
$$g(3) = (3 - 2)^2 + 3 = 4,$$

so $$\text{Average rate of change} = \frac{g(3) - g(0)}{3 - 0} = \frac{4 - 7}{3} = -1.$$

(b) We have

$$g(-1) = (-1 - 2)^2 + 3 = 12$$

$$g(4) = (4 - 2)^2 + 3 \ = 7,$$

so $\quad$ $\begin{array}{c}\text{Average rate}\\\text{of change}\end{array} = \dfrac{g(4) - g(-1)}{4 - (-1)} = \dfrac{7 - 12}{5} = -1.$

Example 2 For the termite colony in Table 4.13,

(a) What is the change in the population the last 6 months? Months 6 to 9?
(b) What is the average rate of change of the population during the last 6 months? Months 6 to 9?

Solution (a) In the last 6 months, we have

$$\text{Change in last 6 months} = \Delta P = f(12) - f(6) = 2800 - 4000 = -1200 \text{ termites}$$
$$\text{Change from months 6 to 9} = \Delta P = f(9) - f(6) \ = 7000 - 4000 = 3000 \text{ termites}.$$

The change in the last 6 months is negative because the colony decreases in size, losing 1200 termites. However, for the first 3 of those months, months 6 to 9, it increases by 3000 termites.

(b) For the last 6 months, we see that

$$\begin{array}{c}\text{Average rate of change}\\\text{in last 6 months}\end{array} = \frac{\Delta P}{\Delta t} = \frac{f(12) - f(6)}{12 - 6} = \frac{-1200 \text{ termites}}{6 \text{ months}} = -200 \frac{\text{termites}}{\text{month}}$$

$$\begin{array}{c}\text{Average rate of change}\\\text{for months 6 to 9}\end{array} = \frac{\Delta P}{\Delta t} = \frac{f(9) - f(6)}{9 - 6} = \frac{3000 \text{ termites}}{3 \text{ months}} = 1000 \frac{\text{termites}}{\text{month}}.$$

The average rate of change over the last 6 months is negative, because the net change is negative, as we saw in (a). In general, if $f(b) < f(a)$ then the numerator in the expression for the average rate of change is negative, so the rate of change is negative (provided $b > a$).

As for months 6 to 9, even though the population change over this period is the same as over the first 6 months (3000 termites), the average rate of change is larger from months 6 to 9. This is because adding 3000 termites in a 3-month period amounts to more termites per month than adding the same number of termites in a 6-month period.

Interpreting the Expression for Average Rate of Change

Suppose a car has traveled $F(t)$ miles t hours after it starts a journey. For example, $F(3)$ is the distance the car has traveled after 3 hours and $F(5)$ is the distance it has traveled after 5 hours. Then the statement $F(5) - F(3) = 140$ means that the car travels 140 miles between 3 and 5 hours after starting. We could also express this as an average rate of change,

$$\frac{F(5) - F(3)}{5 - 3} = 70.$$

Here $5 - 3 = 2$ is the two-hour period from 3 to 5 hours after the start, during which the car travels $F(5) - F(3) = 140$ miles, and the statement tells us that the average velocity of the car during the 2 hour period is 70 mph.

Example 3   Interpret the following statements in terms of the journey of the car.

(a) $F(a+5) - F(a) = 315$ (b) $\dfrac{F(a+5) - F(a)}{5} = 63$

Solution

(a) Since $F(a)$ is the distance the car has traveled after a hours and $F(a+5)$ is the distance it has traveled after $a + 5$ hours (that is, after 5 more hours), $F(a+5) - F(a) = 315$ means it travels 315 miles between a and $a + 5$ hours after starting.

(b) Notice that $(a+5) - a = 5$ is the five-hour period during which the car travels $F(a+5) - F(a) = 315$ miles. The change in input is 5 hours, and the change in output is 315 miles. The statement

$$\frac{F(a+5) - F(a)}{5} = \frac{F(a+5) - F(a)}{(a+5) - a} = 63$$

tells us that the average rate of change of F, or average velocity of the car, during the 5 hour period is 63 mph.

Using Units to Interpret the Average Rate of Change

It if often useful to consider the units of measurement of the input and output quantities when interpreting the average rate of change. For instance, as we saw in Example 3, if the input is time measured in hours and the output is distance measured in miles, then the average rate of change is measured in miles per hour, and represents a velocity.

Example 4   A ball thrown in the air has height $h(t) = 90t - 16t^2$ feet after t seconds.

(a) What are the units of measurement for the average rate of change of h? What does your answer tell you about how to interpret the rate of change in this case?

(b) Find the average rate of change of h between
    (i) $t = 0$ and $t = 2$ (ii) $t = 1$ and $t = 2$ (iii) $t = 2$ and $t = 4$

(c) Use a graph of h to interpret your answers to part (b) in terms of the motion of the ball.

Solution

(a) In the expression

$$\frac{h(b) - h(a)}{b - a}$$

the numerator is measured in feet and the denominator is measured in seconds, so the ratio is measured in ft/sec. Thus it measures a velocity.

(b) We have
    (i) Average rate of change between $t = 0$ and $t = 2$

$$\frac{h(2) - h(0)}{2 - 0} = \frac{(90 \cdot 2 - 16 \cdot 4) - (0 - 0)}{2 - 0} = \frac{116}{2} = 58 \text{ ft/sec.}$$

    (ii) Average rate of change between $t = 1$ and $t = 2$

$$\frac{h(2) - h(1)}{2 - 1} = \frac{(90 \cdot 2 - 16 \cdot 4) - (90 - 16)}{2 - 1} = \frac{42}{1} = 42 \text{ ft/sec.}$$

    (iii) Average rate of change between $t = 2$ and $t = 4$

$$\frac{h(4) - h(2)}{4 - 2} = \frac{(90 \cdot 4 - 16 \cdot 16) - (90 \cdot 2 - 16 \cdot 4)}{4 - 2} = \frac{-12}{2} = -6 \text{ ft/sec.}$$

(c) Figure 4.14 shows the ball rising to a peak somewhere between 2 and 3 seconds, and then starting to fall. The average velocity is positive during the first two seconds because the height is increasing during that time period. The height is also increasing between $t = 1$ and $t = 2$, but since the ball is rising more slowly, the velocity is less. Between $t = 2$ and $t = 4$ the ball rises and falls, experiencing a net loss in height, so its average velocity is negative.

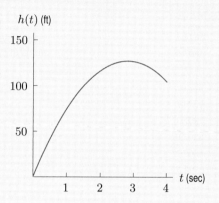

Figure 4.14: Height of a ball

Exercises and Problems for Section 4.4

EXERCISES

Find the average rate of change of $f(x) = x^2 + 3x$ on the intervals indicated in Exercises 1–4.

1. Between 2 and 4.

2. Between -2 and 4.

3. Between -4 and -2.

4. Between 3 and 1.

Find the average rate of change of $g(x) = 2x^3 - 3x^2$ on the intervals indicated in Exercises 5–8.

5. Between 1 and 3.

6. Between -1 and 4.

7. Between 0 and 10.

8. Between -0.1 and 0.1.

9. The value in dollars of an investment t years after 2003 is given by
$$V = 1000 \cdot 2^{t/6}.$$
Find the average rate of change of the investment's value between 2004 and 2007.

10. Atmospheric levels of carbon dioxide (CO_2) have risen from 336 parts per million (ppm) in 1979 to 382 parts per million (ppm) in 2007.[1] Find the average rate of change of CO_2 levels during this time period.

11. Sea levels were most recently at a low point about 22,000 years ago.[2] Since then they have risen approximately 130 meters. Find the average rate of change of the sea level during this time period.

12. Global temperatures may increase by up to $10°$F between 1990 and 2100.[3] Find the average rate of change of global temperatures between 1990 and 2100.

[1] National Oceanic & Atmospheric Administration, http://www.esrl.noaa.gov/gmd/aggi. Page last accessed September 26, 2007.

[2] See http://en.wikipedia.org/wiki/Sea_level_rise, http://en.wikipedia.org/wiki/Last_glacial_maximum, and related links. Pages last accessed September 13, 2006.

[3] See http://en.wikipedia.org/wiki/Global_warming. Page last accessed September 13, 2006.

PROBLEMS

Table 4.14 gives values of $D = f(t)$, the total US debt (in $ billions) t years after 2000.[4] Answer Problems **13–16** based on this information.

Table 4.14

t	D ($ billions)
0	5674.2
1	5807.5
2	6228.2
3	6783.2
4	7379.1
5	7932.7
6	8507.0
7	9007.7
8	10,024.7

13. Evaluate
$$\frac{f(5) - f(1)}{5 - 1}$$
and say what it tells you about the US debt.

14. Which expression has the larger value,
$$\frac{f(5) - f(3)}{5 - 3} \quad \text{or} \quad \frac{f(3) - f(0)}{3 - 0}?$$
Say what this tells you about the US debt.

15. Show that

Average rate of change from 2005 to 2006 $<$ Average rate of change from 2004 to 2005.

Does this mean the US debt is starting to go down? If not, what does it mean?

16. Project the value of $f(10)$ by assuming
$$\frac{f(10) - f(6)}{10 - 6} = \frac{f(6) - f(0)}{6 - 0}.$$
Explain the assumption that goes into making your projection and what your answer tells you about the US debt.

Let $g(t)$ give the market value (in $1000s) of a house in year t. Say what the following statements tell you about the house.

17. $g(5) - g(0) = 30$

18. $\dfrac{g(10) - g(4)}{6} = 3$

19. $\dfrac{g(20) - g(12)}{20 - 12} = -1$

Let $f(t)$ be the population of a town that is growing over time. Say what must be true about a in order for the expressions in Problems **20–22** to be positive.

20. $f(a) - f(3)$

21. $\dfrac{f(3) - f(a)}{3 - a}$

22. $f(t + a) - f(t + b)$

Let $s(t)$ give the number of acres of wetlands in a state in year t. Assuming that the area of wetlands goes down over time, say what the statements in Problems **23–25** tell you about the wetlands.

23. $s(25) - s(0) = -25,000$

24. $\dfrac{s(20) - s(10)}{20 - 10} = -520$

25. $s(30) - s(20) < s(20) - s(10)$

Let $g(t)$ give the amount of electricity (in kWh) used by a manufacturing plant in year t. Assume that, thanks to conservation measures, g is going down over time. Assuming the expressions in Problems **26–28** are defined, say what must be true about h in order for them to be negative.

26. $g(h) - g(10)$

27. $\dfrac{g(5 + h) - g(5)}{h}$

28. $g(h + 1) - g(h - 1)$

Methane is a greenhouse gas implicated as a contributor to global warming. Answer Problems **29–32** based on the table of values of $Q = w(t)$, the atmospheric methane level in parts per billion (ppb) t years after 1980.[5]

Table 4.15

t	0	5	10	15	20	25
Q	1575	1660	1715	1750	1770	1775

---

[4] See http://www.treasurydirect.gov/govt/reports/pd/histdebt/histdebt_histo5.htm. Page last accessed April 25, 2009.

[5] National Oceanic & Atmospheric Administration, http://www.esrl.noaa.gov/gmd/aggi. Page last accessed September 26, 2007.

29. Evaluate
$$\frac{w(10) - w(5)}{10 - 5},$$
and say what it tells you about atmospheric methane levels.

30. Which expression is larger,
$$\frac{w(10) - w(0)}{10 - 0} \quad \text{or} \quad \frac{w(25) - w(10)}{25 - 10}?$$
Say what this tells you about atmospheric methane levels.

31. Show that
$$\text{Average rate of change} \atop \text{from 1995 to 2000} \quad < \quad {\text{Average rate of change} \atop \text{from 2000 to 2005.}}$$

Does this mean the average methane level is going down? If not, what does it mean?

32. Project the value of $w(-20)$ by assuming
$$\frac{w(0) - w(-20)}{20} = \frac{w(10) - w(0)}{10}.$$
Explain the assumption that goes into making your projection and what your answer tells you about atmospheric methane levels.

4.5 FUNCTIONS AND MODELING

When we want to apply mathematics to a real-world situation, we are not always given a function: sometimes we have to find one. Knowledge about the real world can help us to choose a particular type of function, and we then use information about the exact situation we are modeling to select a function of this type. This process is called *modeling*. In this section we consider situations that can be modeled by a proportional relationship between the variables.

Direct Proportionality

Suppose a state sales tax rate is 6%. Then the tax, T, on a purchase of price P is given by the function
$$\text{Tax} = 6\% \times \text{Price} \quad \text{or} \quad T = 0.06P.$$

Rewriting this equation as a ratio, we see that T/P is constant, $T/P = 0.06$, so the tax is proportional to the purchase price. In general

---

A quantity Q is **directly proportional** to a quantity t if

$$Q = k \cdot t,$$

where k is the *constant of proportionality*. We often omit the word "directly" and simply say Q is proportional to t.

In the tax example, the constant of proportionality is $k = 0.06$, because $T = 0.06P$. Similarly, in Example 2 on page 78, the constant of proportionality is $k = 0.2$, and in Example 3 on page 79, the constant of proportionality is 0.44.

Example 1   A car gets 25 miles to the gallon.

(a) How far does the car travel on 1 gallon of gas? 2 gallons? 10 gallons? 20 gallons?

(b) Express the distance, d miles, traveled as a function of the number of gallons, g, of gas used. Explain why d is proportional to g with constant of proportionality 25.

Solution   (a) The car travels 25 miles on each gallon. Thus

$$
\begin{aligned}
\text{Distance on 1 gallon} &= 25 \cdot 1 &=& \quad 25 \text{ miles} \\
\text{Distance on 2 gallons} &= 25 \cdot 2 &=& \quad 50 \text{ miles} \\
\text{Distance on 10 gallons} &= 25 \cdot 10 &=& \quad 250 \text{ miles} \\
\text{Distance on 20 gallons} &= 25 \cdot 20 &=& \quad 500 \text{ miles.}
\end{aligned}
$$

(b) Since an additional gallon of gas enables the car to travel an additional 25 miles, we have

$$d = 25g.$$

Thus d is proportional to g, with constant of proportionality 25, the car's mileage per gallon.

Notice that in Example 1, when the number of gallons doubles from 1 to 2, the number of miles doubles from 25 to 50, and when the number of gallons doubles from 10 to 20, the number of miles doubles from 250 to 500. In general:

The Behavior of Proportional Quantities

If $Q = kt$, then doubling the value of t doubles the value of Q, tripling the value of t triples the value of Q, and so on. Likewise, halving the value of t halves the value of Q, and so on.

Example 2   Vincent pays five times as much for a car as Dominic. Dominic pays \$300 sales tax. How much sales tax does Vincent pay (assuming they pay the same rate)?

Solution   Since Vincent's car costs five times as much as Dominic's car, Vincent's sales tax should be five times as large as Dominic's, or $5 \cdot 300 = \$1500$.

Example 3   For the same car as in Example 1:

(a) How many gallons of gas are needed for a trip of 5 miles? 10 miles? 100 miles?
(b) Find g, the number of gallons needed as a function of d, the number of miles traveled. Explain why g is proportional to d and how the constant of proportionality relates to the constant in Example 1.

Solution   (a) We have

$$
\begin{aligned}
\text{Number of gallons of gas for 5 miles} &= \frac{5}{25} &=& \quad 0.2 \text{ gallon} \\
\text{Number of gallons of gas for 10 miles} &= \frac{10}{25} &=& \quad 0.4 \text{ gallon} \\
\text{Number of gallons of gas for 100 miles} &= \frac{100}{25} &=& \quad 4 \text{ gallons.}
\end{aligned}
$$

(b) Since each mile takes 1/25 gallon of gas,

$$g = \frac{d}{25} = \left(\frac{1}{25}\right) d.$$

Thus g is proportional to d and the constant of proportionality is $1/25$, the reciprocal of the constant in Example 1. We can also see that g is proportional to d by solving for g in $d = 25g$.

Sometimes we have to rewrite the expression for a function to see that it represents a direct proportionality.

Example 4   Does the function represent a direct proportionality? If so, give the constant of proportionality, k.

(a) $f(x) = 19x$ (b) $g(x) = x/53$ (c) $F(a) = 2a + 5a$

(d) $u(t) = \sqrt{5}t$ (e) $A(n) = n\pi^2$ (f) $P(t) = 2 + 5t$

Solution
(a) Here $f(x)$ is proportional to x with constant of proportionality $k = 19$.
(b) We rewrite this as $g(x) = (1/53)x$, so $g(x)$ is proportional to x with constant of proportionality $k = 1/53$.
(c) Simplifying the right-hand side, we get $F(a) = 2a + 5a = 7a$, so $F(a)$ is proportional to a with constant $k = 7$.
(d) Here $u(t)$ is proportional to t with constant $k = \sqrt{5}$.
(e) Rewriting this as $A(n) = \pi^2 n$, we see that it represents a direct proportionality with constant $k = \pi^2$.
(f) Here $P(t)$ is not proportional to t, because of the constant 2 on the right-hand side.

Solving for the Constant of Proportionality

If we do not know the constant of proportionality, we can find it using one pair of known values for the quantities that are proportional.

Example 5   A graduate assistant at a college earns $80 for 10 hours of work. Express her earnings as a function of hours worked. What is the constant of proportionality and what is its practical interpretation?

Solution   We have

$$E = f(t) = kt,$$

where E is the amount earned and t is the number of hours worked. We are given

$$80 = f(10) = 10k,$$

and solving for k we have

$$k = \frac{80 \text{ dollars}}{10 \text{ hours}} = 8 \text{ dollars/hour}.$$

Thus, $k = 8$ dollars/hour, which is her hourly wage.

Example 6   A person's heart mass is proportional to his or her body mass.[6]

(a) A person with a body mass of 70 kilograms has a heart mass of 0.42 kilograms. Find the constant of proportionality, k.

(b) Estimate the heart mass of a person with a body mass of 60 kilograms.

Solution   (a) We have

$$H = k \cdot B.$$

for some constant k. We substitute $B = 70$ and $H = 0.42$ and solve for k:

$$0.42 = k \cdot 70$$
$$k = \frac{0.42}{70} = 0.006.$$

So the formula for heart mass as a function of body mass is

$$H = 0.006B,$$

where H and B are measured in kilograms.

(b) With $B = 60$, we have

$$H = 0.006 \cdot 60 = 0.36 \text{ kg.}$$

Proportionality and Rates

In general the average rate of a change of a function between two points depends on the points we choose. However, the function $f(t) = kt$ describes a quantity which is growing at a constant rate k.

Example 7   Show algebraically that the average rate of change of the function $f(t) = kt$ between any two different values of t is equal to the constant k.

Solution   We have

$$\frac{f(b) - f(a)}{b - a} = \frac{kb - ka}{b - a} = \frac{k(b - a)}{b - a} = k.$$

Thus, if two quantities are proportional, the constant of proportionality is the rate of change of one quantity with respect to the other. Using the units associated with a rate of change often helps us interpret the constant k. For instance, in Example 5 on page 103, the constant of proportionality can be interpreted as a pay rate in dollars per hour.

Example 8   Suppose that the distance you travel, in miles, is proportional to the time spent traveling, in hours:

$$\text{Distance} = k \times \text{Time.}$$

What is the practical interpretation of the constant k?

---

[6]K. Schmidt-Nielsen: *Scaling, Why is Animal Size So Important?* (Cambridge: CUP, 1984).

Solution   To see the practical interpretation of k, we write

$$\text{miles} = \text{Units of } k \times \text{hours.}$$

Therefore,

$$\text{Units of } k = \frac{\text{miles}}{\text{hours}}.$$

The units of k are miles per hour, and it represents speed.

Example 9   The *data rate* of an Internet connection is the rate in bytes per second that data, such as a web page, image, or music file, can be transmitted across the connection.[7] Suppose the data rate is 300 bytes per second. How long does it take to download a file of 42,000 bytes?

Solution   Let T be the time in seconds and N be the number of bytes. Then

$$N = 300T$$
$$42{,}000 \text{ bytes} = 300 \text{ bytes/sec} \cdot T$$
$$T = \frac{42{,}000 \text{ bytes}}{300 \text{ bytes/sec}} = 140 \text{ sec,}$$

so it takes 140 seconds to download the file.

Example 10   Give the units of the constant 25 in Example 1 on page 101 and give a practical interpretation of it.

Solution   In the equation $d = 25g$, the units of g are gallons and the units of distance, d, are miles. Thus

$$\text{miles} = \text{Units of } k \cdot \text{gallons}$$
$$\text{Units of } k = \frac{\text{miles}}{\text{gallons}}$$
$$= \text{miles per gallon, or mpg.}$$

The constant tells us that the car gets 25 miles to a gallon of gas.

Families of Functions

Because the functions $Q = f(t) = kt$ all have the same algebraic form, we think of them as a family of functions, one function for each value of the constant k. Figure 4.15 shows graphs of various functions in this family. All the functions in the family share some common features: their graphs are all lines, and they all pass through the origin. There are also differences between different functions in the family, corresponding to different values of k. Although k is constant for any given member of the family, it changes from one member to the next. We call k a *parameter* for the family.

---

[7] Sometimes mistakenly called "bandwidth." See http://foldoc.doc.ic.ac.uk/foldoc/contents.html.

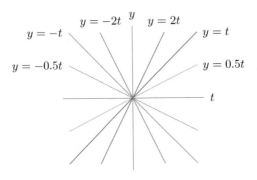

Figure 4.15: The family of functions $y = kt$

Exercises and Problems for Section 4.5

EXERCISES

In Exercises 1–4 is the first quantity proportional to the second quantity? If so, what is the constant of proportionality?

1. d is the distance traveled in miles and t is the time traveled in hours at a speed of 50 mph.

2. P is the price paid in dollars for b barrels of oil at a price of $70.

3. p is the sale price of an item whose original price is p_0 in a 30% off sale.

4. C is the cost of having n drinks at a club, where each drink is $5 and there is a cover charge of $20.

For each of the formulas in Exercises 5–13, is y directly proportional to x? If so, give the constant of proportionality.

5. $y = 5x$

6. $y = x \cdot 7$

7. $y = x \cdot x$

8. $y = \sqrt{5} \cdot x$

9. $y = x/9$

10. $y = 9/x$

11. $y = x + 2$

12. $y = 3(x + 2)$

13. $y = 6z$ where $z = 7x$

In Exercises 14–17, assume the two quantities are directly proportional to each other.

14. If $r = 36$ when $s = 4$, find r when s is 5.

15. If $p = 24$ when $q = 6$, find q when p is 32.

16. If $y = 16$ when $x = 12$, find y when x is 9.

17. If $s = 35$ when $t = 25$, find t when s is 14.

18. The interest paid by a savings account in one year is proportional to the starting balance, with constant of proportionality 0.06. Write a formula for I, the amount of interest earned, in terms of B, the starting balance. Find the interest earned if the starting balance is

    (a) $500   (b) $1000   (c) $5000.

19. If y is directly proportional to x, and $y = 6$ when $x = 4$, find the constant of proportionality, write a formula for y in terms of x, and find x when $y = 8$.

In Exercises 20–28, determine if the graph defines a direct proportion. If it does, estimate the constant of proportionality.

20.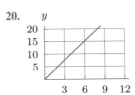

21.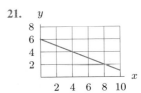

22.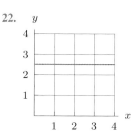

23.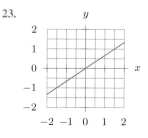

26.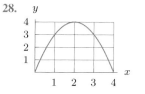

27.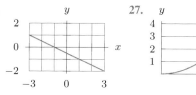

24.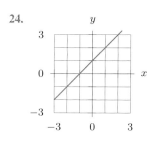

25.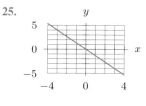

28.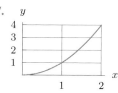

PROBLEMS

29. A factory makes 50 vehicles a week. Is t, the number of weeks, proportional to v, the number of vehicles made? If so, what is the constant of proportionality?

30. One store sells 70 pounds of apples a week, and a second store sells 50 pounds of apples a week. Is the total number of pounds of apples sold, a, proportional to the number of weeks, w? If so, what is the constant of proportionality?

31. The perimeter of a square is proportional to the length of any side. What is the constant of proportionality?

32. A bike shop's revenue is directly proportional to the number of bicycles sold. When 50 bicycles are sold, the revenue is $20,000.

   (a) What is the constant of proportionality, and what are its units?

   (b) What is the revenue if 75 are sold?

33. The total cost of purchasing gasoline for your car is directly proportional to the the number of gallons pumped, and 11 gallons cost $36.63.

   (a) What is the constant of proportionality, and what are its units?

   (b) How much do 15 gallons cost?

34. The required cooling capacity, in BTUs, for a room air conditioner is proportional to the area of the room being cooled. A room of 280 square feet requires an air conditioner whose cooling capacity is 5600 BTUs.

   (a) What is the constant of proportionality, and what are its units?

   (b) If an air conditioner has a cooling capacity of 10,000 BTUs, how large a room can it cool?

35. You deposit $P into a bank where it earns 2% interest per year for 10 years. Use the formula $B = P(1 + r)^t$ for the balance $B, where r is the interest rate (written as a decimal) and t is time in years.

   (a) Explain why B is proportional to P. What is the constant of proportionality?

   (b) Is P proportional to B? If so, what is the constant of proportionality?

36. The length m, in inches, of a model train is proportional to the length r, in inches, of the corresponding real train.

   (a) Write a formula expressing m as a function of r.

   (b) An HO train is $1/87^{\text{th}}$ the size of a real train.[8] What is the constant of proportionality? What is the length in feet of a real locomotive if the HO locomotive is 10.5 inches long?

   (c) A Z scale train is $1/220^{\text{th}}$ the size of a real train. What is the constant of proportionality? What is the length, in inches, of a Z scale locomotive if the real locomotive is 75 feet long?

---

[8] www.internettrains.com, accessed December 11, 2004.

37. The cost of denim fabric is directly proportional to the amount that you buy. Let C be the cost, in dollars, of x yards of denim fabric.

   (a) Write a formula expressing C as a function of x.
   (b) One type of denim costs $28.50 for 3 yards. Find the constant of proportionality and give its units.
   (c) How much does 5.5 yards of this type of denim cost?

38. The distance M, in inches, between two points on a map is proportional to the actual distance d, in miles, between the two corresponding locations.

   (a) If 1/2 inch represents 5 miles, find the constant of proportionality and give its units.

   (b) Write a formula expression M as a function of d.
   (c) How far apart are two towns if the distance between them on the map is 3.25 inches?

39. The blood mass of a mammal is proportional to its body mass. A rhinoceros with body mass 3000 kilograms has blood mass of 150 kilograms. Find a formula for the blood mass of a mammal in terms of the body mass and estimate the blood mass of a human with body mass 70 kilograms.

40. The distance a car travels on the highway is proportional to the quantity of gas consumed. A car travels 225 miles on 5 gallons of gas. Find the constant of proportionality, give units for it, and explain its meaning.

REVIEW EXERCISES AND PROBLEMS FOR CHAPTER 4

EXERCISES

1. Let $f(r)$ be the weight of an astronaut in pounds at the distance r, in thousands of miles from the earth's center. Explain the meaning of the following statements.

   (a) $f(4) = 180$ **(b)** $f(a) = 36$
   (c) $f(36) = b$ **(d)** $w = f(r)$

2. When there are c cars on campus, the number of cars without a parking space is $w(c) = -2000 + c$. Express in function notation the number of cars without parking spaces when there are 8000 cars on campus, and evaluate your expression.

Evaluate the expressions in Exercises 3–8 given that

$$f(x) = \frac{x+1}{2x+1}.$$

3. $f(0)$ **4.** $f(-1)$ **5.** $f(0.5)$

6. $f(-0.5)$ **7.** $f\left(\dfrac{1}{3}\right)$ **8.** $f(\pi)$

9. Let $g(t) = \dfrac{t^2+1}{5+t}$. Evaluate

   (a) $g(3)$ **(b)** $g(-1)$ **(c)** $g(a)$

10. Let $g(s) = \dfrac{5s+3}{2s-1}$. Evaluate

   (a) $g(4)$ **(b)** $g(a)$
   (c) $g(a) + 4$ **(d)** $g(a+4)$

11. Use Table 4.16 to evaluate the expressions.

   (a) $g(10) - 10$ **(b)** $6g(0)$
   (c) $g(10) - g(-10)$ **(d)** $g(5) + 7g(5)$

Table 4.16

y	−10	−5	0	5	10
$g(y)$	−5	0	5	10	−10

12. Table 4.17 gives the number of passenger cars in the US from 1990 to 1994 as a function of year. ($t = 0$ is 1990.)

Table 4.17

t (years)	0	1	2	3	4
$C(t)$ cars (in millions)	133.7	128.3	126.6	127.3	127.9

   (a) Evaluate $C(1)$ and interpret its meaning.
   (b) If $C(t) = 127.3$, find t and interpret its meaning.

13. Census figures for the US population, $P(t)$ (in millions), t years after 1950, are in Table 4.18.

Table 4.18

t (years)	0	10	20	30	40	50
$P(t)$ (millions)	150.7	179.0	205.0	226.5	248.7	281.4

   (a) Evaluate $P(20)$ and interpret its meaning.
   (b) If $P(t) = 281.4$, find t and interpret its meaning.

14. Using Figure 4.16, evaluate

  (a) $f(2) - f(1)$ **(b)** $f(2) - f(-1)$
  (c) $2f(-1) + f(2)$ **(d)** $1/f(-2)$

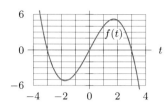

Figure 4.16

15. Using Figure 4.17, estimate

  (a) $d(2)/d(1)$ **(b)** $d(1) \cdot d(1)$
  (c) $4d(-2) - d(-1)$ **(d)** $4d(-2) - (-1)$

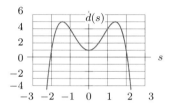

Figure 4.17

In Exercises **16–19**, $g(z) = 4z^2 - 3$. Determine whether the two expressions are equivalent.

16. $2^2 g(z), 4g(z)$ **17.** $4g(z), g(4z)$

18. $g(\sqrt{z}), \sqrt{g(z)}$ **19.** $g(z \cdot z), g(z^2)$

In Exercises **20–27**, find the x value that results in $f(x) = 3$.

20. $f(x) = 5x - 2$ **21.** $f(x) = 5x - 5$

22. $f(x) = \dfrac{2x}{5}$ **23.** $f(x) = \dfrac{5}{2x}$

24. The population of a town, t years after it was founded, is given by $P(t) = 5000 + 350t$.

  (a) Write an equation whose solution is the number of years it takes for the population to reach 12,000.
  (b) Make a plot of the population for $t = 14, 16, 18, 20, 22, 24$ and indicate the solution to the equation in part (a).

25. The number of stamps in a person's passport, t years after the person gets a new job which involves overseas travel, is given by $N(t) = 8 + 4t$.

  (a) Write an equation whose solution is the number of years it takes for the passport to have 24 stamps.
  (b) Make a plot of the number of stamps for $t = 1, 2, 3, 4, 5, 6$, and indicate the solution to the equation in part (a).

26. The number of dirty socks on your roommate's floor, t days after the start of exams, is given by $s(t) = 10 + 2t$.

  (a) Write an equation whose solution is the number of days it takes for the number of socks to reach 26.
  (b) Make a plot of the number of socks for $t = 2, 4, 6, 8, 10, 12$, and indicate the solution to the equation in part (a).

27. For accounting purposes, the value of a machine, t years after it is purchased, is given in dollars by $V(t) = 100{,}000 - 10{,}000t$.

  (a) Write an equation whose solution is the number of years it takes for the machine's value to reach $70,000.
  (b) Make a plot of the value of the machine at $t = 1, 2, 3, 4, 5, 6$, and indicate, on the graph, the solution to the equation in part (a).

Find the average rate of change of $f(x) = x^3 - x^2$ on the intervals indicated in Exercises **28–30**.

28. Between 2 and 4.   **29.** Between -2 and 4.

30. Between -4 and -2.

31. Let $f(N) = 3N$ give the number of glasses a cafe should have if it has an average of N clients per hour.

  (a) How many glasses should the cafe have if it expects an average of 50 clients per hour?
  (b) What is the relationship between the number of glasses and N? What is the practical meaning of the 3 in the expression $f(N) = 3N$?

In Exercises **32–37**, is y directly proportional to x? If so, give the constant of proportionality.

32. $x = 5y$ **33.** $y = 2x - x$

34. $2y = -3x$ **35.** $y - 2 = 3x$

36. $y/3 = \sqrt{5}x$ **37.** $y = ax - 2x$

PROBLEMS

Interpret the expressions in Problems **38–40** in terms of the housing market, where $n = f(p)$ gives the average number of days a house stays on the market before being sold for price p (in \$1000s), and p_0 is the average sale price of houses in the community.

38. $f(p_0)$

39. $f(p_0) + 10$

40. $f(p_0) - f(0.9p_0)$

41. The depth, in inches, of the water in a leaking cauldron after t hours is given by $H(t) = (-0.08t + 6)^2$.

   (a) Find $H(0)$ and interpret its meaning.
   (b) Interpret the meaning of $H(75) = 0$.

42. Using Figure 4.18, find

   (a) $f(4)$
   (b) The values of x such that $f(x) = 4$
   (c) The values of x such that $f(x) = 1$

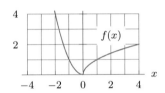

Figure 4.18

43. Consider the graphs in Figure 4.19.

   (a) Evaluate $f(0)$ and $g(0)$.
   (b) Find x such that $f(x) = 0$.
   (c) Find x such that $g(x) = 0$.
   (d) For which values of x is $f(x) = g(x)$?
   (e) For which values of x is $g(x) \geq f(x)$?

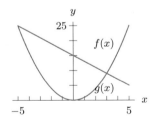

Figure 4.19

Evaluate and simplify the expressions in Problems **44–53**.

44. $f(5)$ given $f(t) = \dfrac{t+1}{t-2}$.

45. $g(3)$ given $g(z) = z\sqrt{z+1} - (z-5)^3$.

46. $g(4)$ given $g(t) = \sqrt{16 - \sqrt{25 - t^2}}$.

47. $g(-2)$ given $g(x) = 2\sqrt{25 - 4\sqrt{10 - 3x}}$.

48. $f(4)$ given $f(x) = \dfrac{4\sqrt{x} - 7}{5 - \sqrt{4x - 7}}$.

49. $g(2t + 1)$ given $g(t) = 4t - 2$.

50. $h\left(4t^2\right)$ given $h(t) = \sqrt{9t - 4}$.

51. $-2g(-x/2)$ given $g(x) = -2x^2$.

52. $h(2x - b - 3)$ given $h(x) = 2x - b - 2$.

53. $v(3z - 2r - s)$ given $v(r) = 2z - 3r - 4s$.

54. Solve $h(r) = -4$ for r given that $h(x) = \dfrac{2 - 3x}{4 - 5x}$.

55. Solve $w(2v + 7) = 3w(v - 1)$ given that $w(v) = 3v + 2$.

Table 4.19 gives $V = f(t)$, the value in Canadian dollars[9] (CAD) of \$1 (USD) t days after November 1, 2007. For instance, 1 USD could be traded for 0.9529 CAD on November 1. Use the table in Problems **56–58**, and say what your answers tell you about the value of the USD.

Table 4.19

t	0	1	2	3	4	5	6
V	0.9529	0.9463	0.9441	0.9350	0.9350	0.9349	0.9295

56. Evaluate $f(3)$.

57. Evaluate $\dfrac{50}{f(5)}$

58. Evaluate and say which value is larger:

$$\frac{f(3) - f(0)}{3 - 0} \quad \text{and} \quad \frac{f(6) - f(3)}{6 - 3}.$$

---

[9]http://www.oanda.com/convert/fxhistory, page accessed November 7, 2007.

■ Answer Problems **59–61** given that a family uses $q(t)$ gallons of gas in week t at an average price per gallon of $p(t)$ dollars.

59. Explain what this expression tells you about the family's use of gasoline:

$$p(t)q(t) - p(t-1)q(t-1).$$

60. Suppose $p(t_2) > p(t_1)$ and $p(t_2)q(t_2) < p(t_1)q(t_1)$. What does this tell you about the family's gasoline usage?

61. Write an expression for the family's average weekly expenditure on gasoline during the four-week period $2 \leq t \leq 5$.

62. Atmospheric levels of carbon dioxide (CO_2) have risen from 336 parts per million (ppm) in 1979 to 382 parts per million (ppm) in 2007.[10] Assuming a constant rate of change of CO_2, predict the level in the year 2020.

63. Methane is a greenhouse gas implicated as a contributor to global warming. The atmospheric methane level[11] in parts per billion (ppb) t years after 1980, $Q = w(t)$, is in Table 4.20. Assuming a constant rate of change in methane levels between years 15 and 20, estimate $w(18)$.

Table 4.20

t	0	5	10	15	20	25
Q	1575	1660	1715	1750	1770	1775

■ The investment portfolio in Problems **64–67** includes stocks and bonds. Let $v(t)$ be the dollar value after t years of the portion held in stocks, and let $w(t)$ be the value held in bonds.

64. Explain what the following expression tells you about the investment:

$$\frac{w(t)}{v(t) + w(t)}.$$

65. The equation $w(t) = 2v(t-1)$ has a solution at $t = 5$. What does this solution tell you about the investment?

66. Write an expression that gives the difference in value of the stock portion of the investment in year t and the bond portion of the investment the preceding year.

67. Write an equation whose solutions are the years in which the value of the bond portion of the investment exceed the value of the stock portion by exactly $3000.

68. The area, in cm^2, of glass used in a door of width w, in cm, is $A(w) = 4500 + w^2$. (See Figure 4.20.)

(a) From the graph, estimate the width of a door using 7000 cm^2 of glass.

(b) Check your answer to part (a) algebraically.

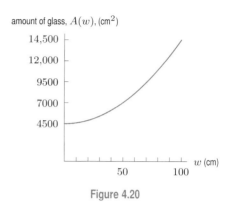

Figure 4.20

■ The *development time* of an insect is how long it takes the insect to develop from egg to adult. Typically, development time goes down as the ambient temperature rises. Table 4.21 gives values of $T = g(H)$, the development time in days at an ambient temperature $H°C$ for the bluebottle blowfly.[12] Answer Problems **69–71** and say what your answers tell you about the development time of blowflies.

Table 4.21

H	10	11	12	13	14	15	16	17	18	19	20
T	68	58	50	43	37	32	29	26	24	22	21

69. Evaluate $g(14)$.

70. Estimate the solution to $g(H) = 23$.

71. Evaluate and say which value is larger:

$$\frac{g(12) - g(10)}{12 - 10} \quad \text{and} \quad \frac{g(15) - g(12)}{15 - 12}.$$

---

[10]National Oceanic & Atmospheric Administration, http://www.esrl.noaa.gov/gmd/aggi. Page last accessed September 26, 2007.

[11]*ibid.*

[12]http://www.sciencebuddies.org/mentoring/project_ideas/Zoo_p023.shtml, accessed October 23, 2007.

72. During the holiday season, a store advertises "Spend $50, save $5. Spend $100, save $10." Assuming that the savings are directly proportional to the amount spent, what is the constant of proportionality? Interpret this in terms of the sale.

73. The distance D, in miles, traveled by a car going at 30 mph is proportional to the time t, in hours, that it has been traveling.

   (a) How far does the car travel in 5 hours?
   (b) What is the constant of proportionality? Show that the units on each side of the proportionality relationship agree.

74. The formula for the circumference of a circle is given by $C = 2\pi r$, where r is the radius of the circle. Is the circumference proportional to the radius?

75. Hooke's law states that the force F required to compress a spring by a distance of x meters is given by $F = -kx$. Is F directly proportional to x?

76. The number of tablespoons of grounds needed to brew coffee is directly proportional to the number of 8 oz cups desired. If 18 tablespoons are needed for 12 cups of coffee, how many cups can be brewed using $4\frac{1}{2}$ tablespoons? What are the units of k?

77. The number of grams of carbohydrates ingested is proportional to the number of crackers eaten. If 3 crackers cause 36 grams of carbohydrates to be ingested, how many grams of carbohydrates are ingested if 8 crackers are eaten? What are the units of k, the constant of proportionality?

78. Which of the lines (A–E) in Figure 4.21 represent a function that is a direct proportion? For those that are, find k.

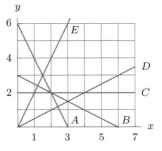

Figure 4.21

79. Observations show that the heart mass H of a mammal is 0.6% of the body mass M, and that the blood mass B is 5% of the body mass.[13]

   (a) Write a formula for M in terms of H.
   (b) Write a formula for M in terms of B.
   (c) Write a formula for B in terms of H. Is this consistent with the statement that the mass of blood in a mammal is about 8 times the mass of the heart?

80. When you convert British pounds (£) into US dollars ($), the number of dollars you receive is proportional to the number of pounds you exchange. A traveler receives $400 in exchange for £250. Find the constant of proportionality, give units for it, and explain its meaning.

81. Three ounces of broiled ground beef contains 245 calories.[14] The number of calories, C, is proportional to the number of ounces of ground beef, b. Write a formula for C in terms of b. How many calories are there in 4 ounces of ground beef?

[13] K. Schmidt-Nielsen, *Scaling, Why is Animal Size so Important?* (Cambridge: CUP, 1984).
[14] *The World Almanac Book of Facts*, 1999, p. 718.

Chapter 5

Linear Functions, Expressions and Equations

CONTENTS

5.1   LINEAR FUNCTIONS

Linear functions describe quantities that have a constant rate of increase (or decrease). For instance, suppose a video store charges \$2.50 for overnight rentals plus a late fee of \$2.99 per day. If $C = f(t)$ is the total cost as a function of t, the number of days late, then

$$f(0) = \text{Rental plus 0 days late fee} = 2.50 \qquad\qquad = 2.50 + 0(2.99) = 2.50$$
$$f(1) = \text{Rental plus 1 day late fee} = 2.50 + 2.99 \qquad\qquad = 2.50 + 1(2.99) = 5.49$$
$$f(2) = \text{Rental plus 2 days late fee} = 2.50 + \underbrace{2.99 + 2.99}_{2} \qquad = 2.50 + 2(2.99) = 8.48$$

$$f(3) = \text{Rental plus 3 days late fee} = 2.50 + \underbrace{2.99 + 2.99 + 2.99}_{3} = 2.50 + 3(2.99) = 11.47$$

$$\vdots \qquad\qquad\qquad\qquad \vdots$$

$$f(t) = \text{Rental plus } t \text{ days late fee} = 2.50 + \underbrace{2.99 + \cdots + 2.99}_{t} \quad = 2.50 + 2.99t.$$

So

$$C = f(t) = 2.50 + 2.99t.$$

See Table 5.1 and Figure 5.1. Notice that the values of C go up by 2.99 each time t increases by 1. This has the effect that the points in Figure 5.1 lie on a line.

Table 5.1

t	C
0	2.50
1	5.49
2	8.48
3	11.47
4	14.46

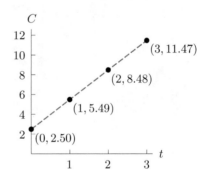

Figure 5.1: Cost, C, of renting a video as a function of t, the number of days late

The Family of Linear Functions

The function $f(t) = 2.50 + 2.99t$ belongs to the family of *linear functions*:

A **linear function** is a function that can be written

$$f(t) = b + mt, \quad \text{for constants } b \text{ and } m.$$

We call the constants b and m the *parameters* for the family.

For instance, the function $f(t) = 2.50 + 2.99t$ has $b = 2.50$ and $m = 2.99$.

What is the Meaning of the Parameters b and m?

In the family of functions representing proportionality, $y = kx$, we found that the constant k has an interpretation as a rate of change. The next example shows that m has the same interpretation for linear functions.

Example 1   The population of a town t years after it is founded is given by the linear function

$$P(t) = 30{,}000 + 2000t.$$

(a) What is the town's population when it is founded?
(b) What is the population of the town one year after it is founded? By how much does the population increase every year?
(c) Sketch a graph of the population.

Solution   (a) The town is founded when $t = 0$, so

$$\text{Initial population} = P(0) = 30{,}000 + 2000(0) = 30{,}000.$$

Thus, the 30,000 in the expression for $P(t)$ represents the starting population of the town.
(b) After one year, we have $t = 1$, and

$$P(1) = 30{,}000 + 2000(1) = 32{,}000,$$

an increase of 2000 over the starting population. After two years we have $t = 2$, and

$$P(2) = 30{,}000 + 2000(2) = 34{,}000,$$

an increase of 2000 over the year 1 population. In fact, the 2000 in the expression for $P(t)$ represents the amount by which the population increases every year.
(c) The population is 30,000 when $t = 0$, so the graph passes through the point $(0, 30{,}000)$. It slopes upward since the population is increasing at a rate of 2000 people per year. See Figure 5.2.

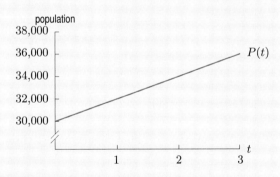

Figure 5.2: Population of a town

The next example gives a decreasing linear function—one whose output value goes down when the input value goes up.

Example 2   The value of a car, in dollars, t years after it is purchased is given by the linear function $V(t) = 18{,}000 - 1700t$.

(a) What is the value of the car when it is new?
(b) What is the value of the car after one year? How much does the value decrease each year?
(c) Sketch a graph of the value.

Solution   (a) As in Example 1, the 18,000 represents the starting value of the car, when it is new:

$$\text{Initial value} = V(0) = 18{,}000 - 1700(0) = \$18{,}000.$$

(b) The value after one year is

$$V(1) = 18{,}000 - 1700(1) = \$16{,}300,$$

which is $1700 less than the value when new. The car decreases in value by $1700 every year until it is worthless. Another way of saying this is that the rate of change in the value of the car is $-\$1700$ per year.

(c) Since the value of the car is 18,000 dollars when $t = 0$, the graph passes through the point $(0, 18{,}000)$. It slopes downward, since the value is decreasing at the rate of 1700 dollars per year. See Figure 5.3.

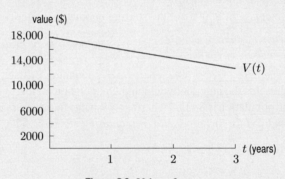

Figure 5.3: Value of a car

Example 1 involves a population given by $P(t) = 30{,}000 + 2000t$, so

$$\text{Current population} = \underbrace{\text{Initial population}}_{\text{30,000 people}} + \underbrace{\text{Growth rate}}_{\text{2000 people per year}} \times \underbrace{\text{Number of years}}_{t}.$$

In Example 2, the value is given by $V(t) = 18{,}000 + (-1700)t$, so

$$\text{Total cost} = \underbrace{\text{Initial value}}_{\$18{,}000} + \underbrace{\text{Change per year}}_{-\$1700 \text{ per year}} \times \underbrace{\text{Number of years}}_{t}.$$

In each case the coefficient of t is a rate of change, and the resulting function is linear, $Q = b + mt$.

Notice the pattern:

$$\underbrace{\text{Output}}_{y} = \underbrace{\text{Initial value}}_{b} + \underbrace{\text{Rate of change}}_{m} \times \underbrace{\text{Input}}_{t}.$$

Graphical and Numerical Interpretation of Linear Functions

We can also interpret the constants b and m in terms of the graph of $f(x) = b + mx$.

Example 3
(a) Make a table of values for the function $f(x) = 5 + 2x$, and sketch its graph.
(b) Interpret the constants 5 and 2 in terms of the table and the graph.

Solution
(a) See Table 5.2 and Figure 5.4.

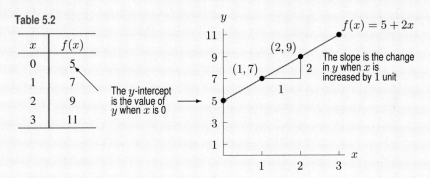

Table 5.2

x	$f(x)$
0	5
1	7
2	9
3	11

The y-intercept is the value of y when x is 0

The slope is the change in y when x is increased by 1 unit

Figure 5.4: Graph of $y = 5 + 2x$

(b) The constant 5 gives the y-value where the graph crosses the y-axis. This is the y-intercept or vertical intercept of the graph. From the table and the graph we see that 2 represents the amount by which y increases when x is increased by 1 unit. This is called the *slope*.

In Example 3, the constant m is positive, so the y-value increases when x is increased by 1 unit, and consequently the graph rises from left to right. If the value of m is negative, the y-value decreases when x is increased by 1 unit, resulting in a graph that falls from left to right.

Example 4
For the function $g(x) = 16 - 3x$,

(a) What does the coefficient of x tell you about the graph?
(b) What is the y-intercept of the graph?

Solution
(a) Since the coefficient of x is negative, the graph falls from left to right. The value of the coefficient, -3, tells us that each time the value of x is increased by 1 unit, the value of y goes down by 3 units. The slope of the graph is -3.
(b) Although we cannot tell from Figure 5.5 where the graph crosses the y-axis, we know that $x = 0$ on the y-axis, so the y-intercept is $y = 16 - 3 \cdot 0 = 16$.

Table 5.3

x	y
3	7
4	4
5	1

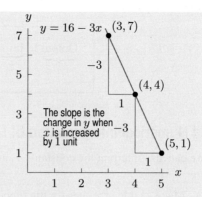

Figure 5.5: Graph of $y = 16 - 3x$

In general, we have

For the linear function
$$Q = f(t) = b + mt,$$

- $b = f(0)$ is the initial value, and gives the vertical intercept of the graph.
- m is the rate of change and gives the slope of the graph.
- If $m > 0$, the graph rises from left to right.
- If $m < 0$, the graph falls from left to right.

Example 5   Find the vertical intercept and slope, and use this information to graph the functions:

(a) $f(x) = 100 + 25x$ (b) $g(t) = 6 - 0.5t$

Solution   (a) The vertical intercept is 100 and the slope is 25, so $f(x)$ starts at 100 when $x = 0$ and increases by 25 each time x increases by 1. See Figure 5.6.

(b) The vertical intercept is 6 and the slope is -0.5, so $g(t)$ starts at 6 when $t = 0$ and decreases by 0.5 each time t increases by 1. See Figure 5.7.

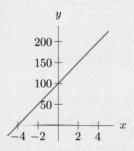

Figure 5.6: Graph of $y = 100 + 25x$

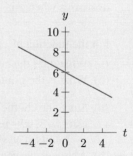

Figure 5.7: Graph of $y = 6 - 0.5t$

Interpreting Linear Functions Using Units

It is often useful to consider units of measurement when interpreting a linear function $Q = b + mt$. Since b is the initial value of Q, the units of b are the same as the units of Q. Since m is the rate of change of Q with respect to t, the units of m are the units of Q divided by the units of t.

Example 6

Worldwide, soda is the third most popular commercial beverage, after tea and milk. The global consumption of soda[1] rose at an approximately constant rate from 150 billion liters in 1995 to 179 billion liters in 2000.

(a) Find a linear function for the quantity of soda consumed, S, in billions of liters, t years after 1995.
(b) Give the units and practical interpretation of the slope and the vertical intercept.

Solution

(a) To find the linear function, we first find the rate of change, or slope. Since consumption increased from $S = 150$ to $S = 179$ over a period of 5 years, we have

$$\text{Rate of change} = \frac{\Delta S}{\Delta t} = \frac{179 - 150}{5} = \frac{29}{5} = 5.8.$$

When $t = 0$ (the year 1995), we have $S = 150$, so the vertical intercept is 150. Therefore

$$S = 150 + 5.8t.$$

(b) Since the rate of change is equal to $\Delta S/\Delta t$, its units are S-units over t-units, or billion liters per year. The slope tells us that world soda production has been increasing at a constant rate of 5.8 billion liters per year.

The vertical intercept 150 is the value of S when t is zero. Since it is a value of S, the units are S-units, or billion liters of soda. The vertical intercept tells us that the global consumption of soda in 1995 was 150 billion liters.

Functions Where the Independent Variable Is Not Time

Units are particularly useful in interpreting functions where the independent variable is not time.

Example 7

A borehole is a hole dug deep in the earth, for oil or mineral exploration. Often temperature gets warmer at greater depths. Suppose that the temperature in a borehole at the surface is $4°C$ and rises by $0.02°C$ with each additional meter of depth. Express the temperature T in $°C$ as a function of depth d in meters.

Solution

The temperature starts at $4°C$, and increases at the rate of 0.02 degrees per meter as you go down. Thus

$$\underbrace{\text{Temperature at } d \text{ meters}}_{T°C} = \underbrace{\text{Temperature at surface}}_{4°C} + \underbrace{\text{Rate of change}}_{0.02 \text{ degrees per meter}} \times \underbrace{\text{Depth}}_{d \text{ meters}}$$
$$T = 4 + 0.02d.$$

[1] The Worldwatch Institute, *Vital Signs 2002* (New York: W.W. Norton, 2002), p. 140.

Exercises and Problems for Section 5.1

EXERCISES

1. A homing pigeon starts 1000 miles from home and flies 50 miles toward home each day. Express distance from home in miles, D, as a function of number of days, d.

2. You buy a saguaro cactus 5 ft high and it grows at a rate of 0.2 inches each year. Express its height in inches, h, as a function of time t in years since the purchase.

3. The temperature of the soil is $30°C$ at the surface and decreases by $0.04°C$ for each centimeter below the surface. Express temperature T as a function of depth d, in centimeters, below the surface.

■Give the values for b and m for the linear functions in Exercises 4–9.

4. $f(x) = 3x + 12$ 5. $g(t) = 250t - 5300$

6. $h(n) = 0.01n + 100$ 7. $v(z) = 30$

8. $w(c) = 0.5c$

9. $u(k) = 0.007 - 0.003k$

10. The cost, $\$C$, of hiring a repairman for h hours is given by $C = 50 + 25h$.

    (a) What does the repairman charge to walk in the door?
    (b) What is his hourly rate?

11. The cost, $\$C$, of renting a limousine for h hours above the 4 hour minimum is given by $C = 300 + 100h$.

    (a) What does the 300 represent?
    (b) What is the hourly rate?

12. The population of a town, t years after it is founded, is given by $P(t) = 5000 + 350t$.

    (a) What is the population when it is founded?
    (b) What is the population of the town one year after it is founded? How much does it increase by during the first year? During the second year?

■In Exercises 13–23, identify the initial value and the rate of change, and explain their meanings in practical terms.

13. An orbiting spaceship releases a probe that travels directly away from Earth. The probe's distance s (in km) from Earth after t seconds is given by $s = 600 + 5t$.

14. After a rain storm, the water in a trough begins to evaporate. The amount in gallons remaining after t days is given by $V = 50 - 1.2t$.

15. The monthly charge of a cell phone is $25 + 0.06n$ dollars, where n is the number of minutes used.

16. The number of people enrolled in Mathematics 101 is $200 - 5y$, where y is the number of years since 2004.

17. On a spring day the temperature in degrees Fahrenheit is $50 + 1.2h$, where h is the number of hours since noon.

18. The value of an antique is $2500 + 80n$ dollars, where n is the number of years since the antique is purchased.

19. A professor calculates a homework grade of $100 - 3n$ for n missing homework assignments.

20. The cost, C, in dollars, of a high school dance attended by n students is given by $C = 500 + 20n$.

21. The total amount, C, in dollars, spent by a company on a piece of heavy machinery after t years in service is given by $C = 20{,}000 + 1500t$.

22. The population, P, of a city is predicted to be $P = 9000 + 500t$ in t years from now.

23. The distance, d, in meters from the shore, of a surfer riding a wave is given by $d = 120 - 5t$, where t is the number of seconds since she caught the wave.

■In Exercises 24–29, identify the slope and y-intercept and graph the function.

24. $f(x) = 2x + 3$ 25. $f(x) = 4 - x$

26. $f(x) = -2 + 0.5x$ 27. $f(x) = 3x - 2$

28. $f(x) = -2x + 5$ 29. $f(x) = -0.5x - 0.2$

PROBLEMS

30. If the tickets for a concert cost p dollars each, the number of people who will attend is $2500 - 80p$. Which of the following best describes the meaning of the 80 in this expression?

    (i) The price of an individual ticket.

    (ii) The slope of the graph of attendance against ticket price.
    (iii) The price at which no one will go to the concert.
    (iv) The number of people who will decide not to go if the price is raised by one dollar.

31. Long Island Power Authority charges its residential customers a monthly service charge plus an energy charge based on the amount of electricity used.[2] The monthly cost of electricity is approximated by the function: $C = f(h) = 36.60 + 0.14h$, where h represents the number of kilowatt hours (kWh) of electricity used in excess of 250 kWh.

   (a) What does the coefficient 0.14 mean in terms of the cost of electricity?

   (b) Find $f(50)$ and interpret its meaning.

32. The following functions describe four different collections of baseball cards. The collections begin with different numbers of cards and cards are bought and sold at different rates. The number, B, of cards in each collection is a function of the number of years, t, that the collection has been held. Describe each of these collections in words.

   (a) $B = 200 + 100t$ **(b)** $B = 100 + 200t$

   (c) $B = 2000 - 100t$ **(d)** $B = 100 - 200t$

33. Match the functions in (a)–(e) with the graphs in Figure 5.8. The constants s and k are the same in each function.

   (a) $f(x) = s$ **(b)** $f(x) = kx$

   (c) $f(x) = kx - s$ **(d)** $f(x) = 2s - kx$

   (e) $f(x) = 2s - 2kx$

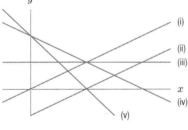

Figure 5.8

34. The velocity of an object tossed up in the air is modeled by the function $v(t) = 48 - 32t$, where t is measured in seconds, and $v(t)$ is measured in feet per second.

   (a) Create a table of values for the function.

   (b) Graph the function.

   (c) Explain what the constants 48 and -32 tell you about the velocity.

   (d) What does a positive velocity indicate? A negative velocity?

35. If a is a constant, does the equation $y = ax + 5a$ define y as a linear function of x? If so, identify the slope and vertical intercept.

36. The graphs of two linear functions have the same slope, but different x-intercepts. Can they have the same y-intercept?

37. The graphs of two linear functions have the same x-intercept, but different slopes. Can they have the same y-intercept?

■ Give the slope and y-intercept for the graphs of the functions in Problems **38–43**.

38. $f(x) = 220 - 12x$ **39.** $f(x) = \dfrac{1}{3}x - 11$

40. $f(x) = \dfrac{x}{7} - 12$ **41.** $f(x) = \dfrac{20 - 2x}{3}$

42. $f(x) = 15 - 2(3 - 2x)$ **43.** $f(x) = \pi x$

44. If n birds eating continuously consume V in^3 of seed in T hours, how much does one bird consume per hour?

45. If n birds eating continuously consume W ounces of seed in T hours, what are the units of $W/(nT)$? What does $W/(nT)$ represent in practical terms?

5.2 WORKING WITH LINEAR EXPRESSIONS

An expression, such as $3 + 2t$, that defines a linear function is called *a linear expression*. When we are talking about the expression, rather than the function it defines, we call b the *constant term* and m the *coefficient*.

Example 1   Identify the constant term and the coefficient in the expression for the following linear functions.

   (a) $u(t) = 20 + 4t$ **(b)** $v(t) = 8 - 0.3t$ **(c)** $w(t) = t/7 + 5$

---
[2] www.lipower.org, accessed December 10, 2004.

Solution   (a) We have constant term 20 and coefficient 4.

(b) Writing $v(t) = 8 + (-0.3)t$, we see that the constant term is 8 and the coefficient is -0.3.

(c) Writing $w(t) = 5 + (1/7)t$, we see that the constant term is 5 and the coefficient is 1/7.

Different forms for linear expressions reveal different aspects of the functions they define.

The Slope-Intercept Form

The form that we have been using for linear functions is called the

---

Slope-Intercept Form

The form
$$f(x) = b + mx \quad \text{or} \quad y = b + mx$$
for expressing a linear function is called *slope-intercept form*, because it shows the slope, m, and the vertical intercept, b, of the graph.

Expressing a function in slope intercept form is helpful in reading its initial value and rate of change.

Example 2   The cost C of a vacation lasting d days consists of the air fare, \$350, plus accommodation expenses of \$55 times the number of days, plus food expenses of \$40 times the number of days.

(a) Give an expression for C as a function of d that shows air fare, accommodation, and food expenses separately.

(b) Express the function in slope-intercept form. What is the significance of the vertical intercept and the slope?

Solution   (a) The cost is obtained by adding together the air fare of \$350, the accommodation, and the food. The accommodation for d days costs $55d$ and the food costs $40d$. Thus

$$C = 350 + 55d + 40d.$$

(b) Collecting like terms, we get
$$C = 350 + 95d$$

dollars, which is linear in d with slope 95 and vertical intercept 350. The vertical intercept represents the initial cost (the air fare) and the slope represents the total daily cost of the vacation, \$95 per day.

The Point-Slope Form

Although slope-intercept form is the simplest form, sometimes another form reveals a different aspect of a function.

Point-Slope Form

The form

$$f(x) = y_0 + m(x - x_0) \quad \text{or} \quad y = y_0 + m(x - x_0)$$

for expressing a linear function is called *point-slope form*, because
- The graph passes through the point (x_0, y_0).
- The slope, or rate of change, is m.

Example 3   The population of a town t years after it is founded is given by

$$P(t) = 16{,}000 + 400(t - 5).$$

(a) What is the practical interpretation of the constants 5 and 16,000 in the expression for P?
(b) Express $P(t)$ in slope-intercept form and interpret the slope and intercept.

Solution   (a) The difference between point-slope form and slope-intercept form is that the coefficient 400 multiplies the expression $t - 5$, rather than just t. Whereas the slope-intercept form tells us the value of the function when $t = 0$, this form tells us the value when $t = 5$:

$$P(5) = 16{,}000 + 400(5 - 5) = 16{,}000 + 0 = 16{,}000.$$

Thus, the population is 16,000 after 5 years.

(b) We have

$$
\begin{aligned}
P(t) &= 16{,}000 + 400(t - 5) && \text{Population in year 5 is 16,000.}\\
&= 16{,}000 + 400t - 400 \cdot 5 && \text{Five years ago, it was 2000 fewer.}\\
&= 14{,}000 + 400t && \text{Thus, the starting value is 14,000.}
\end{aligned}
$$

From the slope-intercept form we see that 400 represents the growth rate of the population per year. When we subtract $400 \cdot 5$ from 16,000 in the above calculation, we are deducting five years of growth in order to obtain the initial population, 14,000.

How Do We Find the Slope?

In Section 4.4 we saw how to find the average rate of change of a function between two points. For a linear function this rate of change is equal to the slope and is the same no matter which two points we choose.

Example 4   Find formulas for the linear functions graphed below.

(a)

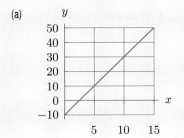

(b)

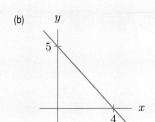

Solution

(a) We use any two points to find the slope. If we use the points $(0, -10)$ and $(5, 10)$, we have

$$m = \frac{\Delta y}{\Delta x} = \frac{10 - (-10)}{5 - 0} = \frac{20}{5} = 4.$$

The slope is $m = 4$. We see in the graph that the y-intercept is -10, so a formula for the function is

$$y = -10 + 4x.$$

(b) We use the two points $(0, 5)$ and $(4, 0)$ to find the slope:

$$m = \frac{\Delta y}{\Delta x} = \frac{0 - 5}{4 - 0} = \frac{-5}{4} = -\frac{5}{4}.$$

The slope is $m = -5/4$. The slope is negative since the graph falls from left to right. We see in the graph that the y-intercept is 5, so a formula for the function is

$$y = 5 - \frac{5}{4}x.$$

Example 5

Find linear functions satisfying:

(a) $f(5) = -8$ and the rate of change is -3
(b) $g(5) = 20$ and $g(8) = 32$.

Solution

(a) The graph passes through $(x_0, y_0) = (5, -8)$ and has slope $m = -3$, so we use point-slope form:

$$\begin{aligned} f(x) &= y_0 + m(x - x_0) \\ &= (-8) + (-3)(x - 5) \\ &= -8 - 3x + 15 \\ &= 7 - 3x. \end{aligned}$$

(b) The graph passes through $(5, 20)$ and $(8, 32)$, so the slope is

$$m = \frac{\Delta y}{\Delta x} = \frac{32 - 20}{8 - 5} = \frac{12}{3} = 4.$$

Using point-slope form with the point $(x_0, y_0) = (5, 20)$, we have

$$\begin{aligned} g(x) &= y_0 + m(x - x_0) \\ &= 20 + 4(x - 5) \\ &= 20 + 4x - 20 \\ &= 4x. \end{aligned}$$

Notice that in either case, we could have left the equation of the line in point-slope form instead of putting it in slope-intercept form.

How Do We Recognize a Linear Expression?

Sometimes a simple rearrangement of the terms is enough to recognize an expression as linear.

Example 6   Is the expression linear in x? If it is, give the constant term and the coefficient.

(a) $5 + 0.2x^2$ (b) $3 + 5\sqrt{x}$ (c) $3 + x\sqrt{5}$

(d) $2(x + 1) + 3x - 5$ (e) $(1 + x)/2$ (f) $ax + x + b$

Solution

(a) This expression is not linear because of the x^2 term.

(b) This expression is not linear because of the $\sqrt{x}$ term.

(c) This expression is linear. Writing this as $y = 3 + \left(\sqrt{5}\right)x$, we see that the constant term is 3 and the coefficient is $\sqrt{5}$.

(d) We distribute the 2 and combine like terms:

$$2(x + 1) + 3x - 5 = 2x + 2 + 3x - 5 = -3 + 5x,$$

so we see that the expression is linear. The constant term is -3, and the coefficient is 5.

(e) Distributing the $1/2$ we get

$$\frac{1 + x}{2} = \frac{1}{2} + \frac{1}{2}x,$$

so we see that the expression is linear. The constant term is $1/2$, and the coefficient is $1/2$.

(f) We collect the x terms to get $ax + x + b = b + (a + 1)x$, so we see that the expression is linear. The constant term is b, and the coefficient is $a + 1$.

How Do We Decide If a Function Is Linear?

To decide if a function $f(x)$ given by an expression is linear, we focus on the form of the expression and see if it can be put in the form $f(x) = b + mx$.

Example 7   Is each function linear?

(a) The share of a community garden plot with area A square feet divided between 5 families is $f(A) = A/5$ square feet per family.

(b) The gasoline remaining in an electric generator running for h hours is $G = 0.75 - 0.3h$.

(c) The circumference of a circle of radius r is $C = 2\pi r$.

(d) The time it takes to drive 300 miles at v mph is $T = 300/v$ hours.

Solution

(a) This function is linear, since we can rewrite it in the form $f(A) = 0 + (1/5)A$. The initial value is 0 and the rate of change is $1/5$.

(b) Since $G = 0.75 + (-0.3)h$, we see this is linear with initial value 0.75 and rate of change -0.3.

(c) This function is linear, with initial value 0 and rate of change 2π.

(d) This function is not linear, since it involves dividing by v, not multiplying v by a constant.

Expressions Involving More Than One Letter

In Example 6(f) there were constants a and b in the expression $ax + x + b$, in addition to the variable x. We could change perspective on this expression, and regard a as the variable and x and b as the constants. Writing it in the form

$$xa + (x + b) = \text{Constant} \cdot a + \text{Constant},$$

we see that it gives a linear function of a. In this case we say that the expression is *linear in a*.

Example 8   (a) Is the expression $xy^2 + 5xy + 2y - 8$ linear in x? In y?
(b) Is the expression $\pi r^2 h$ linear in r? In h?

Solution   (a) To see if the expression is linear in x, we try to match it with the form $b + mx$. We have

$$xy^2 + 5xy + 2y - 8 = (2y - 8) + (y^2 + 5y)x$$

which is linear in x with $b = 2y - 8$ and $m = y^2 + 5y$. Thus the expression is linear in x. It is not linear in y because of the y^2 term.
(b) The expression is not linear in r because of the r^2 term. It is linear in h, with constant term 0 and coefficient πr^2.

Exercises and Problems for Section 5.2

EXERCISES

In Exercises 1–8, is the expression linear?

1. $5t - 3$

2. $5^x + 1$

3. $6r + r - 1$

4. $(3a + 1)/4$

5. $(3a + 1)/a$

6. $5r^2 + 2$

7. $4^2 + (1/3)x$

8. $6A - 3(1 - 3A)$

For each of the linear expressions in x in Exercises 9–14, give the constant term and the coefficient of x.

9. $3x + 4$

10. $5x - x + 5$

11. $w + wx + 1$

12. $x + rx$

13. $mx + mn + 5x + m + 7$

14. $5 - 2(x + 4) + 6(2x + 1)$

In Exercises 15–18, rewrite the function in slope-intercept form.

15. $f(x) = 12 + 3(x - 1)$

16. $f(x) = 1800 + 500(x + 3)$

17. $g(n) = 14 - 2/3(n - 12)$

18. $j(t) = 1.2 + 0.4(t - 5)$

In Exercises 19–22, the form of the expression for the function tells you a point on the graph and the slope of the graph. What are they? Sketch the graph.

19. $f(x) = 3(x - 1) + 5$

20. $f(t) = 4 - 2(t + 2)$

21. $g(s) = (s - 1)/2 + 3$

22. $h(x) = -5 - (x - 1)$

23. Find the slope of each of the lines in Figure 5.9.

(a) 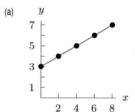   (b)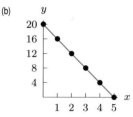

Figure 5.9: Find the slope of each of these lines

24. For working n hours a week, where $n \geq 40$, a personal trainer is paid, in dollars,

$$P(n) = 500 + 18.75(n - 40).$$

What is the practical meaning of the 500 and the 18.75?

25. When n guests are staying in a room, where $n \geq 2$, the Happy Place Hotel charges, in dollars,

$$C(n) = 79 + 10(n - 2).$$

What is the practical meaning of the 79 and the 10?

26. A salesperson receives a weekly salary plus a commission when the weekly sales exceed \$1000. The person's total income in dollars for weekly sales of s dollars (where $s \geq 1000$) is given, in dollars, by

$$T(s) = 600 + 0.15(s - 1000).$$

What is the practical meaning of the 600 and the 0.15?

27. Find a possible formula for the linear function $h(x)$ if $h(-30) = 80$ and $h(40) = -60$.

28. Find a possible formula for the linear function $f(x)$ if $f(20) = 70$ and $f(70) = 10$.

29. Find a possible formula for the linear function $f(x)$ if $f(-12) = 60$ and $f(24) = 42$.

In Exercises 30–33, does the description lead to a linear function? If so, give a formula for the function.

30. The distance traveled is the speed, 45 mph, times the number of hours, t.

31. The area of a circle of radius r is πr^2.

32. The area of a rectangular plot of land w ft wide and 20 ft long is $20w$ ft^2.

33. The area of a square plot of land x ft on a side is x^2 ft^2.

PROBLEMS

34. Greta's Gas Company charges residential customers \$8 per month even if they use no gas, plus 82¢ per therm used. (A therm is a quantity of gas.) In addition, the company is authorized to add a rate adjustment, or surcharge, per therm. The total cost of g therms of gas is given by

$$\text{Total cost } = 8 + 0.82g + 0.109g.$$

(a) Which term represents the rate adjustment? What is the rate adjustment in cents per therm?
(b) Is the expression for the total cost linear?

35. A car trip costs \$1.50 per fifteen miles for gas, 30¢ per mile for other expenses, and \$20 for car rental. The total cost for a trip of d miles is given by

$$\text{Total cost } = 1.5 \left(\frac{d}{15} \right) + 0.3d + 20.$$

(a) Explain what each of the three terms in the expression represents in terms of the trip.
(b) What are the units for cost and distance?
(c) Is the expression for cost linear?

In Problems 36–46, is the given expression linear in the indicated variable? Assume all constants are non-zero.

36. $\dfrac{a+b}{2}, a$

37. $2\pi r^2 + \pi rh, r$

38. $2\pi r^2 + \pi rh, h$

39. $ax^2 + bx + c^3, a$

40. $ax^2 + bx + c^3, x$

41. $2ax + bx + c, a$

42. $2ax + bx + c, x$

43. $3xy + 5x + 2 - 10y, x$

44. $3xy + 5x + 2 - 10y, y$

45. $P(P - c), P$

46. $P(P - c), c$

In Problems 47–49, find the function graphed and give a practical interpretation of the slope and vertical intercept.

47.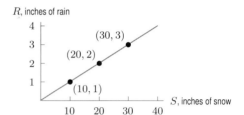

48.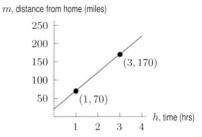

49.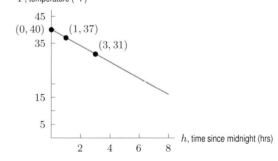

50. A boy's height h, in feet, t years after his 10th birthday, is given by $h = 4 + 0.2t$. Which of the following equivalent expressions for this function shows most clearly his height at age 20? What is that height?

(i) $h = 4 + 0.2t$ (ii) $h = 6 + 0.2(t - 10)$

(iii) $h = 10 + 0.2(t - 30)$

51. The number of books you can afford to buy, b, is a function of the number of CDs, c, you buy and is given by $b = 10 - 0.5c$. Which of the following equivalent expressions for this function most clearly shows the number of books you can afford if you buy 6 CDs?

(i) $b = 10 - 0.5c$ (ii) $b = 6 - 0.5(c - 8)$

(iii) $b = 7 - 0.5(c - 6)$

52. A company's profit after t months of operation is given by $P(t) = 1000 + 500(t - 4)$.

(a) What is the practical meaning of the constants 4 and 1000?

(b) Rewrite the function in slope-intercept form and give a practical interpretation of the constants.

53. After t hours, Liza's distance from home, in miles, is given by $D(t) = 138 + 40(t - 3)$.

(a) What is the practical interpretation of the constants 3 and 138?

(b) Rewrite the function in slope-intercept form and give a practical interpretation of the constants.

54. A cyclist's distance in km from the finish line, t minutes after reaching the flat, is given by $f(t) = 45 - 0.5(t - 12)$.

(a) What is the practical meaning of the constants 12, 45, and 0.5?

(b) Express f in a form that clearly shows the distance from the start of the flat to the finish line.

55. The number of butterflies in a collection x years after 1960 is given by $B(x) = 50 + 2(x - 20)$.

(a) What is the practical interpretation of the constants 20, 50, and 2?

(b) Express B in a form that clearly shows the size of the collection when it started in 1960.

56. The cost, $C(w)$, of mailing a large envelope weighing w ounces, $0 < w \leq 13$, can be modeled by the equation $C(w) = 0.88 + 0.17(w - 1)$. (Note that all fractional ounces are rounded up to the next integer.)

(a) What is the practical interpretation of the constants 0.88 and 1?

(b) What does the 0.17 represent?

(c) How much would it cost to mail a large envelope weighing 9.1 ounces?

■ In Problems **57–59**, why do we expect the situation to be modeled by a linear function? Give an expression for the function.

57. The profit from making q widgets is the revenue minus the cost, where the revenue is the selling price, $27, times the number of widgets, and the cost is $1000 for setting up a production line plus $15 per widget.

58. The construction costs of a road d miles long with 4 toll booths are $500,000 per mile plus $100,000 per toll booth.

59. A farmer builds a fence with two gates, each 4 meters wide, around a square field x meters on a side. The cost of the fence is $10 per meter plus $300 per gate.

■ For Problems **60–62**, write an expression in x representing the result of the given operations on x. Is the expression linear in x?

60. Add 5, multiply by 2, subtract x.

61. Add 5, multiply by x, subtract 2.

62. Add x, multiply by 5, subtract 2.

63. Find a possible formula for the linear function $y = g(x)$ given that:

- The value of the expression $g(100)$ is 30, and
- The solution to the equation $g(x) = 15$ is -50.

5.3 SOLVING LINEAR EQUATIONS

We are often interested in knowing which input values to a linear function give a particular output value. This gives rise to a linear equation.

Example 1   For the town in Example 1 on page 115, the population t years after incorporation is given by $P(t) = 30,000 + 2000t$. How many years does it take for the population to reach 50,000?

Solution   We want to know the value of t that makes $P(t)$ equal to 50,000, so we solve the equation

$$30,000 + 2000t = 50,000$$
$$2000t = 20,000 \qquad \text{subtract 30,000 from both sides}$$
$$t = \frac{20,000}{2000} = 10. \qquad \text{divide both sides by 2000}$$

Thus, it takes 10 years for the population to reach 50,000.

Finding Where Two Linear Functions Are Equal

Linear equations also arise when we want to know what input makes two linear functions $f(x)$ and $g(x)$ have the same output.

Example 2   Incandescent light bulbs are cheaper to buy but more expensive to operate than fluorescent bulbs. The total cost in dollars to purchase a bulb and operate it for t hours is given by

$$f(t) = 0.50 + 0.004t \quad \text{(for incandescent bulbs)}$$
$$g(t) = 5.00 + 0.001t \quad \text{(for fluorescent bulbs).}$$

How many hours of operation gives the same cost with either choice?

Solution   We need to find a value of t that makes $f(t) = g(t)$, so

$$0.50 + 0.004t = 5.00 + 0.001t \quad \text{set expressions for } f(t) \text{ and } g(t) \text{ equal}$$
$$0.50 + 0.003t = 5.00 \qquad \text{subtract } 0.001t$$
$$0.003t = 4.50 \qquad \text{subtract } 0.50$$
$$t = \frac{4.50}{0.003} = 1500 \quad \text{divide by 0.003.}$$

After 1500 hours of use, the cost to buy and operate an incandescent bulb equals the cost to buy and operate a fluorescent bulb.[3] Let's verify our solution:

Cost for incandescent bulb: $f(1500) = 0.50 + 0.004(1500) = 6.50$

Cost for fluorescent bulb: $g(1500) = 5.00 + 0.001(1500) = 6.50.$

We see that the cost of buying and operating either type of bulb for 1500 hours is the same: \$6.50.

Using a Graph to Visualize Solutions

If we graph two functions f and g on the same axes, then the values of t where $f(t) = g(t)$ correspond to points where the two graphs intersect.

---

[3] In fact incandescent bulbs typically last less than 1000 hours, which this calculation does not take into account.

Example 3   In Example 2 we saw that the cost to buy and operate an incandescent bulb for 1500 hours is the same as the cost for a fluorescent bulb.

(a) Graph the cost for each bulb and indicate the solution to the equation in Example 2.
(b) Which bulb is cheaper if you use it for less than 1500 hours? More than 1500 hours?

Solution   See Figure 5.10. The t-coordinate of the point where the two graphs intersect is 1500. At this point, both functions have the same value. In other words, $t = 1500$ is a solution to the equation $f(t) = g(t)$. Figure 5.10 shows that the incandescent bulb is cheaper if you use it for less than 1500 hours. For example, it costs \$4.50 to operate the incandescent bulb for 1000 hours, whereas it costs \$6.00 to operate the fluorescent bulb for the same time. On the other hand, if you operate the bulb for more than 1500 hours, the fluorescent bulb is cheaper.

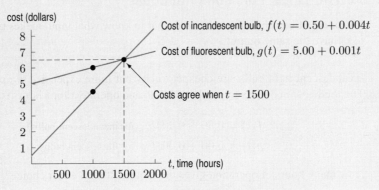

Figure 5.10: Costs of operating different types of bulb

How Many Solutions Does a Linear Equation Have?

In general a linear equation has one solution. If we visualize the equation graphically as in Example 3, we see that the two lines intersect at one point.

Example 4   Solve (a) $3x + 5 = 3(x + 5)$ (b) $3x + 5 = 3(x + 1) + 2$.

Solution   (a) We have

$$3x + 5 = 3(x + 5)$$
$$3x + 5 = 3x + 15$$
$$5 = 15.$$

What is going on here? The last equation is false no matter what the value of x, so this equation has no solution. Figure 5.11 shows why the equation has no solution. The graphs of $y = 3x + 5$ and $y = 3(x + 5)$ have the same slope, so never meet.

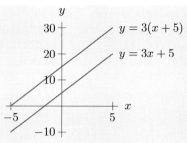

Figure 5.11: Graphs of $y = 3x + 5$ and $y = 3(x + 5)$ do not intersect

(b) We have

$$3x + 5 - 3(x + 1) + 2$$
$$3x + 5 = 3x + 5$$
$$0 = 0$$

You might find this result surprising as well. The last equation is true no matter what the value of x. So this equation has infinitely many solutions; every value of x is a solution. In this case, the graphs of $y = 3x + 5$ and $y = 3(x + 1) + 2$ are the same line, so they intersect everywhere.

In summary

A linear equation can have no solutions, one solution, or infinitely many solutions. See Figure 5.12.

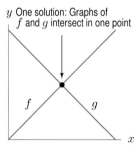
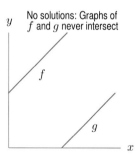
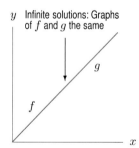

Figure 5.12: Graphs of $f(x)$ and $g(x)$

Looking Ahead When Solving Equations

With foresight, we do not need to solve equations like the ones in Example 4 all the way to the end.

Example 5   At what point during solving the equations in Example 4 could you have stopped and foreseen the number of solutions?

Solution   In part (a), we can see at the second line that no value of x satisfies the equation. The left side has 5 added to $3x$, and the right side has 15 added to the same $3x$. So $3x + 5$ can never equal $3x + 15$, no matter what value $3x$ takes on.

In part (b), we see from the second line that the two sides, $3x+5$ and $3(x+1)+2$ are equivalent expressions, so they are equal for all values of x.

You can also often use the context of an equation to make predictions about the solutions.

Example 6   For the equation comparing light bulb performance in Example 3, explain how you could predict that the solution is a positive number.

Solution   A positive solution means that there is a time in the future (after you buy the bulbs) when the two bulbs end up costing the same. This makes sense, because although the incandescent bulb is cheaper to buy, it uses energy at a greater rate, so the cost of using it will eventually catch up with the fluorescent bulb. This is illustrated in Figure 5.10 on page 130, where the graph for the incandescent bulb starts out lower but rises more steeply than the other graph.

Sometimes you can tell whether the solution is going to be positive or negative just by looking at the sign of the coefficients.

Example 7   Is the solution positive, negative, or zero?

(a) $53x = -29$ (b) $29x + 53 = 53x + 29$
(c) $29x + 13 = 53x + 29$ (d) $29x + 53 = 53x + 53$

Solution   (a) The solution is negative, since it is the ratio of a negative number and a positive number.
(b) When we collect all the x terms on the right and the constant term on the left, we get

$$\text{Positive number} = (\text{Positive number})x,$$

because $53 > 29$, so the solution is positive.
(c) This time the equation is equivalent to

$$\text{Negative number} = (\text{Positive number})x,$$

because $13 < 29$, so the solution is negative.
(d) Since the constant terms are the same on both sides, the equation is equivalent to

$$0 = (\text{Positive number})x,$$

so the solution is 0.

Equations with More Than One Letter

Example 8   Suppose you pay a total of $16,368 for a car whose list price (the price before taxes) is p. Find the list price of the car if you buy it in:

(a) Arizona, where the sales tax is 5.6%.
(b) New York, where the sales tax is 8.25%.
(c) A state where the sales tax is r.

Solution   (a) If p is the list price in dollars then the tax on the purchase is $0.056p$. The total amount paid is $p + 0.056p$, so

$$p + 0.056p = 16{,}368$$
$$(1 + 0.056)p = 16{,}368$$
$$p = \frac{16{,}368}{1 + 0.056} = \$15{,}500.$$

(b) The total amount paid is $p + 0.0825p$, so

$$p + 0.0825p = 16{,}368$$
$$(1 + 0.0825)p = 16{,}368$$
$$p = \frac{16{,}368}{1 + 0.0825} = \$15{,}120.55.$$

(c) The total amount paid is $p + rp$, so

$$p + rp = 16{,}368$$
$$(1 + r)p = 16{,}368$$
$$p = \frac{16{,}368}{1 + r} \text{ dollars.}$$

In part (c) of Example 8, there is a constant r in the equation, in addition to the variable p. Since the equation is linear in p, we can solve it. The only difference is that the solution contains the constant r.

Equations with more than one letter can be nonlinear in some of the letters, but still be linear in the letter that you are trying to solve for.

Example 9   If P dollars is invested at an annual interest rate r compounded monthly, then its value in dollars after T years is

$$P\left(1 + \frac{r}{12}\right)^{12T}.$$

What amount must be invested to produce a balance of $10,000 after T years?

Solution   We must solve the equation

$$P\left(1 + \frac{r}{12}\right)^{12T} = 10{,}000$$

for P. If we treat r and T as constants, then the expression on the left is linear in P, because it is P multiplied by an expression that involves only r and T. So it is of the form

$$\overbrace{P\left(1 + \frac{r}{12}\right)^{12T}}^{\text{constant}} = P \times \text{constant}.$$

Thus, we can solve for P by dividing through by the constant:

$$P = \frac{10{,}000}{\left(1 + \frac{r}{12}\right)^{12T}}.$$

Exercises and Problems for Section 5.3

EXERCISES

1. The tuition cost for part-time students taking C credits at Stonewall College is given by $300 + 200C$ dollars.

   (a) Find the tuition cost for eight credits.
   (b) If the tuition cost is $1700, how many credits are taken ?

2. A car's value t years after it is purchased is given by $V(t) = 18,000 - 1700t$. How long does it take for the car's value to drop to $2000?

3. For the function $f(t) = \dfrac{2t + 3}{5}$,

   (a) Evaluate $f(11)$ (b) Solve $f(t) = 2$.

4. Solve $g(x) = 7$ given that $g(x) = \dfrac{5x}{2x - 3}$.

■ In Exercises 5–9, a company offers three formulas for the weekly salary of its sales people, depending on the number of sales, s, made each week:

(a) $100 + 0.10s$ dollars   (b) $150 + 0.05s$ dollars
(c) 175 dollars

5. How many sales must be made under option (a) to receive $200 a week?

6. How many sales must be made under option (c) to receive $200 a week?

7. At what sales level do options (a) and (b) produce the same salary?

8. At what sales level do options (b) and (c) produce the same salary?

9. At what sales level do options (a) and (c) produce the same salary?

■ Solve the equations in Exercises 10–16.

10. $7 - 3y = -17$ 11. $13t + 2 = 49$

12. $3t + \dfrac{2(t - 1)}{3} = 4$

13. $2(r + 5) - 3 = 3(r - 8) + 20$

14. $2x + x = 27$

15. $4t + 2(t + 1) - 5t = 13$

16. $\dfrac{9}{x - 3} - \dfrac{5}{1 - x} = 0$

■ Without solving them, say whether the equations in Exercises 17–28 have a positive solution, a negative solution, a zero solution, or no solution. Give a reason for your answer.

17. $3x = 5$ 18. $3a + 7 = 5$

19. $5z + 7 = 3$ 20. $3u - 7 = 5$

21. $7 - 5w = 3$ 22. $4y = 9y$

23. $4b = 9b + 6$ 24. $6p = 9p - 4$

25. $8r + 3 = 2r + 11$ 26. $8 + 3t = 2 + 11t$

27. $2 - 11c = 8 - 3c$ 28. $8d + 3 = 11d + 3$

■ In Exercises 29–35, does the equation have no solution, one solution, or an infinite number of solutions?

29. $4x + 3 = 7$

30. $4x + 3 = -7$

31. $4x + 3 = 4(x + 1) - 1$

32. $4x + 3 = 4(x + 1) + 1$

33. $4x + 3 = 3$

34. $4x + 3 = 4(x - 1) + 5$

35. $4x + 3 = 4(x - 1) + 7$

36. Solve $t(t + 3) - t(t - 5) = 4(t - 5) - 7(t - 3)$.

PROBLEMS

37. You have a coupon worth $20 off the purchase of a scientific calculator. At the same time the calculator is offered with a discount of 20%, and no further discounts apply. For what tag price on the calculator do you pay the same amount for each discount?

38. Apples are 99 cents a pound, and pears are $1.25 a pound. If I spend $4 and the weight of the apples I buy is twice the weight of the pears, how many pounds of pears do I buy?

39. A car rental company charges $37 per day and $0.25 per mile.

    (a) Compute the cost of renting the car for one day, assuming the car is driven 100 miles.
    (b) Compute the cost of renting the car for three days, assuming the car is driven 400 miles.
    (c) Andy rented a car for five days, but he did not keep track of how many miles he drove. He gets a bill for $385. How many miles did he drive?

40. Using Figure 5.13, determine the value of s.

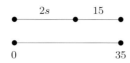

Figure 5.13

41. The floor plan for a room is shown in Figure 5.14. The total area is 144 ft^2. What is the missing length?

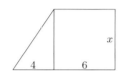

Figure 5.14: Not drawn to scale

42. You drive 100 miles. Over the first 50 miles you drive 50 mph, and over the second 50 miles you drive V mph.

(a) Calculate the time spent on the first 50 miles and on the second 50 miles.

(b) Calculate the average speed for the entire 100 mile journey.

(c) If you want to average 75 mph for the entire journey, what is V?

(d) If you want to average 100 mph for the entire journey, what is V?

In Problems 43–50, decide for what value(s) of the constant A (if any) the equation has

(a) The solution $x = 0$ **(b)** A positive solution in x

(c) No solution in x.

43. $3x = A$

44. $Ax = 3$

45. $3x + 5 = A$

46. $3x + A = 5$

47. $3x + A = 5x + A$

48. $Ax + 3 = Ax + 5$

49. $\dfrac{7}{x} = A$

50. $\dfrac{A}{x} = 5$

In Problems 51–57, $f(t) = 2t + 7$. Does the equation have no solution, one solution, or an infinite number of solutions?

51. $f(t) = 7$

52. $2f(t) = f(2t)$

53. $f(t) = f(t + 1) - 2$

54. $f(t) = f(-t)$

55. $f(t) = -f(t)$

56. $f(t) + 1 = f(t + 1)$

57. $f(t) + f(3t) - 2f(2t) = 0$

58. If $b_1 + m_1 x = b_2 + m_2 x$, what can be said about the constants $b_1, m_1, b_2,$ and m_2 if the equation has

(a) One solution?

(b) No solutions?

(c) An infinite number of solutions?

5.4 EQUATIONS FOR LINES IN THE PLANE

The formula $y = b + mx$ expresses y as a linear function of x, whose graph is a line. But it can also be thought of in a different way, as a linear equation in two variables, whose solution is a line in the xy-plane. In this section we use this point of view to analyze the geometry of lines.

Slope-Intercept Form and Point-Slope Form Revisited

In Section 5.2 we used slope-intercept form and point-slope form to interpret linear functions. We can also use them to understand lines.

Example 1   For each of the following equations, find the vertical intercept and slope of its graph by putting the equation in slope-intercept form. Use this information to match the equations with the graphs.

(a) $y - 5 = 8(x + 1)$ (b) $3x + 4y = 20$ (c) $6x - 15 = 2y - 3$

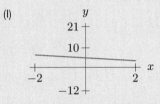
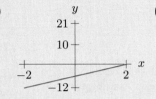
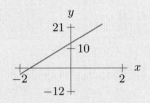

Figure 5.15: Which graph goes with which equation?

Solution   We solve for y in each case:

(a)
$$y - 5 = 8(x + 1)$$
$$y - 5 = 8x + 8$$
$$y = 8x + 8 + 5 = 13 + 8x.$$

The y-intercept is 13 and the slope is 8. This matches graph (III).

(b)
$$3x + 4y = 20$$
$$4y = 20 - 3x$$
$$y = \frac{20 - 3x}{4} = \frac{20}{4} - \frac{3}{4}x = 5 - \frac{3}{4}x.$$

The y-intercept is 5 and the slope is $-3/4$. Since the slope is negative, the y-values decrease as the x-values increase, and the graph falls. This matches graph (I).

(c)
$$6x - 15 = 2y - 3$$
$$2y = 6x - 15 + 3 = 6x - 12$$
$$y = \frac{6x - 12}{2} = \frac{6}{2}x - \frac{12}{2} = -6 + 3x.$$

The y-intercept is -6 and the slope is 3. This matches graph (II). Notice that this graph rises more gently than graph (III) because its slope is 3, which is less than 8, the slope of graph (III).

Example 2   Explain how the form of the equation

$$y = 5 + 3(x - 2)$$

enables you to see without calculation that

(a) its graph passes through the point $(2, 5)$ (b) the slope is 3.

Solution   (a) When we put $x = 2$ in $5 + 3(x - 2)$ the second term is zero, so we know the value is 5:

$$y = 5 + 3(2 - 2) = 5 + 0 = 5.$$

(b) The coefficient 3 in $y = 5 + 3(x - 2)$ becomes the coefficient of x when we distribute it over the $x - 2$:

$$y = 5 + 3(x - 2) = \text{constant} + 3x + \text{constant}.$$

Combining the constants, we see the graph has slope 3. See Figure 5.16.

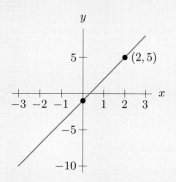

Figure 5.16: Graph of $y = 5 + 3(x - 2)$

Equations of Horizontal and Vertical Lines

A line with positive slope rises and one with negative slope falls as we move from left to right. What about a line with slope $m = 0$? Such a line neither rises nor falls, but is horizontal.

Example 3   Explain why the equation $y = 4$ represents a horizontal line and the equation $x = 4$ represents a vertical line, when regarded as equations in two variables x and y.

Solution   We can think of $y = 4$ as an equation in x and y by rewriting it

$$y = 4 + 0 \cdot x.$$

The value of y is 4 for all values of x, so all points with y-coordinate 4 lie on the graph. As we see in Figure 5.17, these points lie on a horizontal line. Similarly, the equation $x = 4$ means that x is 4 no matter what the value of y is. Every point on the vertical line in Figure 5.18 has x equal to 4.

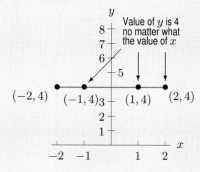

Figure 5.17: The horizontal line $y = 4$ has slope 0

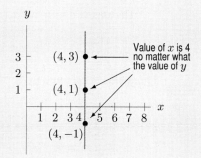

Figure 5.18: The vertical line $x = 4$ has an undefined slope

In Example 3 the equation $x = 4$ cannot be put into slope-intercept form. This is because the slope is defined as $\Delta y / \Delta x$ and Δx is zero. A vertical line does not have an equation of the form $y = mx + b$, and its slope is undefined.

In summary,

For any constant k:
- The graph of the equation $y = k$ is a horizontal line, and its slope is zero.
- The graph of the equation $x = k$ is a vertical line, and its slope is undefined.

The Standard Form of a Linear Equation

The slope-intercept and point-slope forms for linear equations have one of the variables isolated on the left side of the equation. Linear equations can also be written in a form like $5x + 4y = 20$, where neither x nor y is isolated on one side. This is called *standard form*.

Standard Form for a Linear Equation

Any line, including horizontal and vertical lines, has an equation of the form

$$Ax + By = C, \qquad \text{where } A, B, \text{ and } C \text{ are constants.}$$

Example 4   What values of A, B, and C in the standard form give the horizontal and vertical lines in Example 3?

Solution   Writing $y = 4$ as

$$0 \cdot x + 1 \cdot y = 4$$

we have $A = 0$, $B = 1$, and $C = 4$. Similarly, writing $x = 4$ as

$$1 \cdot x + 0 \cdot y = 4$$

we have $A = 1$, $B = 0$, and $C = 4$.

The standard form makes it easy to find the places where a line intercepts the axes.

Example 5   Find the intercepts of the line $5x + 4y = 20$ and graph the line.

Solution   The y-axis is the line $x = 0$, so we put $x = 0$ and get

$$5 \cdot 0 + 4y = 4y = 20, \quad y = \frac{20}{4} = 5.$$

So the y-intercept is 5. Similarly, the x-intercept is 4, because putting $y = 0$ we get

$$5x + 4 \cdot 0 = 5x = 20, \quad x = \frac{20}{5} = 4.$$

Slopes of Parallel and Perpendicular Lines

Figure 5.19 shows two parallel lines. These lines are parallel because they have equal slopes.

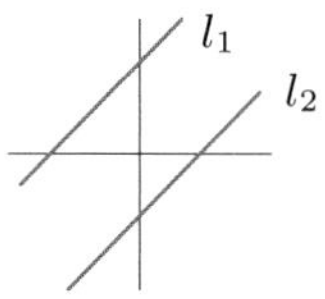

Figure 5.19: Parallel lines: l_1 and l_2 have equal slopes

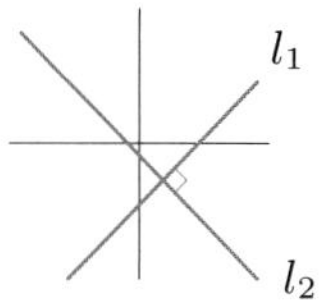

Figure 5.20: Perpendicular lines: l_1 has a positive slope and l_2 has a negative slope

What about perpendicular lines? Two perpendicular lines are graphed in Figure 5.20. We can see that if one line has a positive slope, then any line perpendicular to it must have a negative slope. Perpendicular lines have slopes with opposite signs.

We show (on page 141) that if l_1 and l_2 are two perpendicular lines with slopes m_1 and m_2, then m_1 is the negative reciprocal of m_2. If m_1 and m_2 are not zero, we have the following result:

Let l_1 and l_2 be two lines having slopes m_1 and m_2, respectively. Then:
- These lines are parallel if and only if $m_1 = m_2$.
- These lines are perpendicular if and only if $m_1 = -\dfrac{1}{m_2}$.

In addition, any two horizontal lines are parallel and $m_1 = m_2 = 0$. Any two vertical lines are parallel and m_1 and m_2 are undefined. A horizontal line is perpendicular to a vertical line. See Figures 5.21–5.23.

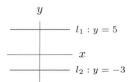

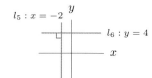

Figure 5.21: Any two horizontal lines are parallel

Figure 5.22: Any two vertical lines are parallel

Figure 5.23: A horizontal line and a vertical line are perpendicular

---

Example 6   Show that the three lines are parallel:

$$3x + 2y = 20$$
$$3x + 2y = 5$$
$$12x + 8y = 5.$$

Solution   We find the slope of line $3x + 2y = 20$ by putting it in slope-intercept form.

$$3x + 2y = 20$$
$$2y = 20 - 3x$$

$$y = 10 - \left(\frac{3}{2}\right)x.$$

So the slope of the first line is $-3/2$.

The only difference between first and second equation is the 5 instead of 20 on the right-hand side. This affects the intercept, but not the slope. The equation is

$$y = \frac{5}{2} - \left(\frac{3}{2}\right)x.$$

For the third equation, the coefficients of x and y have changed, but notice they are both multiplied by 4. The equation is

$$y = \frac{5}{8} - \left(\frac{12}{8}\right)x$$
$$= \frac{5}{8} - \left(\frac{3}{2}\right)x,$$

and once again the slope is $-3/2$.

Example 7   Find an equation for

(a) The line parallel to the graph of $y = 12 - 3x$ with a y-intercept of 7.
(b) The line parallel to the graph of $5x + 3y = -6$ containing the point $(9, 4)$.
(c) The line perpendicular to the graph of $y = 5x - 20$ that intersects the graph at $x = 6$.

Solution

(a) Since the lines are parallel, the slope is $m = -3$. The y-intercept is $b = 7$, and from the slope-intercept form we get $y = 7 - 3x$.
(b) To find the slope of $5x + 3y = -6$, we put it in slope-intercept form $y = -2 - (5/3)x$. So the slope is $m = -5/3$. Since the parallel line has the same slope and contains the point $(9, 4)$, we can use point-slope form $y = y_0 + m(x - x_0)$ to get

$$y = 4 + \left(-\frac{5}{3}\right)(x - 9)$$
$$= 4 + \left(-\frac{5}{3}\right)x + 15$$
$$= 19 - \frac{5}{3}x.$$

(c) The slope of the original line is $m_1 = 5$, so the slope of the perpendicular line is

$$m_2 = -\frac{1}{5} = -0.2.$$

The lines intersect at $x = 6$. From the original equation, this means

$$y = 5x - 20 = 5 \cdot 6 - 20 = 10,$$

so the lines intersect at the point $(6, 10)$. Using the point-slope form, we have

$$y = 10 + (-0.2)(x - 6)$$
$$= 10 - 0.2x + 1.2$$
$$= 11.2 - 0.2x.$$

Justification of the Formula for Slopes of Perpendicular Lines

Figure 5.24 shows l_1 and l_2, two perpendicular lines through the origin with slope m_1 and m_2. Neither line is horizontal or vertical, so m_1 and m_2 are both defined and nonzero.

We measure a distance 1 along the x-axis and mark the point on l_1 above it. This point has coordinates $(1, m_1)$, since $x = 1$ and since the equation for the line is $y = m_1 x$:

$$y = m_1 x = m_1 \cdot 1 = m_1.$$

Likewise, we measure a distance 1 down along the y-axis and mark the point on l_2 to the right of it. This point has coordinates $(-1/m_2, -1)$, since $y = -1$ and the equation for the line is $y = m_2 x$:

$$-1 = m_2 x \qquad \text{so} \qquad x = -\frac{1}{m_2}.$$

By drawing in the dashed lines shown in Figure 5.24, we form two triangles. These triangles have the same shape and size, meaning they have three equal angles and three equal sides. (Such triangles are called *congruent* triangles.) We know this is true because both triangles have a side of length 1 along one of the axes and the angle between that side and and the longest side (the *hypotenuse*) is the same in both triangles, since the axes are perpendicular to each other, and l_1 and l_2 are also perpendicular to each other. Therefore the sides drawn with dashed lines are also equal. From the coordinates of the two labeled points, we see that these two dashed lines measure m_1 and $-1/m_2$, respectively, so

$$m_1 = \frac{-1}{m_2}.$$

A similar argument works for lines that do not intersect at the origin.

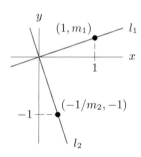

Figure 5.24: Perpendicular lines

Exercises and Problems for Section 5.4

EXERCISES

Write the linear equations in Exercises 1–4 in slope-intercept form $y = b + mx$. What are the values of m and b?

1. $y = 100 - 3(x - 20)$
2. $80x + 90y = 100$
3. $\dfrac{x}{100} + \dfrac{y}{300} = 1$
4. $x = 30 - \dfrac{2}{3}y$

In Exercises 5–10, graph the equation.

5. $y = 3x - 6$
6. $y = 5$
7. $2x - 3y = 24$
8. $x = 7$
9. $y = -\dfrac{2}{3}x - 4$
10. $y = 200 - 4x$

11. Without a calculator, match the equations (a)–(g) to the graphs (I)–(VII).

(a) $y = x - 3$ **(b)** $-3x + 2 = y$

(c) $2 = y$ **(d)** $y = -4x - 3$

(e) $y = x + 2$ **(f)** $y = x/3$

(g) $4 = x$

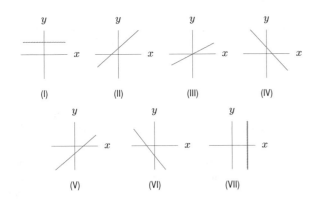

(I)   (II)   (III)   (IV)

(V)   (VI)   (VII)

In Exercises **12–22**, write an equation in point-slope form for the line.

12. Through $(2, 3)$ with slope $m = 5$

13. Through $(-1, 7)$ with slope $m = 6$

14. Through $(8, 10)$ with slope $m = -3$

15. Through $(2, -9)$ with slope $m = -2/3$

16. Through $(4, 7)$ and $(1, 1)$

17. Through $(6, 5)$ and $(7, 1)$

18. Through $(-2, -8)$ and $(2, 4)$

19. Through $(6, -7)$ and $(-6, -1)$

20. Through $(-1, -8)$ and parallel to $y = 5x - 2$

21. Through $(3, -6)$ and parallel to $y = 5/4(x + 10)$

22. Through $(12, 20)$ and perpendicular to $y = -4x - 3$.

For Exercises **23–32**, put the equation in standard form.

23. $x = 3y - 2$ **24.** $y = 2 + 4(x - 3)$

25. $5x = 7 - 2y$ **26.** $y - 6 = 5(x + 2)$

27. $x + 4 = 3(y - 1)$ **28.** $6(x + 4) = 3(y - x)$

29. $9(y + x) = 5$

30. $3(2y + 4x - 7) = 5(3y + x - 4)$

31. $y = 5x + 2a$, with a constant

32. $5b(y + bx + 2) = 4b(4 - x + 2b)$, with b constant

33. Find an equation for the line parallel to the graph of

(a) $y = 3 + 5x$ with a y-intercept of 10.
(b) $4x + 2y = 6$ with a y-intercept of 12.
(c) $y = 7x + 2$ and containing the point $(3, 22)$.
(d) $9x + y = 5$ and containing the point $(5, 15)$.

In Exercises **34–41**, are the lines parallel?

34. $y = 12 + ax$; $y = 20 + ax$, where a is a constant

35. $y = 1 + x$; $y = 1 + 2x$

36. $y = 5 + 4(x - 2)$; $y = 2 + 4x$

37. $y = 2 + 3(x + 5)$; $y = 2 + 4(x + 5)$

38. $2x + 3y = 5$; $4x + 6y = 7$

39. $qx + ry = 3$; $qx + ry = 4$, where q and r are nonzero constants

40. $y = 7 + 4(x - 2)$; $y = 8 + 2(2x + 3)$

41. $y = 5 + 6(x + 2)$ $y = 5 + 6(3x - 1)$

PROBLEMS

Match the statements in Problems **42–43** with the lines I–VI.

I. $y = 2(x - 4) + 9$ II. $y - 9 = -3(x - 4)$

III. $y + 9 = -2(x - 4)$ IV. $y = 4x + 9$

V. $y = 9 - 2(4 - x)$ VI. $y = 9 - \dfrac{4 - 8x}{4}$

42. These three lines pass through the same point.

43. These three lines have the same slope.

Match the statements in Problems **44–46** with equations I–VI.

I. $y = 20 + 2(x - 8)$ II. $y = 20 - 2(x - 8)$

III. $y = 5x + 30$ IV. $y = -5(6 - x)$

V. $y = \dfrac{2x + 90}{3}$ VI. $y = -\dfrac{2}{3}(x - 8) + 20$

44. These three lines pass through the same point.

45. These two lines have the same y-intercept.

46. These two lines have the same slope.

47. Find the equation of the line intersecting the graph of $y = x^3 - x + 3$ at $x = -2$ and $x = 2$.

◼Find possible equations for the straight lines in Exercises **48–50.**

48. The line is perpendicular to the graph of $5x - 3y = 6$, and the two lines intersect at $x = 15$.

49. The line is perpendicular to the graph of $y = 0.7 - 0.2x$ and intersects it at $x = 1.5$.

50. The line is perpendicular to the graph of $y = 400 + 25x$ and intersects it at $x = 12$.

◼In Problems **51–55**, is the point-slope form or slope-intercept form the easier form to use when writing an equation for the line?

51. Slope $= 3$, Intercept $= -6$

52. Passes through $(2, 3)$ and $(-6, 7)$

53. Passes through $(-5, 10)$ and has slope 6

54. Is parallel to the line $y = 0.4x - 5.5$ and has the same y-intercept as the line $y = -2x - 3.4$

55. Is parallel to the line $y = 4x - 6$ and contains the point $(2, -3)$

◼In Problems **56–59**, write the equation in the form $y = b + mx$, and identify the values of b and m.

56. $y - y_0 = r(x - x_0)$ 57. $y = \beta - \dfrac{x}{\alpha}$

58. $Ax + By = C$, if $B \neq 0$

59. $y = b_1 + m_1x + b_2 + m_2x$

60. Give an equation for the line parallel to $y = 20 - 3x$ and passing through the point $(\sqrt{3}, \sqrt{8})$.

◼In Problems **61–66**, which line has the greater

(a) Slope?   **(b)** y-intercept?

61. $y = 3 + 6x, \quad y = 5 - 3x$

62. $y = \frac{1}{5}x, \quad y = 1 - 6x$

63. $2x = 4y + 3, \quad y = -x - 2$

64. $3y = 5x - 2, \quad y = 2x + 1$

65. $y + 2 = 3(x - 1), \quad y = 6 - 50x$

66. $y - 3 = -4(x + 2), \quad -2x + 5y = -3$

67. Which equation, (a)–(d), has the graph that crosses the y-axis at the highest point?

   (a) $y = 3(x - 1) + 5$ **(b)** $x = 3y + 2$
   (c) $y = 1 - 6x$ **(d)** $2y = 3x + 1$

68. Which of the following equations has a graph that slopes down the most steeply as you move from left to right?

   (a) $y + 4x = 5$ **(b)** $y = 5x + 3$
   (c) $y = 10 - 2x$ **(d)** $y = -3x + 2$

69. Explain the differences between the graphs of the equations $y = 14x - 18$ and $y = -14x + 18$.

70. Put the equation $y = 3xt + 2xt^2 + 5$ in the form $y = b + mx$. What are the values of b and m? [Note: Your answers could include t.]

71. Show that the points $(0, 12), (3, 0)$, and $(17/3, 2/3)$ form the corners of a right triangle (that is, a triangle with a right angle).

72. **(a)** Find the equation of the line with intercepts

   (i) $(2, 0)$ and $(0, 5)$

   (ii) Double those in part (i)

   (b) Are the two lines in part (a) parallel? Justify your answer.

   (c) In words, generalize your conclusion to part (b). (There are many ways to do this; pick one. No justification is necessary.)

5.5 MODELING WITH LINEAR FUNCTIONS AND EQUATIONS

We have seen that linear functions model situations in which one quantity is varying at a constant rate of change with respect to another. How do we know if a linear model is appropriate?

Deciding When to Use a Linear Model

Table 5.4 gives the temperature-depth profile measured in a borehole in 1988 in Belleterre, Quebec.[4] How can we decide whether a linear function models the data in Table 5.4? There are two ways to answer this question. We can plot the data to see if the points fall on a line, or we can calculate the slope to see if it is constant.

---

[4]Hugo Beltrami of St. Francis Xavier University and David Chapman of the University of Utah posted this data at http://geophysics.stfx.ca/public/borehole/borehole.html. Page last accessed on March 3, 2003.

Table 5.4   *Temperature in a borehole at different depths*

d, depth (m)	150	175	200	225	250	275	300
H, temp (°C)	5.50	5.75	6.00	6.25	6.50	6.75	7.00

Example 1   Is the temperature data in Table 5.4 linear with respect to depth? If so, find a formula for temperature, H, as a function of depth, d, for depths ranging from 150 m to 300 m.

Solution   Since the temperature rises by 0.25°C for every 25 additional meters of depth, the rate of change of temperature with respect to depth is constant. Thus, we use a linear function to model this relationship. The slope is the constant rate of change:

$$\text{Slope} = \frac{\Delta H}{\Delta d} = \frac{0.25}{25} = 0.01 \text{ °C/m}.$$

To find the linear function, we substitute this value for the slope and any point from Table 5.4 into the point-slope form. For example, using the first entry in the table, $(d, H) = (150, 5.50)$, we get

$$H = 5.50 + 0.01(d - 150)$$
$$H = 5.50 + 0.01d - 1.5$$
$$H = 4.0 + 0.01d.$$

The linear function $H = 4.0 + 0.01d$ gives temperature as a function of depth.

Recognizing Values of a Linear Function

Values of x and y in a table could be values of a linear function $f(x) = b + mx$ if the same change in x-values always produces the same change in the y-values.

Example 2   Which of the following tables could represent values of a linear function?

(a)

x	20	25	30	35
y	17	14	11	8

(b)

x	2	4	6	8
y	10	20	28	34

Solution   (a) The x-values go up in steps of 5, and the corresponding y-values go down in steps of 3:

$$14 - 17 = -3 \quad \text{and} \quad 11 - 14 = -3 \quad \text{and} \quad 8 - 11 = -3.$$

Since the y-values change by the same amount each time, the table satisfies a linear equation.

(b) The x-values go up in steps of 2. The corresponding y-values do not go up in steps of constant size, since

$$20 - 10 = 10 \quad \text{and} \quad 28 - 20 = 8 \quad \text{and} \quad 34 - 28 = 6.$$

Thus, even though the value of Δx is the same for consecutive entries in the table, the value of Δy is not. This means the slope changes, so the table does not satisfy a linear equation.

Using Models to Make Predictions

In Example 1 we found a function for the temperature in a borehole based on temperature data for depths ranging from 150 meters to 300 meters. We can use our function to make predictions.

Example 3   Use the function from Example 1 to predict the temperature at the following depths. Do you think these predictions are reasonable?

(a)   260 m   (b)   350 m   (c)   25 m

Solution   (a) We have $H = 4 + 0.01(260) = 6.6$, so the function predicts a temperature of $6.6°C$ at a depth of 260 m. This is a reasonable prediction, because it is a little warmer than the temperature at 250 m and a little cooler than the temperature at 275 m:

	(shallower)	(in between)	(deeper)
Depth (m)	250	**260**	275
Temp (°C)	6.50	**6.60**	6.75
	(cooler)	(in between)	(warmer)

(b) We have $H = 4 + 0.01(350) = 7.5$, so the function predicts a temperature of $7.5°C$ at a depth of 350 m. This seems plausible, since the temperature at 300 m is $7°C$. However, unlike in part (a), the value 350 is outside the range 150–300 represented in Table 5.4 on the preceding page. It is possible that the linear trend fails to hold as we go deeper.

(c) We have $H = 4 + 0.01(25) = 4.25$, so the function predicts a temperature of $4.25°C$ at a depth of 25 m. Since this prediction is using a value of d that is well outside the range of our original data set, we should treat it with some caution. (In fact it turns out that at depths less than 100 m, the temperature actually rises as you get closer to the surface, instead of falling as our function predicts, so this is not a good prediction.)

Modeling With Linear Equations: Constraint Equations

The standard form of a linear equation is useful for describing *constraints*, or situations involving limited resources.

Example 4   A newly designed motel has S small rooms measuring 250 ft^2 and L large rooms measuring 400 ft^2. The designers have 10,000 ft^2 of available space. Write an equation relating S and L.

Solution   Altogether, we have

$$\text{Total room space} = \underbrace{\text{space for small rooms}}_{250S} + \underbrace{\text{space for large rooms}}_{400L}$$
$$= 250S + 400L.$$

We know that the total available space is 10,000 ft^2, so

$$250S + 400L = 10{,}000.$$

We graph this by determining the axis intercepts:

$$\text{If } S = 0 \text{ then } L = \frac{10{,}000}{400} = 25.$$

$$\text{If } L = 0 \text{ then } S = \frac{10{,}000}{250} = 40.$$

See Figure 5.25.

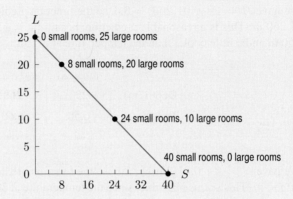

Figure 5.25: Different combinations of small and large rooms for the motel

The equation $250S + 400L = 10{,}000$ in Example 4 on the preceding page is called a *constraint equation* because it describes the constraint that floor space places on the number of rooms built. Constraint equations are usually written in standard form.

Example 5   Revised plans for the motel in Example 4 on the previous page provide for a total floor space of $16{,}000 \text{ ft}^2$. Find the new constraint equation. Sketch its graph together with the graph of the original constraint. How do the two graphs compare?

Solution   The constraint in this case is the total floor area of $16{,}000 \text{ ft}^2$. Since small rooms have an area of 250 ft^2, the total area of S small rooms is $250S$. Similarly, the total area of L large rooms is $400L$. Since

$$\text{Area of small rooms} + \text{Area of large rooms} = \text{Total area,}$$

we have a constraint equation

$$250S + 400L = 16{,}000.$$

Since the coefficients of S and L stay the same as in Example 4, the new constraint line is parallel to the old one. See Figure 5.26. However, at every value of $S \le 40$, the value of L is 15 units larger than before. Since one large room uses 400 ft^2, we see that 6000 ft^2 provides for the additional $6000/400 = 15$ large rooms.

Likewise, for every value of $L \le 25$, the value of S is 24 units larger than before. Again, this is a consequence of the extra space: at 250 ft^2 each, there is space for $6000/250 = 24$ additional small rooms.

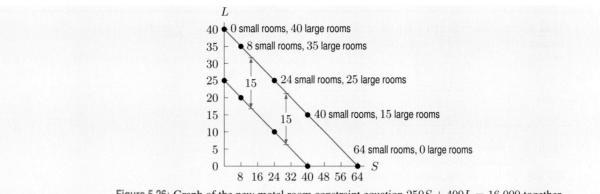

Figure 5.26: Graph of the new motel room constraint equation $250S + 400L = 16,000$ together with the old constraint equation $250S + 400L = 10,000$

Exercises and Problems for Section 5.5

EXERCISES

In Exercises 1–5, could the table represent the values of a linear function?

1.

x	-6	-3	0	3	6
y	12	8	4	0	-4

2.

x	7	9	11	13	15
y	43	46	49	52	55

3.

x	2	4	8	16
y	5	10	15	20

4.

x	2	4	8	16	32
y	5	7	11	19	35

5.

x	-2	0	4	10
y	3	2	0	-3

In Exercises 6–8, could the table represent the values of a linear function? Give a formula if it could.

6.

t	0	1	2	3	4	5
Q	100.00	95.01	90.05	85.11	80.20	75.31

7.

t	3	7	19	21	26	42
Q	5.79	6.67	9.31	9.75	10.85	14.37

8.

x	0	2	10	20
y	50	58	90	130

For Exercises 9–12,
  (a) Write a constraint equation.
  (b) Choose two solutions.
  (c) Graph the equation and mark your solutions.

9. The relation between quantity of chicken and quantity of steak if chicken costs $1.29/lb and steak costs $3.49/lb, and you have $100 to spend on a barbecue.

10. The relation between the time spent walking and driving if you walk at 3 mph then hitch a ride in a car traveling at 75 mph, covering a total distance of 60 miles.

11. The relation between the volume of titanium and iron in a bicycle weighing 5 kg, if titanium has a density of 4.5g/cm^3 and iron has a density of 7.87 g/cm^3 (ignore other materials).

12. The relation between the time spent walking and the time spent canoeing on a 30 mile trip if you walk at 4 mph and canoe at 7 mph.

PROBLEMS

13. Table 5.5 shows the readings given by an instrument for measuring weight when various weights are placed on it.[5]

Table 5.5

w, weight (lbs)	0.0	0.5	1.0	1.5	2.0	2.5
I, reading	46.0	272.8	499.6	726.4	953.2	1180.0

   (a) Is the reading, I, a linear function of the weight, w? If so, find a formula for it.

   (b) Give possible interpretations for the slope and vertical intercept in your formula from part (a). [Hint: 1 lb = 453.6 g.]

14. Table 5.6 shows the air temperature T as a function of the height h above the earth's surface.[6] Is T a linear function of h? Give a formula if it is.

Table 5.6

h, height (m)	0	2000	4000	6000	8000	10,000
T, temperature (°C)	15	2	−11	−24	−37	−50

15. A gram of fat contains 9 dietary calories, whereas a gram of carbohydrates contains only 4.[7]

   (a) Write an equation relating the amount f, in grams, of fat and the amount c, in grams, of carbohydrates that one can eat if limited to a total of 2000 calories/day.

   (b) The USDA recommends that calories from fat should not exceed 30% of all calories. What does this tell you about f?

16. Put the equation $250S + 400L = 10,000$ from Example 4 into slope-intercept form in two different ways by solving for **(a)** L **(b)** S.
   Which form fits best with Figure 5.25?

17. The graph in Figure 5.27 shows the relationship between hearing ability score h and age a.

   (a) What is the expected hearing ability score for a 40 year old?

   (b) What age is predicted to have a hearing ability score of 40?

   (c) Give an equation that relates hearing ability and age.

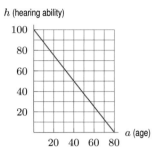

Figure 5.27

18. The final plans for the motel in Example 4 call for a total floor space of 16,000 ft², and for less spacious small rooms measuring 200 ft² instead of 250 ft². The large rooms are to remain 400 ft². Sketch a graph of the resulting constraint equation together with the constraint from Example 5. How do the two graphs compare?

■ The coffee variety *Arabica* yields about 750 kg of coffee beans per hectare, while *Robusta* yields about 1200 kg/hectare. In Problems **19–20**, suppose that a plantation has a hectares of *Arabica* and r hectares of *Robusta*. [8]

19. Write an equation relating a and r if the plantation yields 1,000,000 kg of coffee.

20. On August 14, 2003, the world market price of coffee was \$1.42/kg for *Arabica* and \$0.73/kg for *Robusta*.[9] Write an equation relating a and r if the plantation produces coffee worth \$1,000,000.

5.6 SYSTEMS OF LINEAR EQUATIONS

The solutions of a linear equation in two variables x and y are pairs of numbers. For any given x-value, we can solve for y to find a corresponding y-value, and vice versa.

---

[5] Adapted from J.G. Greeno, *Elementary Theoretical Psychology* (Massachusetts: Addison-Wesley, 1968).

[6] Adapted from H. Tennekes, *The Simple Science of Flight* (Cambridge: MIT Press, 1996).

[7] Food and Nutrition Information Center, USDA, www.nalusda.gov/fnic/Dietary/9dietgui.htm, accessed October 6, 2007.

[8] http://www.da.gov.ph/tips/coffee.html, accessed October 6, 2007.

[9] http://www.cafedirect.co.uk/about/gold_prices.php, accessed August 14, 2003.

Example 1   Find solutions to the equation $3x + y = 14$ given

(a) $x = 1, 2, 3$ (b) $y = 1, 2, 3$.

Solution   (a) We substitute the given values for x, then solve for y:

$$x = 1: \quad 3(1) + y = 14 \quad \rightarrow \quad y = 11$$
$$x = 2: \quad 3(2) + y = 14 \quad \rightarrow \quad y = 8$$
$$x = 3: \quad 3(3) + y = 14 \quad \rightarrow \quad y = 5.$$

The solutions are the (x, y)-pairs $(1, 11), (2, 8), (3, 5)$.

(b) We substitute the given values for y, then solve for x:

$$y = 1: \quad 3x + 1 = 14 \quad \rightarrow \quad x = 13/3$$
$$y = 2: \quad 3x + 2 = 14 \quad \rightarrow \quad x = 4$$
$$y = 3: \quad 3x + 3 = 14 \quad \rightarrow \quad x = 11/3.$$

The solutions are the (x, y)-pairs $(13/3, 1), (4, 2), (11/3, 3)$.

Sometimes we are interested in finding pairs that satisfy more than one equation at the same time.

Example 2   The total cost of tickets to a play for one adult and two children is \$11. The total for two adults and three children is \$19. What are the individual ticket prices for adults and children?

Solution   If A is the adult ticket price and C is the child ticket price, then

$$A + 2C = 11 \quad \text{and} \quad 2A + 3C = 19.$$

We want a pair of values for A and C that satisfies both of these equations simultaneously. Because adults are often charged more than children, we might first guess that $A = 9$ and $C = 1$. While these values satisfy the first equation, they do not satisfy the second equation, since the left-hand side evaluates to

$$2 \cdot 9 + 3 \cdot 1 = 21,$$

whereas the right-hand side is 19. However, the values $A = 5$ and $C = 3$ satisfy both equations:

$$5 + 2 \cdot 3 = 11 \quad \text{and} \quad 2 \cdot 5 + 3 \cdot 3 = 19.$$

Thus, an adult ticket costs \$5 and a child ticket costs \$3.

What is a System of Equations?

In Example 2, we saw that the (A, C)-pair $(5, 3)$ is a solution to the equation $A + 2C = 11$ and also to the equation $2A + 3C = 19$. Since this pair of values makes both equations true, we say that it is a solution to the *system of equations*

$$\begin{cases} A + 2C = 11 \\ 2A + 3C = 19. \end{cases}$$

Systems of Equations

A **system of equations** is a set of two or more equations. A solution to a system of equations is a set of values for the variables that makes all of the equations true.

We write the system of equations with a brace to indicate that a solution must satisfy both equations.

Example 3   Find a solution to the system of equations

$$\begin{cases} 3x + y = 14 & \text{(equation 1)} \\ 2x + y = 11 & \text{(equation 2).} \end{cases}$$

Solution   In Example 1 we saw that solutions to equation 1 include the pairs $(1, 11), (2, 8)$, and $(3, 5)$. We can also find pairs of solutions to equation 2:

$$\begin{aligned} x = 1: & \quad 2(1) + y = 11 & \rightarrow & \quad y = 9 \\ x = 2: & \quad 2(2) + y = 11 & \rightarrow & \quad y = 7 \\ x = 3: & \quad 2(3) + y = 11 & \rightarrow & \quad y = 5 \\ & \quad \vdots \end{aligned}$$

As before, the solutions are pairs: $(1, 9), (2, 7), (3, 5), \ldots$. Notice that one of these pairs, $(3, 5)$, is also a solution to the first equation. Since this pair of values solves both equations at the same time, it is a solution to the system.

In Example 3, we find the solution by trial and error. We now consider two algebraic methods for solving systems of equations. The first method is known as *substitution*.

Solving Systems of Equations Using Substitution

In this approach we solve for one of the variables and then substitute the solution into the other equation.

Example 4   Solve the system

$$\begin{cases} 3x + y = 14 & \text{(equation 1)} \\ 2x + y = 11 & \text{(equation 2).} \end{cases}$$

Solution   Solving equation 1 for y gives

$$y = 14 - 3x \quad \text{(equation 3).}$$

Since a solution to the system satisfies both equations, we substitute this expression for y into equation 2 and solve for x:

$$2x + y = 11$$
$$2x + \underbrace{14 - 3x}_{y} = 11 \quad \text{using equation 3 to substitute for } y \text{ in equation 2}$$

$$2x - 3x = 11 - 14$$
$$-x = -3$$
$$x = 3.$$

To find the corresponding y-value, we can substitute $x = 3$ into either equation. Choosing equation 1, we have $3(3) + y = 14$. Thus $y = 5$, so the (x, y)-pair $(3, 5)$ satisfies both equations, as we now verify:

$$3(3) + 5 = 14 \longrightarrow \quad (3, 5) \text{ solves equation 1}$$
$$2(3) + 5 = 11 \longrightarrow \quad (3, 5) \text{ solves equation 2.}$$

To summarize:

Method of Substitution

- Solve for one of the variables in one of the equations.
- Substitute into the other equation to get an equation in one variable.
- Solve the new equation, then find the value of the other variable by substituting back into one of the original equations.

Deciding Which Substitution Is Simplest

When we use the method of substitution, we can choose which equation to start with and which variable to solve for. For instance, in Example 4, we began by solving equation 1 for y, but this is not the only possible approach. We might instead solve equation 1 for x, obtaining

$$x = \frac{14 - y}{3} \quad \text{(equation 4).}$$

We can use this equation to substitute for x in equation 2, obtaining

$$2 \underbrace{\left(\frac{14 - y}{3} \right)}_{x} + y = 11 \quad \text{using equation 4 to substitute for } x \text{ in equation 2}$$

$$\frac{2}{3}(14 - y) + y = 11$$
$$2(14 - y) + 3y = 33 \quad \text{multiply both sides by 3}$$
$$28 - 2y + 3y = 33$$
$$y = 33 - 28 = 5.$$

From equation 4, we have $x = (14 - 5)/3 = 3$, so we get the solution $(x, y) = (3, 5)$. This is same pair of values as before, only this time the fractions make the algebra messier. The moral is that we should make our substitution as simple as possible. One way to do this is to look for terms having a coefficient of 1. This is why we chose to solve for y in Example 4.

Solving Systems of Equations Using Elimination

Another technique for solving systems of equations is known as *elimination*. As the name suggests, the goal of elimination is to eliminate one of the variables from one of the equations.

Example 5   We solve the system from Example 4 using elimination instead of substitution:

$$\begin{cases} 3x + y = 14 & \text{(equation 1)} \\ 2x + y = 11 & \text{(equation 2).} \end{cases}$$

Solution   Since equation 2 tells us that $2x + y = 11$, we subtract $2x + y$ from both sides of equation 1:

$$3x + y - (2x + y) = 14 - (2x + y)$$
$$3x + y - (2x + y) = 14 - 11 \quad \text{(equation 3)} \quad \text{because } 2x + y \text{ equals } 11$$
$$3x + y - 2x - y = 3$$
$$x = 3.$$

We have eliminated y from the first equation by subtracting the second equation, allowing us to solve for x, obtaining $x = 3$. As before, we substitute $x = 3$ into either equation to find the corresponding value of y. Our solution, $(x, y) = (3, 5)$, is the same answer that we found in Example 4.

Adding and Subtracting Equations

In Example 5, we subtracted $2x + y$ from the left-hand side of equation 1 and subtracted 11 from the right-hand side. We can do this because, if equation 2 is true, then $2x + y$ has the value 11. Notice that we have, in effect, subtracted equation 2 from equation 1. We write

$$\text{Equation 3} = \text{Equation 1} - \text{Equation 2}.$$

Similarly, we say that we add two equations when we add their left- and right-hand sides, respectively. We also say that we multiply an equation by a constant when we multiply both sides by a constant.

Example 6   Solve the following system using elimination.

$$\begin{cases} 2x + 7y = -3 & \text{(equation 1)} \\ 4x - 2y = 10 & \text{(equation 2).} \end{cases}$$

Solution   In preparation for eliminating x from the two equations, we multiply equation 1 by -2 in order to make the coefficient of x into -4, the negative of the coefficient in equation 2. This gives equation 3:

$$\underbrace{-4x - 14y}_{-2(2x+7y)} = \underbrace{6}_{-2(-3)} \quad \text{(equation 3} = -2 \times \text{equation 1)}$$

Notice that equation 2 has a $4x$ term and equation 3 has a $-4x$ term. We can eliminate the variable

x by adding these two equations:

	$4x$	$-2y$	$=$	10	(equation 2)
$+$	$-4x$	$-14y$	$=$	6	(equation 3)
		$-16y$	$=$	16	(equation 4 = equation 2 + equation 3)

Solving equation 4, we obtain $y = -1$. We substitute $y = -1$ into equation 1 to solve for x:

$$2x + 7(-1) = -3 \quad \text{solving for } x \text{ using equation 1}$$
$$2x = 4$$
$$x = 2.$$

Thus, $(2, -1)$ is the solution. We can verify this solution using equation 2:

$$4 \cdot 2 - 2(-1) = 8 + 2 = 10.$$

In summary:

Method of Elimination

- Multiply one or both equations by constants so that the coefficients of one of the variables are either equal or are negatives of each other.
- Add or subtract the equations to eliminate that variable and solve the resulting equation for the other variable.
- Substitute back into one of the original equations to find the value of the first variable.

Solving Systems of Equations Using Graphs

The number of solutions to a system of linear equations is related to whether their graphs are parallel lines, overlapping lines, or lines that intersect at a single point.

One Solution: Intersecting Lines

In Example 5 we solved the system of equations:

$$\begin{cases} 3x + y = 14 \\ 2x + y = 11. \end{cases}$$

We can also solve this system graphically. In order to graph the lines, we rewrite both equations in slope-intercept form:

$$\begin{cases} y = 14 - 3x \\ y = 11 - 2x. \end{cases}$$

The graphs are shown in Figure 5.28. Each line shows all solutions to one of the equations. Their point of intersection is the common solution of both equations. It appears that these two lines intersect at the point $(x, y) = (3, 5)$. We check that these values satisfy both equations, so the solution to the system of equations is $(x, y) = (3, 5)$.

If it exists, the solution to a system of two linear equations in two variables is the point at which their graphs intersect.

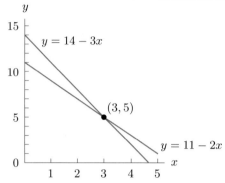

Figure 5.28: The solution to the system $y = 14 - 3x$ and $y = 11 - 2x$ is the point of intersection of these two lines

No Solutions: Parallel Lines

The system in the next example has no solutions.

Example 7   Solve

$$\begin{cases} 6x - 2y = 8 & \text{(equation 1)} \\ 9x - 3y = 6 & \text{(equation 2).} \end{cases}$$

Solution   We can use the process of elimination to eliminate the x terms.

$$18x - 6y = 24 \quad \text{(equation 3 = 3 × equation 1)}$$
$$18x - 6y = 12 \quad \text{(equation 4 = 2 × equation 2).}$$

There are no values of x and y that can make $18x - 6y$ equal to both 12 and 24, so there is *no* solution to this system of equations. To see this graphically, we rewrite each equation in slope-intercept form:

$$\begin{cases} y = -4 + 3x \\ y = -2 + 3x. \end{cases}$$

Since their slopes, m, are both equal to 3, the lines are parallel and do not intersect. Therefore, the system of equations has no solution. See Figure 5.29.

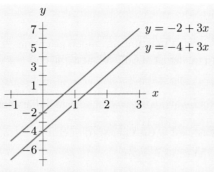

Figure 5.29: The system $y = -4 + 3x$ and $y = -2 + 3x$ has no solution, corresponding to the fact that the graphs of these equations are parallel (non-intersecting) lines

Many Solutions: Overlapping Lines

The system in the next example has many solutions.

Example 8   Solve:

$$\begin{cases} x - 2y = 4 & \text{(equation 1)} \\ 2x - 4y = 8 & \text{(equation 2).} \end{cases}$$

Solution   Notice that equation 2 is really equation 1 in disguise: Multiplying equation 1 by 2 gives equation 2. Therefore, our system of equations is really just the same equation written in two different ways. The set of solutions to the system of equations is the same as the set of solutions to either one of the equations separately. To see this graphically, we rewrite each equation in slope-intercept form:

$$\begin{cases} y = -2 + 0.5x \\ y = -2 + 0.5x. \end{cases}$$

We get the same equation twice, so the two equations represent the same line. The graph of this system is the single line $y = -2 + 0.5x$, and any (x, y) point on the line is a solution to the system. See Figure 5.30. The system of equations has infinitely many solutions.

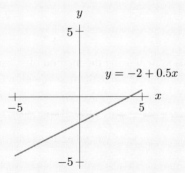

Figure 5.30: The system $x - 2y = 4$ and $2x - 4y = 8$ has infinitely many solutions, since both equations give the same line and every point on the line is a solution

Applications of Systems of Linear Equations

Example 9   At a fabric store, silk costs three times as much as cotton. A customer buys 4 yards of cotton and 1.5 yards of silk, for a total cost of $55. What is the cost per yard of cotton and what is the cost per yard of silk?

Solution   Let x be the cost of cotton and y the cost of silk, in dollars per yard. Then

$$\text{Total cost} = 4 \times \text{Cost per yard of cotton} + 1.5 \times \text{Cost per yard of silk}$$
$$55 = 4x + 1.5y.$$

Also, since the price of silk is three times the price of cotton, we have $y = 3x$. Thus we have the system of equations:

$$\begin{cases} y = 3x \\ 55 = 4x + 1.5y. \end{cases}$$

We use the method of substitution, since the first equation is already in the right form to substitute into the second:

$$55 = 4x + 1.5(3x)$$
$$55 = 4x + 4.5x$$
$$55 = 8.5x$$
$$x = \frac{55}{8.5} = 6.47.$$

Thus cotton costs $6.47 per yard and silk costs $3(6.47) = \$19.41$ per yard.

Example 10   A farmer raises chickens and pigs. His animals together have a total of 95 heads and a total of 310 legs. How many chickens and how many pigs does the farmer have?

Solution   We let x represent the number of chickens and y represent the number of pigs. Each animal has one head, so we know $x + y = 95$. Since chickens have 2 legs and pigs have 4 legs, we know $2x + 4y = 310$. We solve the system of equations:

$$\begin{cases} x + y = 95 \\ 2x + 4y = 310. \end{cases}$$

We can solve this system using the method of substitution or the method of elimination. Using the method of substitution, we solve the first equation for y:

$$y = 95 - x.$$

We substitute this for y in the second equation and solve for x:

$$2x + 4y = 310$$
$$2x + 4(95 - x) = 310$$
$$2x + 380 - 4x = 310$$

$$-2x = 310 - 380$$
$$-2x = -70$$
$$x = 35.$$

Since $x = 35$ and $x + y = 95$, we have $y = 60$. The farmer has 35 chickens and 60 pigs.

Exercises and Problems for Section 5.6

EXERCISES

■ Solve the systems of equations in Exercises **1–20**.

1. $\begin{cases} x + y = 5 \\ x - y = 7 \end{cases}$

2. $\begin{cases} 3x + y = 10 \\ x + 2y = 15 \end{cases}$

3. $\begin{cases} 3x - 4y = 7 \\ y = 4x - 5 \end{cases}$

4. $\begin{cases} x = y - 9 \\ 4x - y = 0 \end{cases}$

5. $\begin{cases} 2a + 3b = 4 \\ a - 3b = 11 \end{cases}$

6. $\begin{cases} 3w - z = 4 \\ w + 2z = 6 \end{cases}$

7. $\begin{cases} 2p + 3r = 10 \\ -5p + 2r = 13 \end{cases}$

8. $\begin{cases} 5d + 4e = 2 \\ 4d + 5e = 7 \end{cases}$

9. $\begin{cases} 8x - 3y = 7 \\ 4x + y = 11 \end{cases}$

10. $\begin{cases} 4w + 5z = 11 \\ z - 2w = 5 \end{cases}$

11. $\begin{cases} 20n + 50m = 15 \\ 70m + 30n = 22 \end{cases}$

12. $\begin{cases} r + s = -3 \\ s - 2r = 6 \end{cases}$

13. $\begin{cases} y = 20 - 4x \\ y = 30 - 5x \end{cases}$

14. $\begin{cases} 2p + 5q = 14 \\ 5p - 3q = 4 \end{cases}$

15. $\begin{cases} 9x + 10y = 21 \\ 7x + 11y = 26 \end{cases}$

16. $\begin{cases} 7x + 5y = -1 \\ 11x + 8y = -1 \end{cases}$

17. $\begin{cases} 5x + 2y = 1 \\ 2x - 3y = 27 \end{cases}$

18. $\begin{cases} 11v + 7w - 2 \\ 13v + 8w = 1 \end{cases}$

19. $\begin{cases} 3e + 2f = 4 \\ 4e + 5f = -11 \end{cases}$

20. $\begin{cases} 4p - 7q = 2 \\ 5q - 3p = -1 \end{cases}$

■ Determine the points of intersection for Exercises **21–24**.

21.
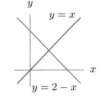
$y = x$, $y = 2 - x$

22.
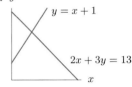
$y = x + 1$, $2x + 3y = 13$

23.

$y = 2x$, $2x + y = 16$

24.
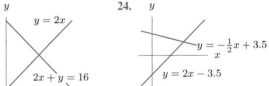
$y = -\frac{1}{2}x + 3.5$, $y = 2x - 3.5$

■ In Exercises **25–28**, solve the system of equations graphically.

25. $\begin{cases} y = 6x - 7 \\ y = 3x + 2 \end{cases}$

26. $\begin{cases} y = -2x + 7 \\ y = 4x + 1 \end{cases}$

27. $\begin{cases} 2x + 5y = 7 \\ -3x + 2y = 1 \end{cases}$

28. $\begin{cases} y = 22 + 4(x - 8) \\ y = 11 - 2(x + 6) \end{cases}$

PROBLEMS

■ In Problems **29–33**, without solving the equations, decide how many solutions the system has.

29. $\begin{cases} x - 2y = 7 \\ x + y = 9 \end{cases}$ **30.** $\begin{cases} x + 3y = 1 \\ 3x + 9y = 3 \end{cases}$

31. $\begin{cases} 3y = 2 - x \\ x = 2 + 3y \end{cases}$ **32.** $\begin{cases} 4x - y = 2 \\ 12x - 3y = 2 \end{cases}$

33. $\begin{cases} 4x - 3 = y \\ \dfrac{4}{y} - \dfrac{1}{x} = \dfrac{3}{xy} \end{cases}$

■ Solve the systems of equations in Problems **34–41**.

34. $\begin{cases} 2x + 5y = 1 \\ 2y - 3x = 8 \end{cases}$ **35.** $\begin{cases} 7x - 3y = 24 \\ 4y + 5x = 11 \end{cases}$

36. $\begin{cases} 5x - 7y = 31 \\ 2x + 3y = -5 \end{cases}$ **37.** $\begin{cases} 11\alpha - 7\beta = 31 \\ 4\beta - 3\alpha = 2 \end{cases}$

38. $\begin{cases} 3\alpha + \beta = 32 \\ 2\beta - 3\alpha = 1 \end{cases}$ **39.** $\begin{cases} 3x - 2y = 4 \\ 3y - 5x = -5 \end{cases}$

40. $\begin{cases} 3(e + f) = 5e + f + 2 \\ 4(f - e) = e + 2f - 4 \end{cases}$

41. $\begin{cases} 7\kappa - 9\psi = 23 \\ 2\kappa + 3\psi = 1 \end{cases}$

42. You want to build a patio. Builder A charges \$3 a square foot plus a \$500 flat fee, and builder B charges \$2.50 a square foot plus a \$750 flat fee. For each builder, write an expression relating the cost C to the area s square feet of the patio. Which builder is cheaper for a 200 square foot patio? Which is cheaper for a 1000 square foot patio? For what size patio will both builders charge the same?

43. Which of the following systems of equations have the same solution? Give reasons for your answers that do not depend on solving the equations.

I. $\begin{cases} x - y = 9 \\ x + y = 26 \end{cases}$ II. $\begin{cases} x + y = 9 \\ x - y = 26 \end{cases}$

III. $\begin{cases} 2x + 2y = 52 \\ x - y = 9 \end{cases}$ IV. $\begin{cases} \dfrac{x + y}{2} = 13 \\ 9 + y = x \end{cases}$

V. $\begin{cases} 2x - y = 18 \\ 2x + y = 52 \end{cases}$

44. Consider two numbers x and y satisfying the equations $x + y = 4$ and $x - y = 2$.

(a) Describe in words the conditions that each equation places on the two numbers.

(b) Find two numbers x and y satisfying both equations.

45. Find two numbers with sum 17 and difference 12.

46. A motel plans to build small rooms of size 250 ft² and large rooms of size 500 ft², for a total area of 16,000 ft². Also, local fire codes limit the legal occupancy of the small rooms to 2 people and of the large rooms to 5 people, and the total occupancy of the entire motel is limited to 150 people.

(a) Use linear equations to express the constraints imposed by the size of the motel and by the fire code.

(b) Solve the resulting system of equations. What does your solution tell you about the motel?

47. A fast-food fish restaurant serves meals consisting of fish, chips, and hush-puppies.

- One fish, one order of chips, and one pair of hush-puppies costs \$2.27.
- Two fish, one order of chips, and one pair of hush-puppies costs \$3.26.
- One fish, one order of chips, and two pairs of hush-puppies costs \$2.76.

(a) How much should a meal of two fish, two orders of chips, and one pair of hush-puppies cost?

(b) Show that

$$2x + 2y + z = 3(x + y + z) - (x + y + 2z)$$

is an identity in x, y, and z.

(c) Use the identity in part (b) to solve part (a).

(d) Someone says that you do not need the information that "Two fish, one order of chips, and one pair of hush-puppies costs \$3.26" to solve part (a). Is this true?

48. For the system

$$\begin{cases} 2x + 3y = 5 \\ 4x + 6y = n, \end{cases}$$

what must be true about n in order for there to be many solutions?

49. Solve the system of equations

$$\begin{cases} 3x + 2y + 5z = 11 \\ 2x - 3y + z = 7 \\ z = 2x. \end{cases}$$

Hint: Use the third equation to substitute for z in the other two.

REVIEW EXERCISES AND PROBLEMS FOR CHAPTER 5

EXERCISES

In Exercises **1–4**, give the initial value and the slope and explain their meaning in practical terms.

1. The rental charge, C, at a video store is given by $C = 4.29 + 3.99n$, where n is the number of days greater than 2 for which the video is kept.

2. While driving back to college after spring break, a student realizes that his distance from home, D, in miles, is given by $D = 55t + 30$, where t is the time in hours since the student made the realization.

3. The cost, C, in dollars, of making n donuts is given by $C = 250 + n/36$.

4. The number of people, P, remaining in a lecture hall m minutes after the start of a very boring lecture is given by $P = 300 - 19m/3$.

In Exercises **5–7**, describe in words how the temperature changes with time.

5.

T, temperature ($^\circ$F)

```
70
60
50   T = 50 + 2h
40
30
20
10
        2   4   6   8
                    h, hours since midnight
```

6.

T, temperature ($^\circ$F)

```
70
60   T = 50 − 2h
50
40
30
20
10
        2   4   6   8
                    h, hours since midnight
```

7.

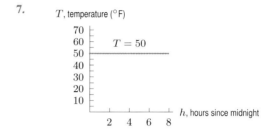

T, temperature ($^\circ$F)

```
70
60     T = 50
50
40
30
20
10
        2   4   6   8
                    h, hours since midnight
```

Match functions I–VI, which describe the value of investments after t years, to the statements in Exercises **8–10**.

I. $V = 5000 + 200t$ II. $V = 500t$
III. $V = 8000$ IV. $V = 10{,}000 + 200(t - 10)$
V. $V = 200t + 7000$ VI. $V = 8500 - 550t$

8. These investments begin with the same amount of money.

9. This investment's value never changes.

10. This investment begins with the most money.

In Exercises **11–19**, express the quantity as a linear function of the indicated variable.

11. The cost for h hours if a handyman charges \$50 to come to the house and \$45 for each hour that he works.

12. The population of a town after y years if the initial population is 23,400 and decreases by 200 people per year.

13. The value of a computer y years old if it costs \$2400 when new and depreciates \$500 each year.

14. A company's cumulative operating costs after m months if it needs \$7600 for equipment, furniture, etc. and if the rent is \$3500 for each month.

15. A student's test score if he answers p extra-credit problems for 2 points each and has a base score of 80.

16. A driver's distance from home after h hours if she starts 200 miles from home and drives 50 miles per hour away from home.

17. The cumulative cost of a gym membership after m months if the membership fee is \$350 and the monthly rate is \$30.

18. $P = h(t)$ gives the size of a population at time t in years, that begins with 12,000 members and grows by 225 members each year.

19. In year $t = 0$, a previously uncultivated region is planted with crops. The depth d of topsoil is initially 77 cm and begins to decline by 3.2 cm each year.

20. The number of viewers for a new TV show is 8 million for the first episode. It drops to half this level by the fifth episode. Assuming viewership drops at a constant rate, find a formula for $p(n)$, the number of viewers (in millions) for the nth episode.

■ Identify the constants b and m in the expression $b + mt$ for the linear functions in Exercises **21–26**.

21. $f(t) = 200 + 14t$ 22. $g(t) = 77t - 46$

23. $h(t) = t/3$ 24. $p(t) = 0.003$

25. $q(t) = \dfrac{2t + 7}{3}$ 26. $r(t) = \sqrt{7} - 0.3t\sqrt{8}$

■ Evaluate the expressions in Exercises **27–30** given that $w(x)$ is a linear function defined by

$$w(x) = 9 + 4x.$$

Simplify your answers.

27. $w(-4)$ 28. $w(x - 4)$

29. $w(x + h) - w(x)$ 30. $w(9 + 4x)$

■ In Exercises **31–42**, is the expression linear in the indicated variable? Assume all constants are nonzero.

31. $5x - 7 + 2x,\ x$ 32. $3x - 2x^2,\ x$

33. $mx + b + c^3,\ x$ 34. $mx + b + c^3x^2,\ x$

35. $P(P - b)(c - P),\ P$ 36. $P(P - b)(c - P),\ c$

37. $P(2 + P) - P^2,\ P$ 38. $P(2 + P) - 2P^2,\ P$

39. $xy + ax + by + ab,\ x$ 40. $xy + ax + by + ab,\ y$

41. $xy + ax + by + ab,\ a$ 42. $xy + ax + by + ab,\ b$

■ Without solving them, say whether the equations in Exercises **43–54** have a positive solution, a negative solution, a zero solution, or no solution. Give a reason for your answer.

43. $7x = 5$ 44. $3x + 5 = 7$

45. $5x + 3 = 7$ 46. $5 - 3x = 7$

47. $3 - 5x = 7$ 48. $9x = 4x + 6$

49. $9x = 6 - 4x$ 50. $9 - 6x = 4x - 9$

51. $8x + 11 = 2x + 3$ 52. $11 - 2x = 8 - 4x$

53. $8x + 3 = 8x + 11$ 54. $8x + 3x = 2x + 11x$

■ In Exercises **55–58**, solve for the indicated variable. Assume all constants are nonzero.

55. $I = Prt$, for r.   56. $F = \dfrac{9}{5}C + 32$, for C.

57. $\dfrac{a - cy}{b + dy} + a = 0$, for y, if $c \neq ad$.

58. $\dfrac{Ax - B}{C - B(1 - 2x)} = 3$, for x, if $A \neq 6B$.

■ In Exercises **59–60**, which line has the greater

(a) Slope?   (b) y-intercept?

59. $y = 5 - 2x,\quad y = 8 - 4x$

60. $y = 7 + 3x,\quad y = 8 - 10x$

■ In Exercises **61–62**, graph the equation.

61. $y = \dfrac{2}{3}x - 4$ 62. $4x + 5y = 2$

■ Write the equations in Exercises **63–68** in the form $y = b + mx$, and identify the values of b and m.

63. $y = \frac{x}{8} - 14$ 64. $y - 8 = 3(x - 5)$

65. $\dfrac{y - 4}{5} = \dfrac{2x - 3}{3}$ 66. $4x - 7y = 12$

67. $y = 90$ 68. $y = \sqrt{8}x$

■ For Exercises **69–70**, put the equation in standard form.

69. $y = 3x - 2$ 70. $3x = 2y - 1$

In Exercises **71–73**, write an equation for each of the lines whose graph is shown.

71.

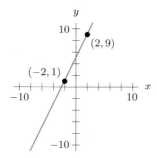

72.

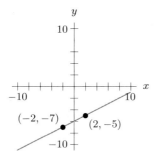

73.

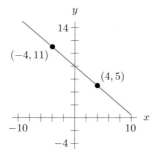

In Exercises **74–85**, write an equation in slope-intercept form for the line that

74. Passes through $(4, 6)$ with slope $m = 2$

75. Passes through $(-6, 2)$ with slope $m = -3$

76. Passes through $(-4, -8)$ with slope $m = 1/2$

77. Passes through $(9, 7)$ with slope $m = -2/3$

78. Contains $(6, 8)$ and $(8, 12)$

79. Contains $(-3, 5)$ and $(-6, 4)$

80. Contains $(4, -2)$ and $(-8, 1)$

81. Contains $(5, 6)$ and $(10, 3)$

82. Contains $(4, 7)$ and is parallel to $y = (1/2)x - 3$

83. Contains $(-10, -5)$ and is parallel to $y = -(4/5)x$

84. Passes through $(0, 5)$ and is parallel to $3x + 5y = 6$

85. Passes through $(-10, -30)$ and is perpendicular to $12y - 4x = 8$.

Find formulas for the linear functions described in Exercises **86–94**.

86. The graph intercepts the x-axis at $x = 30$ and the y-axis at $y = -80$.

87. $P = s(t)$ describes a population that begins with 8200 members in year $t = 0$ and reaches 12,700 members in year $t = 30$.

88. The total cost C of an international call lasting n minutes if there is a connection fee of \$2.95 plus an additional charge of \$0.35/minute.

89. $w(x)$, where $w(4) = 20, w(12) = -4$

90. The graph of $y = p(x)$ contains the points $(-30, 20)$ and $(70, 140)$.

91. $w(0.3) = 0.07, w(0.7) = 0.01$

92. $g(x)$, where $g(5) = 50, g(30) = 25$

93. This function's graph is parallel to the line $y = 20 - 4x$ and contains the point $(3, 12)$.

94. The graph of $f(x)$ passes through $(-1, 4)$ and $(2, -11)$.

Exercises **95–100** give data from a linear function. Find a formula for the function.

95.

Temperature, $y = f(x)$ (°C)	0	5	20
Temperature, x (°F)	32	41	68

96.

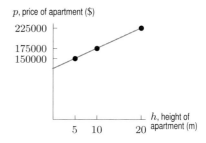

97.

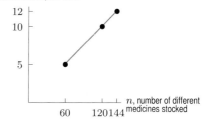

98.

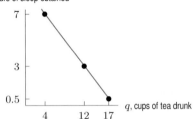

s, hours of sleep obtained

q, cups of tea drunk

99.

Temperature, $y = f(x)$, (°R)	459.7	469.7	489.7
Temperature, x (°F)	0	10	30

100.

Price per bottle, p ($)	0.50	0.75	1.00
Number of bottles sold, $q = f(p)$	1500	1000	500

In the tables in Exercises **101–106** could the second row be a linear function of the first?

101.

t	1	2	3	4	5
v	5	4	3	2	1

102.

p	−4	−2	0	2	4
q	15	15.5	16	16.5	17

103.

x	0	1	2	3	4
y	0	1	4	9	16

104.

n	3	6	9	12	15
C	5	4	3	4	5

105.

x	1	5	11	19
y	10	8	5	1

106.

x	1	3	9	27
y	−5	−10	−15	−20

Solve the systems in Exercises **107–110**.

107. $\begin{cases} 2a - 3b = 22 \\ 3a + 4b = -1. \end{cases}$

108. $\begin{cases} 2n + 7m = 1 \\ 3n + 10m = 3. \end{cases}$

109. $\begin{cases} 10x + 4y = -3 \\ 6x - 5y = 13. \end{cases}$

110. $\begin{cases} 2v + 3w = 11 \\ 2w - 3v = 29. \end{cases}$

PROBLEMS

111. Water is added to a barrel at a constant rate for 25 minutes, after which it is full. The quantity of water in the barrel after t minutes is $100 + 4t$ gallons. What do the 100 and 4 mean in practical terms? How much water can the barrel hold?

112. A party facility charges $500 for a banquet room and $20 per person. In addition there is a 20% surcharge on the entire fee. Why would you expect the total cost of the party to be given by a linear function of the number of people, P? Give the function.

113. A design shop offers ceramics painting classes. The fee for the class is $30 and each item that is painted costs $12. There is also a firing fee of $3 for each item and a 7% tax on the cost of the item. Why would you expect the total cost of the class to be given by a linear function of the number of items, i? Give the function.

114. The total cost of ownership for an Epson Stylus Photo R320 printer is given by

$$\text{Total cost} = 200 + 1.00n,$$

where n is the number of 8×10 photos printed. Interpret the constants 200 and 1.00 in practical terms.

115. A ski shop rents out skis for an initial payment of I dollars plus r dollars per day. A skier spends $I + 9r$ dollars on skis for a vacation. What is the meaning of the 9 in this formula?

Find formulas for the linear functions described in Problems **116–117**.

116. The graph of $y = v(x)$ intersects the graph of $u(x) = 1 + x^3$ at $x = -2$ and $x = 3$.

117. The graph of f intersects the graph of $y = 0.5x^3 - 4$ at $x = -2$ and $x = 4$.

■ If two populations have the same constant rate of change, then the graphs that describe them are parallel lines. Are the graphs of the populations in Problems **118–121** parallel lines?

118. Towns A and B are each growing by 1000 people per decade.

119. Bacteria populations C and D are each growing at 30% each hour.

120. Country E is growing by 1 million people per decade. Country F is growing by 100 thousand people per year.

121. Village G has 200 people and is growing by 2 people per year. Village H has 100 people and is growing by 1 person per year.

122. A Chinese restaurant charges $3 per dish. You leave a 15% tip. Express your cost as a function of the number, d, of dishes ordered. Is it a linear function?

123. You buy a $15 meal and leave a t% tip. Express your your total cost as a function of t. Is it a linear function?

124. Your personal income taxes are 25% of your adjusted income. To calculate your adjusted income you subtract $1000 for each child you have from your income, I. You have five children. Find an expression for the tax in terms of I and say whether it is linear in I.

125. Two parents take their N children to the movies. Adult tickets are $9 each and child tickets are $7. They buy a $3 box of popcorn for each child and pay $5 parking. Find an expression for the cost of the excursion and say whether it is linear in N.

126. A person departs Tucson driving east on I-10 at 65 mph. One hour later a second person departs in the same direction at 75 mph. Both vehicles stop when the second person catches up.

 (a) Write an expression for the distance, D, in miles, between the two vehicles as a function of the time t in hours since the first one left Tucson.

 (b) For what values of t does the expression make sense in practical terms?

127. A pond with vertical sides has a depth of 2 ft and a surface area of 10 ft^2. If the pond is full of water and evaporation causes the water level to drop at the rate of 0.3 inches/day, write an expression that represent the volume of water in the pond after d days. For what values of d is your expression valid?

128. A 200 gallon container contains 100 gallons of water. Water pours in at 1 gallon per minute, but drains out at the rate of 5 gallon per minute. How much water is in the container after t minutes? Is the container emptying or filling? When is the container empty or full?

129. While on a European vacation, a student wants to be able to convert from degrees Celsius ($°C$) to degrees Fahrenheit ($°F$). One day the temperature is $20°C$, which she is told is equivalent to $68°F$. Another day, the temperature is $25°C$, which she is told is equivalent to $77°F$.

 (a) Write a linear function that converts temperatures from Celsius to Fahrenheit.

 (b) Using the function from part (a), convert the temperature of

 (i) $10°C$ to Fahrenheit.

 (ii) $86°F$ to Celsius.

130. The road to the summit of Mt. Haleakala, located in Haleakala National Park in Hawaii, holds the world's record for climbing to the highest elevation in the shortest distance.[10] You can drive from sea level to the 10,023 ft summit over a distance of 35 miles, passing through five distinct climate zones. The temperature drops about $3°F$ per 1000 foot rise. The average temperature at Park Headquarters (7000 foot elevation) is $53°F$.[11]

 (a) Express the average temperature T, in $°F$, in Haleakala National Park as a function of the elevation E in ft.

 (b) What is the average temperature at

 (i) Sun Visitor Center, at 9745 feet?

 (ii) Sea level?

131. When the carnival comes to town, a group of students wants to attend. The cost of admission and going on 3 rides is $12.50, while the cost of admission and going on 6 rides is $17.

 (a) Write a linear function to represent the cost, C, of entering and the carnival and going on n rides.

 (b) Express the function in slope-intercept form.

 (c) What is the meaning of the slope and the intercept in part (b)?

132. A small band would like to sell CDs of its music. The business manager says that it costs $320 to produce 100 CDs and $400 to produce 500 CDs.

 (a) Write a linear function to represent the cost, C, of producing n CDs.

 (b) Find the slope and the intercept of the graph of the function, and interpret their meaning

 (c) How much does it cost to produce 750 CDs?

 (d) If the band can afford to spend $500, how many CDs can it produce?

---

[10] www.hawaiiweb.com/maui/html/sites/haleakala_national_park.html.

[11] www.mauidownhill.com/haleakala/facts/haleakalaweather.html.

133. The cost of owning a timeshare consists of two parts: the initial cost of buying the timeshare, and the annual cost of maintaining it (the maintenance fee), which is paid each year in advance. Figure 5.31 shows the cumulative cost of owning a particular California timeshare over a 10-year period from the time it was bought. The first maintenance fee was paid at the time of purchase. What was the maintenance fee? How much did the timeshare cost to buy?

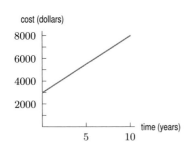

Figure 5.31

■ Without solving, decide which of the following statements applies to the solution to each of the equations in Problems 134–153:

(i) $x \leq -1$ (ii) $-1 < x < 0$ (iii) $x = 0$
(iv) $0 < x < 1$ (v) $x \geq 1$ (vi) No solution

Give a reason for your answer.

134. $7x = 4$

135. $4x + 2 = 7$

136. $2x + 8 = 7$

137. $11x + 7 = 2$

138. $11x - 7 = 5$

139. $8 - 2x = 7$

140. $5x + 3 = 7x + 5$

141. $5x + 7 = 2x + 5$

142. $3 - 4x = -4x - 3$

143. $5x + 3 = 8x + 3$

144. $\dfrac{x}{2} = \dfrac{1}{5}$

145. $\dfrac{2}{x} = \dfrac{3}{5}$

146. $\dfrac{x + 8}{2} = 4$

147. $\dfrac{x + 8}{3} = 2$

148. $\dfrac{10}{5 + x} = 1$

149. $\dfrac{10}{5 + x} = 2$

150. $\dfrac{2x + 5}{2x + 6} = 1$

151. $\dfrac{2x + 2}{3x + 5} = 1$

152. $\dfrac{1}{x} - 2 = 2$

153. $3 - \dfrac{1}{x} = 3$

■ For Problems **154–156**, use the equation

$$F = \frac{9}{5}C + 32,$$

which relates temperatures in degrees Celsius (°C) to degrees Fahrenheit (°F).

154. (a) By solving the equation for C, find a formula which converts temperatures in degrees Fahrenheit (°F) to degrees Celsius (°C).
(b) Typical oven cooking temperatures are between 350°F and 450°F. What are these temperatures in °C?

155. A rule of thumb used by some cooks[12] is: For the range of temperatures used for cooking, the number of degrees Fahrenheit is twice the number of degrees Celsius.

(a) Write an equation, in terms of F and C, that represents this rule of thumb.
(b) For what values of F and C does this rule of thumb give the correct conversion temperature?
(c) Typical oven cooking temperatures are between 350°F (176.7°C) and 450°F (232.2°C). Use graphs to show that the maximum error in using this rule of thumb occurs at the upper end of the temperature range.

156. A convenient way to get a feel for temperatures measured in Celsius is to know[13] that room temperature is about 21°C.

(a) Show that, according to this, room temperature is about 70°F.
(b) Someone says: "So if 21°C is about 70°F, then 42°C is about 140°F." Is this statement accurate?

■ Problems **157–159** concern laptop computer batteries. As the computer runs, the total charge on the battery, measured in milliampere-hours or mAh, goes down. A typical new laptop battery has a capacity of 4500 mAh when fully charged, and under normal use will discharge at a constant rate of about 0.25 mAh/s.

157. Find an expression for $f(t)$, the amount of charge (in mAh) remaining on a new battery after t seconds of normal use, assuming that the battery is charged to capacity at time $t = 0$.

158. If fully charged at time $t = 0$, the charge remaining on an older laptop battery after t seconds of use is given by $g(t) = 3200 - 0.4t$. What do the parameters of this function tell you about the old laptop battery as compared to a new battery?

---

[12]Graham and Rosemary Haley, *Haley's Hints*, Rev. Ed. (Canada: 3H Productions, 1999), p. 355.
[13]Graham and Rosemary Haley, *Haley's Hints*, Rev. Ed. (Canada: 3H Productions, 1999), p. 351.

159. Given that $f(t_1) = 0$ and $g(t_2) = 0$, evaluate the expression t_1/t_2. What does the expression tells you about the two batteries?

■ Problems 160–162 involve a simple mathematical model of the behavior of songbirds. This model relates d, the amount of time in hours a songbird spends each day defending its territory, to s, the amount of time it spends singing (looking for mates). The more time a bird devotes to one of these behaviors, the less time it can devote to the other.[14]

160. Suppose that a particular species must consume 5 calories for each hour spent singing and 10 calories for each hour spent defending its territory. If this species consumes 60 calories per day on these activities, find a linear equation relating s and d and sketch its graph, placing d on the vertical axis. Say what the s- and d-intercepts of the graph tell you about the bird.

161. A second species must consume 4 calories for each hour spent singing and 12 calories for each hour spent defending its territory. If this species also consumes 60 calories per day on these activities, find a linear equation relating s and d and sketch its graph, placing d on the vertical axis. Say what the intercepts tell you about the bird.

162. What does the point of intersection of the graphs you drew for Problems 160 and 161 tell you about the two bird species?

■ Italian coffee costs \$10/lb, and Kenyan costs \$15/lb. A workplace has a \$60/week budget for coffee. If it buys I lbs of Italian coffee and K pounds of Kenyan coffee, which of the following equations best matches the statements in Problems 163–166?
   (a) The total amount spent each week on coffee equals \$60.
   (b) The amount of Kenyan coffee purchased can be found by subtracting the amount spent on Italian coffee from the total and then dividing by the price per pound of Kenyan.
   (c) The amount of Italian coffee that can be purchased is proportional to the amount of Kenyan coffee *not* purchased (as compared to the largest possible amount of Kenyan coffee that can be purchased).
   (d) For every two pounds of Kenyan purchased, the amount of Italian purchased goes down by 3 lbs.

163. $I = 6 - \frac{3}{2}K$

164. $K = \dfrac{60 - 10I}{15}$

165. $I = 1.5(4 - K)$

166. $10I + 15K = 60$

■ Solve the systems of equations in Problems 167–170 for x and y.

167. $\begin{cases} 2x + 4y = 44 \\ y = \frac{3}{4}x + 6 \end{cases}$

168. $\begin{cases} 3x - y = 20 \\ -2x - 3y = 5 \end{cases}$

169. $\begin{cases} 2(x + y) = 5 \\ x = y + 3(x - 3) \end{cases}$

170. $\begin{cases} bx + y = 2b \\ x + by = 1 + b^2, \quad \text{if } b \neq \pm 1 \end{cases}$

171. Two companies sell and deliver sand used as a base for building patios. Company A charges \$10 a cubic yard and a flat \$40 delivery fee for any amount of sand up to 12 cubic yards. Company B charges \$8 a cubic yard and a flat \$50 delivery fee for any amount of sand up to 12 cubic yards. You wish to have x cubic yards of sand delivered, where $0 < x \leq 12$. Which company should you buy from?

■ Measures in US recipes are not the same size as in UK recipes. Suppose that a UK cup is u US cups, a UK tablespoon is t US cups, and a UK dessertspoon is d US cups. Problems 172–173 use information from the same web page.[15]

172. According to the web, 3/4 of a US cup is 1/2 a UK cup plus 2 UK tablespoons, and 2/3 of a US cup is 1/2 a UK cup plus 1 UK tablespoon, so

$$\frac{3}{4} = \frac{1}{2}u + 2t$$
$$\frac{2}{3} = \frac{1}{2}u + t.$$

Solve this system and interpret your answer in terms of the relation between UK and US cups, and between the UK tablespoon and the US cup.

173. According to the web, 1 US cup is 3/4 of a UK cup plus 2 UK dessertspoons, and 1/4 of a US cup is 1/4 of a UK cup minus 1 UK dessertspoon, so

$$1 = \frac{3}{4}u + 2d$$
$$\frac{1}{4} = \frac{1}{4}u - d.$$

Solve this system and interpret your answer in terms of the relation between UK and US cups, and the UK dessertspoon and the US cup.

---

[14] This model neglects other behaviors such as foraging, nest building, caring for young, and sleeping.
[15] http://allrecipes.com/advice/ref/conv/conversions_brit.asp, accessed on May 14, 2003.

SOLVING DRILL

In Problems **1–25**, solve the equation for the indicated variable.

1. $5x + 2 = 12$; for x
2. $7 - 3x = 25$; for x
3. $5 + 9t = 72$; for t
4. $10 = 25 - 3r$; for r
5. $5x + 2 = 3x - 7$; for x
6. $7 + 5t = 10 - 3t$; for t
7. $100 - 24w = 5w - 30$; for w
8. $1.25 + 0.07x = 3.92$; for x
9. $0.5t - 13.4 = 25.8$; for t
10. $12.53 + 5.67x = 45.1x - 125$; for x
11. $3(x - 2) + 15 = 5(x + 4)$; for x
12. $3t - 5(t + 2) = 3(2 - 4t) + 8$; for t
13. $5(2p - 6) + 10 = 6(p + 3) + 4(2p - 1)$; for p
14. $2.3(2x + 5.9) = 0.1(24.7 + 54.2x) + 2.4x$; for x
15. $10.8 - 3.5(40 - 5.1t) = 3.2t + 4.5(25.4 - 5.6t)$; for t
16. $ax + b = c$; for x
17. $rt + s = pt + q$; for t
18. $rsw - 0.2rw + 0.1sw = 5r - 1.8s$; for s
19. $rsw - 0.2rw + 0.1sw = 5r - 1.8s$; for r
20. $r(As - Bt + Cr) = 25t + A(st + Br)$; for t
21. $r(As - Bt + Cr) = 25t + A(st + Br)$; for s
22. $0.2PQ^2 + RQ(P - 1.5Q + 2R) = 2.5$; for P
23. $x^2y^2 + 2xyy' + x^2y' - 5x + 2y' + 10 = 0$; for y'
24. $25V_0S^2[T] + 10(H^2 + V_0) = A_0(3V + V_0)$; for V_0
25. $25V_0S^2[T] + 10(H^2 + V_0) = A_0(3V + V_0)$; for $[T]$

Chapter 6

Rules for Exponents and the Reasons for Them

CONTENTS

6.1 INTEGER POWERS AND THE EXPONENT RULES

Repeated addition can be expressed as a product. For example,

$$\underbrace{2+2+2+2+2}_{\text{5 terms in sum}} = 5 \times 2.$$

Similarly, repeated multiplication can be expressed as a power. For example,

$$\underbrace{2 \times 2 \times 2 \times 2 \times 2}_{\text{5 factors in product}} = 2^5.$$

Here, 2 is called the *base* and 5 is called the *exponent*. Notice that 2^5 is not the same as 5^2, because $2^5 = 2 \times 2 \times 2 \times 2 \times 2 = 32$ but $5^2 = 5 \times 5 = 25$.

In general, if a is any number and n is a positive integer, then we define

$$a^n = \underbrace{a \cdot a \cdot a \cdots a}_{n \text{ factors}}.$$

Notice that $a^1 = a$, because here we have only 1 factor of a. For example, $5^1 = 5$. We call a^2 the *square* of a and a^3 the *cube* of a.

Multiplying and Dividing Powers with the Same Base

When we multiply powers with the same base, we can add the exponents to get a more compact form. For example, $5^2 \cdot 5^3 = (5 \cdot 5) \cdot (5 \cdot 5 \cdot 5) = 5^{2+3} = 5^5$. In general,

$$a^n \cdot a^m = \underbrace{a \cdot a \cdot a \cdots a}_{n \text{ factors}} \cdot \underbrace{a \cdot a \cdot a \cdots a}_{m \text{ factors}} = \underbrace{a \cdot a \cdot a \cdots a}_{n+m \text{ factors}} = a^{n+m}.$$

Thus,

$$a^n \cdot a^m = a^{n+m}.$$

Example 1	Write with a single exponent:

(a) $q^5 \cdot q^7$ (b) $6^2 \cdot 6^3$ (c) $2^n \cdot 2^m$
(d) $3^n \cdot 3^4$ (e) $(x+y)^2(x+y)^3$.

Solution   Using the rule $a^n \cdot a^m = a^{n+m}$ we have

(a) $q^5 \cdot q^7 = q^{5+7} = q^{12}$ (b) $6^2 \cdot 6^3 = 6^{2+3} = 6^5$
(c) $2^n \cdot 2^m = 2^{n+m}$ (d) $3^n \cdot 3^4 = 3^{n+4}$
(e) $(x+y)^2(x+y)^3 = (x+y)^{2+3} = (x+y)^5$.

Just as we applied the distributive law from left to right as well as from right to left, we can use the rule $a^n \cdot a^m = a^{n+m}$ written from right to left as $a^{n+m} = a^n \cdot a^m$.

Example 2   Write as a product:

(a) 5^{2+a} (b) x^{r+4} (c) y^{t+c} (d) $(z+2)^{z+2}$.

Solution   Using the rule $a^{n+m} = a^n \cdot a^m$ we have

(a) $5^{2+a} = 5^2 \cdot 5^a = 25 \cdot 5^a$ (b) $x^{r+4} = x^r \cdot x^4$

(c) $y^{t+c} = y^t \cdot y^c$ (d) $(z+2)^{z+2} = (z+2)^z \cdot (z+2)^2$.

When we divide powers with a common base, we subtract the exponents. For example, when we divide 5^6 by 5^2, we get

$$\frac{5^6}{5^2} = \frac{\overbrace{5 \cdot 5 \cdot 5 \cdot 5 \cdot 5 \cdot 5}^{6 \text{ factors of } 5}}{\underbrace{5 \cdot 5}_{2 \text{ factors of } 5}} = \frac{\overbrace{\not5 \cdot \not5}^{2 \text{ factors of } 5 \text{ cancel}} \cdot \overbrace{5 \cdot 5 \cdot 5 \cdot 5}^{6 - 2 = 4 \text{ factors of } 5 \text{ are left after canceling}}}{\underbrace{\not5 \cdot \not5}_{2 \text{ factors of } 5 \text{ cancel}}} = \underbrace{5 \cdot 5 \cdot 5 \cdot 5}_{6 - 2 = 4 \text{ factors}} = 5^{6-2} = 5^4.$$

More generally, if $n > m$,

$$\frac{a^n}{a^m} = \frac{\overbrace{a \cdot a \cdot a \cdot a \cdots a}^{n \text{ factors of } a}}{\underbrace{a \cdot a \cdots a}_{m \text{ factors of } a}} = \frac{\overbrace{\not a \cdot \not a \cdots \not a}^{m \text{ factors of } a \text{ cancel}} \cdot \overbrace{a \cdots a}^{n - m \text{ factors of } a \text{ are left after canceling}}}{\underbrace{\not a \cdot \not a \cdots \not a}_{m \text{ factors of } a \text{ cancel}}} = \underbrace{a \cdot a \cdot a \cdots a}_{n - m \text{ factors}} = a^{n-m}.$$

Thus,

$$\frac{a^n}{a^m} = a^{n-m}, \text{ if } n > m.$$

Example 3   Write with a single exponent:

(a) $\dfrac{q^7}{q^5}$ (b) $\dfrac{6^7}{6^3}$ (c) $\dfrac{3^n}{3^4}$, where $n > 4$

(d) $\dfrac{\pi^5}{\pi^3}$ (e) $\dfrac{(c+d)^8}{(c+d)^2}$.

Solution   Since $\dfrac{a^n}{a^m} = a^{n-m}$ we have

(a) $\dfrac{q^7}{q^5} = q^{7-5} = q^2$ (b) $\dfrac{6^7}{6^3} = 6^{7-3} = 6^4$

(c) $\dfrac{3^n}{3^4} = 3^{n-4}$ (d) $\dfrac{\pi^5}{\pi^3} = \pi^{5-3} = \pi^2$

(e) $\dfrac{(c+d)^8}{(c+d)^2} = (c+d)^{8-2} = (c+d)^6$.

Just as with the products, we can write $\dfrac{a^n}{a^m} = a^{n-m}$ in reverse as $a^{n-m} = \dfrac{a^n}{a^m}$.

Example 4   Write as a quotient:

(a) 10^{2-k} (b) e^{b-4} (c) z^{w-s} (d) $(p+q)^{a-b}$.

Solution   Since $a^{n-m} = \dfrac{a^n}{a^m}$ we have

(a) $10^{2-k} = \dfrac{10^2}{10^k} = \dfrac{100}{10^k}$ (b) $e^{b-4} = \dfrac{e^b}{e^4}$

(c) $z^{w-s} = \dfrac{z^w}{z^s}$ (d) $(p+q)^{a-b} = \dfrac{(p+q)^a}{(p+q)^b}$.

Raising a Power to a Power

When we take a number written in exponential form and raise it to a power, we multiply the exponents. For example,

$$(5^2)^3 = 5^2 \cdot 5^2 \cdot 5^2 = 5^{2+2+2} = 5^{2 \cdot 3} = 5^6.$$

More generally,

$$(a^m)^n = \underbrace{(a \cdot a \cdot a \cdots a)}_{m \text{ factors of } a}{}^n = \overbrace{\underbrace{(a \cdot a \cdot a \cdots a)}_{m \text{ factors of } a}\underbrace{(a \cdot a \cdot a \cdots a)}_{m \text{ factors of } a} \cdots \underbrace{(a \cdot a \cdot a \cdots a)}_{m \text{ factors of } a}}^{\substack{\text{The } m \text{ factors of } a \text{ are multiplied } n \text{ times,} \\ \text{giving a total of } m \cdot n \text{ factors of } a}} = a^{m \cdot n}.$$

Thus,

$$(a^m)^n = a^{m \cdot n}.$$

Example 5   Write with a single exponent:

(a) $(q^7)^5$ (b) $(7^p)^3$ (c) $(y^a)^b$
(d) $(2^x)^x$ (e) $\left((x+y)^2\right)^3$ (f) $\left((r-s)^t\right)^z$.

Solution   Using the rule $(a^m)^n = a^{m \cdot n}$ we have

(a) $(q^7)^5 = q^{7 \cdot 5} = q^{35}$ (b) $(7^p)^3 = 7^{3p}$
(c) $(y^a)^b = y^{ab}$ (d) $(2^x)^x = 2^{x^2}$
(e) $\left((x+y)^2\right)^3 = (x+y)^{2 \cdot 3} = (x+y)^6$ (f) $\left((r-s)^t\right)^z = (r-s)^{tz}$.

Example 6   Write as a power raised to a power:

(a) $2^{3 \cdot 2}$ (b) 4^{3x} (c) e^{4t} (d) 6^{z^2}.

Solution   Using the rule $a^{m \cdot n} = (a^m)^n$ we have

(a) $2^{3 \cdot 2} = (2^3)^2$. This could also have been written as $(2^2)^3$.

(b) $4^{3x} = (4^3)^x$, which simplifies to 64^x. This could also have been written as $(4^x)^3$.

(c) $e^{4t} = (e^4)^t$. This could also have been written as $(e^t)^4$.

(d) $6^{z^2} = (6^z)^z$.

Products and Quotients Raised to the Same Exponent

When we multiply $5^2 \cdot 4^2$ we can change the order of the factors and rewrite it as $5^2 \cdot 4^2 = (5 \cdot 5) \cdot (4 \cdot 4) = 5 \cdot 5 \cdot 4 \cdot 4 = (5 \cdot 4) \cdot (5 \cdot 4) = (5 \cdot 4)^2 = 20^2$. Sometimes, we want to use this process in reverse: $10^2 = (2 \cdot 5)^2 = 2^2 \cdot 5^2$.

In general,

$$(a \cdot b)^n = \underbrace{(a \cdot b)(a \cdot b)(a \cdot b) \cdots (a \cdot b)}_{n \text{ factors of } (a \cdot b)} = \underbrace{(\overbrace{a \cdot a \cdot a \cdots a}^{n \text{ factors of } a}) \cdot (\overbrace{b \cdot b \cdot b \cdots b}^{n \text{ factors of } b})}_{\substack{\text{since we can rearrange the order using the} \\ \text{commutative property of multiplication}}} = a^n \cdot b^n.$$

Thus,

$$(ab)^n = a^n b^n.$$

Example 7   Write without parentheses:

(a) $(qp)^7$ (b) $(3x)^n$ (c) $(4ab^2)^3$ (d) $(2x^{2n})^{3n}$.

Solution   Using the rule $(ab)^n = a^n b^n$ we have

(a) $(qp)^7 = q^7 p^7$

(b) $(3x)^n = 3^n x^n$

(c) $(4ab^2)^3 = 4^3 a^3 (b^2)^3 = 64 a^3 b^6$

(d) $(2x^{2n})^{3n} = (2^{3n})(x^{2n})^{3n} = (2^3)^n (x)^{2n \cdot 3n} = 8^n x^{6n^2}$.

Example 8   Write with a single exponent:

(a) $c^4 d^4$ (b) $2^n \cdot 3^n$ (c) $4x^2$

(d) $a^4 (b+c)^4$ (e) $(x^2 + y^2)^5 (c - d)^5$.

Solution   Using the rule $a^n b^n = (ab)^n$ we have

(a) $c^4 d^4 = (cd)^4$ (b) $2^n \cdot 3^n = (2 \cdot 3)^n = 6^n$

(c) $4x^2 = 2^2 x^2 = (2x)^2$ (d) $a^4 (b+c)^4 = (a(b+c))^4$

(e) $(x^2 + y^2)^5 (c - d)^5 = \left((x^2 + y^2)(c - d) \right)^5$.

Division of two powers with the same exponent works the same way as multiplication. For example,

$$\frac{6^4}{3^4} = \frac{6 \cdot 6 \cdot 6 \cdot 6}{3 \cdot 3 \cdot 3 \cdot 3} = \frac{6}{3} \cdot \frac{6}{3} \cdot \frac{6}{3} \cdot \frac{6}{3} = \left(\frac{6}{3} \right)^4 = 2^4 = 16.$$

Or, reversing the process,

$$\left(\frac{4}{5}\right)^3 = \frac{4}{5} \cdot \frac{4}{5} \cdot \frac{4}{5} = \frac{4 \cdot 4 \cdot 4}{5 \cdot 5 \cdot 5} = \frac{4^3}{5^3}.$$

More generally,

$$\left(\frac{a}{b}\right)^n = \underbrace{\left(\frac{a}{b}\right) \cdot \left(\frac{a}{b}\right) \cdot \left(\frac{a}{b}\right) \cdots \left(\frac{a}{b}\right)}_{n \text{ factors of } a/b} = \frac{\overbrace{a \cdot a \cdot a \cdots a}^{n \text{ factors of } a}}{\underbrace{b \cdot b \cdot b \cdots b}_{n \text{ factors of } b}} = \frac{a^n}{b^n}.$$

Thus,

$$\left(\frac{a}{b}\right)^n = \frac{a^n}{b^n}.$$

Example 9   Write without parentheses:

(a) $\left(\dfrac{4}{5}\right)^2$ (b) $\left(\dfrac{c}{d}\right)^{12}$ (c) $\left(\dfrac{y}{z^3}\right)^4$ (d) $\left(\dfrac{2u}{3v}\right)^3.$

Solution   Using the rule $\left(\dfrac{a}{b}\right)^n = \dfrac{a^n}{b^n}$ we have

(a) $\left(\dfrac{4}{5}\right)^2 = \dfrac{4^2}{5^2} = \dfrac{16}{25}$ (b) $\left(\dfrac{c}{d}\right)^{12} = \dfrac{c^{12}}{d^{12}}$

(c) $\left(\dfrac{y}{z^3}\right)^4 = \dfrac{(y)^4}{(z^3)^4} = \dfrac{y^4}{z^{12}}$ (d) $\left(\dfrac{2u}{3v}\right)^3 = \dfrac{(2u)^3}{(3v)^3} = \dfrac{2^3 u^3}{3^3 v^3} = \dfrac{8u^3}{27v^3}.$

Example 10   Write with a single exponent:

(a) $\dfrac{3^5}{7^5}$ (b) $\dfrac{q^7}{p^7}$ (c) $\dfrac{9x^2}{y^2}$

(d) $\dfrac{(p+q)^4}{z^4}$ (e) $\dfrac{(x^2+y^2)^5}{(a+b)^5}$

Solution   Using the rule $\dfrac{a^n}{b^n} = \left(\dfrac{a}{b}\right)^n$ we have

(a) $\dfrac{3^5}{7^5} = \left(\dfrac{3}{7}\right)^5$ (b) $\dfrac{q^7}{p^7} = \left(\dfrac{q}{p}\right)^7$ (c) $\dfrac{9x^2}{y^2} = \left(\dfrac{3x}{y}\right)^2$

(d) $\dfrac{(p+q)^4}{z^4} = \left(\dfrac{p+q}{z}\right)^4$ (e) $\dfrac{(x^2+y^2)^5}{(a+b)^5} = \left(\dfrac{x^2+y^2}{a+b}\right)^5.$

Zero and Negative Integer Exponents

We have seen that 4^5 means 4 multiplied by itself 5 times, but what is meant by 4^0, 4^{-1} or 4^{-2}? We choose definitions for exponents like $0, -1, -2$ that are consistent with the exponent rules.

If $a \neq 0$, the exponent rule for division says

$$\frac{a^2}{a^2} = a^{2-2} = a^0.$$

But $\dfrac{a^2}{a^2} = 1$, so we define $a^0 = 1$ if $a \neq 0$. The same idea tells us how to define negative powers. If $a \neq 0$, the exponent rule for division says

$$\frac{a^0}{a^1} = a^{0-1} = a^{-1}.$$

But $\dfrac{a^0}{a^1} = \dfrac{1}{a}$, so we define $a^{-1} = 1/a$. In general, we define

$$a^{-n} = \frac{1}{a^n}, \quad \text{if } a \neq 0.$$

Note that a negative exponent tells us to take the reciprocal of the base and change the sign of the exponent, *not* to make the number negative.

Example 11   Evaluate:

(a) 5^0 (b) 3^{-2} (c) 2^{-1} (d) $(-2)^{-3}$ (e) $\left(\dfrac{2}{3}\right)^{-1}$

Solution   (a)   Any nonzero number to the zero power is one, so $5^0 = 1$.

(b)   We have

$$3^{-2} = \frac{1}{3^2} = \frac{1}{9}.$$

(c)   We have

$$2^{-1} = \frac{1}{2^1} = \frac{1}{2}.$$

(d)   We have

$$(-2)^{-3} = \frac{1}{(-2)^3} = \frac{1}{(-2)\cdot(-2)\cdot(-2)} = \frac{1}{-8} = -\frac{1}{8}.$$

(e)   We have

$$\left(\frac{2}{3}\right)^{-1} = \frac{1}{\left(\frac{2}{3}\right)} = \frac{3}{2}.$$

With these definitions, we have the exponent rule for division, $\dfrac{a^n}{a^m} = a^{n-m}$ where n and m are integers.

Example 12   Rewrite with only positive exponents. Assume all variables are positive.

(a) $\dfrac{1}{3x^{-2}}$ (b) $\left(\dfrac{x}{y}\right)^{-3}$ (c) $\dfrac{3r^{-2}}{(2r)^{-4}}$ (d) $\dfrac{(a+b)^{-2}}{(a+b)^{-5}}$

Solution   (a) We have

$$\frac{1}{3x^{-2}} = \frac{1}{3 \cdot \frac{1}{x^2}} = \frac{1}{\frac{3}{x^2}} = \frac{x^2}{3}.$$

(b) We have

$$\left(\frac{x}{y}\right)^{-3} = \frac{1}{\left(\frac{x}{y}\right)^3} = \frac{1}{\frac{x^3}{y^3}} = \frac{y^3}{x^3}.$$

(c) We have

$$\frac{3r^{-2}}{(2r)^{-4}} = \frac{3 \cdot \frac{1}{r^2}}{\frac{1}{(2r)^4}} = \frac{\frac{3}{r^2}}{\frac{1}{16r^4}} = \frac{3}{r^2} \cdot \frac{16r^4}{1} = \frac{48r^4}{r^2} = 48r^2.$$

(d) We have

$$\frac{(a+b)^{-2}}{(a+b)^{-5}} = (a+b)^{-2+5} = (a+b)^3.$$

In part (a) of Example 12, we saw that the x^{-2} in the denominator ended up as x^2 in the numerator. In general:

$$\frac{1}{a^{-n}} = a^n \quad \text{and} \quad \left(\frac{a}{b}\right)^{-n} = \left(\frac{b}{a}\right)^n.$$

Example 13   Write each of the following expressions with only positive exponents. Assume all variables are positive.

(a) $\dfrac{1}{4^{-2}}$ (b) $\dfrac{3}{4m^{-4}}$ (c) $\left(\dfrac{4}{5}\right)^{-2}$ (d) $\left(\dfrac{2x}{3y}\right)^{-3}$ (e) $\left(\dfrac{a+b}{a-b}\right)^{-1}$

Solution   (a) $\dfrac{1}{4^{-2}} = 4^2 = 16.$

(b) $\dfrac{3}{4m^{-4}} = \dfrac{3m^4}{4}.$

(c) $\left(\dfrac{4}{5}\right)^{-2} = \left(\dfrac{5}{4}\right)^2 = \dfrac{25}{16}.$

(d) $\left(\dfrac{2x}{3y}\right)^{-3} = \left(\dfrac{3y}{2x}\right)^3 = \dfrac{27y^3}{8x^3}.$

(e) $\left(\dfrac{a+b}{a-b}\right)^{-1} = \dfrac{a-b}{a+b}.$

Summary of Exponent Rules

We summarize the results of this section as follows.

Expressions with a Common Base

If m and n are integers,

1. $a^n \cdot a^m = a^{n+m}$ 2. $\dfrac{a^n}{a^m} = a^{n-m}$ 3. $(a^m)^n = a^{m \cdot n}$

Expressions with a Common Exponent

If n is an integer,

1. $(ab)^n = a^n b^n$ 2. $\left(\dfrac{a}{b}\right)^n = \dfrac{a^n}{b^n}$

Zero and Negative Exponents

If a is any nonzero number and n is an integer, then:
- $a^0 = 1$
- $a^{-n} = \dfrac{1}{a^n}$

Common Mistakes

Be aware of the following notations that are sometimes confused:

$$ab^n = a(b^n), \qquad \text{but, in general, } ab^n \neq (ab)^n,$$
$$-b^n = -(b^n), \qquad \text{but, in general, } -b^n \neq (-b)^n,$$
$$-ab^n = (-a)(b^n).$$

For example, $-2^4 = -(2^4) = -16$, but $(-2)^4 = (-2)(-2)(-2)(-2) = 16$.

Example 14   Evaluate the following expressions for $x = -2$ and $y = 3$:

(a) $(xy)^4$ (b) $-xy^2$ (c) $(x+y)^2$
(d) x^y (e) $-4x^3$ (f) $-y^2$.

Solution   (a) $(-2 \cdot 3)^4 = (-6)^4 = (-6)(-6)(-6)(-6) = 1296$.
(b) $-(-2) \cdot (3)^2 = 2 \cdot 9 = 18$.
(c) $(-2+3)^2 = (1)^2 = 1$.
(d) $(-2)^3 = (-2)(-2)(-2) = -8$.
(e) $-4(-2)^3 = -4(-2)(-2)(-2) = 32$.
(f) $-(3)^2 = -9$.

Exercises and Problems for Section 6.1

EXERCISES

Evaluate the expressions in Exercises **1–12** without using a calculator.

1. $3 \cdot 2^3$

2. -3^2

3. $(-2)^3$

4. $5^1 \cdot 1^4 \cdot 3^2$

5. $5^2 \cdot 2^2$

6. $\dfrac{-1^3 \cdot (-3)^4}{9^2}$

7. $(-5)^3 \cdot (-2)^2$

8. $-5^3 \cdot -2^2$

9. $-1^4 \cdot (-3)^2 \left(-2^3\right)$

10. $\left(\dfrac{-4^3}{-2^3}\right)^2$

11. 3^0

12. 0^3

In Exercises **13–22**, evaluate the following expressions for $x = 2, y = -3$, and $z = -5$.

13. $-xyz$

14. y^x

15. $-y^x$

16. $\left(\dfrac{y}{z}\right)^x$

17. $\left(\dfrac{x}{z}\right)^{-y}$

18. x^{-z}

19. $-x^{-z}$

20. $\left(\dfrac{x^3 y}{2z}\right)^2$

21. $\left(\dfrac{3y}{2z}\right)^3$

22. $\left(\dfrac{x+y}{x-z}\right)^x$

In Exercises **23–31**, write the expression in the form x^n, assuming $x \neq 0$.

23. $x^3 \cdot x^5$

24. $\dfrac{x^3 \cdot x^4}{x^2}$

25. $(x^4 \cdot x)^2$

26. $\left(\dfrac{x^5}{x^2}\right)^4$

27. $\dfrac{x^5 \cdot x^3}{x^4 \cdot x^2}$

28. $\dfrac{x^7}{x^4} \cdot \dfrac{x^5}{x}$

29. $(x^3)^5$

30. $\dfrac{x \cdot x^6}{(x^3)^2}$

31. $\dfrac{\left(x^4 \cdot x^6\right)^2}{\left(x^2 \cdot x^3\right)^3}$

In Exercises **32–45**, write with a single exponent.

32. $4^2 \cdot 4^n$

33. $2^n 2^2$

34. $a^5 b^5$

35. $\dfrac{a^x}{b^x}$

36. $\dfrac{2^a 3^a}{6^b}$

37. $\dfrac{4^n}{2^m}$

38. $A^{n+3} B^n B^3$

39. $B^a B^{a+1}$

40. $(x^2 + y)^3 (x + y^2)^3$

41. $\left((x+y)^4\right)^5$

42. $16^2 y^8$

43. $\dfrac{(g+h)^6 (g+h)^5}{((g+h)^2)^4}$

44. $\left((a+b)^2\right)^5$

45. $\dfrac{(a+b)^5}{(a+b)^2}$

Without a calculator, decide whether the quantities in Exercises **46–59** are positive or negative.

46. $(-4)^3$

47. -4^3

48. $(-3)^4$

49. -3^4

50. $(-23)^{42}$

51. -31^{66}

52. 17^{-1}

53. $(-5)^{-2}$

54. -5^{-2}

55. $(-4)^{-3}$

56. $(-73)^0$

57. -48^0

58. $(-47)^{-15}$

59. $(-61)^{-42}$

In Exercises **60–72**, write each expression without parentheses. Assume all variables are positive.

60. $\left(\dfrac{a}{b}\right)^5$

61. $\left(\dfrac{c^3}{d}\right)^4$

62. $\left(\dfrac{2p}{q^3}\right)^5$

63. $\left(\dfrac{4r^2}{5s^4}\right)^3$

64. $\left(\dfrac{3}{w^4}\right)^4$

65. $\left(\dfrac{6g^5}{7h^7}\right)^2$

66. $(cf)^9$

67. $(2p)^5$

68. $\left(\dfrac{2}{3}\right)^4$

69. $\left(4b^2\right)^{2t}$

70. $3\left(10e^{3t}\right)^2$

71. $3\left(2^x e^x\right)^4$

72. $\left(3x^2\right)^{2n}$

PROBLEMS

In Problems **73–77**, decide which expressions are equivalent. Assume all variables are positive.

73. (a) 3^{-2} **(b)** $\dfrac{1}{3^{-2}}$ **(c)** $\dfrac{1}{3^2}$

    (d) $\left(\dfrac{1}{3}\right)^2$ **(e)** $\left(\dfrac{1}{3}\right)^{-2}$

74. (a) $\left(\dfrac{2}{3}\right)^{-n}$ **(b)** $\left(\dfrac{1}{\frac{2}{3}}\right)^n$ **(c)** $\left(\dfrac{3}{2}\right)^n$

    (d) $-\left(\dfrac{2}{3}\right)^n$ **(e)** $\dfrac{2^{-n}}{3^{-n}}$

75. (a) $\dfrac{1}{x^{-r}}$ **(b)** $\dfrac{1}{x^r}$ **(c)** $\left(\dfrac{1}{x}\right)^{-r}$

    (d) x^{-r} **(e)** $\dfrac{1}{\frac{1}{x^r}}$

76. (a) $\dfrac{1}{\left(\frac{r}{s}\right)^{-t}}$ **(b)** $\left(\dfrac{s}{r}\right)^{-t}$ **(c)** $\dfrac{1}{\left(\frac{r}{s}\right)^t}$

    (d) $\left(r^{-t}\right)\dfrac{1}{s^{-t}}$ **(e)** $\left(rs^{-1}\right)^t$

77. **(a)** $\left(\dfrac{p}{q}\right)^{-m}$ **(b)** $\left(\dfrac{\frac{1}{p}}{\frac{1}{q}}\right)^{m}$ **(c)** $\left(\dfrac{p^{-1}}{q^{-1}}\right)^{-m}$

   (d) $\left(p^{-1}q\right)^{-m}$ **(e)** $\left(\dfrac{\frac{1}{q^{-1}}}{\frac{1}{p^{-1}}}\right)^{-m}$

■ In Problems 78–91, write each expression as a product or a quotient. Assume all variables are positive.

78. 3^{2+3} 79. a^{4+1}

80. e^{2+r} 81. 10^{4-z}

82. k^{a-b} 83. 4^{p+3}

84. 6^{a-1} 85. $(-n)^{a+b}$

86. x^{a+b+1} 87. $p^{1-(a+b)}$

88. $(r-s)^{t+z}$ 89. $(p+q)^{a-b}$

90. $e^{t-1}(t+1)$ 91. $(x+1)^{ab+c}$

■ In Problems 92–98, write each expression as a power raised to a power. There may be more than one correct answer.

92. $4^{2\cdot4}$ 93. 2^{3x}

94. 5^{2y} 95. 3^{4a}

96. 5^{x^2} 97. $3e^{2t}$

98. $(x+3)^{2w}$

99. If $3^a = w$, express 3^{3a} in terms of w.

100. If $3^x = y$, express 3^{x+2} in terms of y.

101. If $4^b = c$, express 4^{b-3} in terms of c.

102. If $x^a = \dfrac{y}{z}$, $y = x^b$, and $z = x^c$, what is a?

6.2 FRACTIONAL EXPONENTS AND RADICAL EXPRESSIONS

A *radical expression* is an expression involving roots. For example, $\sqrt{a}$ is the positive number whose square is a. Thus, $\sqrt{9} = 3$ since $3^2 = 9$, and $\sqrt{625} = 25$ since $25^2 = 625$. Similarly, the cube root of a, written $\sqrt[3]{a}$, is the number whose cube is a. So $\sqrt[3]{64} = 4$, since $4^3 = 64$. In general, the nth root of a is the number b, such that $b^n = a$.

If $a < 0$, then the nth root of a exists when n is odd, but is not a real number when n is even. For example, $\sqrt[3]{-27} = -3$, because $(-3)^3 = -27$, but $\sqrt{-9}$ is not defined, because there is no real number whose square is -9.

Example 1   Evaluate:

   (a) $\sqrt{121}$ **(b)** $\sqrt[5]{-32}$ **(c)** $\sqrt{-4}$ **(d)** $\sqrt[4]{\dfrac{81}{625}}$

Solution   **(a)** Since $11^2 = 121$, we have $\sqrt{121} = 11$.
   (b) The fifth root of -32 is the number whose fifth power is -32. Since $(-2)^5 = -32$, we have $\sqrt[5]{-32} = -2$.
   (c) Since the square of a real number cannot be negative, $\sqrt{-4}$ does not exist.
   (d) Since
$$\left(\frac{3}{5}\right)^4 = \frac{81}{625}, \quad \text{we have} \quad \sqrt[4]{\frac{81}{625}} = \frac{3}{5}.$$

Using Fractional Exponents to Describe Roots

The laws of exponents suggest an exponential notation for roots involving fractional exponents. For instance, applying the exponent rules to the expression $a^{1/2}$, we get
$$(a^{1/2})^2 = a^{(1/2)\cdot2} = a^1 = a.$$

Thus, $a^{1/2}$ should be the number whose square is a, so we define

$$a^{1/2} = \sqrt{a}.$$

Similarly, we define

$$a^{1/3} = \sqrt[3]{a} \quad \text{and} \quad a^{1/n} = \sqrt[n]{a}.$$

The Exponent Laws Work for Fractional Exponents

The exponent laws on page 175 also work for fractional exponents.

Example 2   Evaluate

(a) $25^{1/2}$ (b) $9^{-1/2}$ (c) $8^{1/3}$ (d) $27^{-1/3}$.

Solution   (a) We have $25^{1/2} = \sqrt{25} = 5$.

(b) Using the rules about negative exponents, we have

$$9^{-1/2} = \frac{1}{9^{1/2}} = \frac{1}{\sqrt{9}} = \frac{1}{3}.$$

(c) Since $2^3 = 8$, we have $8^{1/3} = \sqrt[3]{8} = 2$.

(d) Since $3^3 = 27$, we have

$$27^{-1/3} = \frac{1}{27^{1/3}} = \frac{1}{\sqrt[3]{27}} = \frac{1}{3}.$$

Fractional Exponents with Numerators Other Than One

We can also use the exponent rule $(a^n)^m = a^{nm}$ to define the meaning of fractional exponents in which the numerator of the exponent is not 1.

Example 3   Find

(a) $64^{2/3}$ (b) $9^{-3/2}$ (c) $(-216)^{2/3}$ (d) $(-625)^{3/4}$

Solution   (a) Writing $2/3 = 2 \cdot (1/3)$ we have

$$64^{2/3} = 64^{2 \cdot (1/3)} = \left(64^2\right)^{1/3} = \sqrt[3]{64^2} = \sqrt[3]{4096} = 16.$$

We could also do this the other way around, and write $2/3 = (1/3) \cdot 2$:

$$64^{2/3} = 64^{(1/3) \cdot 2} = \left(64^{1/3}\right)^2 = \left(\sqrt[3]{64}\right)^2 = 4^2 = 16.$$

(b) We have

$$9^{-3/2} = \frac{1}{9^{3/2}} = \frac{1}{(\sqrt{9})^3} = \frac{1}{3^3} = \frac{1}{27}.$$

(c) To find $(-216)^{2/3}$, we can first evaluate $(-216)^{1/3} = -6$, and then square the result. This gives $(-216)^{2/3} = (-6)^2 = 36$.

(d) Writing $(-625)^{3/4} = \left((-625)^{1/4}\right)^3$, we conclude that $(-625)^{3/4}$ is not a real number since $(-625)^{1/4}$ is an even root of a negative number.

Example 4   Write each of the following as an equivalent expression in the form x^n and give the value for n.

(a) $\dfrac{1}{x^3}$

(b) $\sqrt[5]{x}$

(c) $(\sqrt[3]{x})^2$

(d) $\sqrt{x^5}$

(e) $\dfrac{1}{\sqrt[4]{x}}$

(f) $\left(\dfrac{1}{\sqrt{x}}\right)^3$

Solution

(a) We have
$$\frac{1}{x^3} = x^{-3} \quad \text{so} \quad n = -3.$$

(b) We have $\sqrt[5]{x} = x^{1/5}$, so $n = 1/5$.

(c) We have $(\sqrt[3]{x})^2 = (x^{1/3})^2$. According to the exponent laws we multiply the exponents in this expression, so $(x^{1/3})^2 = x^{2/3}$, so $n = 2/3$.

(d) We have $\sqrt{x^5} = (x^5)^{1/2} = x^{5/2}$, so $n = 5/2$.

(e) We have
$$\frac{1}{\sqrt[4]{x}} = \frac{1}{x^{1/4}} = x^{-1/4} \quad \text{so} \quad n = -1/4.$$

(f) We have
$$\left(\frac{1}{\sqrt{x}}\right)^3 = \left(\frac{1}{x^{1/2}}\right)^3 = (x^{-1/2})^3 = x^{-3/2} \quad \text{so} \quad n = -3/2.$$

Example 5   Simplify each expression, assuming all variables are positive.

(a) $\sqrt[3]{2}\,\sqrt[3]{4}$

(b) $\sqrt[3]{z^6 w^9}$

Solution

(a) Using the fact that $\sqrt[n]{a} \cdot \sqrt[n]{b} = \sqrt[n]{ab}$, we have $\sqrt[3]{2} \cdot \sqrt[3]{4} = \sqrt[3]{8} = 2$.

(b) We can write $\sqrt[3]{z^6 w^9}$ as $(z^6 w^9)^{1/3} = z^2 w^3$.

Working with Radical Expressions

Since roots are really fractional powers, our experience with integer powers leads us to expect that expressions involving roots of sums and differences will not be as easy to simplify as expressions involving roots of products and quotients.

Example 6   Do the expressions have the same value?

(a) $\sqrt{9 + 16}$ and $\sqrt{9} + \sqrt{16}$

(b) $\sqrt{100 - 64}$ and $\sqrt{100} - \sqrt{64}$

(c) $\sqrt{(9)(4)}$ and $\sqrt{9} \cdot \sqrt{4}$

(d) $\sqrt{\dfrac{100}{4}}$ and $\dfrac{\sqrt{100}}{\sqrt{4}}$

Solution

(a) We have $\sqrt{9 + 16} = \sqrt{25} = 5$, but $\sqrt{9} + \sqrt{16} = 3 + 4 = 7$, so the expressions have different values. In general, $\sqrt{a + b} \neq \sqrt{a} + \sqrt{b}$.

(b) Here, $\sqrt{100 - 64} = \sqrt{36} = 6$. However, $\sqrt{100} - \sqrt{64} = 10 - 8 = 2$. In general, $\sqrt{a - b} \neq \sqrt{a} - \sqrt{b}$.

(c) We have $\sqrt{(9)(4)} = \sqrt{36} = 6$. Also, $\sqrt{9} \cdot \sqrt{4} = 3 \cdot 2 = 6$. In general, the square root of a product is equal to the product of the square roots, or $\sqrt{ab} = \sqrt{a} \cdot \sqrt{b}$.

(d) We have $\sqrt{\dfrac{100}{4}} = \sqrt{25} = 5$. Also, $\dfrac{\sqrt{100}}{\sqrt{4}} = \dfrac{10}{2} = 5$. In general, $\sqrt{\dfrac{a}{b}} = \dfrac{\sqrt{a}}{\sqrt{b}}$.

Simplifying Radical Expressions That Contain Sums and Differences

Although the laws of exponents do not tell us how to simplify expressions involving roots of sums and differences, there are other methods that work. Again, we start with a numerical example.

Example 7

Are the expressions equivalent?

(a) $\sqrt{100} + \sqrt{100}$ and $\sqrt{200}$

(b) $\sqrt{100} + \sqrt{100}$ and $2\sqrt{100}$

Solution

(a) We know that $\sqrt{100} + \sqrt{100} = 10 + 10 = 20$. But, the square root of 200 cannot be 20, since $20^2 = 400$, so they are not equivalent.

(b) They are equivalent, since $\sqrt{100} + \sqrt{100} = 10 + 10 = 20$ and $2\sqrt{100} - 2(10) = 20$. Another way to see this without evaluating is to use the distributive law to factor out the common term $\sqrt{100}$:

$$\sqrt{100} + \sqrt{100} = \sqrt{100}(1 + 1) = 2\sqrt{100}.$$

Expressing roots as fractional powers helps determine the principles for combining like terms in expressions involving roots. The principles are the same as for integer powers: we can combine terms involving the same base and the same exponent.

Example 8

Combine the radicals if possible.

(a) $3\sqrt{7} + 6\sqrt{7}$

(b) $4\sqrt{x} + 3\sqrt{x} - \sqrt{x}$

(c) $9\sqrt[3]{2} - 5\sqrt[3]{2} + \sqrt{2}$

(d) $-5\sqrt{5} + 8\sqrt{20}$

(e) $5\sqrt{4x^3} + 3x\sqrt{36x}$

Solution

(a) We have $3\sqrt{7} + 6\sqrt{7} = 9\sqrt{7}$.

(b) We have $4\sqrt{x} + 3\sqrt{x} - \sqrt{x} = 6\sqrt{x}$.

(c) Expressing the roots in exponential notation, we see that not all the exponents are the same:

$$9\sqrt[3]{2} - 5\sqrt[3]{2} + \sqrt{2} = 9 \cdot 2^{1/3} - 5 \cdot 2^{1/3} + 2^{1/2} = 4 \cdot 2^{1/3} + 2^{1/2} = 4\sqrt[3]{2} + \sqrt{2}.$$

(d) Here there is a term involving $\sqrt{5} = 5^{1/2}$ and a term involving $\sqrt{20} = 20^{1/2}$. Thus the exponents are the same in both terms, but the bases are different. However, $8\sqrt{20} = 8\sqrt{(4)(5)} = 8\sqrt{4}\sqrt{5} = 8 \cdot 2\sqrt{5} = 16\sqrt{5}$. Therefore,

$$-5\sqrt{5} + 8\sqrt{20} = -5\sqrt{5} + 16\sqrt{5} = 11\sqrt{5}.$$

(e) Expressing each term in exponential form, we get

$$5\sqrt{4x^3} + 3x\sqrt{36x} = 5(4x^3)^{1/2} + 3x(36x)^{1/2} = 5 \cdot 4^{1/2}(x^3)^{1/2} + 3x(36)^{1/2}x^{1/2}$$
$$= 10x^{3/2} + 18x^{3/2} = 28x^{3/2}.$$

Rationalizing the Denominator

In Section 2.3 we saw how to simplify fractions by taking a common factor out of the numerator and denominator. Sometimes it works the other way around: we can simplify a fraction with a radical expression in the denominator, such as

$$\frac{\sqrt{2}}{2+\sqrt{2}},$$

by multiplying the numerator and denominator by a carefully chosen common factor.

Example 9   Simplify $\dfrac{\sqrt{2}}{2+\sqrt{2}}$ by multiplying the numerator and denominator by $2-\sqrt{2}$.

Solution   You might think that multiplying $2+\sqrt{2}$ by $2-\sqrt{2}$ will result in a complicated expression, but in fact the result is very simple because the product is a difference of squares, and squaring undoes the square root:

$$(2+\sqrt{2})(2-\sqrt{2}) = 2^2 - (\sqrt{2})^2 = 4 - 2 = 2.$$

So multiplying the numerator and denominator of $\dfrac{\sqrt{2}}{2+\sqrt{2}}$ by $2-\sqrt{2}$ gives

$$\frac{\sqrt{2}}{2+\sqrt{2}} \cdot \frac{2-\sqrt{2}}{2-\sqrt{2}} = \frac{2\sqrt{2}-2}{2} = \sqrt{2}-1.$$

The process in Example 9 is called *rationalizing the denominator*. In general, to rationalize a sum of two terms, one or more of which is a radical, we multiply it by the sum obtained by changing the sign of one of the radicals. The resulting sum is a *conjugate* of the original sum.

Example 10   Multiply each expression by a conjugate.

(a) $1+\sqrt{2}$ (b) $-6-2\sqrt{7}$ (c) $3\sqrt{5}-2$ (d) $\sqrt{3}+3\sqrt{6}$

Solution   (a) The conjugate of $1+\sqrt{2}$ is $1-\sqrt{2}$, and the product is

$$(1+\sqrt{2})(1-\sqrt{2}) = 1^2 - (\sqrt{2})^2 = 1 - 2 = -1.$$

(b) The conjugate of $-6-2\sqrt{7}$ is $-6+2\sqrt{7}$, and the product is

$$(-6-2\sqrt{7})(-6+2\sqrt{7}) = (-6)^2 - (2\sqrt{7})^2 = 36 - 28 = 8.$$

(c) The conjugate of $3\sqrt{5}-2$ is $-3\sqrt{5}-2$. We rewrite the product so it is easier to see the difference of two squares:

$$(3\sqrt{5}-2)(-3\sqrt{5}-2) = (-2+3\sqrt{5})(-2-3\sqrt{5}) = (-2)^2 - (3\sqrt{5})^2 = 4 - 45 = -41.$$

(d) Here there are two radicals, so there are two possible conjugates, $\sqrt{3}-3\sqrt{6}$ and $-\sqrt{3}+3\sqrt{6}$. We choose the first one:

$$(\sqrt{3}+3\sqrt{6})(\sqrt{3}-3\sqrt{6}) = (\sqrt{3})^2 - (3\sqrt{6})^2 = 3 - 54 = -51.$$

Exercises and Problems for Section 6.2

EXERCISES

■Evaluate the expressions in Exercises 1–4 without using a calculator.

1. $4^{1/2}$

2. $25^{-1/2}$

3. $\left(\dfrac{4}{9}\right)^{-1/2}$

4. $\left(\dfrac{64}{27}\right)^{-1/3}$

■In Exercises 5–6, write each expression without parentheses. Assume all variables are positive.

5. $\left(2^{4x}5^{4x}\right)^{1/2}$

6. $\left(\dfrac{10^{6a}}{5^{6a}}\right)^{1/3}$

■Simplify the expressions in Exercises 7–10, assuming all variables are positive.

7. $\sqrt[5]{\dfrac{x^{10}}{y^5}}$

8. $\sqrt[4]{4x^3}\,\sqrt[4]{4x^5}$

9. $\sqrt{48a^3b^7}$

10. $\dfrac{\sqrt[3]{96x^7y^8}}{\sqrt[3]{3x^4y}}$

■In Exercises 11–16, write the expression as an equivalent expression in the form x^n and give the value for n.

11. $\dfrac{1}{\sqrt{x}}$

12. $\dfrac{1}{x^5}$

13. $\sqrt{x^3}$

14. $\left(\sqrt[3]{x}\right)^5$

15. $1/(1/x^{-2})$

16. $\dfrac{(x^3)^2}{(x^2)^3}$

■In Exercises 17–26, combine radicals, if possible.

17. $12\sqrt{11} - 3\sqrt{11} + \sqrt{11}$

18. $5\sqrt{9} - 2\sqrt{144}$

19. $2\sqrt{3} + \dfrac{\sqrt{3}}{2}$

20. $\sqrt[3]{4x} + 6\sqrt[3]{4x} - 2\sqrt[3]{4x}$

21. $8\sqrt[3]{3} - 2\sqrt[3]{3} - 2\sqrt{3}$

22. $-6\sqrt{98} + 4\sqrt{8}$

23. $6\sqrt{48a^4} + 2a\sqrt{27a^2} - 3a^2\sqrt{75}$

24. $5\sqrt{12t^3} + 2t\sqrt{128t} - 3t\sqrt{48t}$

25. $\dfrac{\sqrt{45}}{5} - \dfrac{2\sqrt{20}}{5} + \dfrac{\sqrt{80}}{\sqrt{25}}$

26. $4xy\sqrt{90xy} + 2\sqrt{40x^3y^3} - 3xy\sqrt{50xy}$

■In Exercises 27–32, find a conjugate of each expression and the product of the expression with the conjugate.

27. $4 + \sqrt{6}$

28. $\sqrt{13} - 10$

29. $-\sqrt{5} - \sqrt{6}$

30. $7\sqrt{2} - 2\sqrt{7}$

31. $b\sqrt{a} + a\sqrt{b}$

32. $1 - \sqrt{r+1}$

■In Exercises 33–37, rewrite each expression by rationalizing the denominator.

33. $\dfrac{2}{\sqrt{3}+1}$

34. $\dfrac{\sqrt{5}}{5 - \sqrt{5}}$

35. $\dfrac{10}{\sqrt{6}-1}$

36. $\dfrac{\sqrt{3}}{3\sqrt{2}+\sqrt{3}}$

37. $\dfrac{\sqrt{6}+\sqrt{2}}{\sqrt{6}-\sqrt{2}}$

PROBLEMS

38. The surface area (not including the base) of a right circular cone of radius r and height $h > 0$ is given by

$$\pi r \sqrt{r^2 + h^2}.$$

Explain why the surface area is always greater than πr^2

(a) In terms of the structure of the expression.
(b) In terms of geometry.

■By giving specific values for $a, b,$ and c, explain how the exponent rule

$$\left(a^b\right)^c = a^{bc}$$

is used to rewrite the expressions in Problems 39–40.

39. $\left(2m^2n^4\right)^{3r+3} = \left(8m^6n^{12}\right)^{r+1}$

40. $\sqrt{(x+1)(x^2+2x+1)} = (x+1)^{3/2}$

REVIEW EXERCISES AND PROBLEMS FOR CHAPTER 6

EXERCISES

■ Evaluate the expressions in Exercises 1–6 without using a calculator.

1. $-2^3 \cdot -3^2$

2. $(-2)^3 \cdot (-3)^2$

3. $-4^3 \cdot -1^5$

4. $-2(-3)^4$

5. 5^{-2}

6. 9^{-1}

■ In Exercises 7–13, write with a single exponent.

7. $\dfrac{a^b}{3^b}$

8. $\dfrac{9^2}{q^4}$

9. $\dfrac{64x^3 y^6}{y^3}$

10. $27^2 (x^2)^3$

11. $(a+b)^2 (a+b)^5$

12. $\left((p+q)^3\right)^4 \left((r+s)^6\right)^2$

13. $(x+y+z)^{21}(u+v+w)^{21}$

■ Without a calculator, decide whether the quantities in Exercises 14–19 are positive or negative.

14. -16^{33}

15. $(-12)^{15}$

16. 14^{-5}

17. -7^{-3}

18. $(-44)^0$

19. $(-20)^{-13}$

■ Evaluate the expressions in Exercises 20–24 using $r = -1$ and $s = 7$.

20. $(5r)^{-3}$

21. $-\dfrac{rs^{-2}}{5^0}$

22. -5^r

23. $2s^r$

24. $(174s^4 r^{12})^0$

■ In Exercises 25–30, write each expression without parentheses. Assume all variables are positive.

25. $\left(\dfrac{k}{g}\right)^8$

26. $(5a^5)^2$

27. $\left(\dfrac{r}{s^4}\right)^5$

28. $(6m^5 n^3)^3$

29. $-5(ab^2)^6$

30. $\left(\dfrac{3p^3}{-2q^2}\right)^4$

■ In Exercises 31–50, write each expression with only positive exponents. Assume all variables are positive.

31. $\dfrac{6}{x^{-5}}$

32. $\dfrac{k^{-3}}{5}$

33. $9d^{-2}$

34. $\dfrac{3m^{-3}}{4}$

35. $\left(\dfrac{r}{s}\right)^{-n}$

36. $\left(\dfrac{3x}{y^2}\right)^{-4}$

37. $\dfrac{x^a x^b}{2x^{-1}}$

38. $\left(\dfrac{a^3}{b^{-2}}\right)^{-2}$

39. $\dfrac{4c^{-4}}{6c^{-6}}$

40. $(-2a^3)^{-2}$

41. $-5\left(x^4 y^{-2}\right)^{-3}$

42. $\dfrac{3^2 x^4 y^{-3}}{3^{-1} x^{-4} y^{-3}}$

43. $\dfrac{(2bc^2)^2}{2(bc)^{-1}}$

44. $\dfrac{3\left(p^2 r^{-3}\right)^{-1}}{p^{-3}}$

45. $\dfrac{x\left(a^4\right)^{1/2}}{(ax^{-1})^3}$

46. $\left(27x^{-9} y^6\right)^{1/3}$

47. $\dfrac{3(cx)^{-2}}{(3c)^3 x^{-4}}$

48. $4\left(\dfrac{a}{b}\right)^{-1}\left(\dfrac{b}{a^2}\right)^{-2}$

49. $\dfrac{5^{3t+2} \cdot 5^{1-2t}}{5^{t+4}}$

50. $\dfrac{4^{-1}\left(3p^2 q^{-3}\right)^0}{r^{-4}}$

■ In Exercises 51–54, find a conjugate of each expression and the product of the expression with the conjugate.

51. $-2\sqrt{3} + \sqrt{5}$

52. $\sqrt{m} - 4$

53. $\sqrt{4+h} - 2$

54. $\sqrt{x+h} + \sqrt{x-h}$

■ In Exercises 55–58, rewrite each expression by rationalizing the denominator.

55. $\dfrac{\sqrt{a}}{\sqrt{a}+b}$

56. $\dfrac{\sqrt{pq}}{\sqrt{p}+\sqrt{q}}$

57. $\dfrac{1-x}{1-\sqrt{x}}$

58. $\dfrac{r-4}{\sqrt{r-3}-1}$

PROBLEMS

■ If $a > 0, b > 0, c < 0$, and $d < 0$, decide whether the quantities in Problems **59–66** are positive or negative.

59. $(ab)^5$

60. $-c^3$

61. $(bc)^{21}$

62. $(-abc)^{17}$

63. $(-c)^3$

64. $-(bc)^{21}$

65. $\dfrac{(-ac)^5}{-(bd)^2}$

66. $\dfrac{(-ab)^5}{-(cd)^2}$

■ In Problems **67–70**, write each expression as a product or a quotient. Assume all variables are positive.

67. $(a + b)^{(2+s)-r}$

68. $2(m + n)^{a+3}$

69. $p^{(2+t)}(p + q)^{2-t}$

70. $(x + y)^{2a-1}$

Chapter 7

Power Functions, Expressions, and Equations

CONTENTS

7.1 POWER FUNCTIONS

The area of a circle, A, is a function of its radius, r, given by

$$A = \pi r^2.$$

Here A is proportional to r^2, with constant of proportionality π. In general,

---

A **power function** is a function that can be given by

$$f(x) = kx^p, \qquad \text{for constants } k \text{ and } p.$$

We call k the coefficient and p the exponent.

Example 1   The braking distance in feet of an Alfa Romeo on dry road is proportional to the square of its speed, v mph, at the time the brakes are applied, and is given by the power function

$$f(v) = 0.036v^2.$$

(a) Identify the constants k and p.
(b) Find the braking distances of an Alfa Romeo going at 35 mph and at 140 mph (its top speed). Which braking distance is larger? Explain your answer in algebraic and practical terms.

Solution   (a) This is a power function $f(v) = kv^p$ with $k = 0.036$ and $p = 2$.
(b) At 35 mph the braking distance is

$$f(35) = 0.036 \cdot 35^2 = 44 \text{ ft}.$$

At 140 mph it is

$$f(140) = 0.036 \cdot 140^2 = 706 \text{ ft}.$$

This is larger than the braking distance for 35 mph; it is almost as long as two football fields. This makes sense algebraically, because $140 > 35$, so $140^2 > 35^2$. It also makes sense in practical terms, since a car takes longer to stop if it is going faster. See Figure 7.1.

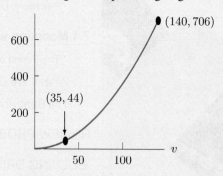

Figure 7.1: Braking distance of an Alfa Romeo

Interpreting the Coefficient

If we want to compare two power functions with the same exponent, we look at the coefficient.

Example 2

The braking distance of an Alfa Romeo on wet road going at v mph is given by

$$g(v) = 0.045v^2.$$

Find the braking distance at 35 mph. Is it greater than or less than the braking distance on dry road, from Example 1? Explain your answer in algebraic and practical terms.

Solution

The braking distance on wet road at 35 mph is

$$g(35) = 0.045(35)^2 = 55 \text{ ft}.$$

This is greater than the stopping distance of 44 ft at the same speed on dry road, from Example 1. This makes sense algebraically, since

$$0.045 > 0.036, \quad \text{so} \quad 0.045(35^2) > 0.036(35^2).$$

It also makes sense in practical terms, since it takes a greater distance to stop on a wet road than on a dry one.

Power functions arise in geometry, where the exponent is related to the dimension of the object.

Example 3

The solid objects in Figure 7.2 have volume, V, in cm^3, given by

$$\text{Cube:} \qquad V = w^3$$

$$\text{Sphere:} \qquad V = \frac{\pi}{6}w^3$$

$$\text{Cylinder:} \qquad V = \frac{\pi}{4}w^3.$$

(a) What is the exponent?
(b) Which has the largest coefficient? Which the smallest?

Interpret your answers in practical terms.

 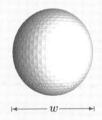

Figure 7.2: Three solid objects

Solution   (a) Each expression has the form $V = kw^3$, so the exponent is 3. This makes sense since w is a length, measured in cm, and V is a volume, measured in cm^3.

(b) For the cube we have $V = w^3 = 1 \cdot w^3$, so $k = 1$. For the sphere we have $k = \pi/6$ and for the cylinder we have $k = \pi/4$. Since $\pi/6 \approx 0.5$ and $\pi/4 \approx 0.8$, the cube has the largest coefficient and the sphere has the smallest. This makes sense, since the cube has the largest volume and the sphere has the smallest. The sphere can fit inside the cylinder, which in turn can fit inside the cube.

Comparing Different Powers

Figures 7.3 and 7.4 show comparisons of the first three powers, x, x^2, and x^3. Notice that all the graphs pass through the points $(0, 0)$ and $(1, 1)$, because 0 to any power is 0 and 1 to any power is 1. What is the relationship between different powers when x is not 0 or 1?

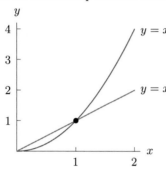

Figure 7.3: Comparison of x^2 with x

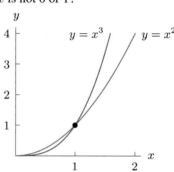

Figure 7.4: Comparison of x^3 with x^2

Example 4   In Figure 7.3, the graph of $y = x^2$ is above the graph of $y = x$ when x is greater than 1, and it is below it when x is between 0 and 1.

(a) Express these facts algebraically using inequalities.

(b) Explain why they are true.

Solution   (a) If one graph is above another then the y-values on the first are greater than the corresponding y-values on the second. So Figure 7.3 suggests that

$$x^2 > x \quad \text{when} \quad x > 1,$$
$$x^2 < x \quad \text{when} \quad 0 < x < 1.$$

(b) We write $x^2 = x \cdot x$ and $x = x \cdot 1$. So we want to compare

$$x \cdot x \quad \text{and} \quad x \cdot 1.$$

On the left we have x multiplied by itself, and on the right we have x multiplied by 1. So if $x > 1$ the quantity on the left is larger, and if $x < 1$ (and positive) then the quantity on the right is larger.

Similarly, Figure 7.4 illustrates that $x^3 > x^2$ when $x > 1$ and $x^3 < x^2$ when $0 < x < 1$. In general,

- When $x > 1$, a higher power of x is greater than a lower power.
- When $0 < x < 1$, a higher power of x is less than a lower power.

Powers with Coefficients

Example 5   Which of the expressions
$$6x^2 \quad \text{or} \quad x^3$$

is larger if

(a) $x = 2$ (b) $x = 5$ (c) $x = 10$ (d) $x = 100$.

Solution   (a) When $x = 2$, the expression $6x^2$ is $6 \cdot 2 \cdot 2$ and the expression x^3 is $2 \cdot 2 \cdot 2$. Since 6 is larger than 2, we have

$$6 \cdot 2 \cdot 2 > 2 \cdot 2 \cdot 2$$
$$6 \cdot 2^2 > 2^3.$$

When $x = 2$, we have $6x^2 > x^3$ and the expression $6x^2$ is larger.

(b) Similarly, when $x = 5$, we have

$$6 \cdot 5 \cdot 5 > 5 \cdot 5 \cdot 5.$$

When $x = 5$, we have $6x^2 > x^3$ and the expression $6x^2$ is larger.

(c) Since $6 < 10$, we have

$$6 \cdot 10 \cdot 10 < 10 \cdot 10 \cdot 10.$$

When $x = 10$, we have $6x^2 < x^3$ and the expression x^3 is larger.

(d) Since $6 < 100$, we have

$$6 \cdot 100 \cdot 100 < 100 \cdot 100 \cdot 100.$$

When $x = 100$, we have $6x^2 < x^3$ and the expression x^3 is larger.

Example 6   We see in Example 5 that for small values of x, the expression $6x^2$ is larger and for large values of x the expression x^3 is larger. At what value of x does x^3 become larger than $6x^2$?

Solution   We have
$$x \cdot x \cdot x > 6 \cdot x \cdot x \quad \text{when} \quad x > 6.$$

We see that

$$x^3 > 6x^2 \quad \text{if} \quad x > 6,$$
$$x^3 < 6x^2 \quad \text{if} \quad x < 6.$$

A higher power of x will always eventually be larger than a lower power of x. The value of x at which it becomes larger depends on the expressions and the coefficients.

> When x is large enough, a power function with a higher power of x is greater than a power function with a lower power.

Example 7

(a) For large values of x, is $2x^3$ or $100x^2$ larger?
(b) At what value of x does it become larger?
(c) Use a graph of $y = 2x^3$ and $y = 100x^2$ to illustrate this relationship.

Solution

(a) Since the exponent 3 is greater than the exponent 2, we know that $2x^3$ is larger for large values of x.
(b) Since $2x^3 = 2 \cdot x \cdot x \cdot x$ and $100x^2 = 100 \cdot x \cdot x$, we want to know what value of x makes

$$2 \cdot x \cdot x \cdot x > 100 \cdot x \cdot x.$$
$$\text{We want} \quad 2 \cdot x > 100,$$
$$\text{so} \quad x > 50.$$

For all values greater than 50, the expression $2x^3$ will be larger and for all values less than 50, the expression $100x^2$ will be larger.

(c) See Figure 7.5.

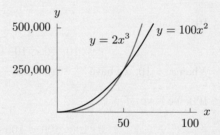

Figure 7.5: Comparison of $2x^3$ with $100x^2$

Figure 7.5 looks very similar to Figure 7.4. In both cases, the expression with the higher power becomes larger than the expression with the lower power. This general rule always holds. Coefficients determine where the graphs intersect, but not the overall behavior of the functions.

Powers of Negative Numbers

When x is negative, we have

$$x^2 = x \cdot x = \text{Negative} \cdot \text{Negative} = \text{Positive}.$$

Since x^2 is also positive when x is positive, we see that x^2 is always positive, except when $x = 0$. On the other hand, if x is negative,

$$x^3 = x \cdot x \cdot x = \text{Negative} \cdot \text{Negative} \cdot \text{Negative} = \text{Negative} \cdot \text{Positive} = \text{Negative}.$$

In general, if you multiply a negative number by itself an even number of times, you get a positive number, since the factors can be paired up, and if you multiply it by itself an odd number of times, you get a negative number. See Figures 7.6 and 7.7. Note that $y = x^2$ and $y = x^4$ are entirely above the x-axis, except at $x = 0$, whereas $y = x^3$ and $y = x^5$ are below the axis when x is negative.

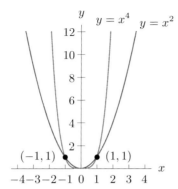

Figure 7.6: Graphs of positive even powers of x

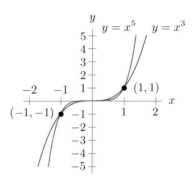

Figure 7.7: Graphs of positive odd powers of x

Zero and Negative Powers

In Example 1 we use the fact that squaring a larger number gives a larger value. In the next example we consider what happens when the exponent p is negative.

Example 8   The weight, w, in pounds, of an astronaut r thousand miles from the center of the earth is given by

$$w = 2880r^{-2}.$$

(a) Is this a power function? Identify the constants k and p.
(b) How much does the astronaut weigh at the earth's surface, 4000 miles from the center? How much does the astronaut weigh 1000 miles above the earth's surface? Which weight is smaller?

Solution   (a) This is a power function $w = kr^p$ with $k = 2880$ and $p = -2$.
(b) At the earth's surface we have $r = 4$, since r is measured in thousands of miles, so

$$w = 2880 \cdot 4^{-2} = \frac{2880}{4^2} = 180 \text{ lb.}$$

So the astronaut weighs 180 lb at the earth's surface. At 1000 miles above the earth's surface we have $r = 5$, so

$$w = 2880 \cdot 5^{-2} = \frac{2880}{5^2} = 115.2 \text{ lb.}$$

So the astronaut weighs about 115 lb when she is 1000 miles above the earth's surface. Notice that this is smaller than her weight at the surface. For this function, larger input values give smaller output values.

Example 9   Rewrite the expression for the function in Example 8 in a form that explains why larger inputs give smaller outputs.

Solution   If we write

$$w = 2880r^{-2} = \frac{2880}{r^2},$$

we see that the r^2 is in the denominator, so inputting a larger value of r means dividing by a larger number, which results in a smaller output. This can be seen from the graph in Figure 7.8, which shows the function values decreasing as you move from left to right.

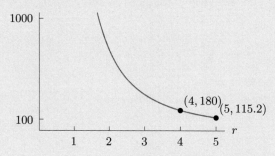

Figure 7.8: Weight of an astronaut

Inverse Proportionality

We often describe a power function $y = kx^p$ by saying that y is proportional to x^p. When the exponent is negative, there is another way of describing the function that comes from expressing it as a fraction. For example, if

$$y = 2x^{-3} = \frac{2}{x^3},$$

we say that y is *inversely proportional* to x^3. In general, if

$$y = kx^{-n} = \frac{k}{x^n}, \quad n \text{ positive},$$

we say y is inversely proportional to x^n.

Graphs of Negative Integer Powers

In Example 9 we saw that the value of $w = 2880r^{-2}$ gets smaller as r gets larger. This is true in general for negative powers. See Table 7.1, and Figures 7.9 and 7.10.

Table 7.1   *Negative powers of x get small as x gets large*

x	0	10	20	30	40	50
$x^{-1} = 1/x$	Undefined	0.1	0.05	0.033	0.025	0.02
$x^{-2} = 1/x^2$	Undefined	0.01	0.0025	0.0011	0.0006	0.0004

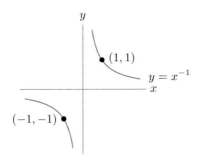

Figure 7.9: Graph of $y = x^{-1} = 1/x$

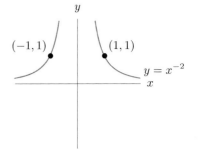

Figure 7.10: Graph of $y = x^{-2} = 1/x^2$

Figures 7.9 and 7.10 also illustrate what happens when $x < 0$. Just as with positive exponents, the graph is above the x axis if the exponent is even and below it if the exponent is odd.

The Special Cases $p = 0$ and $p = 1$

In general a power function is not linear. But if $p = 0$ or $p = 1$, then the power functions are linear. For example, the graph of $y = x^0 = 1$ is a horizontal line through the point $(1, 1)$. The graph of $y = x^1 = x$ is a line through the origin with slope 1. See Figures 7.11 and 7.12.

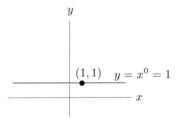

Figure 7.11: Graph of $y = x^0 = 1$

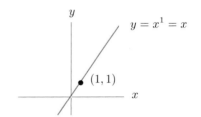

Figure 7.12: Graph of $y = x^1 = x$

Fractional Powers

Power functions where the exponent p is not an integer arise in biological applications.

Example 10   For a certain species of animal, the bone length, L, in cm, is given by

$$L = 7\sqrt{A},$$

where A is the cross-sectional area of the bone in cm^2.

(a) Is L a power function of A? If so, identify the values of k and p.
(b) Find the bone lengths for $A = 36$ cm^2 and $A = 100$ cm^2. Which bone length is larger? Explain your answer in algebraic terms.

Solution   (a) Since $\sqrt{A} = A^{1/2}$, we have

$$L = 7A^{1/2}.$$

So this is a power function of A with $k = 7$ and $p = 1/2$. See Figure 7.13.

(b) For $A = 36$ cm^2, the bone length is

$$L = 7\sqrt{36} = 42 \text{ cm}.$$

For $A = 100$ cm^2, the bone length is

$$L = 7\sqrt{100} = 70 \text{ cm}.$$

The bone length for $A = 100$ cm^2 is larger. This makes sense algebraically, because $100 > 36$, so $\sqrt{100} > \sqrt{36}$.

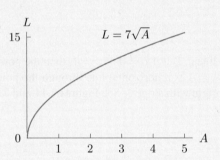

Figure 7.13: Bone length as a function of cross-sectional area

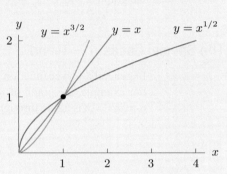

Figure 7.14: Comparison of $y = x^{1/2}$, $y = x$, and $y = x^{3/2}$

Comparing Different Fractional Powers

We can compare the size of fractional powers using the same ideas we used for integer powers. As before, the relative size of powers of x depends on whether x is greater than or less than 1. The box on page 189 tells us that if $x > 1$, a higher power of x is greater than a lower power. Figure 7.14 shows us that this result holds for fractional exponents as well. Notice that the function $y = x$ in Figure 7.14 is the power function $y = x^1$. We see that when $x > 1$, we have $x^{3/2} > x^1$ and $x^1 > x^{1/2}$. In each case, the higher power of x is greater than the lower power. For $0 < x < 1$, the situation is reversed, and the higher power of x is less than the lower power.

Example 11   For each pair of numbers, use the exponents to determine which is greater.

(a)   3 or $3^{1/2}$ (b)   0.3 or $(0.3)^{1/2}$

Solution   (a) Since the base 3 is larger than 1, we know that a higher power is bigger than a lower power. The two exponents are 1 and $1/2$ and 1 is larger than $1/2$, so we have $3^1 > 3^{1/2}$. The number 3 is greater than the number $3^{1/2}$.

(b) Since the base 0.3 is less than 1, the situation is reversed, and a higher power is smaller than a lower power. Since the exponent 1 is larger than the exponent $1/2$, we have $(0.3)^1 < (0.3)^{1/2}$. The number $(0.3)^{1/2}$ is greater than the number 0.3.

Figure 7.14 illustrates these calculations. It shows that the graph of $y = x^{1/2}$ lies above the graph of $y = x$ when $0 < x < 1$, and below it when $x > 1$.

Example 12   For each pair of numbers, use the exponents to determine which is greater.

(a)   3 or $3^{3/2}$ (b)   0.3 or $(0.3)^{3/2}$

Solution   (a) Since the base 3 is larger than 1, we know that a higher power is bigger than a lower power. The two exponents are 1 and 3/2 and 3/2 is larger than 1, so we have $3^{3/2} > 3^1$. The number $3^{3/2}$ is greater than the number 3^1.

(b) Since the base 0.3 is less than 1, the situation is reversed and a higher power is smaller than a lower power. Since the exponent 3/2 is larger than the exponent 1, we have $(0.3)^{3/2} < (0.3)^1$. The number 0.3 is greater than the number $(0.3)^{3/2}$.

Again, Figure 7.14 illustrates these calculations. It shows that the graph of $y = x^{3/2}$ lies below the graph of $y = x$ when $0 < x < 1$, and above it when $x > 1$.

Exercises and Problems for Section 7.1

EXERCISES

In Exercises 1–5, identify the exponent and the coefficient for each power function.

1. The area of a square of side x is $A = x^2$.

2. The perimeter of a square of side x is $P = 4x$.

3. The side of a cube of volume V is $x = \sqrt[3]{V}$.

4. The circumference of a circle of radius r is $C = 2\pi r$.

5. The surface area of a sphere of radius r is $S = 4\pi r^2$.

6. The area, A, of a rectangle whose length is 3 times its width is given by $A = 3w^2$, where w is its width.

   (a) Identify the coefficient and exponent of this power function.
   (b) If the width is 5 cm, what is the area of the rectangle?

7. The volume, V, of a cylinder whose radius is 5 times its height is given by $V = \frac{1}{5}\pi r^3$, where r is the radius.

   (a) Identify the coefficient and exponent of this power function.
   (b) If the radius is 2 cm, what is the volume?
   (c) If the height is 0.8 cm, what is the volume?

8. A ball dropped into a hole reaches a depth $d = 4.9t^2$ meters, where t is the time in seconds since it was dropped.

   (a) Identify the coefficient and exponent of this power function.
   (b) How deep is the ball after 2 seconds?
   (c) If the ball hits the bottom of the hole after 4 seconds, how deep is the hole?

For the graphs of power functions $f(x) = kx^p$ in Exercises 9–13, is

(a) $p > 1$?   (b) $p = 1$?   (c) $0 < p < 1$?
(d) $p = 0$?   (e) $p < 0$?

9.

10.
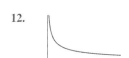

11.

12.

13.

In Exercises 14–17,

(a) Is y proportional, or is it inversely proportional, to a positive power of x?

(b) Make a table of values showing corresponding values for y when x is 1, 10, 100, and 1000.

(c) Use your table to determine whether y increases or decreases as x gets larger.

14. $y = 2x^2$ 15. $y = 3\sqrt{x}$

16. $y = \dfrac{1}{x}$ 17. $y = \dfrac{5}{x^2}$

PROBLEMS

■Problems **18–20** describe power functions of the form $y = kx^p$, for p an integer. Is p

(a) Even or odd?   **(b)** Positive or negative?

18.   **19.**

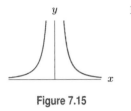

Figure 7.15

Table 7.2

x	y
-2	-24
-1	-3
0	0
1	3
2	24

20. The graph of $y = f(x)$ gets closer to the x-axis as x gets large. For $x < 0$, $y < 0$, and for $x > 0$, $y > 0$.

■In Problems **21–24**, a and b are positive constants. If $a > b$ then which is larger?

21. a^4, b^4

22. $a^{1/4}, b^{1/4}$

23. a^{-4}, b^{-4}

24. $a^{-1/4}, b^{-1/4}$

25. (a) In Figure 7.7, for which x-values is the graph of $y = x^5$ above the graph of $y = x^3$, and for which x-values is it below?

(b) Express your answers in part (a) algebraically using inequalities.

26. (a) In Figure 7.16, for which x-values is the graph of $y = x^{1/2}$ above the graph of $y = x^{1/4}$, and for which x-values is it below?

(b) Express your answers in part (a) algebraically using inequalities.

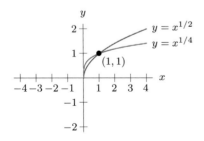

Figure 7.16

27. Figure 7.6 shows that the graph of $y = x^4$ is above the graph of $y = x^2$ when x is greater than 1, and it is below it when x is between 0 and 1. Express these facts algebraically using inequalities.

28. When a car's tires are worn, its braking distance increases by 30%. Use this information and Example 1 to find $h(v)$, the braking distance of an Alfa Romeo with worn tires going at a speed of v mph.

29. A student takes a part-time job to earn $2400 for summer travel. The number of hours, h, the student has to work is inversely proportional to the wage, w, in dollars per hour, and is given by

$$h = \frac{2400}{w}.$$

(a) How many hours does the student have to work if the job pays $4 an hour? What if it pays $10 an hour?

(b) How do the number of hours change as the wage goes up from $4 an hour to $10 an hour? Explain your answer in algebraic and practical terms.

(c) Is the wage, w, inversely proportional to the number hours, h? Express w as a function of h.

30. If a ball is dropped from a high window, the distance, D, in feet, it falls is proportional to the square of the time, t, in seconds, since it was dropped and is given by

$$D = 16t^2.$$

How far has the ball fallen after three seconds and after five seconds? Which distance is larger? Explain your answer in algebraic terms.

31. A city's electricity consumption, E, in gigawatt-hours per year, is given by

$$E = \frac{0.15}{p^{3/2}},$$

where p is the price in dollars per kilowatt-hour charged.

(a) Is E a power function of p? If so, identify the exponent and the constant of proportionality.

(b) What is the electricity consumption at a price of $0.16 per kilowatt-hour? At a price of $0.25 per kilowatt hour? Explain the change in electricity consumption in algebraic terms.

32. The surface area of a mammal is given by $f(M) = kM^{2/3}$, where M is the body mass, and the constant of proportionality k is a positive number that depends on the body shape of the mammal. Is the surface area larger for a mammal of body mass 60 kilograms or for a mammal of body mass 70 kilograms? Explain your answer in algebraic terms.

33. The radius, r, in cm, of a sphere of volume V cm^3 is approximately $r = 0.620 \sqrt[3]{V}$.

   (a) Graph the radius function, r, for volumes from 0 to 40 cm^3.

   (b) Use your graph to estimate the volume of a sphere of radius 2 cm.

34. Plot the expressions $x^2 \cdot x^3$, x^5, and x^6, on three separate graphs in the window $-1 < x < 1$, $-1 < y < 1$. Does it appear from the graphs that $x^2 \cdot x^3 = x^5 = x^{2+3}$ or $x^2 \cdot x^3 = x^6 = x^{2 \cdot 3}$?

35. Plot the expressions $-x^4$ and $(-x)^4$ on the same graph in the window $-1 < x < 1$, $-1 < y < 1$. Is $-x^4 = (-x)^4$?

7.2 WORKING WITH POWER EXPRESSIONS

Recognizing Powers

In practice, an expression for a power function is often not given to us directly in the form kx^p, and we need to do some algebraic simplification to recognize the form.

Example 1   Is the given function a power function? If so, identify the coefficient k and the exponent p.

   (a) $f(x) = \dfrac{2}{x^3}$ **(b)** $g(x) = 4x + 2$ **(c)** $h(x) = \dfrac{5x}{2}$ **(d)** $j(x) = \dfrac{3x^5}{12x^6}$

Solution   **(a)** We have

$$f(x) = \frac{2}{x^3} = 2x^{-3},$$

so this is a power function with $k = 2$ and $p = -3$.

   (b) We cannot rewrite $4x + 2$ in the form kx^p, so this is not a power function.

   (c) We have

$$h(x) = \frac{5x}{2} = \frac{5}{2}x^1,$$

so $k = 5/2$ and $p = 1$.

   (d) We have

$$j(x) = \frac{3x^5}{12x^6} = \frac{1}{4x} = \frac{1}{4}x^{-1},$$

so $k = 1/4$ and $p = -1$.

It is important to remember the order of operations in a power function $f(x) = kx^p$: first raise x to the pth power, then multiply by k.

Example 2   Which of the expressions define the same function?

   (a) $2x^{-3}$ **(b)** $\dfrac{1}{2x^3}$ **(c)** $\dfrac{2}{x^3}$ **(d)** $0.5x^{-3}$

Solution   We have

$$x^{-3} = \frac{1}{x^3}, \quad \text{so} \quad 2x^{-3} = 2\left(\frac{1}{x^3}\right) = \frac{2}{x^3}.$$

Thus (a) and (c) define the same function. From (a) we see that it is a power function with $k = 2$

and $p = -3$. Also,

$$\frac{1}{2x^3} = \left(\frac{1}{2}\right)\left(\frac{1}{x^3}\right) = 0.5x^{-3}.$$

So (b) and (d) define the same function, and from (d) we see that it is a power function with $k = 0.5$ and $p = -3$. So it is different from the power function defined by (a) and (c).

Power functions arising in applications often have constants represented by letters. Recognizing which letter in an expression stands for the independent variable and which letters stand for constants is important if we want to understand the behavior of a function. Often a quite complicated-looking expression defines a simple function.

Example 3   For each power function, identify the coefficient k, and the exponent, p. (Assume that all constants are positive.) Which graph (I)–(IV) gives the best fit?

(a) The volume of a sphere as a function of its radius r:

$$V = \frac{4}{3}\pi r^3.$$

(b) The gravitational force that the Earth exerts on an object as a function of the object's distance r from the earth's center:

$$F = \frac{GmM}{r^2}, \quad G, m, M \text{ constants.}$$

(c) The period of a pendulum as a function of its length l:

$$P = 2\pi\sqrt{\frac{l}{g}}, \quad g \text{ a constant.}$$

(d) The pressure of a quantity of gas as a function of its volume V (at a fixed temperature T):

$$P = \frac{nRT}{V}, \quad n, R, T \text{ constants.}$$

(e) Energy as a function of mass m:

$$E = mc^2, \quad c \text{ a constant (the speed of light).}$$

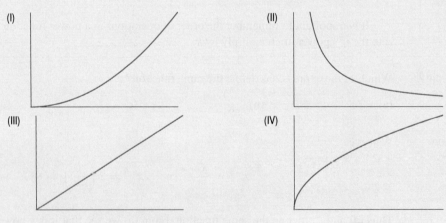

Figure 7.17: Which graph fits?

Solution   (a) The volume V is a power function of r, with $k = (4/3)\pi$ and exponent $p = 3$. Recall that the graph of a power function with p a positive odd integer resembles Figure 7.7 on page 191. In the first quadrant, a graph such as (I) satisfies this requirement.

(b) Writing the equation as

$$F = \frac{GmM}{r^2} = GmMr^{-2},$$

we see that F is a constant times r^{-2}, so it is a power function of r with $k = GmM$ and exponent $p = -2$. Since the exponent is negative, the function's output decreases as its input increases, and graph (II) fits best.

(c) Writing the equation as

$$P = 2\pi\sqrt{\frac{l}{g}} = 2\pi\frac{\sqrt{l}}{\sqrt{g}} = \frac{2\pi}{\sqrt{g}}l^{1/2},$$

we see that P is a power function of l with $k = \frac{2\pi}{\sqrt{g}}$ and exponent $p = 1/2$. Graph (IV) looks most like a square root function, so it fits best.

(d) Writing the equation as

$$P = \frac{nRT}{V} = nRTV^{-1},$$

we see that P is a constant times V^{-1}, so it is a power function of V with $k = nRT$ and exponent $p = -1$. Since the exponent is negative, as in part (b), graph (II) again fits best.

(e) Writing the equation as

$$E = c^2 m^1,$$

we see that E is a power function of m with coefficient $k = c^2$ and exponent $p = 1$. We recognize this as an equation of a line. Thus graph (III) fits best.

Writing Expressions for Power Functions

Example 4   In Example 3 on page 187 we gave formulas for the volume of a sphere of diameter w, and of a cylinder of diameter w and height w (see Figure 7.2). Derive these formulas from the usual formulas for the volume of a sphere and a cylinder.

Solution   A sphere of radius r has volume $V = (4/3)\pi r^3$. The width of the sphere is twice the radius, so $w = 2r$. Therefore $r = w/2$, and

$$V = \frac{4}{3}\pi r^3 = \frac{4}{3}\pi\left(\frac{w}{2}\right)^3 = \frac{4}{3}\pi\left(\frac{1}{2}\right)^3 w^3 = \frac{\pi}{6}w^3.$$

A cylinder with radius r and height h has volume $V = \pi r^2 h$. As before we have $r = w/2$, and, since the height is equal to the diameter, we also have $h = w$, so

$$V = \pi r^2 h = \pi\left(\frac{w}{2}\right)^2 w = \frac{\pi}{4}w^3.$$

Example 5   The surface area of a closed cylinder of radius r and height h is

$$S = 2\pi r^2 + 2\pi rh.$$

(a) Write the expression as a power function of r or h, if possible.

(b) If the height is twice the radius, write S as a power function of r.

Solution   (a) We cannot write this expression as a power function of h or r, because there are two terms that cannot be combined into either the form $S = kr^p$ or the form $S = kh^p$.

(b) Since $h = 2r$, we have

$$S = 2\pi r^2 + 2\pi r \cdot 2r$$
$$= 2\pi r^2 + 4\pi r^2$$
$$= 6\pi r^2.$$

Thus, S is a power function of r, with $k = 6\pi$, and $p = 2$.

Interpreting Expressions for Power Functions

We saw in Section 7.1 that power functions with positive exponents have very different behavior from power functions with negative exponents. Sometimes all we need to know in order to answer a question about a power function is whether the exponent is positive or negative.

Example 6   Which is larger, $f(10)$ or $f(5)$?

(a) $f(A) = 7A^3$ (b) $f(x) = \sqrt{\dfrac{25}{x^3}}$ (c) $f(u) = \sqrt{3u^3}$

Solution   If p is positive, then $10^p > 5^p$, and if p is negative, then $10^p < 5^p$. Multiplying both sides of these inequalities by a positive coefficient k does not change them.

(a) Here exponent $p = 3$ and the coefficient $k = 7$ are both positive, so $f(10) > f(5)$.

(b) We have

$$f(x) = \sqrt{\frac{25}{x^3}} = \frac{\sqrt{25}}{\sqrt{x^3}} = \frac{5}{x^{3/2}} = 5x^{-3/2}.$$

Since the exponent is negative, we have $10^{-3/2} < 5^{-3/2}$, so $f(10) < f(5)$.

(c) We have

$$f(u) = \sqrt{3u^3} = \sqrt{3}\sqrt{u^3} = 3^{1/2}u^{3/2}.$$

Since both the coefficient and the exponent are positive, $f(10) > f(5)$.

The following example shows how to interpret power functions that have a negative coefficient.

Example 7   Which is larger, $f(0.5)$ or $f(0.25)$?

(a) $f(t) = \dfrac{-4}{\sqrt[3]{t}}$ (b) $f(r) = \dfrac{-r^2}{5r^{3/2}}$

Solution   Since $0.5 > 0.25$, we have $0.5^p > 0.25^p$ if p is positive, and $0.5^p < 0.25^p$ if p is negative. However, these inequalities are reversed if we multiply both sides by a negative coefficient k.

(a) We have
$$f(t) = -4t^{-1/3},$$

with coefficient -4 and exponent $-1/3$. Since the exponent is negative, we have $0.5^{-1/3} < 0.25^{-1/3}$. Multiplying both sides by the negative number -4 reverses the direction of this inequality, so $f(0.5) > f(0.25)$ (both numbers are negative, and $f(0.5)$ is closer to zero).

(b) We have
$$f(r) = \frac{-r^2}{5r^{3/2}} = \frac{-1}{5}r^{2-3/2} = -\frac{1}{5}r^{1/2}.$$

Since the exponent $1/2$ is positive, we have $0.5^{1/2} > 0.25^{1/2}$. Multiplying both sides by $-1/5$ reverses the inequality, so $f(0.5) < f(0.25)$.

Although the functions in the next examples are not strictly speaking power functions, the same principles apply in interpreting them.

Example 8   After ten years, the value, V, in dollars, of a $\$5000$ certificate of deposit earning annual interest x is given by
$$V = 5000(1 + x)^{10}.$$

What is the value of the certificate after 10 years if the interest rate is 5%? 10%?

Solution   If the interest rate is 5% then $x = 0.05$, so
$$V = 5000(1 + 0.05)^{10} = 5000 \cdot 1.05^{10} = 8144.47 \text{ dollars.}$$

When the interest rate is 10%, we have $x = 0.10$, so
$$V = 5000(1 + 0.10)^{10} = 5000 \cdot 1.10^{10} = 12{,}968.71 \text{ dollars.}$$

Note that since $1.10 > 1.05$, we have $1.10^{10} > 1.05^{10}$, so the certificate with the higher interest rate has a greater value after 10 years.

Example 9   The amount you need to invest for 10 years at an annual interest rate x if you want to have a final balance of $\$10{,}000$ is given by
$$P = 10{,}000(1 + x)^{-10} \text{ dollars.}$$

How much do you have to invest if the interest rate is 5%? 10%?

Solution   For an interest rate of 5% we have
$$P = 10{,}000(1 + 0.05)^{-10} = 10{,}000 \cdot 1.05^{-10} = 6139.13 \text{ dollars.}$$

When the interest rate is 10% we have
$$P = 10{,}000(1 + 0.10)^{-10} = 10{,}000 \cdot 1.10^{-10} = 3855.43 \text{ dollars.}$$

In this case, $1.10 > 1.05$, we have $1.10^{-10} < 1.05^{-10}$, so the certificate with the higher interest rate requires a lower initial investment.

Exercises and Problems for Section 7.2

EXERCISES

■ Are the functions in Exercises **1–6** power functions?

1. $y = 14x^{12}$

2. $y = 12 \cdot 14^x$

3. $y = 3x^3 + 2x^2$

4. $y = 2/(x^3)$

5. $y = x^3/2$

6. $y = \sqrt{4x^4}$

■ In Exercises **7–16**, write the expression as a constant times a power of a variable. Identify the coefficient and the exponent.

7. $3\sqrt{p}$

8. $\sqrt{2b}$

9. $\dfrac{4}{\sqrt[4]{z}}$

10. $\dfrac{\sqrt{x}}{\sqrt[3]{x}}$

11. $\dfrac{x^4}{4\sqrt{x^2}}, x > 0$

12. $\left(\dfrac{1}{5\sqrt{x}}\right)^3$

13. $\sqrt[3]{\dfrac{8}{x^6}}$

14. $(2\sqrt{x})x^2$

15. $\sqrt{\dfrac{\sqrt{3}}{4}s}$

16. $\sqrt{\sqrt{4t^3}}$

■ In Exercises **17–30**, can the expression be written in the form kx^p? If so, give the values of k and p.

17. $\dfrac{2}{3\sqrt{x}}$

18. $\dfrac{1}{3x^2}$

19. $(7x^3)^2$

20. $\sqrt{\dfrac{49}{x^5}}$

21. $(3x^2)^3$

22. $\left(\dfrac{2}{\sqrt{x}}\right)^3$

23. $y = \dfrac{1}{5x}$

24. $(x^2 + 3x^2)^2$

25. $3 + x^2$

26. $\sqrt{9x^5}$

27. $\left(\dfrac{1}{2\sqrt{x}}\right)^3$

28. $(x^2 + 4x)^2$

29. $\left((-2x)^2\right)^3$

30. $\sqrt[3]{x/8}$

■ In Exercises **31–36**, a and b are positive numbers and $a > b$. Which is larger, $f(a)$ or $f(b)$?

31. $f(x) = \dfrac{x^3}{3}$

32. $f(t) = \dfrac{\sqrt{5}}{t^7}$

33. $f(r) = \sqrt{\dfrac{12}{r}}$

34. $f(t) = -3t^2\left(-2t^3\right)^5$

35. $f(v) = -5\left((-2v)^2\right)^3$

36. $f(x) = 2\sqrt[3]{x} + 3\sqrt[9]{x^3}$

PROBLEMS

■ In Problems **37–48**, what is the exponent of the given power function? Which of (I)–(IV) in Figure 7.18 best fits its graph? Assume all constants are positive.

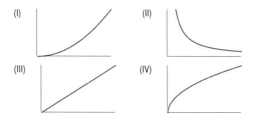

Figure 7.18

37. The heart mass, H, of a mammal as a function of its body mass, B:
$$H = kB.$$

38. The number of species, N, on an island as a function of the area, A, of the island:
$$N = k\sqrt[3]{A}.$$

39. The energy, E, of a swimming dolphin as a function of the speed, v, of the dolphin:
$$E = av^3.$$

40. The strength, S, of a beam as a function of its thickness, h:
$$S = bh^2.$$

41. The average velocity, v, on a trip over a fixed distance d as a function of the time of travel, t:

$$v = \frac{d}{t}.$$

42. The surface area, S, of a mammal as a function of the body mass, B:

$$S = kB^{2/3}.$$

43. The number of animal species, N, of a certain body length as a function of the body length, L:

$$N = \frac{A}{L^2}.$$

44. The circulation time, T, of a mammal as a function of its body mass, B:

$$T = M\sqrt[4]{B}.$$

45. The weight, W, of plaice (a type of fish) as a function of the length, L, of the fish:

$$W = \frac{a}{b} \cdot L^3.$$

46. The surface area, s, of a person with weight w and height h as a function of w, if height h is fixed:

$$s = 0.01w^{0.25}h^{0.75}.$$

47. The judged loudness, J, of a sound as a function of the actual loudness L:

$$J = aL^{0.3}.$$

48. The blood mass, M, of a mammal as a function of its body mass, B:

$$M = kB.$$

■ Given that each expression in Problems 49–52 is defined and not equal to zero, state its sign (positive or negative).

49. $-(b-1)^2$

50. $c\sqrt{-c}$

51. $\left(1+r^2\right)^2 - 1$

52. $w\sqrt{-v} + v\sqrt{vw}$

53. The perimeter of a rectangle of length l and width w is

$$P = 2l + 2w.$$

   (a) Write the expression as a power function of l or w, if possible.
   (b) If the length is three times the width, write P as a power function of w and give the values of the coefficient k and the exponent p.

54. A window is in the shape of a square with a semicircle on top. If the side of the square is l feet long then the area of the glass sheet in the window is

$$A = l^2 + \frac{\pi}{2}\left(\frac{l}{2}\right)^2 \text{ ft}^2.$$

   (a) Is A a power function of l? If so, identify the coefficient k and the exponent p.
   (b) Without computing the area, say whether the area is larger when $l = 1.5$ feet or when $l = 2.5$ feet. Explain your answer in algebraic terms.

55. The side length of an equilateral triangle of area A is

$$s = \sqrt{\frac{4}{\sqrt{3}}A}.$$

Write s as a power function of A and identify the coefficient and the exponent.

■ In Problems 56–65, a and x are positive. What is the effect of increasing a on the value of the expression? Does the value increase, decrease, or remain unchanged?

56. $\dfrac{(ax)^2}{a^2}$

57. $(ax)^2 + a^2$

58. $\dfrac{(ax)^2}{a^3}$

59. a^x

60. $\dfrac{1}{a^{-x}}$

61. $x + a^{-1}$

62. $a^0 x$

63. $x + \dfrac{1}{a^0}$

64. $\dfrac{a^4}{(ax)^3}$

65. $\dfrac{(ax)^{1/3}}{\sqrt{a}}$

■ A certificate of deposit is worth $P(1+r)^t$ dollars after t years, where r is the annual interest rate expressed as a decimal, and P is the amount initially deposited. In Problems 66–68, state which investment will be worth more.

66. Investment A, in which $P = \$7000$, $r = 4\%$, and $t = 5$ years or investment B, in which $P = \$6500$, $r = 6\%$ and $t = 7$ years.

67. Investment A, in which $P = \$10,000$, $r = 2\%$, and $t = 10$ years or investment B, in which $P = \$5000$, $r = 4\%$, and $t = 10$ years.

68. Investment A, in which $P = \$10,000$, $r = 3\%$, and $t = 30$ years or investment B, in which $P = \$15,000$, $r = 4\%$, and $t = 15$ years.

69. A town's population in thousands in 20 years is given by $15(1 + x)^{20}$, where x is the growth rate per year. What is the population in 20 years if the growth rate is

(a)   2%?   (b)   7%?   (c) -5%?

70. An astronaut r thousand miles from the center of the earth weighs $2880/r^2$ lbs, and the surface of the earth is 4000 miles from the center.

(a)   If the astronaut is h miles above the surface of the earth, express r as a function of h.

(b)   Express her weight w in pounds as a function of h.

7.3   SOLVING POWER EQUATIONS

In Example 1 on page 186 we considered the braking distance of an Alfa Romeo traveling at v mph, which is given, in feet, by

$$\text{Braking distance} = 0.036v^2.$$

Suppose we know the braking distance and want to find v. Then we must solve an equation.

Example 1   At the scene of an accident involving an Alfa Romeo, skid marks show that the braking distance is 270 feet. How fast was the car going when the brakes were applied?

Solution   Substituting 270 for the braking distance, we have the equation

$$270 = 0.036v^2.$$

To find the speed of the car, we solve for v. First, we isolate v^2 by dividing both sides by 0.036:

$$v^2 = \frac{270}{0.036} = 7500.$$

Taking the square root of both sides, we get

$$v = \pm\sqrt{7500}.$$

Since the speed of the car is a positive number, we choose the positive square root as the solution:

$$v = \sqrt{7500} = 86.603 \text{ mph}.$$

We see that the car was going nearly 87 mph at the time the brakes were applied.

Notice that in Example 1, we introduced a new operation for solving equations, taking the square root of both sides. Taking the square root of both sides of an equation is different from the other operations we have looked at since it leads to *two* equations.

> Taking the square root of both sides of an equation leads to two equations,
>
> $$\sqrt{\text{Left side}} = \sqrt{\text{Right side}} \quad \text{and} \quad \sqrt{\text{Left side}} = -\sqrt{\text{Right side}}.$$
>
> The same is true of any even root.

For example, taking the square root of both sides of the equation

$$x^2 = 9$$

leads to the two equations

$$x = +\sqrt{9} = +3 \quad \text{and} \quad x = -\sqrt{9} = -3.$$

As in Example 1, we often combine the two possibilities using the $\pm$ symbol, and write $x = \pm 3$.

Example 2   For the cube, sphere, and cylinder in Example 3 on page 187, find the width w if the volume is 10 ft^3. See Figure 7.19.

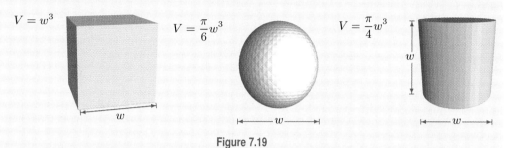

$V = w^3$ $V = \dfrac{\pi}{6}w^3$ $V = \dfrac{\pi}{4}w^3$

Figure 7.19

Solution   For the cube the volume is $V = w^3$, so we must solve the equation

$$w^3 = 10.$$

Taking the cube root of both sides, we get

$$w = 10^{1/3} = 2.15 \text{ ft.}$$

For the sphere, the volume is $V = (\pi/6)w^3$, so we must solve

$$\frac{\pi}{6}w^3 = 10$$

$$w^3 = \frac{6}{\pi}\cdot 10 = \frac{60}{\pi}$$

$$w = \sqrt[3]{\frac{60}{\pi}} = 2.67 \text{ ft.}$$

Similarly, for the cylinder we solve

$$\frac{\pi}{4}w^3 = 10$$

$$w^3 = \frac{4}{\pi}10 = \frac{40}{\pi}$$

$$w = \sqrt[3]{\frac{40}{\pi}} = 2.34.$$

Taking the cube root of both sides of an equation does not create the same problems as taking the square root, since every number has exactly one cube root.

Example 3   An astronaut's weight in pounds is inversely proportional to the square of her distance, r, in thousands of miles, from the earth's center and is given by

$$\text{Weight} = f(r) = \frac{2880}{r^2}.$$

Find the distance from the earth's center at the point when the astronaut's weight is 100 lbs.

Solution

Since the weight is 100 lbs, the distance from the earth's center is the value of r such that $100 = f(r)$, so

$$100 = \frac{2880}{r^2}$$
$$100r^2 = 2880$$
$$r^2 = 28.8$$
$$r = \pm\sqrt{28.8} = \pm 5.367 \text{ thousand miles.}$$

We choose the positive square root since r is a distance. When the astronaut weighs 100 lbs, she is approximately 5367 miles from the earth's center (which is about 1367 miles from the earth's surface).

How Many Solutions Are There?

Notice that in Examples 1 and 3 the equation had two solutions, and in Example 2 the equation had only one solution. In general, how many solutions can we expect for the equation $x^p = a$? In what follows we assume the exponent p can be either a positive or negative integer, but not zero.

Two Solutions

When we have an equation like $x^2 = 5$ or $x^4 = 16$, in which an even power of x is set equal to a positive number, we have two solutions. For instance, $x^4 = 16$ has two solutions, $x = 2$ and $x = -2$, because $(-2)^4 = 2^4 = 16$. See Figure 7.20.

In general, if p is an even integer (other than zero) and a is positive, the equation

$$x^p = a$$

has two solutions, $x = a^{1/p}$ and $x = -a^{1/p}$.

Example 4

Find all solutions for each equation.

(a) $x^2 = 64$ (b) $x^4 = 64$ (c) $x^{-2} = 64$ (d) $x^{-4} = 64$

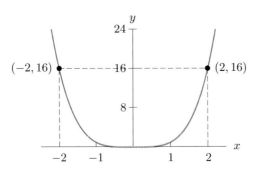

Figure 7.20: Solutions to $x^4 = 16$

Solution   (a) We take square roots of both sides. Because there are two numbers whose square is 64, there are two solutions, 8 and -8.

(b) We take the fourth root of both sides and get

$$x = \pm(64)^{1/4} = \pm(2^6)^{1/4} = \pm 2^{6/4} = \pm 2^{4/4} \cdot 2^{2/4} = \pm 2\sqrt{2}.$$

(c) Multiplying both sides by x^2, we get

$$1 = 64x^2$$
$$\frac{1}{64} = x^2$$
$$x = \pm\frac{1}{8}.$$

(d) We can use the same method as in part (c), or we can simply raise both sides to the $-1/4$ power:

$$x = \pm 64^{-1/4} = \pm\frac{1}{64^{1/4}} = \pm\frac{1}{2\sqrt{2}}.$$

One Solution

When we have an equation like $x^3 = 8$ or $x^5 = -2$, in which an odd power of x is set equal to a number (positive or negative), we have one solution. For instance, $x^3 = 8$ has the solution $x = 2$, because $2^3 = 8$, and $x^3 = -8$ has the solution $x = -2$, because $(-2)^3 = -8$. See Figure 7.21.

In general, if p is odd,

$$x^p = a$$

has one solution, $x = a^{1/p}$. There is another situation when the equation $x^p = a$ has one solution, and that is if $a = 0$. In that case there is just the one solution $x = 0$.

Example 5   Find all solutions for each equation.

(a) $x^5 = -773$ (b) $x^4 = 773$ (c) $x^8 = 0$

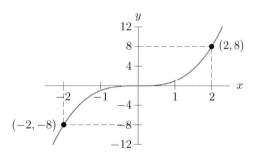

Figure 7.21: Solutions to $x^3 = 8$ and $x^3 = -8$

Solution   (a) Since $p = 5$ is odd, there is one solution: $x = \sqrt[5]{-773} = -3.781$.
   (b) Since $p = 4$ is even and $a = 773$ is positive, there are two solutions, $x = \sqrt[4]{773} = 5.273$ and $x = -\sqrt[4]{773} = -5.273$.
   (c) Since $a = 0$, there is one solution, $x = 0$.

No Solutions

Since we cannot take the square root of a negative number, the equation $x^2 = -64$ has no solutions. In general, if p is even and a is negative, then the equation

$$x^p = a$$

has no solutions. There is another situation when the equation $x^p = a$ has no solution, and that is if p is negative and $a = 0$.

Example 6   How many solutions are there to each of the following equations?
   (a) $-773 = 22x^3$ (b) $-773 = 22x^2$
   (c) $773 = 22x^2$ (d) $0 = 22x^2$

Solution   (a) Since the exponent is odd, there is one solution.
   (b) Since the exponent is even and $-773/22$ is a negative number, there are no solutions.
   (c) Since the exponent is even and $773/22$ is a positive number, there are two solutions.
   (d) Since $0/22$ is 0, there is one solution.

Example 7   A \$5000 certificate of deposit at an annual interest rate of r yields $5000(1 + r)^{10}$ after 10 years. What does the equation $5000(1 + r)^{10} = 10{,}000$ represent? How many solutions are there? Which solutions make sense in practical terms?

Solution   We know that $5000(1 + r)^{10}$ represents the value of the certificate after 10 years. So the solution to the equation

$$5000(1 + r)^{10} = 10{,}000$$

represents the interest rate at which the certificate will grow to \$10,000 in 10 years. To solve the equation, we divide both sides by 5000 to get

$$(1 + r)^{10} = \frac{10{,}000}{5000} = 2,$$

then take 10th roots to get

$$1 + r = \pm \sqrt[10]{2}$$
$$r = -1 \pm \sqrt[10]{2}.$$

Since $\sqrt[10]{2} = 1.0718$, there are two possible values for r,

$$r = -1 + 1.0718 = 0.0718 \quad \text{and} \quad r = -1 - 1.0718 = -2.0718.$$

Since r represents an interest rate, it must be positive, so only the first solution makes sense. The interest rate needed to double the value in 10 years is $r = 0.0718$, or 7.18%.

Equations Involving Fractional Powers

In previous chapters we considered the operations of adding a constant to both sides and multiplying both sides by a nonzero constant. These operations produce an equivalent equation, because they do not change whether two numbers are equal or not.

In this chapter we have introduced new operations, such as squaring both sides, which *can* change the equality or inequality of two numbers. For example, the two numbers 2 and -2 are unequal before squaring, but both equal to 4 after squaring. The same problem can arise in raising both sides of an equation to any even power. These operations can lead to *extraneous solutions*: solutions to the new equation obtained by squaring both sides which are not solutions to the original equation. Whenever we use these operations, we have to check all solutions by substituting in the original equation.

Example 8   Solve each of the following equations:

(a) $\sqrt{t} + 9 = 21$ (b) $\sqrt{t} + 21 = 9$

(c) $2A^{1/5} = 10$

Solution   (a) We solve for $\sqrt{t}$ and then square both sides:

$$\sqrt{t} + 9 = 21$$
$$\sqrt{t} = 12$$
$$(\sqrt{t})^2 = (12)^2$$
$$t = 144.$$

We check that 144 is a solution: $\sqrt{144} + 9 = 12 + 9 = 21$.

(b) Proceeding as before, we solve for $\sqrt{t}$ and then square both sides:

$$\sqrt{t} + 21 = 9$$
$$\sqrt{t} = -12$$
$$(\sqrt{t})^2 = (-12)^2$$
$$t = 144.$$

However, in this case 144 is not a solution: $\sqrt{144} + 21 = 12 + 21 = 33 \neq 9$. We could have noticed this without solving the equation, since $\sqrt{t}$ is always positive, so $\sqrt{t} + 21$ cannot be 9.

(c) The first step is to isolate the $A^{1/5}$. We then raise both sides of the equation to the fifth power:

$$2A^{1/5} = 10$$
$$A^{1/5} = 5$$
$$\left(A^{1/5}\right)^5 = 5^5$$
$$A = 3125.$$

Checking $A = 3125$ in the original equation, we get

$$2(3125)^{1/5} = 2(5^5)^{1/5} = 2 \cdot 5 = 10.$$

In Example 8(b), something strange happened. Although we started out with an equation that had no solutions, we ended up with $t = 144$. This is because we used the operation of squaring both sides, which produced an extraneous solution. When we raise both sides of an equation to an even power, it is possible to get an equation that is not equivalent to the original one.

Summary: Using Powers and Roots to Solve Equations

We offer the following summary about solving equations involving powers and roots. In general:

- Taking the square root of both sides of an equation leads to two equations,

$$\sqrt{\text{Left side}} = \sqrt{\text{Right side}} \quad \text{and} \quad \sqrt{\text{Left side}} = -\sqrt{\text{Right side}}.$$

  Similarly, taking any even root of both sides of an equation leads to two equations.

- Squaring both sides of an equation can lead to extraneous solutions, so you have to check your solutions in the original equation. The same rules apply for any even exponent, such as $4, 6, \ldots$.

- For odd exponents such as $3, 5, \ldots$, or odd roots, there is no problem. Taking the cube root of or cubing both sides of an equation leads to an equivalent equation.

Exercises and Problems for Section 7.3

EXERCISES

In Exercises **1–21**, solve the equation for the variable.

1. $x^3 = 50$
2. $2x^2 = 8.6$
3. $4 = x^{-1/2}$
4. $4w^3 + 7 = 0$
5. $z^2 + 5 = 0$
6. $2b^4 - 11 = 81$
7. $\sqrt{a} - 2 = 7$
8. $3\sqrt[3]{x} + 1 = 16$
9. $\sqrt{y-2} = 11$
10. $\sqrt{2y-1} = 9$
11. $\sqrt{3x-2} + 1 = 10$
12. $(x+1)^2 + 4 = 29$
13. $(3c-2)^3 - 50 = 100$
14. $2x = 54x^{-2}$
15. $2p^5 + 64 = 0$
16. $\frac{1}{4}t^3 = \frac{4}{t}$
17. $16 - \frac{1}{L^2} = 0$
18. $\sqrt{r^2 + 144} = 13$
19. $\frac{1}{\sqrt[3]{x}} = -3$
20. $4\sqrt{x} - 2\sqrt{x} = \frac{2}{3}x$
21. $12 = \sqrt{\frac{z}{5\pi}}$

In Exercises **22–26**, solve the equation for the indicated variable. Assume all other letters represent nonzero constants.

22. $y = kx^2$, for x
23. $A = \frac{1}{2}\pi r^2$, for r
24. $L = kB^2 D^3$, for D
25. $y^2 x^2 = (3y^2)^2$, for x
26. $w = 4\pi \sqrt{\dfrac{x}{t}}$, for x.

Without solving them, say whether the equations in Exercises **27–42** have

(i)   One positive solution
(ii) One negative solution
(iii) One solution at $x = 0$
(iv) Two solutions
(v)   Three solutions
(vi) No solution

Give a reason for your answer.

27. $x^3 = 5$
28. $x^5 = 3$
29. $x^2 = 5$
30. $x^2 = 0$
31. $x^2 = -9$
32. $x^3 = -9$
33. $x^4 = -16$
34. $x^7 = -9$
35. $x^{1/3} = 2$
36. $x^{1/3} = -2$
37. $x^{1/2} = 12$
38. $x^{1/2} = -12$
39. $x^{-1} = 4$
40. $x^{-2} = 4$
41. $x^{-3} = -8$
42. $x^{-6} = -\frac{1}{64}$

In Exercises 43–48, what operation transforms the first equation into the second? Identify any extraneous solutions and any solutions that are lost in the transformation.

43. $\sqrt{x+4} = x-2$
$x+4 = (x-2)^2$

44. $t+1 = 1$
$(t+1)^2 = 1$

45. $r^2 + 3r = 7r$
$r+3 = 7$

46. $(2x)^2 = 16$
$2x = 4$

47. $3 - \dfrac{1}{p} = \dfrac{p-1}{p}$
$3p - 1 = p - 1$

48. $\dfrac{2x}{x+1} = 1 - \dfrac{2}{x+1}$
$2x = x + 1 - 2$

PROBLEMS

49. The equation
$$x\sqrt{8-x} = 5$$
has two solutions. Are they positive, zero, or negative? Give an algebraic reason why this must be the case. You need not find the solutions.

50. The volume of a cone of height 2 and radius r is $V = \frac{2}{3}\pi r^2$. What is the radius of such a cone whose volume is 3π?

51. Let $V = s^3$ give the volume of a cube of side length s centimeters. For what side length is the cube's volume 27 cm^3?

52. A city's electricity consumption, E, in gigawatt-hours per year, is given by $E = 0.15p^{-3/2}$, where p is the price in dollars per kilowatt-hour charged. What does the solution to the equation $0.15p^{-3/2} = 2$ represent? Find the solution.

53. The surface area, S, in cm^2, of a mammal of mass M kg is given by $S = kM^{2/3}$, where k depends on the body shape of the mammal. For people, assume that $k = 1095$.

 (a) Find the body mass of a person whose surface area is 21,000 cm^2.
 (b) What does the solution to the equation $1095M^{2/3} = 30,000$ represent?
 (c) Express M in terms of S.

54. Solve each of the following geometric formulas for the radius r.

 (a) The circumference of a circle of radius r: $C = 2\pi r$.
 (b) The area of a circle of radius r: $A = \pi r^2$.
 (c) The volume of a sphere of radius r: $V = (4/3)\pi r^3$.
 (d) The volume of a cylinder of radius r and height h: $V = \pi r^2 h$.
 (e) The volume of a cone of base radius r and height h: $V = (1/3)\pi r^2 h$.

55. The volume of a cone of base radius r and height h is $(1/3)\pi r^2 h$, and the volume of a sphere of radius r is $(4/3)\pi r^3$. Suppose a particular sphere of radius r has the same volume as a particular cone of base radius r.

 (a) Write an equation expressing this situation.
 (b) What is the height of the cone in terms of r?

56. When P dollars is invested at an annual interest rate r compounded once a year, the balance, A, after 2 years is given by $A = P(1 + r)^2$.

 (a) Evaluate A when $r = 0$, and interpret the answer in practical terms.
 (b) If r is between 5% and 6%, what can you conclude about the percentage growth after 2 years?
 (c) Express r as a function of A. Under what circumstances might this function be useful?
 (d) What interest rate is necessary to obtain an increase of 25% in 2 years?

57. The balance in a bank account earning interest at r% per year doubles every 10 years. What is r?

58. How can you tell immediately that the equation $x + 5\sqrt{x} = -4$ has no solutions?

In Problems 59–70, decide for what values of the constant A the equation has

 (a) The solution $t = 0$
 (b) A positive solution
 (c) A negative solution

59. $t^3 = A$

60. $t^4 = A$

61. $(-t)^3 = A$

62. $(-t)^4 = A$

63. $t^3 = A^2$

64. $t^4 = A^2$

65. $t^4 = -A^2$

66. $t^3 = -A^2$

67. $t^3 + 1 = A$

68. $At^2 + 1 = 0$

69. $At^2 = 0$

70. $A^2 t^2 + 1 = 0$

In Problems 71–74, decide for what value(s) of the constant A (if any) the equation has

 (a) The solution $x = 1$
 (b) A solution $x > 1$
 (c) No solution

71. $4x^2 = A$

72. $4x^3 = A$

73. $-4x^2 = A$

74. $4x^{-2} = A$

75. Figure 7.22 shows two points of intersection of the graphs of $y = 1.3x^3$ and $y = 120x$.

(a) Use a graphing calculator to find the x-coordinate of the nonzero point of intersection accurate to one decimal place.

(b) Use algebra to find the nonzero point of intersection accurate to three decimal places.

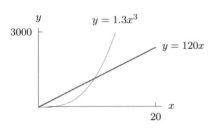

Figure 7.22

76. Figure 7.23 shows two points of intersection of the graphs of $y = 0.2x^5$ and $y = 1000x^2$.

(a) Use a graphing calculator to find the x-coordinate of the nonzero point of intersection accurate to one decimal place.

(b) Use algebra to find the nonzero point of intersection accurate to three decimal places.

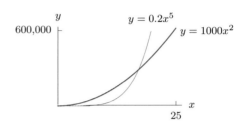

Figure 7.23

7.4 MODELING WITH POWER FUNCTIONS

In Example 1 on page 186, an Alfa Romeo's braking distance as a function of velocity is given by $f(v) = 0.036v^2$. How does the engineer testing the car find the function? Often we model a phenomenon with a family of functions such as $f(v) = kv^2$, and find the parameter k using experimental data. For example, if a test drive shows that the braking distance at 35 mpg is 44 ft, then we know $f(35) = 44$, so we can solve $k(35)^2 = 44$ for k and get $k = 44/35^2 = 0.036$.

Example 1   Find a formula for y in terms of x.

(a) The quantity y is proportional to the fourth power of x, and $y = 150$ when $x = 2$.

(b) The quantity y is inversely proportional to the cube of x, and $y = 5$ when $x = 3$.

Solution   (a) We have $y = kx^4$. We substitute $y = 150$ and $x = 2$ and solve for k:

$$y = kx^4$$
$$150 = k \cdot (2^4)$$
$$k = 150/(2^4) = 9.375.$$

The formula is

$$y = 9.375x^4.$$

(b) We have $y = k/x^3$. We substitute $y = 5$ and $x = 3$ and solve for k:

$$y = \frac{k}{x^3}$$
$$5 = \frac{k}{3^3}$$
$$k = 5 \cdot 3^3 = 135.$$

The formula is

$$y = \frac{135}{x^3}.$$

The Period of a Pendulum

The period of a pendulum is the time it takes to make one full swing back and forth. As long as the pendulum is not swinging too wildly the period depends only on the length of the pendulum.

Example 2   A pendulum's period P, in seconds, is proportional to the square root of its length L, in feet. If a 3 foot pendulum has a period of 1.924 seconds, find the constant of proportionality and write $P = f(L)$ as a power function.

Solution   Since P is proportional to the square root of L, we have

$$P = f(L) = k\sqrt{L} = kL^{1/2}.$$

To find k, we substitute $L = 3$ and $P = 1.924$:

$$P = kL^{1/2}$$
$$1.924 = k(3^{1/2})$$
$$k = \frac{1.924}{3^{1/2}} = 1.111.$$

So

$$P = f(L) = 1.111L^{1/2}.$$

Once we have obtained a model using data, we can apply it to other situations.

Example 3   The length of Foucault's pendulum, built in 1851 in the Panthéon in Paris, is 220 feet. Find the period of Foucault's pendulum.

Solution   We substitute $L = 220$ and see that $P = 1.111(220^{1/2}) = 16.48$ seconds.

What If We Do Not Know the Power?

In Example 2 we are given the power of L that P is proportional to, $L^{1/2}$. But what if we are looking for a power function without knowing what the power is? For example, in biology, scientists often look for power functions to describe the relationship between the body dimensions of individuals from a given species. Table 7.3 relates the weight, y, of plaice (a type of fish) to its length, x.[1]

Table 7.3   *Are length and weight of fish proportional?*

Length (cm), x	33.5	34.5	35.5	36.5	37.5	38.5	39.5	40.5	41.5
Weight (gm), y	332	363	391	419	455	500	538	574	623

We want to model the relationship between y and x by a power function, $y = kx^p$.

[1] Adapted from R. J. H. Beverton and S. J. Holt, "On the Dynamics of Exploited Fish Populations", in *Fishery Investigations*, Series II, 19, 1957.

Example 4   Show that the data in Table 7.3 does not support the hypothesis that y is directly proportional to x.

Solution   If one quantity y is directly proportional to another quantity x, then $y = kx$, and so $k = y/x$. This tells us that the *ratio* of the first quantity to the second quantity is the constant value k. So to test the hypothesis we see if the ratios y/x are approximately constant:

$$\frac{332}{33.5} = 9.910, \qquad \frac{363}{34.5} = 10.522, \qquad \frac{391}{35.5} = 11.014.$$

Since the ratios do not appear to be approximately constant, we conclude that y is not proportional to x.

Similarly, if one quantity y is proportional to a power of another quantity, x, then we have

$$y = kx^p, \qquad \text{for some } p.$$

We can solve for k to obtain

$$k = \frac{y}{x^p}.$$

In this case, the ratio of y to the pth power of x is a constant.

> If y is proportional to x^p, then the ratio y/x^p is constant.

Example 5   Determine whether the data in Table 7.3 supports the hypothesis that the weight of a plaice is approximately proportional to the cube of its length.

Solution   To see if weight y is proportional to the cube of length x, we see if the ratios y/x^3 are approximately constant. To three decimal places, we have

$$\frac{332}{(33.5)^3} = 0.009, \qquad \frac{363}{(34.5)^3} = 0.009, \qquad \frac{391}{(35.5)^3} = 0.009, \qquad \frac{419}{(36.5)^3} = 0.009,$$

and so on for the other ratios. To three decimal places, all the ratios y/x^3 are the same, so weight is approximately proportional to the cube of its length, with constant of proportionality 0.009.

The same method can be used to determine if one quantity is inversely proportional to a positive power of another quantity.

Example 6   Using Table 7.4, show that y could be inversely proportional to x^2.

Table 7.4

x	1	2	3	4	5	6
y	21,600	5400	2400	1350	864	600

Solution   If y is inversely proportional to x^2 then it is proportional to x^{-2}, so $y = kx^{-2}$. We calculate y/x^{-2} to see if this ratio is constant. We have:

$$\frac{21{,}600}{1^{-2}} = 21{,}600 \quad \frac{5400}{2^{-2}} = 21{,}600 \quad \frac{2400}{3^{-2}} = 21{,}600$$

$$\frac{1350}{4^{-2}} = 21{,}600 \quad \frac{864}{5^{-2}} = 21{,}600 \quad \frac{600}{6^{-2}} = 21{,}600.$$

We see that the ratio is constant: $21{,}600$. Therefore $y = 21{,}600x^{-2}$.

The Behavior of Power Functions

Understanding power functions can help us predict what happens to the value of the output variable when we change the input. If y is directly proportional to x, then doubling the value of x doubles the value of y, tripling the value of x triples the value of y, and so on. What happens if y is proportional to a power of x?

Example 7   What happens to the braking distance for the Alfa Romeo in Example 1 on page 186:

(a) If its speed is doubled from $v = 20$ mph to 40 mph? From 30 mph to 60 mph?
(b) If the speed is tripled from 10 mph to 30 mph? From 25 mph to 75 mph?

Solution   (a) If the speed doubles from $v = 20$ mph to 40 mph, then the braking distance increases from $0.036(20)^2 = 14.4$ ft to $0.036(40)^2 = 57.6$ ft. Taking ratios, we see that the braking distance increases by a factor of 4:

$$\frac{57.6 \text{ ft}}{14.4 \text{ ft}} = 4.$$

Likewise, if the speed doubles from $v = 30$ mph to 60 mph, the braking distance increases from 32.4 ft to 129.6 ft. Once again, this is a fourfold increase:

$$\frac{129.6 \text{ ft}}{32.4 \text{ ft}} = 4.$$

(b) If the speed triples from $v = 10$ mph to 30 mph, the braking distance increases from 3.6 ft to 32.4 ft. If the speed triples from 25 mph to 75 mph, the braking distance increases from 22.5 ft to 202.5 ft. In both cases the braking distance increases by a factor of 9:

$$\frac{32.4 \text{ ft}}{3.6 \text{ ft}} = \frac{202.5 \text{ ft}}{22.5 \text{ ft}} = 9.$$

Notice that in each case in Example 7(a), multiplying the speed by 2 multiplies the braking distance by $4 = 2^2$, and in each case in Example 7(b), multiplying the speed by 3 multiplies the braking distance by $9 = 3^2$. Doubling the speed quadruples the braking distance, while tripling the speed multiplies the braking distance by 9. Examples 8 and 9 show why this is true algebraically.

Example 8   For the Alfa Romeo in Example 1 on page 186, what does the expression $0.036(2v)^2$ represent in terms of braking distance? Write this expression as a constant times $0.036v^2$. What does your answer tell you about how braking distance changes if you double the car's speed?

Solution   The expression $0.036v^2$ represents the braking distance of an Alfa Romeo traveling at v mph, so the expression $0.036(2v)^2$ represents the braking distance if the car's speed is doubled from v to $2v$. We have

$$\text{New braking distance} = 0.036(2v)^2 = 0.036 \cdot 2^2 v^2 = 4(0.036v^2)$$
$$= 4 \cdot \text{Old braking distance.}$$

So if the car's speed is doubled, its braking distance is multiplied by 4.

Example 9   What happens to the braking distance if the speed of the Alfa Romeo in Example 1 on page 186 is tripled?

Solution   If the original speed is v, then after being tripled the speed is $3v$, so

$$\text{New braking distance} = 0.036(3v)^2 = 0.036 \cdot 3^2 v^2 = 9(0.036v^2)$$
$$= 9 \cdot \text{Old braking distance.}$$

So if the car's speed is tripled, its braking distance is multiplied by 9.

Exercises and Problems for Section 7.4

EXERCISES

■ In Exercises **1–4**, write a formula for y in terms of x if y satisfies the given conditions.

1. Proportional to the 5th power of x, and $y = 744$ when $x = 2$.

2. Proportional to the cube of x, with constant of proportionality -0.35.

3. Proportional to the square of x, and $y = 1000$ when $x = 5$.

4. Proportional to the 4th power of x, and $y = 10.125$ when $x = 3$.

5. Find a formula for s in terms of t if s is proportional to the square root of t, and $s = 100$ when $t = 50$.

6. If A is inversely proportional to the cube of B, and $A = 20.5$ when $B = -4$, write A as a power function of B.

7. Suppose c is directly proportional to the square of d. If $c = 50$ when $d = 5$, find the constant of proportionality and write the formula for c in terms of d. Use your formula to find c when $d = 7$.

8. Suppose c is inversely proportional to the square of d. If $c = 50$ when $d = 5$, find the constant of proportion-

ality and write the formula for c in terms of d. Use your formula to find c when $d = 7$.

■ In Exercises **9–12**, write a formula representing the function.

9. The strength, S, of a beam is proportional to the square of its thickness, h.

10. The energy, E, expended by a swimming dolphin is proportional to the cube of the speed, v, of the dolphin.

11. The radius, r, of a circle is proportional to the square root of the area, A.

12. Kinetic energy, K, is proportional to the square of velocity, v.

■ In Exercises **13–16**, what happens to y when x is doubled? Here k is a positive constant.

13. $y = kx^3$

14. $y = \dfrac{k}{x^3}$

15. $xy = k$

16. $\dfrac{y}{x^4} = k$

PROBLEMS

17. The thrust, T, in pounds, of a ship's propeller is proportional to the square of the propeller speed, R, in rotations per minute, times the fourth power of the propeller diameter, D, in feet.[2]

(a) Write a formula for T in terms of R and D.

(b) If $R = 300D$ for a certain propeller, is T a power function of D?

(c) If $D = 0.25\sqrt{R}$ for a different propeller, is T a power function of R?

18. Poiseuille's Law gives the rate of flow, R, of a gas through a cylindrical pipe in terms of the radius of the pipe, r, for a fixed drop in pressure between the two ends of the pipe.

(a) Find a formula for Poiseuille's Law, given that the rate of flow is proportional to the fourth power of the radius.

(b) If $R = 400$ cm^3/sec in a pipe of radius 3 cm for a certain gas, find a formula for the rate of flow of that gas through a pipe of radius r cm.

(c) What is the rate of flow of the gas in part (b) through a pipe with a 5 cm radius?

19. The circulation time of a mammal (that is, the average time it takes for all the blood in the body to circulate once and return to the heart) is proportional to the fourth root of the body mass of the mammal.

(a) Write a formula for the circulation time, T, in terms of the body mass, B.

(b) If an elephant of body mass 5230 kilograms has a circulation time of 148 seconds, find the constant of proportionality.

(c) What is the circulation time of a human with body mass 70 kilograms?

20. When an aircraft takes off, it accelerates until it reaches its takeoff speed V. In doing so it uses up a distance R of the runway, where R is proportional to the square of the takeoff speed. If V is measured in mph and R is measured in feet, then 0.1639 is the constant of proportionality.[3]

(a) A Boeing 747-400 aircraft has a takeoff speed of about 210 miles per hour. How much runway does it need?

(b) What would the constant of proportionality be if R was measured in meters, and V was measured in meters per second?

21. Biologists estimate that the number of animal species of a certain body length is inversely proportional to the square of the body length.[4] Write a formula for the number of animal species, N, of a certain body length in terms of the length, L. Are there more species at large lengths or at small lengths? Explain.

In Problems **22–25**, find possible formulas for the power functions.

22.

x	0	1	2	3
y	0	2	8	18

23.

x	1	2	3	4
y	4	16	36	64

24.

x	-2	-1	1	2
y	-16	-1	-1	-16

25.

x	-2	-1	1	2
y	$8/5$	$1/5$	$-1/5$	$-8/5$

26. The volume V of a sphere is a function of its radius r given by

$$V = f(r) = \frac{4}{3}\pi r^3.$$

(a) Find $\dfrac{f(2r)}{f(r)}$. (b) Find $\dfrac{f(r)}{f(\frac{1}{2}r)}$.

(c) What do you notice about your answers to (a) and (b)? Explain this result in terms of sphere volumes.

27. The gravitational force F exerted on an object of mass m at a distance r from the Earth's center is given by

$$F = g(r) = kmr^{-2}, \qquad k, m \text{ constant}.$$

(a) Find $\dfrac{g\left(\frac{r}{10}\right)}{g(r)}$. (b) Find $\dfrac{g(r)}{g(10r)}$.

(c) What do you notice about your answers to (a) and (b)? Explain this result in terms of gravitational force.

---

[2] Thomas C. Gillner, *Modern Ship Design* (US Naval Institute Press, 1972).

[3] Adapted from H. Tennekes, *The Simple Science of Flight* (Cambridge: MIT Press, 1996).

[4] *US News & World Report*, August 18, 1997, p. 79.

28. A square of side x has area x^2. By what factor does the area change if the length is

(a) Doubled?   (b) Tripled?

(c) Halved?   (d) Multiplied by 0.1?

29. A cube of side x has volume x^3. By what factor does the volume change if the length is

(a) Doubled?   (b) Tripled?

(c) Halved?   (d) Multiplied by 0.1?

30. If the radius of a circle is halved, what happens to its area?

31. If the side length of a cube is increased by 10%, what happens to its surface area?

32. If the side length of a cube is increased by 10%, what happens to its volume?

33. Figure 7.24 shows the life span of birds and mammals in captivity as a function of their body size.[5]

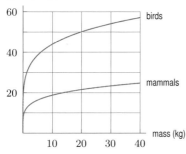

Figure 7.24

(a) Which mammals live longer in captivity, large ones or small ones? If a bird and a mammal have the same size, which has the greater life span in captivity?

(b) What would you expect the life span to be for a body size of 0 kg? For each of the body sizes 10, 20, 30, 40 kg, estimate the life span for birds and mammals, and then use these estimates to plot bird life span as a function of mammal life span. Use this graph to support the statement that in captivity "birds tend to live more than twice as long as mammals of the same size".

(c) The graphs come from the experimentally derived estimates $L_M = 11.8W^{0.20}$ and $L_B = 28.3W^{0.19}$, where L_M and L_B are the life spans of the mammals and birds, and W is body size.

(i) Solve each formula for W and show that $L_B = 28.3\left(\dfrac{L_M}{11.8}\right)^{0.95}$. What does this have to do with the graph you produced in part (b)?

(ii) Clearly $L_M = L_B$ when $W = 0$. Is there another value of W that makes $L_M = L_B$? What does this mean in terms of the life spans of birds and mammals? Is this realistic?

34. If z is proportional to a power of y and y is proportional to a power of x, is z proportional to a power of x?

35. If z is proportional to a power of x and y is proportional to the same power of x, is $z + y$ proportional to a power of x?

36. If z is proportional to a power of x and y is proportional to a power of x, is zy proportional to a power of x?

37. If z is proportional to a power of x and y is proportional to a different power of x, is $z + y$ proportional to a power of x?

REVIEW EXERCISES AND PROBLEMS FOR CHAPTER 7

EXERCISES

In Exercises **1–12**, is the function a power function? If it is a power function, write it in the form $y = kx^p$ and give the values of k and p.

1. $y = \dfrac{3}{x^2}$

2. $y = 5\sqrt{x}$

3. $y = \dfrac{3}{8x}$

4. $y = 2^x$

5. $y = \dfrac{5}{2\sqrt{x}}$

6. $y = (3x^5)^2$

7. $y = \dfrac{2x^2}{10}$

8. $y = 3 \cdot 5^x$

9. $y = (5x)^3$

10. $y = \dfrac{8}{x}$

11. $y = \dfrac{x}{5}$

12. $y = 3x^2 + 4$

In Exercises **13–24**, write each expression as a constant times a power of the variable, and state the base, exponent and coefficient.

13. $\dfrac{-1}{7w}$

14. $(-2t)^3$

15. $\dfrac{125v^5}{25v^3}$

---

[5] Adapted from K. Schmidt-Nielsen, *Scaling, Why is Animal Size So Important?* (Cambridge: CUP, 1984), p. 147.

16. $\dfrac{4}{3}\pi r^3$ **17.** $(4x^3)(3x^{-2})$ **18.** $-z^4$

19. $\dfrac{3}{x^2}$ **20.** $\dfrac{8}{-2/x^6}$ **21.** $3(-4r)^2$

22. $\dfrac{\frac{5}{2}}{10t^5}$ **23.** $(\pi a)(\pi a)$ **24.** $\pi a + \pi a$

29. $\dfrac{100}{(x-2)^2} = 4$ **30.** $\dfrac{A}{Bx^n} = C$

31. Solve the equation $\dfrac{2\pi\sqrt{L}}{C^2} = R$

    (a) For L **(b)** For C

■ In Exercises 32–37, say without solving how many solutions the equation has. Assume that a is a positive constant.

■ In Exercises 25–30, solve for x, assuming all constants are nonzero.

25. $\dfrac{50}{x^3} = 2.8$ **26.** $\dfrac{5}{x^2} = \dfrac{8}{x^3}$

27. $\dfrac{12}{\sqrt{x}} = 3$ **28.** $\dfrac{1}{\sqrt{x-3}} = \dfrac{5}{4}$

32. $x^3 = a$ **33.** $2t^2 = a$

34. $x^{1/2} = a$ **35.** $s^6 = -a$

36. $3t^{1/5} = a$ **37.** $x^5 = -a$

PROBLEMS

38. The energy, E, in foot-pounds, delivered by an ocean wave is proportional[6] to the length, L, of the wave times the square of its height, h.

    (a) Write a formula for E in terms of L and h.

    (b) A 30-foot high wave of length 600 feet delivers 4 million foot-pounds of energy. Find the constant of proportionality and give its units.

    (c) If the height of a wave is one-fourth the length, find the energy E in terms of the length L.

    (d) If the length is 5 times the height, find the energy E in terms of the height h.

39. Poiseuille's Law tells us that the rate of flow, R, of a gas through a cylindrical pipe is proportional to the fourth power of the radius, r, of the pipe, given a fixed drop in pressure between the two ends of the pipe. For a certain gas, if the rate of flow is measured in cm^3/sec and the radius is measured in cm, the constant of proportionality is 4.94.

    (a) If the rate of flow of this gas through a pipe is 500 cm^3/sec, what is the radius of the pipe?

    (b) Solve for the radius r in terms of the rate of flow R.

    (c) Is r proportional to a power of R? If so, what power?

40. The energy, E, in foot-pounds, delivered by an ocean wave is proportional[7] to the length, L, in feet, of the wave times the square of its height, h, in feet, with constant of proportionality 7.4.

    (a) If a wave is 50 ft long and delivers 40,000 ft-lbs of energy, what is its height?

    (b) For waves that are 20 ft long, solve for the height of the wave in terms of the energy. Put the answer in the form $h = kE^p$ and give the values of the coefficient k and the exponent p.

41. A quantity P is inversely proportional to the cube of a quantity R. Solve for R in terms of P. Is R inversely proportional or proportional to a positive power of P? What power?

42. The thrust, T, in pounds, of a ship's propeller is proportional to the square of the propeller speed, R, in rotations per minute, times the fourth power of the propeller diameter, D, in feet.[8]

    (a) Write a formula for T in terms of R and D.

    (b) Solve for the propeller speed R in terms of the thrust T and the diameter D. Write your answer in the form $R = CT^nD^m$ for some constants C, n, and m. What are the values of n and m?

    (c) Solve for the propeller diameter D in terms of the thrust T and the speed R. Write your answer in the form $D = CT^nR^m$ for some constants C, n, and m. What are the values of n and m?

[6]Thomas C. Gillner, *Modern Ship Design* (US Naval Institute Press, 1972).
[7]Thomas C. Gillner, *Modern Ship Design*, (US Naval Institute Press, 1972).
[8]Thomas C. Gillner, *Modern Ship Design*, (US Naval Institute Press, 1972).

Without solving them, say whether the equations in Problems 43–56 have a positive solution $x = a$ such that

(i) $a > 1$ (ii) $a = 1$
(iii) $0 < a < 1$ (iv) No positive solution

Give a reason for your answer.

43. $x^3 = 2$

44. $x^2 = 5$

45. $x^7 = \frac{1}{8}$

46. $x^3 = 0.6$

47. $2x^3 = 5$

48. $5x^2 = 3$

49. $2x^5 = 3$

50. $3x^5 = 3$

51. $x^{-1} = 9$

52. $x^{-3} = \frac{1}{4}$

53. $8x^{-5} = 3$

54. $5x^{-8} = -3$

55. $3x^{-5} = 8$

56. $-8x^{-5} = -8$

In Problems 57–59, demonstrate a sequence of operations that could be used to solve $4x^2 = 16$. Begin with the step given.

57. Take the square root of both sides.

58. Divide both sides of the equation by 4.

59. Divide both sides of the equation by 16.

60. Which of the following steps is the appropriate next step to solve the equation $x^3 + 8 = 64$?

(a) Take the cube root of both sides of the equation
(b) Subtract 8 from both sides of the equation.

61. Which of the following equations have the same solutions as the equation $9x^2 = 81$?

(a) $3x = 9$ (b) $9x = \pm 9$
(c) $3x = \pm 9$ (d) $x^2 = 9$

62. Table 7.5 shows the weight and diameter of various different sassafras trees.[9] The researchers who collect the data theorize that weight, w, is related to diameter, d, by a power model

$$w = kd^s$$

for some constants k and s.

Table 7.5   *Weight and diameter of sassafras trees*

Diameter, d (cm)	5.6	6.5	11.8	16.7	23.4
Weight, w (kg)	5.636	7.364	30.696	76.730	169.290

(a) Plot the data with w on the vertical axis and d on the horizontal axis.
(b) Does your plot support the hypothesis that $s = 1$? Why or why not?

(c) What would a graph of the relationship look like if $s = 2$? What would it look like if $s = 3$? Does it look possible that $s = 2$ or 3 from your data plot?
(d) Instead of looking at the plot to decide whether $s = 2$ is a possibility, look at the data. Add a row to Table 7.5 that shows the ratios w/d^2. What is the overall trend for the numbers in this table: increasing, decreasing, or roughly the same? What should the trend be if $s = 2$ is the correct value for the hypothesis?
(e) To decide if $s = 3$ is a possibility, add a row to Table 7.5 that shows the ratio w/d^3. How does this table behave differently from the one in part (d)? What do you conclude about the correct value of s for the hypothesis?
(f) Make your best estimate of the exponent s and the constant of proportionality k.

63. One of Kepler's three laws of planetary motion states that the square of the period, P, of a body orbiting the sun is proportional to the cube of its average distance, d, from the sun. The earth has a period of 365 days and its distance from the sun is approximately 93,000,000 miles.

(a) Find a formula that gives P as a function of d.
(b) The planet Mars has an average distance from the sun of 142,000,000 miles. What is the period in earth days for Mars?

64. Heap's Law says that the number of different vocabulary words V in a typical English text of length n words is approximately

$$V = Kn^{\beta}, \quad \text{for constants } K \text{ and } \beta.$$

Figure 7.25 shows a typical graph of this function for specific values of K and β.

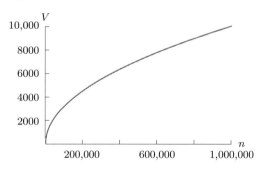

Figure 7.25: Heap's Law

(a) For the function graphed, which of the following is true: $\beta < 0$ or $0 < \beta < 1$ or $\beta > 1$?
(b) Assuming $\beta = 0.5$ in Figure 7.25, is K closer to 1, 10, or 100?

---

[9]http://www.yale.edu/fes519b/totoket/allom/allom.htm, accessed April 19, 2008.

SOLVING DRILL

In Problems 1–30, solve the equation for the indicated variable.

1. $x^3 = 10$; for x
2. $5x^2 = 32$; for x
3. $4x + 1 = 8$; for x
4. $2t^5 = 74$; for t
5. $5p - 12 = 3p + 8$; for p
6. $1.3t + 10.9 = 6.2$; for t
7. $25\sqrt{t} = 8$; for t
8. $8w^3 + 5 = 30$; for w
9. $1.2p^4 = 60$; for p
10. $5.2x - 17.1 = 3.9x + 15.8$; for x
11. $2.3x^2 - 4.5 = 6.8$; for x
12. $25.6t^{2.6} = 83.1$; for $t > 0$
13. $6q^5 = 15q^2$; for q
14. $25.4x^{3.7} = 4.6x^{2.4}$; for $x > 0$
15. $25t^{5.2} = 4.9t^3$; for $t > 0$
16. $5(x^3 - 31) = 27$; for x
17. $3.1(t^5 + 12.4) = 10$; for t
18. $4(s + 3) - 7(s + 10) = 25$; for s
19. $10(r^2 - 4) = 18$; for r
20. $5.2(t^2 + 3.1) = 12$; for t
21. $ax^3 = b$; for x
22. $p^2 r^3 - 5p^3 = 100$; for r
23. $5p^2 + 6q + 17 = 8q$; for p
24. $5p^2 + 6q + 17 = 8q$; for q
25. $5x(2x + 6y) = 2y(3x + 10)$; for y
26. $6r^2(s^3 + 5rt + 2) = 8rt - 15$; for s
27. $6r^2(s^3 + 5rt + 2) = 8rt - 15$; for t
28. $6\sqrt{xy} - 32y^4 + 56y^2 = 3y(6y^3 + 5y)$; for x, assuming $x > 0$ and $y > 0$
29. $A + \sqrt[3]{Bx + C} = D$; for x
30. $25V_0 S^2[T] + 10(H^2 + V_0) = A_0(3V + V_0)$; for S

Chapter 8

More on Functions

CONTENTS

8.1 DOMAIN AND RANGE

Often there is a restriction on the numbers a function can take as inputs. For example, the function

$$P = f(L) = 1.111\sqrt{L}$$

from Example 2 on page 213 is not defined if $L < 0$, since the square root of a negative number is not real. Thus the inputs for f are values with $L \geq 0$.

If $Q = f(t)$, then
- the **domain** of f is the set of input values, t, which yield an output value.
- the **range** of f is the corresponding set of output values, Q.

For $f = 1.111\sqrt{L}$, the domain is all non-negative numbers.

Finding the Domain: Evaluating Expressions

If f is defined by an algebraic expression and no other information is given, we often take the domain to be all inputs x for which $f(x)$ is defined. For example, we exclude any input that leads to dividing by zero or taking the square root of a negative number.

Example 1   For each of the following functions, find the domain.

(a) $f(x) = \dfrac{1}{x-2}$ (b) $g(x) = \dfrac{1}{x} - 2$ (c) $h(x) = \sqrt{1-x}$ (d) $k(x) = \sqrt{x-1}$.

Solution   (a) Since the expression for $f(x)$ is an algebraic fraction, we watch out for inputs that lead to dividing by zero. The denominator is zero when $x = 2$,

$$f(2) = \frac{1}{2-2} = \frac{1}{0} = \text{Undefined}.$$

Any other value of x is allowable, so the domain is all real numbers except $x = 2$. We sometimes write this as

Domain: all real $x \neq 2$.

(b) Here we have an x in the denominator, so $x = 0$ is a problem:

$$g(0) = \frac{1}{0} - 2 = \text{Undefined}.$$

Any other input is allowable, so

Domain: all real $x \neq 0$.

(c) Here we need to make sure that we are taking the square root of a number that is positive or zero. So we must have $1 - x \geq 0$, which means $x \leq 1$. So

Domain: $x \leq 1$.

(d) This time the order of subtraction under the square root is reversed, so we want $x - 1 \geq 0$, which means $x \geq 1$. So

Domain: $x \geq 1$.

Finding the Range: Solving Equations

If the equation $f(t) = -3$ has a solution, then the number -3 is an output value of the function f. The range of f is the set of all outputs, so it is the set of numbers k for which the equation $f(t) = k$ has a solution.

Example 2   Find the range of (a) $h(x) = 5 - x$ (b) $f(t) = 3t + 5$ (c) $g(x) = 5$

Solution   (a) We try to solve the equation $h(x) = k$, where k is an arbitrary number:

$$5 - x = k$$
$$5 = k + x$$
$$x = 5 - k.$$

Since k can be any number, we have

Range: all real numbers.

(b) The equation

$$f(t) = 3t + 5 = k$$

always has the solution $t = (k - 5)/3$, no matter what the value of k. So we have

Range: all real numbers.

(c) The output is 5 no matter what the input, so the range consists of the single number 5.

In general, if m is not zero, the range of a linear function $f(x) = b + mx$ is all real numbers.

Visualizing the Domain and Range on the Graph

An input value a for a function f corresponds to a point on the graph with x-coordinate a, and an output value b corresponds to a point with y-coordinate b. So the domain of f is the set of all values on the input axis for which there is a corresponding point on the graph. For example, Figure 8.1 shows the graph of $k(x) = \sqrt{x - 1}$, which we saw in Example 1(d) has domain $x \geq 1$. The graph has no portion to the left of $x = 1$, starts at $(1, 0)$, and has points corresponding to every number to the right of $x = 1$. Similarly, the range is the set of all values on the output axis with a corresponding point on the graph, or all real $y \geq 0$.

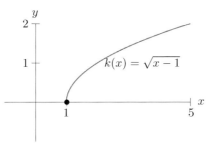

Figure 8.1: The domain of $k(x) = \sqrt{x - 1}$ is $x \geq 1$

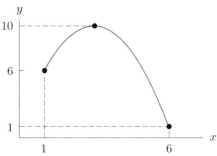

Figure 8.2: Find the domain and range of this function

Example 3   Estimate the domain and range of the function in Figure 8.2, assuming that the entire graph is shown.

Solution   Since the graph has points with x-coordinates between 1 and 6, these are the x-values for which $f(x)$ is defined, so the domain is $1 \leq x \leq 6$. To estimate the range, we look for the highest and lowest points on the graph. The highest point has $y = 10$, the lowest point has $y = 1$, and the graph appears to have points with every y-value in between, so the range is $1 \leq y \leq 10$.

Example 4   For the function

$$g(x) = \frac{1}{x - 1},$$

(a) Find the domain and range from its graph
(b) Verify your answer algebraically.

Solution   (a) In Figure 8.3, we see that there is no point (x, y) on the graph with $x = 1$, because the vertical line $x = 1$ does not intersect the graph. Assuming that the graph keeps leveling out along the parts of the x-axis that are not visible, it appears that every other x-value has an associated y-value, so the domain is all real numbers $x \neq 1$.

Similarly, the only y-value that does not correspond to a point on the graph seems to be $y = 0$, since the horizontal line $y = 0$ does not intersect the graph anywhere. For any other value $k \neq 0$, the horizontal line intersects the graph, so there is a point on the graph with $y = k$, and therefore k is an output of the function. So the range is all real numbers $y \neq 0$.

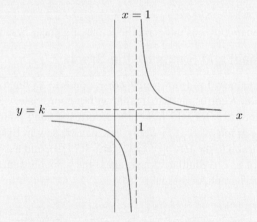

Figure 8.3: The graph of $g(x) = 1/(x - 1)$

(b) The expression for $g(x)$ is undefined when $x = 1$, because $1/(1 - 1) = 1/0$ and division by 0 is undefined. It is defined for all other values of x, so

Domain: all real x, $\quad x \neq 1$.

To find the range, we check to see if $y = k$ is an output of the function by seeing if the equation $1/(x - 1) = k$ can be solved for x. Provided $k \neq 0$, the equation can be solved as follows:

$$\frac{1}{x - 1} = k$$

$$x - 1 = \frac{1}{k}$$

$$x = 1 + \frac{1}{k}.$$

If $k = 0$, then the equation

$$\frac{1}{x - 1} = 0$$

has no solution, since 1 divided by a number is never zero. Thus we have

Range: all real y, $\quad y \neq 0$.

Example 5   Graph

$$f(x) = \frac{5}{2 - \sqrt{x - 2}},$$

then find its domain algebraically.

Solution   The graph in Figure 8.4 lies to the right of the line $x = 2$, and appears to have a break at $x = 6$. Thus it appears that the domain is $x \geq 2$, $x \neq 6$.

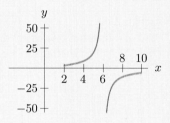

Figure 8.4

To confirm this algebraically, we need to know for what values of x the expression for f is defined, so we consider one by one the operations used in forming the expression.

(a) Subtract 2 from x. This does not restrict the domain, since we can subtract 2 from any number.
(b) Take the square root of $x - 2$. We cannot take the square root of a negative number, so the result of step (a) cannot be negative. Thus, $x \geq 2$.
(c) Subtract $\sqrt{x - 2}$ from 2. This does not restrict the domain, since we can subtract any number from 2.
(d) Divide 5 by $2 - \sqrt{x - 2}$. We cannot divide by 0, so $\sqrt{x - 2}$ cannot equal 2. Therefore, $x - 2$ cannot equal 4, and x cannot equal 6.
(e) Putting steps (c) and (d) together, we see that the domain is $x \geq 2$, $x \neq 6$, as the graph suggests.

Determining the Domain from the Context

If a function is being used to model a particular situation, we only allow inputs that make sense in the situation. For instance, in the function $P = 0.111L^{1/2}$ for the period of a pendulum, a real pendulum would have to have positive length, so we should exclude $L = 0$, leaving $L > 0$.

Example 6   Find the domain of

(a) The function $f(A) = A/350$ used to calculate the number of gallons of paint needed to cover an area A.

(b) The function f giving the cost of purchasing n stamps,

$$f(n) = 0.44n.$$

Solution   (a) Although the algebraic expression for f has all real numbers A as its domain, in this case we say f has domain $A \geq 0$ because A is the area painted, and an area cannot be negative.

(b) We can only purchase a whole number of stamps, or possibly no stamps at all, so the domain is all integers greater than or equal to 0.

Example 7   Find the domain of the function

$$G = 0.75 - 0.3h$$

giving the amount of gasoline, G gallons, in a portable electric generator h hours after it starts.

Solution   Considered algebraically, the expression for G makes sense for all values of h. However, in the context, the only values of h that make sense are those from when the generator starts, at $h = 0$, until it runs out of gas, where $G = 0$. Figure 8.5 shows that the allowable values of h are $0 \leq h \leq 2.5$.

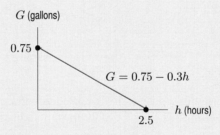

Figure 8.5: Gasoline runs out when $h = 2.5$

We can also calculate algebraically the value of h that makes $G = 0$ by solving the equation $0.3h = 0.75$, so when $h = 0.75/0.3 = 2.5$.

Domain and Range of a Power Function

The domain of a power function $f(x) = kx^p$, with $k \neq 0$, is the values of x for which x^p is defined.

Example 8   Graph the following functions and give their domain.

(a) $g(x) = x^3$ (b) $h(x) = x^{-2}$ (c) $f(x) = x^{1/2}$ (d) $k(x) = x^{-1/3}$

Solution   See Figure 8.6.

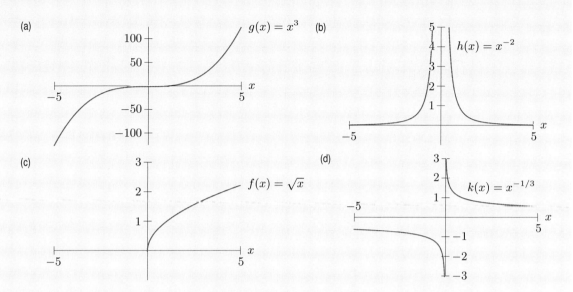

Figure 8.6: Domains of power functions

(a) Since we can cube any number, the domain of g is all real numbers.

(b) We have

$$h(x) = x^{-2} = \frac{1}{x^2}.$$

Since we cannot divide by 0, we cannot input 0 into h. Any other number produces an output, so the domain of h is all numbers except 0. Notice that on the graph of h, there is no point with $x = 0$.

(c) The domain of f is all non-negative numbers. Notice that the graph of f has no points with negative x-coordinates.

(d) We can take the cube root of any real number, but we cannot divide by 0, so the domain of k is all real numbers except 0.

In general, we have the following rules for determining the domain of a power function:

- x^p is defined if x and p are positive
- Negative powers of x are not defined at $x = 0$
- Fractional powers $x^{n/m}$, where n/m is a fraction in lowest terms and m is even, are not defined if $x < 0$.

The next examples show how we can also use these rules to decide whether a number a is in the range of a power function $f(x) = kx^p$.

Example 9   Find the range, and explain your answer in terms of equations.

(a) $f(x) = 2x^2$ (b) $g(x) = 5x^3$ (c) $h(x) = \dfrac{1}{x}$ (d) $k(x) = \dfrac{-3}{x^2}$.

Solution   (a) The equation $2x^2 = a$ is equivalent to $x^2 = a/2$, which has no solutions if a is negative, and has the solution $x = \sqrt{a/2}$ if $a \geq 0$, so

$$\text{Range} = \text{all } a \text{ such that } a \geq 0.$$

We can see this from the graph of $y = 2x^2$ in Figure 8.7. There is a point on the graph corresponding to every non-negative y-value.

(b) The equation $5x^3 = a$ has the solution $x = \sqrt[3]{a/5}$ for all values of a, so

$$\text{Range} = \text{all real numbers}.$$

We also see this from the graph of $y = 5x^3$ in Figure 8.8. There is a point on the graph corresponding to every number on the vertical axis.

(c) The equation $x^{-1} = a$ has the solution $x = a^{-1}$ for all a except $a = 0$, so the range is all real numbers except 0. Figure 8.9 shows that there is a point on the graph for every y-value except 0.

(d) The equation $-3/x^{-2} = a$ is equivalent to $x^2 = -3/a$. The right-hand side must be positive for this to have solutions, so a must be negative. There is no solution if $a = 0$, since a is in the denominator. So the range is all negative numbers. Figure 8.10 shows that there are points on the graph for all negative y-values, and no points if $y \geq 0$.

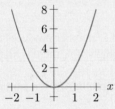

Figure 8.7: Graph of $f(x) = 2x^2$

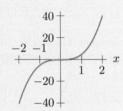

Figure 8.8: Graph of $g(x) = 5x^3$

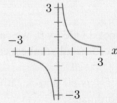

Figure 8.9: Graph of $h(x) = x^{-1}$

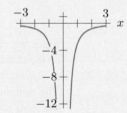

Figure 8.10: Graph of $k(x) = -3x^{-2}$

Exercises and Problems for Section 8.1

EXERCISES

In Exercises 1–13, find

(a) The domain.   (b) The range.

1. $m(x) = 9 - x$
2. $y = x^2$
3. $y = 7$
4. $y = x^2 - 3$
5. $f(x) = x - 3$
6. $y = 5x - 1$
7. $f(x) = \dfrac{1}{\sqrt{x-4}}$
8. $y = \sqrt{x} + 1$
9. $y = \sqrt{x+1}$
10. $y = \dfrac{1}{x-2}$
11. $f(x) = \dfrac{1}{x+1} + 3$
12. $y = \sqrt{2x-4}$
13. $h(z) = 5 + \sqrt{z - 25}$.

14. A restaurant is open from 2 pm to 2 am each day, and a maximum of 200 clients can fit inside. If $f(t)$ is the number of clients in the restaurant t hours after 2 pm each day,

(a) What is reasonable domain for f?

(b) What is a reasonable range for f?

15. A car's average gas mileage, G, is a function $f(v)$ of the average speed driven, v. What is a reasonable domain for $f(v)$?

In Exercises **16–19**, assume the entire graph is shown. Estimate:

(a) The domain   (b) The range

16.

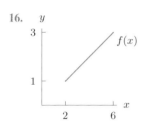

17.

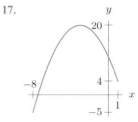

18.

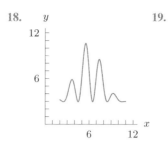

19.

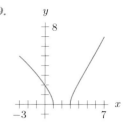

20. The value, V, of a car that is a years old is given by $V = f(a) = 18{,}000 - 3000a$. Find and interpret:

(a) The domain   (b) The range

21. The cost, $\$C$, of producing x units of a product is given by the function $C = 2000 + 4x$, up to a cost of $\$10{,}000$. Find and interpret:

(a) The domain   (b) The range

22. For what values of k does the equation $5 - 3x = k$ have a solution? What does your answer say about the range of the function $f(x) = 5 - 3x$?

23. For what values of k does the equation $-1 = k$ have a solution? What does your answer tell you about the range of the function $f(x) = -1$?

PROBLEMS

The range of the function $y = 9 - (x - 2)^2$ is $y \leq 9$. Find the range of the functions in Problems **24–27**.

24. $y = 10 - (x - 2)^2$ **25.** $y = (x - 2)^2 - 9$

26. $y = 18 - 2(x - 2)^2$ **27.** $y = \sqrt{9 - (x - 2)^2}$

28. A movie theater is filled to capacity with 550 people. After the movie ends, people start leaving at the rate of 100 each minute.

(a) Write an expression for N, the number of people in the theater, as a function of t, the number of minutes after the movie ends.

(b) For what values of t does the expression make sense in practical terms?

Give the domain and range of the functions described in Problems **29–32**.

29. Let $d = g(q)$ give the distance a certain car can travel on q gallons of gas without stopping. Its fuel economy is 24 mpg, and its gas tank holds a maximum of 14 gallons.

30. Let $N = f(H)$ given the number of days it takes a certain kind of insect to develop as a function of the temperature H (in °C). At 40°C—the maximum it can tolerate—the insect requires 10 full days to develop. An additional day is required for every 2°C drop, and it cannot develop in temperatures below 10° C.

31. Let $P = v(t)$ give the total amount earned (in dollars) in a week by an employee at a store as a a function of the number of hours worked. The employee earns $\$7.25$ per hour and must work 4 to 6 days per week, from 7 to 9 hours per day.

32. Let $T = w(r)$ give the total number of minutes of radio advertising bought in a month by a small company as a function of the rate r in $\$$/minute. Rates vary depending on the time of day the ad runs, ranging from $\$40$ to $\$100$ per minute. The company's monthly radio advertising budget is $\$2500$.

■For the functions in Problems **33–38**,

(a) List the algebraic operations in order of evaluation. What restrictions does each operation place on the domain of the function?

(b) Give the function's domain.

33. $y = \dfrac{2}{x - 3}$

34. $y = \sqrt{x - 5} + 1$

35. $y = 4 - (x - 3)^2$

36. $y = \dfrac{7}{4 - (x - 3)^2}$

37. $y = 4 - (x - 3)^{1/2}$

38. $y = \dfrac{7}{4 - (x - 3)^{1/2}}$

■In Problems **39–42**, find the range of f by finding the values of a for which $f(x) = a$ has a solution.

39. $f(x) = \dfrac{5x + 7}{2}$

40. $f(x) = \dfrac{2}{5x + 7}$

41. $f(x) = 2(x + 3)^2$

42. $f(x) = 5 + 2(4x + 3)^2$

■In Problems **43–44** $z(t) = 2a + \sqrt{b - 2t}$.

43. Find b if the domain of $y = z(t)$ is $t \le 7$.

44. Find a given that the range of $y = z(t)$ is $y \ge -8$.

■Judging from their graphs, find the domain and range of the functions in Problems **45–48**.

45. $y = 100 \cdot 2^{-x^2}$

46. $y = 20x \cdot 2^{-x}$

47. $y = \sqrt{10x - x^2 - 9}$

48. $y = \dfrac{5}{(x - 3)^2} + 1$

8.2 COMPOSING AND DECOMPOSING FUNCTIONS

Sometimes it helps in understanding how a function works if we break it down into a series of simpler operations.

Example 1 Most countries measure temperature in degrees Celsius (°C), different from the degrees Fahrenheit (°F) used in the United States. The freezing point of water is 0°C and 32°F. Also, an increase of 1°C corresponds to an increase of 1.8°F. If the temperature is T in degrees Fahrenheit, then corresponding temperature t in degrees Celsius is given by

$$t = \frac{T - 32}{1.8}.$$

Describe the steps in evaluating this function and explain the significance of each step in terms of temperature scales.

Solution Starting with the input T, we compute the output in the following two steps:

- subtract 32, giving $T - 32$
- divide by 1.8, giving $(T - 32)/1.8$

The first step subtracts the freezing point, and gives the number of degrees Fahrenheit above freezing. The second step converts that into the number of degrees Celsius above freezing. Since freezing point is 0°C, this is the temperature in degrees Celsius.

Composition of Functions

Each of the two steps in Example 1 is itself a function:

- the "subtract 32" function, $u = T - 32$
- the "divide by 1.8" function, $t = \dfrac{u}{1.8}$.

To obtain t, we take the output of the first function and make it the input of the second. This process is called *composing* the two functions. In general, given two functions

$$u = g(x) \quad \text{and} \quad y = f(u),$$

the composition of f with g is

$$y = f(g(x)).$$

We call g the *inside function* and f the *outside function*.

The process of composing two functions is similar to the process of evaluating a function, except that instead of substituting a specific number for the independent variable, you are substituting a function $u = g(x)$.

Example 2   If $f(x) = 5x + 1$ and $g(x) = x^2 - 4$, find

(a) $f(2)$ (b) $f(a)$ (c) $f(a-2)$ (d) $f(x+3)$ (e) $f(g(x))$

Solution   (a) We have $f(2) = 5(2) + 1 = 10 + 1 = 11$.
(b) We have $f(a) = 5a + 1$.
(c) We have $f(a - 2) = 5(a - 2) + 1 = 5a - 10 + 1 = 5a - 9$.
(d) Using $x + 3$ as the input to f, we have

$$f(x + 3) = 5(x + 3) + 1 = 5x + 15 + 1 = 5x + 16.$$

(e) Using $g(x)$ as the input to f, we have

$$f(g(x)) = f(x^2 - 4) = 5(x^2 - 4) + 1 = 5x^2 - 20 + 1 = 5x^2 - 19.$$

The order in which you compose two functions makes a difference, as the next example shows.

Example 3   If $f(x) = 2 + 3x$ and $g(x) = 5x + 1$, find (a) $f(g(x))$ (b) $g(f(x))$

Solution   (a) We have
$$f(g(x)) = f(5x + 1) = 2 + 3(5x + 1) = 5 + 15x.$$

(b) We have
$$g(f(x)) = g(2 + 3x) = 5(2 + 3x) + 1 = 11 + 15x.$$

Notice that $f(g(x)) \neq g(f(x))$ in this case.

Expressing a Function as a Composition

In Example 1, we saw how to make sense of the function converting from Fahrenheit to Celsius by breaking it down into its component operations. For functions modeling a real-world situation, expressing the function as a composition of smaller functions can often help make sense of it.

Example 4   After ten years, the value, V, in dollars, of a \$5000 certificate of deposit earning annual interest x is given by
$$V = 5000(1 + x)^{10}.$$

Express V as a composite of simpler functions and interpret each function in terms of finance.

Solution   We choose $u = 1 + x$ as the inside function, so that $V = 5000u^{10}$ is the outside function. The inside function can be interpreted as the factor the balance is multiplied by each year. For example, if the interest rate is 5% then $x = 0.05$. Each year the balance is multiplied by $u = 1 + x = 1.05$. The outside function $V = 5000u^{10}$ describes the effect of multiplying a balance of \$5000 by a factor of u every year for 10 years.

If a function is given by a complicated expression, we can try to express it as a composition of simpler functions in order to better understand it.

Example 5   Express each of the following as a composition of two simpler functions.

(a) $y = (2x + 1)^5$ (b) $y = \dfrac{1}{\sqrt{x^2 + 1}}$ (c) $y = 2(5 - x^2)^3 + 1$

Solution   (a) We define $u = 2x + 1$ as the inside function and $y = u^5$ as the outside function. Then

$$y = u^5 = (2x + 1)^5.$$

(b) We define $u = x^2 + 1$ as the inside function and $y = 1/\sqrt{u}$ as the outside function. Then

$$y = \frac{1}{\sqrt{u}} = \frac{1}{\sqrt{x^2 + 1}}.$$

Notice that we instead could have defined $u = \sqrt{x^2 + 1}$ and $y = 1/u$. There are other possibilities as well. There are multiple ways to decompose a function.

(c) One possibility is to define $u = 5 - x^2$ as the inside function and $y = 2u^3 + 1$ as the outside function. Then

$$y = 2u^3 + 1 = 2(5 - x^2)^3 + 1.$$

Again, there are other possible answers.

By looking for ways to express functions as compositions of simpler functions, we can often discover important properties of the function.

Example 6   Explain why the functions

$$y = (2x + 1)^2, \qquad y = (x^2 - 1)^2, \qquad y = (5 - 2x)^2$$

have no negative numbers in their range.

Solution   Each function is of the form

$$y = (\text{expression})^2,$$

so we can write the function as a composition in which the outside function is $y = u^2$. Since a square can never be negative, the values of y can never be negative.

There are usually many different ways to express a function as a composition. Choosing the right way for a given real-world situation can reveal new information about the situation.

Example 7   As a spherical balloon inflates, its volume in cubic inches after t seconds is given by

$$V = \frac{32}{3}\pi t^3.$$

(a) Express V as a composition in which the outside function is the formula for the volume in terms of the radius, r, in inches:

$$V = \frac{4}{3}\pi r^3.$$

(b) Use your answer to (a) to determine how fast the radius is increasing.

Solution   (a) Comparing the formula for the volume in terms of t with the formula in terms of r, we see that they differ by a factor of 8, which is 2^3. So we can rewrite

$$V = \frac{32}{3}\pi t^3 \quad \text{as} \quad V = 8\frac{4}{3}\pi t^3 = 2^3 \frac{4}{3}\pi t^3 = \frac{4}{3}\pi(2t)^3.$$

This is the composition of

$$V = \frac{4}{3}\pi r^3 \quad \text{with} \quad r = 2t.$$

(b) The inside function, $r = 2t$, tells us that the radius is growing at a rate of 2 in/sec.

Exercises and Problems for Section 8.2

EXERCISES

In Exercises 1–8, use substitution to compose the two functions.

1. $y = u^4$ and $u = x + 1$
2. $y = 5u^3$ and $u = 3 - 4x$
3. $w = r^2 + 5$ and $r = t^3$
4. $p = 2q^4$ and $D = 5p - 1$
5. $w = 5s^3$ and $q = 3 + 2w$
6. $P = 3q^2 + 1$ and $q = 2r^3$
7. $y = u^2 + u + 1$ and $u = x^2$
8. $y = 2u^2 + 5u + 7$ and $u = 3x^3$

In Exercises 9–11, express the function as a composition of two simpler functions.

9. $y - \sqrt{x^2 + 1}$
10. $y = 5(x - 2)^3$

11. $y = 3x^3 - 2$

Write the functions in Exercises 12–19 in the form $y = k \cdot (h(x))^p$ for some function $h(x)$.

12. $y = \dfrac{(2x + 1)^5}{3}$
13. $y = \dfrac{17}{(1 - x^3)^4}$
14. $y = \sqrt{5 - x^3}$
15. $y = \dfrac{2}{\sqrt{1 + \dfrac{1}{x}}}$
16. $y = 100\left(1 + \sqrt{x}\right)^4$
17. $y = \left(1 + x + x^2\right)^3$
18. $y = 5\sqrt{12 - \sqrt[3]{x}}$
19. $y = 0.5(x + 3)^2 + 7(3 + x)^2$

PROBLEMS

In Problems 20–21, the two functions share either an inside function or an outside function. Which is it? Describe the shared function.

20. $y = (2x + 1)^3$ and $y = \dfrac{1}{\sqrt{2x + 1}}$

21. $y = \sqrt{5x - 2}$ and $y = \sqrt{x^2 + 4}$

22. If we compose the two functions $w = f(s)$ and $q = g(w)$ using substitution, what is the input variable of the resulting function? What is the output variable?

23. Evaluate and simplify $g(0.6c)$ given that

$$f(x) = x^{-1/2}$$
$$g(v) = f\left(1 - v^2/c^2\right).$$

24. Evaluate and simplify $p(2)$ given that

$$V(r) = \frac{4}{3}\pi r^3$$
$$p(t) = V(3t).$$

25. Find a possible formula for f given that

$$f\left(x^2\right) = 2x^4 + 1$$
$$f(2x) = 8x^2 + 1$$
$$f(x+1) = 2x^2 + 4x + 3.$$

26. Give the composition of any two functions such that

   (a) The outside function is a power function and the inside function is a linear function.
   (b) The outside function is a linear function and the inside function is a power function.

In Problems 27–28, find

   (a) $f(g(x))$ **(b)** $g(f(x))$

27. $f(x) = x^3$ and $g(x) = 5 + 2x$

28. $f(x) = x^3 + 1$ and $g(x) = \sqrt{x}$

29. Using $f(t) = 3t^2$ and $g(t) = 2t + 1$, find

   (a) $f(g(t))$ **(b)** $g(f(t))$ **(c)** $f(f(t))$ **(d)** $g(g(t))$

30. If $f(g(x)) = 5(x^2 + 1)^3$ and $g(x) = x^2 + 1$, find $f(x)$.

31. Give three different composite functions with the property that the outside function raises the inside function to the third power.

32. Give a formula for a composite function with the property that the outside function takes the square root and the inside function multiplies by 5 and adds 2.

33. Give a formula for a composite function with the property that the inside function takes the square root and the outside function multiplies by 5 and adds 2.

34. The amount, A, of pollution in a certain city is a function of the population P, with $A = 100P^{0.3}$. The population is growing over time, and $P = 10000 + 2000t$, with t in years since 2000. Express the amount of pollution A as a function of time t.

35. The rate R at which the drug level in the body changes when an intravenous line is used is a function of the amount Q of the drug in the body. For a certain drug, we have $R = 25 - 0.08Q$. The quantity Q of the drug is a function of time t with $Q = \sqrt{t}$ over a fixed time period. Express the rate R as a function of time t.

8.3 SHIFTING AND SCALING

In the previous section we saw how to break the expression for a function into simpler components. It is also useful to be able to add components to a function. When we use functions to model real-world situations, we often want to adjust an expression for a function that describes a situation in order to be able to describe a different but related situation.

Adding a Constant to the Output: Vertical Shifts

Example 1   Two balls are thrown in the air at the same time. After t seconds, the first has height $h(t) = 90t - 16t^2$ feet and the second has height $g(t) = 20 + 90t - 16t^2$. How do the expressions for the two functions compare with each other, and what does this say about the motion of the two balls?

Solution   Looking at the expressions for the two functions, we see that

$$\underbrace{20 + 90t - 16t^2}_{\text{expression for } g(t)} = \underbrace{90t - 16t^2}_{\text{expression for } f(t)} + 20$$

The expression for the height of the second ball is obtained by adding the constant 20 to the expression for the height of the first ball. This means that the second ball must be 20 feet higher than the first ball at all times. In particular, it was thrown from a position 20 feet higher. See Figure 8.11.

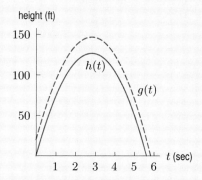

Figure 8.11: Height of two balls thrown from different heights

Notice that the effect of adding a constant to the output is simply to shift the graph of the function up by a certain amount. This also affects the range of the function, as the next example illustrates.

Example 2   Sketch the graph and give the range of each function.

(a) $y = x^2$ (b) $y = x^2 + 1$ (c) $y = x^2 - 3$

Solution   (a) See Figure 8.12(a). The expression x^2 is never negative, but can take on any positive value, so we have

$$\text{Range: all real } y \geq 0.$$

(b) For every x-value, the y-value for $y = x^2 + 1$ is one unit larger than the y-value for $y = x^2$. Since all the y-coordinates are increased by 1, the graph shifts vertically up by 1 unit. See Figure 8.12(b). We have

$$\text{Range: all real } y \geq 1.$$

(c) The y-coordinates are 3 units smaller than the corresponding y-coordinates of $y = x^2$, so the graph is shifted *down* by 3 units. See Figure 8.12(c). The range is

$$\text{Range: all real } y \geq -3.$$

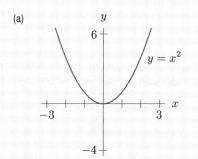
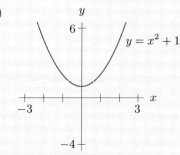
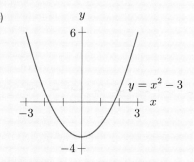

Figure 8.12: Vertical shifts of $y = x^2$

In general, we have

> For a positive constant k, the graph of
>
> $$y = f(x) + k$$
>
> is the graph of $y = f(x)$ shifted up by k units, and the graph of
>
> $$y = f(x) - k$$
>
> is the graph of $y = f(x)$ shifted down by k units.

Adding a Constant to the Input: Horizontal Shifts

Another adjustment we can make to a function is to add or subtract a constant to the input.

Example 3   Two balls are thrown in the air one after the other. At time t seconds after the first ball is thrown, it has height $h(t) = 90t - 16t^2$ feet and the second ball has height $g(t) = 90(t-2) + 16(t-2)^2$. How do the expressions for the two functions compare with each other, and what does this say about the motion of the two balls?

Solution   The expression for $g(t)$ is obtained by replacing t with $t - 2$ in the expression for $h(t)$. Thus the second ball always has the same height as the first ball had 2 seconds earlier. The second ball was thrown two seconds after the first ball, but after that its motion is exactly the same (see Figure 8.13).

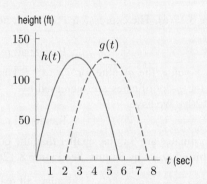

Figure 8.13: Height of two balls thrown at different times

Notice that in Example 3, the effect of subtracting 2 from the input was to describe the motion of a ball that was lagging 2 seconds behind the first ball. Since this makes the motion of the ball occur 2 seconds later, its graph is shift to the right by 2.

Example 4   Sketch the graph of each function and give its domain.

(a) $y = \dfrac{1}{x}$ (b) $y = \dfrac{1}{x-1}$ (c) $y = \dfrac{1}{x+3}$

Solution   (a) See Figure 8.14(a). Since $1/x$ is defined for all x except $x = 0$, we have

$$\text{Domain: all real } x, \ x \neq 0.$$

(b) See Figure 8.14(b). Notice that it is the same shape as the graph of $y = 1/x$, shifted 1 unit to the right. Since the original graph has no y-value at $x = 0$, the new graph has no y value at $x = 1$, which makes sense because $1/(x - 1)$ is defined for all x except $x = 1$, so

$$\text{Domain: all real } x, \ x \neq 1.$$

(c) See Figure 8.14(c), which has the same shape as the graph of $y = 1/x$, shifted 3 units to the left. This time there is no y-value for $x = -3$, and $1/(x+3)$ is defined for all x except $x = -3$, so

$$\text{Domain: all real } x, \ x \neq -3.$$

(a)
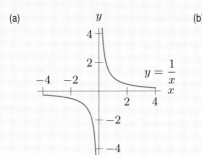

(b)
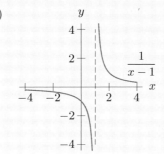

(c)
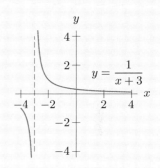

Figure 8.14: Horizontal shifts of $y = 1/x$

In general, we have

For a positive constant k, the graph of

$$y = f(x - k)$$

is the graph of $y = f(x)$ shifted k units to the right, and the graph of

$$y = f(x + k)$$

is the graph of $y = f(x)$ shifted k units to the left.

Vertical and Horizontal Shifts on Graphs

Example 5   Figure 8.15 shows the graph of $y = \sqrt{x}$. Use shifts of this function to graph each of the following:

(a) $y = \sqrt{x} + 1$ (b) $y = \sqrt{x - 1}$ (c) $y = \sqrt{x} - 2$ (d) $y = \sqrt{x + 3}$

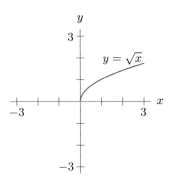

Figure 8.15: Graph of $y = \sqrt{x}$

Solution   (a) Here the output values of $y = \sqrt{x}$ are shifted up by 1. See Figure 8.16(a).

(b) Since $x - 1$ is substituted for x in the expression $\sqrt{x}$, the graph is a horizontal shift of the graph of $y = \sqrt{x}$. Substituting $x = 1$ gives $\sqrt{1 - 1} = \sqrt{0}$, so we see that $x = 1$ in the new graph gives the same value as $x = 0$ in the old graph. The graph is shifted to the right 1 unit. See Figure 8.16(b).

(c) Since 2 is subtracted from the expression $\sqrt{x}$, the graph is a vertical shift down 2 units of the graph of $y = \sqrt{x}$. See Figure 8.16(c).

(d) Since $x + 3$ is substituted for x in the expression $\sqrt{x}$, the graph is a horizontal shift of the graph of $y = \sqrt{x}$. Substituting $x = -3$ gives $\sqrt{-3 + 3} = \sqrt{0}$, so we see that $x = -3$ in the new graph gives the same value as $x = 0$ in the old graph. The graph is shifted to the left 3 units. See Figure 8.16(d).

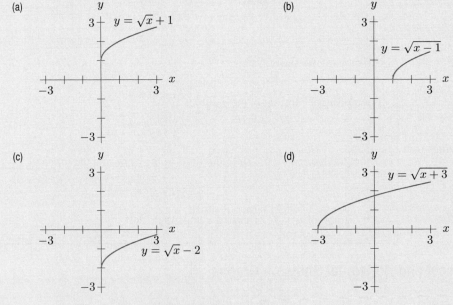

Figure 8.16: Horizontal and vertical shifts of $y = \sqrt{x}$

Multiplying the Outside of a Function by a Constant: Vertical Scaling

In Example 1, we observed the effect of composing a function with a vertical shift by 20. In the next example, we observe the effect of composing a function with multiplication by k. Multiplying by a constant is also called *scaling*.

Example 6   Let $f(x) = x^3$ and $g(x) = 2x^3$ and $h(x) = \frac{1}{2}x^3$. Compare f and g and h using their output values and graphs.

Solution   Each output value of g is double the corresponding output value of f. For example, $f(3) = 27$ and $g(3) = 54$. Since the output values are displayed on the vertical axis, the graph of g is stretched vertically away from the x-axis by a factor of 2. Each output value of h is half the corresponding output value of f. For example, $f(3) = 27$ and $h(3) = 13.5$. Since the output values are displayed on the vertical axis, the graph of h is compressed vertically towards the x-axis by a factor of $\frac{1}{2}$. See Figure 8.17.

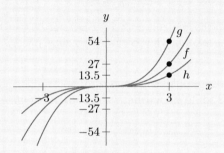

Figure 8.17: Effect of k for $k \cdot f(x)$

Example 7   Let $r(t) = \sqrt{t}$ and $s(t) = -\sqrt{t}$. Compare r and s using their output values and graphs.

Solution   Multiplying the output values of r by the constant, -1, changes the sign of corresponding output value of s. For example, $r(4) = 2$ and $s(4) = -2$. The graph of s resembles the graph of r with positive outputs becoming negative and negative outputs becoming positive. See Figure 8.18. The graph of s is a reflection of the graph of r across the x-axis.

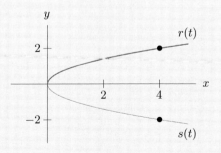

Figure 8.18: Effect of negative multiplication factor

Example 8   Let $f(t)$ give the altitude in yards of a weather balloon t minutes after it is released. Find an expression for $g(t)$, the balloon's altitude in feet.

Solution   For every yard there are 3 feet, so if the altitude in yards is $f(t)$, then the altitude in feet is $3f(t)$. Thus $g(t) = 3f(t)$.

Multiplying the Inside of a Function by a Constant: Horizontal Scaling

Previously when we multiplied the outside function by a constant we changed the output value corresponding to a certain input. Now if we multiply the inside of a function by a constant it is the input value that changes.

Example 9   If $F(x) = \dfrac{1}{x+1}$, find a formula for

   (a) $5F(x)$ (b) $F(5x)$ (c) $F\left(\dfrac{x}{5}\right)$

Solution   (a) $5F(x) = 5\left(\dfrac{1}{x+1}\right) = \dfrac{5}{x+1}.$

   (b) $F(5x) = \dfrac{1}{(5x)+1} = \dfrac{1}{5x+1}.$

   (c) $F\left(\dfrac{x}{5}\right) = \dfrac{1}{\left(\frac{x}{5}\right)+1} = \dfrac{5}{5}\cdot\left(\dfrac{1}{\left(\frac{x}{5}\right)+1}\right) = \dfrac{5}{x+5}.$

Example 10   Let $f(t)$ give the altitude in yards of a weather balloon t minutes after it is released. Find an expression for $w(r)$, the balloon's altitude r hours after it is released.

Solution   The number of minutes in r hours is $t = 60r$, so the height after r hours is the same as the height after $60r$ minutes, namely $f(60r)$. Thus $w(r) = f(60r)$.

Combining Inside and Outside Changes

Just as the order in which a sequence of operations is performed can matter in arithmetic, the same is true with transformations of functions.

Example 11   Let $f(x) = 2x$. Write a formula for a new function $g(x)$ resulting from the following sequence of operations.

   (a) Add 2 units to the output, then multiply by a factor of 5.
   (b) Multiply the output by a factor of 5, then add 2 units to the result.

Solution   (a) The first operation changes f to $2x + 2$, then the multiplication by 5 gives $g(x) = 5(2x+2) = 10x + 10$
   (b) The first operation changes f to $10x$, then the adding 2 units gives $g(x) = 10x + 2$

Example 12   A new function $g(x)$ is formed from $f(x)$ by a sequence of operations. List the operations in order for the following:

(a) $g(x) = 5f(2x) - 4$ (b) $g(x) = 5(f(2x) - 4)$

Solution   (a) First the input of f is multiplied by a factor of 2, then that result is multiplied by a factor of 5. Finally the function values are shifted down by 4 units.

(b) Although the operations are the same as those in Part (a), the order in which they are performed is different. First the input of f is multiplied by a factor of 2, then 4 units are subtracted from that result. The final operation is multiplication by a factor of 5.

Example 13   If $H = f(d)$ is the Celsius temperature (°C) at an ocean depth of d meters, find a formula for $g(s)$, the Fahrenheit temperature (°F) at an ocean depth of s feet, given that:

- The Fahrenheit temperature, T, for a given Celsius temperature of H is $T = \dfrac{9}{5} \cdot H + 32$.
- The depth d in meters at s feet is $d = 0.3048s$.

Solution   We have

$$f(d) = \text{Temperature in °C at } d \text{ meters}$$
$$f(0.3048s) = \text{Temperature in °C at } s \text{ feet} \qquad \text{since } d = 0.3048s$$
$$\frac{9}{5} \cdot f(0.3048s) + 32 = \text{Temperature in °F at } s \text{ feet} \qquad \text{since } T = \frac{9}{5} \cdot H + 32.$$

Thus,

$$g(s) = \frac{9}{5} \cdot f(0.3048s) + 32.$$

We see that g involves scaling the input by a factor of 0.3048, scaling the output by a factor of $9/5$, and shifting the output up by 32 units.

Exercises and Problems for Section 8.3

EXERCISES

In Exercises 1–6, the graph of the function $g(x)$ is a horizontal and/or vertical shift of the graph of $f(x) = x^3$, shown in Figure 8.19. For each of the shifts described, sketch the graph of $g(x)$ and find a formula for $g(x)$.

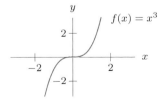

$f(x) = x^3$

Figure 8.19

1. Shifted vertically up 3 units.

2. Shifted vertically down 2 units.

3. Shifted horizontally to the left 1 unit.

4. Shifted horizontally to the right 2 units.

5. Shifted vertically down 3 units and horizontally to the left 1 unit.

6. Shifted vertically up 2 units and horizontally to the right 4 units.

In Exercises **7–10**, the graph of the function $g(x)$ is a horizontal and/or vertical shift of the graph of $f(x) = 5 - x$, shown in Figure 8.20. For each of the shifts described, sketch the graph of $g(x)$ and find a formula for $g(x)$.

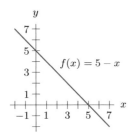

Figure 8.20

7. Shifted vertically down 3 units.

8. Shifted vertically up 1 unit.

9. Shifted horizontally to the right 2 units.

10. Shifted horizontally to the left 4 units.

The graphs in Exercises **11–16** are horizontal and/or vertical shifts of the graph of $y = x^2$. Find a formula for each function graphed.

11.

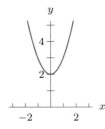

12.

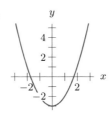

13.

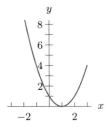

14.

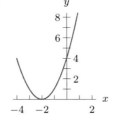

15.

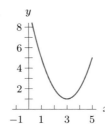

16.

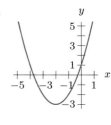

Exercises **17–22** refer to the graph of $y = f(x)$ in Figure 8.21. Sketch the graph of each function.

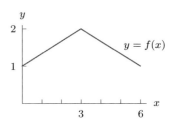

Figure 8.21

17. $y = f(x) - 3$ 18. $y = f(x) + 2$

19. $y = f(x - 1)$ 20. $y = f(x + 3)$

21. $y = f(x + 1) + 2$ 22. $y = f(x - 3) - 1$

In Exercises **23–28**, find a formula for n in terms of m where:

23. n is a weight in oz and m is the weight in lbs.

24. n is a length in feet and m is the length in inches.

25. n is a distance in km and m the distance in meters.

26. n is an age in days and m the age in weeks.

27. n is an amount in dollars and m the amount in cents.

28. n is an elapsed time in hours and m the time in minutes.

PROBLEMS

29. The function $H = f(t)$ gives the temperature, $H°$F, of an object t minutes after it is taken out of the refrigerator and left to sit in a room. Write a new function in terms of $f(t)$ for the temperature if:

    (a) The object is taken out of the refrigerator 5 minutes later. (Give a reasonable domain for your function.)
    (b) Both the refrigerator and the room are $10°$F colder.

30. The height, h in cm, of an eroding sand dune as a function of year, t, is given by $h = f(t)$. Describe the difference between this sand dune and a second one one whose height is given by

(a) $h = f(t + 30)$
(b) $h = f(t) + 50$.

31. A line has equation $y = x$.

(a) Find the new equation if the line is shifted vertically up by 5 units.
(b) Find the new equation if the line is shifted horizontally to the left by 5 units.
(c) Compare your answers to (a) and (b) and explain your result graphically.

32. The growth rate of a colony of bacteria at temperature $T°$F is $P(T)$. The Fahrenheit temperature T for $H°$C is

$$T = \frac{9}{5} \cdot H + 32.$$

Find an expression for $Q(H)$, the growth rate as a function of H.

■ In Problems 33–38, find a formula for g by scaling the output of f.

33. Let $f(t)$ give the snowfall in feet t hours after a blizzard begins, and $g(t)$ the snowfall in inches.

34. Let $f(t)$ give the number of liters of fuel oil burned in t hours, and $g(t)$ the number of gallons burned. Use the fact that 1 gal equals 3.785 liters.

35. Let $f(n)$ give the average time in seconds required for a computer to process n megabytes (MB) of data, and $g(n)$ the time in microseconds. Use the fact that 1 s equals 1,000,000 μs.

36. Let $f(t)$ give the speed in mph of a jet at time t, and $g(t)$ the speed in kilometers per hour (kph). Use the fact that 1 kph is 0.621 mph.

37. Let $f(t)$ give the distance in light years to a receding star in year t, and $g(t)$ the distance in parsecs. Use the fact that 1 parsec equals 3.262 light years.

38. Let $f(t)$ give the area in square miles (mi^2) of a town in year t, and $g(t)$ the area in square kilometers (km^2). Use the fact that 1 mi^2 equals 2.59 km^2.

■ In Problems 39–44, find a formula for w by scaling the input of f.

39. Let $f(t)$ give the snowfall in feet t hours after a blizzard begins, and $w(r)$ the snowfall after r minutes.

40. Let $f(t)$ give the number of liters of fuel oil burned in t days, and $w(r)$ the liters burned in r weeks.

41. Let $f(n)$ give the average time in seconds required for a computer to process n megabytes (MB) of data, and $w(r)$ the time required for r gigabytes (GB). Use the fact that 1 GB equals 1024 MB.

42. Let $f(u)$ give the maximum speed of a jet at a thrust of u pounds-force (lbs) and $w(v)$ the maximum speed at a thrust of v newtons (N). Use the fact that 1 lb is 4 448 N.

43. Let $f(r)$ be the average yield in bushels from r acres of corn, and $w(s)$ be the yield from s hectares. Use the fact that one hectare is 2.471 acres.

44. Let $f(p)$ be the number of songs downloaded from an online site at a price per song of p dollars, and $w(q)$ be the number downloaded at a price per song of q cents.

■ In Problems 45–50, find a formula for g by scaling the input and/or output of f.

45. Let $f(t)$ give the measured precipitation in inches on day t, and $g(t)$ give the precipitation in centimeters. Use the fact that 1 in equals 2.54 cm.

46. Let $f(t)$ give the area in acres of a county's topsoil lost to erosion in year t, and $g(t)$ give the area in hectares lost. Use the fact that 1 acre equals 0.405 hectares.

47. Let $f(s)$ give the volume of water in liters in a reservoir when the depth measures s meters, and $g(s)$ give the volume of water in gallons. Use the fact that 1 gal equals 3.785 l.

48. Let $f(s)$ give the volume of water in liters in a reservoir when the depth measures s meters, and $g(d)$ give the volume of water in liters when the depth measures d centimeters. Use the fact that 1 m equals 100 cm.

49. Let $f(w)$ give the expected weight in lbs of an infant at one month's age if its weight at birth is w lbs, and $g(m)$ give the expected weight in kg if the weight at birth is m kg. Use the fact that 1 lb equals 0.454 kg.

50. Let $f(t)$ give the number of acres lost to a wildfire t days after it is set, and $g(n)$ give the number of hectares lost after n hours. Use the fact that 1 acre equals 0.405 hectares.

8.4   INVERSE FUNCTIONS

In Section 8.2 we saw how to break a function down into a sequence of simpler operations. In this section we consider how to use this idea to undo the operation of a function.

Inverse Operations

An inverse operation is used to undo an operation. Performing an operation followed by its inverse operation gets us back to where we started. For example, the inverse operation of "add 5" is "subtract 5." If we start with any number x and perform these operations in order, we get back to where we started:

$$x \quad \to \quad x+5 \quad \to \quad (x+5)-5 \quad \to \quad x$$
$$\text{(add 5)} \qquad\qquad \text{(subtract 5)}$$

Example 1   State in words the inverse operation.

    (a)   Add 12   (b)   Multiply by 7
    (c)   Raise to the 5th power   (d)   Take the cube root

Solution   (a) To undo adding 12, we subtract 12. The inverse operation is "Subtract 12".
    (b) To undo multiplying by 7, we divide by 7. The inverse operation is "Divide by 7". (We could also say that the inverse operation is to multiply by $1/7$, which is equivalent to dividing by 7.)
    (c) To undo raising an expression to the 5th power, we raise to the $1/5^{\text{th}}$ power. The inverse operation is to "Raise to the $1/5^{\text{th}}$ power" or, equivalently, "Take the fifth root".
    (d) To undo taking a cube root, we cube the result. The inverse operation is "Cube" or "Raise to the 3rd power".

Example 2   Find the sequence of operations to undo "multiply by 2 and then add 5".

Solution   First we undo "add 5" by performing the operation "subtract 5". Then we undo "multiply by 2" by performing the operation "divide by 2." Thus the sequence that undoes "multiply by 2 and add 5" is "subtract 5 and divide by 2." Notice that we have not only inverted each operation, but we performed them in reverse order, since we had to undo the most recently performed operation first.

Not every operation has an inverse operation.

Example 3   Explain why the operation of squaring a number does not have an inverse operation.

Solution   You might think that the inverse of squaring a number would be to take the square root. However, if we start with a negative number, such as -2, we have

$$-2 \quad \to \quad (-2)^2 = 4 \quad \to \quad \sqrt{4} \quad \to \quad 2$$
$$\text{(square)} \qquad\qquad \text{(take square root)}$$

Notice that we do not get back to the number we started with. There is no inverse operation for squaring a number that works for all numbers.

Inverse Operations and Solving Equations

We use inverse operations routinely in solving equations. For example, in the equation

$$2x + 5 = 9,$$

the sequence of operations "multiply by 2 and add 5" from Example 2 is applied to x and the result set equal to 9. To solve the equation we undo this sequence by performing the sequence "subtract 5 and divide by 2."

$$2x + 5 = 9 \qquad \text{subtract 5 from both sides}$$
$$2x = 9 - 5 \qquad \text{divide both sides by 2}$$
$$x = \frac{9 - 5}{2} = 2.$$

Example 4   Describe the operations you would use to solve the following equations:

(a) $3x - 2 = 7$ (b) $3(x - 2) = 7$ (c) $3x^2 = 7$

Solution   (a) The left-hand side is formed by first multiplying x by 3 and then subtracting 2. To reverse this we add 2 and divide by 3. So the solution is

$$x = \frac{7 + 2}{3} = 3.$$

(b) This equation uses the same operations as the previous one but in the other order. The left-hand side is formed by first subtracting 2 from x and then multiplying by 3. To reverse this we divide by 3 and then add 2. So

$$x = \frac{7}{3} + 2 = 4\frac{1}{3}.$$

(c) Here the left-hand side is formed by squaring x and then multiplying by 3. We can partially undo this by dividing by 3, which gives

$$x^2 = \frac{7}{3}.$$

Unlike the equations in (a) and (b), this equation has two solutions, $x = \pm\sqrt{7/3}$. There is no single operation that is the inverse of squaring.

Inverse Functions

For any function f, in order to solve the equation

$$f(x) = k,$$

we want to be able to "undo" f, that is, find a function g so that when we compose it with f we get back to where we started. This leads to the following definition:

Inverse Functions

Given a function $f(x)$, we say that $g(x)$ is the inverse function to $f(x)$ if

$$g(f(x)) = x \quad \text{for all } x \text{ in the domain of } f$$

and

$$f(g(x)) = x \quad \text{for all } x \text{ in the domain of } g.$$

Example 5   A town's population t years after incorporation is given by $P(t) = 30{,}000 + 2000t$.

(a) Find the time it takes for the population to reach a level of k.
(b) Find the inverse function to $P(t)$.

Solution   (a) We want to know the value of t that makes $P(t)$ equal to k, so we solve the equation

$$30{,}000 + 2000t = k$$
$$2000t = k - 30{,}000 \qquad \text{subtract 30,000 from both sides}$$
$$t = \frac{k - 30{,}000}{2000} \qquad \text{divide both sides by 2000.}$$

Thus, it takes $(k - 30{,}000)/2000$ years for the population to reach a level of k.
(b) The formula for t obtained in part (a) is also the formula for the inverse function to P:

$$Q(k) = \text{Number of years to reach a population of } k = \frac{k - 30{,}000}{2000}.$$

The output of a function becomes the input for its inverse function, if it has one. Furthermore, the output of the inverse function is the input of the original function.

Example 6   Let $s = f(t)$ give the altitude in yards of a weather balloon t minutes after it is released. Describe in words the inverse function $t = g(s)$.

Solution   Since the input of f is the output of g, and the output of f is the input of g, we see that $t = g(s)$ gives the number of minutes required for the balloon to reach an altitude of s.

Finding an Inverse

If a function is built up out of a series of simple operations, we can find the inverse function by finding the inverse operations.

Example 7   In Example 2 we saw that the inverse to "multiply by 2 and add 5" is "subtract 5 and divide by 2." Express both of these as functions and show that they are inverses.

Solution   The sequence of operations "multiply by 2 and add 5" takes x to $2x$ and then takes $2x$ to $2x + 5$, so it expresses the function

$$f(x) = 2x + 5.$$

The inverse sequence of operations is "subtract 5 and divide by 2", which takes x to $x - 5$ and then takes $x - 5$ to $(x - 5)/2$, and so describes the function

$$g(x) = \frac{x - 5}{2}.$$

We check that these two functions are inverses by composing them in both orders. We have

$$f(g(x)) = 2g(x) + 5 = 2\left(\frac{x - 5}{2}\right) + 5 \quad \text{multiplying by 2 undoes dividing by 2}$$
$$= (x - 5) + 5 \qquad \text{adding 5 undoes subtracting 5}$$
$$= x. \qquad \text{We are back to where we started.}$$

Also,

$$g(f(x)) = \frac{f(x) - 5}{2} = \frac{(2x + 5) - 5}{2} \quad \text{subtracting 5 undoes adding 5}$$
$$= \frac{2x}{2} \qquad \text{dividing by 2 undoes multiplying by 2}$$
$$= x. \qquad \text{We are back to where we started.}$$

We have seen that inverse functions can be used to solve equations. We can turn this around and use the process of solving an equation to find the inverse function.

Example 8   Find the inverses of the linear functions (a) $y = 6 - 2t$ (b) $p = 2(n + 1)$

Solution   (a) The equation

$$y = 6 - 2t$$

expresses y as a function of t. To get the inverse function we solve this equation for t. This has the effect of reversing the operations that produce y from t, but enables us to keep track of the process. We have

$$y = 6 - 2t$$
$$y - 6 = -2t$$
$$\frac{y - 6}{-2} = t$$
$$t = \frac{y - 6}{-2} = -\frac{1}{2}y + 3.$$

This expresses t as a function of y, which is the inverse of the original function.
(b) We solve $p = 2(n + 1)$ for n.

$$p = 2(n + 1)$$
$$\frac{p}{2} = n + 1$$

$$\frac{p}{2} - 1 = n$$

$$n = \frac{p}{2} - 1.$$

This expresses n as a function of p, which is the inverse of the original function.

Example 9   Find the inverse of the function $f(x) = 5x^3 - 2$.

Solution   In this case there is no variable representing the dependent variable, but we can choose one, say k, for the purpose of finding the inverse. So we want to solve

$$f(x) = k.$$

We have

$5x^3 - 2 = k$	add 2 to both sides
$5x^3 = k + 2$	divide both sides by 5
$x^3 = \dfrac{k + 2}{5}$	take the cube root of both sides
$x = \sqrt[3]{\dfrac{k + 2}{5}}.$	

So the inverse function is the function g that outputs $\sqrt[3]{(k + 2)/5}$ whenever you input k. Using x to stand for the input, we get

$$g(x) = \sqrt[3]{\frac{x + 2}{5}}.$$

Exercises and Problems for Section 8.4

EXERCISES

■ In Exercises 1–4, state in words the inverse operation.

1. Subtract 8.
2. Divide by 10.
3. Raise to the 7th power.
4. Take the ninth root.

■ In Exercises 5–8, find the sequence of operations to undo the sequence given.

5. Multiply by 5 and then subtract 2.
6. Add 10 and then multiply the result by 3.
7. Raise to the 5th power and then multiply by 2.
8. Multiply by 6, add 10, then take the cube root.

■ In Exercises 9–12, show that composing the functions in either order gets us back to where we started.

9. $y = 7x - 5$ and $x = \dfrac{y + 5}{7}$

10. $y = 8x^3$ and $x = \sqrt[3]{\dfrac{y}{8}}$

11. $y = x^5 + 1$ and $x = \sqrt[5]{y - 1}$

12. $y = \dfrac{10 + x}{3}$ and $x = 3y - 10$

In Exercises **13–14**,

(a) Write a function of x that performs the operations described.

(b) Find the inverse and describe in words the sequence of operations in the inverse.

13. Raise x to the fifth power, multiply by 8, and then add 4.

14. Subtract 5, divide by 2, and take the cube root.

In Exercises **15–20**, let g be the inverse of f. Describe in words the function g.

15. Let $T = f(n)$ give the average time in seconds required for a computer to process n megabytes (MB) of data.

16. Let $v = f(r)$ give the maximum speed of a jet at a thrust of r pounds-force.

17. Let $Y = f(r)$ give the average yield in bushels from r acres of corn.

18. Let $P = f(N)$ give the cost in dollars to cater a wedding reception attended by N people.

19. Let $P = f(s)$ give the atmospheric pressure in kilopascals (kPa) at an altitude of s km.

20. Let $T = f(c)$ give the boiling point in $^\circ$C of salt water at a concentration (in moles/liter) of c.

PROBLEMS

In Problems **21–25**, check that the functions are inverses.

21. $f(x) = 2x - 7$ and $g(t) = \dfrac{t}{2} + \dfrac{7}{2}$

22. $f(x) = 6x^7 + 4$ and $g(t) = \left(\dfrac{t-4}{6}\right)^{1/7}$

23. $f(x) = 32x^5 - 2$ and $g(t) = \dfrac{(t+2)^{1/5}}{2}$

24. $f(x) = \dfrac{x}{4} - \dfrac{3}{2}$ and $g(t) = 4\left(t + \dfrac{3}{2}\right)$

25. $f(x) = 1 + 7x^3$ and $g(t) = \sqrt[3]{\dfrac{t-1}{7}}$

Solve the equations in Problems **26–29** exactly. Use an inverse function when appropriate.

26. $2x^3 + 7 = -9$

27. $\dfrac{3\sqrt{x} + 5}{4} = 5$

28. $\sqrt[3]{30 - \sqrt{x}} = 3$

29. $\sqrt{x^3 - 2} = 5$

In Problems **30–34**, find the inverse function.

30. $h(x) = 2x + 4$

31. $h(x) = 9x^5 + 7$

32. $h(x) = \sqrt[3]{x + 3}$

33. $p(x) = \dfrac{5 - \sqrt{x}}{3 + 2\sqrt{x}}.$

34. $q(x) = \dfrac{4 - 3x}{7 - 9x}.$

REVIEW EXERCISES AND PROBLEMS FOR CHAPTER 8

EXERCISES

1. Tuition cost T, in dollars, for part-time students at a college is given by $T = 300 + 200C$, where C represents the number of credits taken. Part-time students cannot take more than 10 credit hours. Give a reasonable domain and range for this function.

In Exercises **2–9**, give the domain and range of the power function.

2. $y = x^4$

3. $y = x^{-3}$

4. $y = 2x^{3/2}$

5. $A = 5t^{-1}$

6. $P = 6x^{2/3}$

7. $Q = r^{5/3}$

8. $y = x^{-1/2}$

9. $M = 6n^3$

In Exercises **10–11**, give the domain and range.

10. $y = \sqrt{4 - \sqrt{9 - x}}.$

11. $y = 4 + \dfrac{2}{x - 3}.$

In Exercises **12–19**, find the range.

12. $m(x) = 9 - x$

13. $f(x) = \dfrac{1}{\sqrt{x - 4}}$

14. $f(x) = x - 3$

15. $h(x) = \sqrt[3]{\dfrac{1}{x}}$

16. $y = \dfrac{3 + \sqrt{x}}{2}$

17. $y = \left(2 + \sqrt{x}\right)^3$

18. $y = \dfrac{5}{x - 3} + 9$

19. $y = (x - \sqrt{2})^3$

In Exercises **20–21**, assume the entire graph is shown. Estimate:

(a) The domain   (b) The range

20.

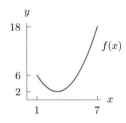

21.

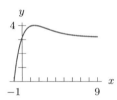

In Exercises **22–24**, express the function as a composition of two simpler functions.

22. $y = \dfrac{5}{x^2 + 1}$

23. $y = 1 + 2(x - 1) + 5(x - 1)^2$

24. $y = 25(3x - 2)^5 + 100$

In Exercises **25–26**, find

(a) $f(g(x))$ (b) $g(f(x))$

25. $f(x) = 5x^2$ and $g(x) = 3x$

26. $f(x) = \dfrac{1}{x}$ and $g(x) = x^2 + 1$

In Exercises **27–30**, the graph of the function $g(x)$ is a horizontal and/or vertical shift of the graph of $f(x) = 2x^2 - 1$, shown in Figure 8.22. For each of the shifts described, sketch the graph of $g(x)$ and find a formula for $g(x)$.

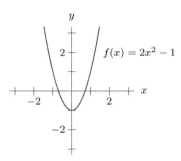

Figure 8.22

27. Shifted horizontally to the left 1 unit.

28. Shifted vertically down 3 units.

29. Shifted vertically up 2 units and horizontally to the right 3 units.

30. Shifted vertically down 1 unit and horizontally to the left 2 units.

The function $f(t)$ gives the distance in miles traveled by a car after t hours. Express the functions in Exercises **31–34** in terms of f.

31. $g(s)$, the distance in miles traveled in s seconds.

32. $F(t)$, the distance in feet traveled in t hours.

33. $h(m)$, the distance in kilometers traveled in m minutes.

34. $r(s)$, the distance in meters traveled in s seconds.

In Exercises **35–38**, find the inverse function.

35. $h(x) = 7x + 5$

36. $h(x) = 11x^3 - 2$

37. $h(x) = \dfrac{7x^5 + 2}{7x^5 + 3}$

38. $h(x) = \dfrac{4x}{5x + 4}$

In Exercises **39–42**, state in words the inverse operation.

39. Multiply by 77.

40. Add -10.

41. Raise to the 1/7th power.

42. Take the 7th root.

PROBLEMS

43. A movie theater seats 200 people. For any particular show, the amount of money the theater makes is a function of the number of people, n, in attendance. If a ticket costs $4.00, find the domain and range of this function. Sketch its graph.

44. The gravitational force (in Newtons) exerted by the Earth on the Space Shuttle depends on r, the Shuttle's distance in km from the Earth's center, and is given by

$$F = GmM_E r^{-2},$$

where[1]

$$M_E = 5.97 \times 10^{24} \qquad \text{Mass of Earth in kg}$$
$$m = 1.05 \times 10^{5} \qquad \text{Mass of Shuttle orbiter in kg}$$
$$G = 6.67 \times 10^{-11} \qquad \text{Gravitational constant}$$
$$R_E = 6380 \qquad \text{Radius of Earth in km}$$
$$A = 1000 \text{ km} \qquad \text{Max. altitude of Orbiter.}$$

Based on this information, give the domain and range of this function.

45. **(a)** What is the domain of the function $P = -100,000 + 50,000s$?

(b) If P represents the profit of a silver mine at price s dollars per ounce, and if the silver mine closes if profits fall below zero, what is the domain?

■ The range of the function $y = 2^{-x^2}$ is $0 < y \le 1$. Find the range of the functions in Problems **46–49**.

46. $y = 5 - 2^{-x^2}$

47. $y = 3 \cdot 2^{-x^2}$

48. $y = 7 \cdot 2^{-x^2} + 3$

49. $y = 2^{2-x^2}$

■ In Problems **50–51**, find the range.

50. $g(x) = \sqrt[3]{x^2 - 8}$

51. $p(x) = \sqrt[3]{x^2 - 1} + \sqrt{x^2 - 1}$

52. For what values of k does the equation

$$\frac{5x}{x-1} = k$$

have a solution? What does your answer say about the range of the function $f(x) = 5x/(x-1)$?

53. Find the range of the function $G = 0.75 - 0.3h$ in Example 7 on page 228 giving the amount of gasoline, G gallons, in a portable electric generator h hours after it starts.

54. **(a)** Give the domain and range of the linear function $y = 100 - 25x$.

(b) If x and y represent quantities which cannot be negative, give the domain and range.

■ In Problems **55–56**, the two functions share either an inside function or an outside function. Which is it? Describe the shared function.

55. $y = 5(1 - 3x)^2$ and $y = 5(x^2 + 1)^2$

56. $y = \sqrt{x^2 + 1}$ and $y = (x^2 + 1)^3 + 5$

■ In Problems **57–58**, find

(a) $g(h(x))$ **(b)** $h(g(x))$

57. $g(x) = x^5, h(x) = x + 3$

58. $g(x) = \sqrt{x}, h(x) = 2x + 1$

59. Using $f(t) = 2 - 5t$ and $g(t) = t^2 + 1$, find

(a) $f(g(t))$ **(b)** $g(f(t))$
(c) $f(f(t))$ **(d)** $g(g(t))$

60. If $f(g(x)) = \sqrt{5x + 1}$ and $f(x) = \sqrt{x}$, find $g(x)$.

61. Give three different composite functions with the property that the outside function takes the square root of the inside function.

62. Give a formula for a function with the property that the inside function raises the input to the 5th power, and the outside function multiplies by 2 and subtracts 1.

63. Give a formula for a function with the property that the outside function raises the input to the 5th power, and the inside function multiplies by 2 and subtracts 1.

64. If we compose the functions $U = f(V)$ and $V = g(W)$ using substitution, what is the input variable of the resulting function? What is the output variable?

65. A line has equation $y = b + mx$.

(a) Find the new equation if the line is shifted vertically up by k units. What is the y-intercept of this line?

(b) Find the new equation if the line is shifted horizontally to the right by k units. What is the y-intercept of this line?

---

[1] Figures from Google calculator (see http://www.google.com/intl/en/help/features.html#calculator) and from http://en.wikipedia.org/wiki/Space_Shuttle accessed June 9, 2007. The mass given for the Shuttle is for the Orbiter alone; at liftoff, the total mass of the Shuttle, including the plane-like Orbiter, the external solid rocket boosters, and the external liquid-fuel tank, is closer to 2 million kg than to 105,000 kg.

66. The cost, C in dollars, for an amusement park includes an entry fee and a certain amount per ride. We have $C = f(r)$ where r represents the number of rides. Give a formula in terms of $f(r)$ for the cost as a function of number of rides if the situation is modified as described:

   (a) The entry fee is increased by $5.
   (b) The entry fee includes 3 free rides.

■ In Exercises **67–68**, check that the functions are inverses.

67. $g(x) = 1 - \dfrac{1}{x-1}$ and $f(t) = 1 + \dfrac{1}{1-t}$

68. $h(x) = \sqrt{2x}$ and $k(t) = \dfrac{t^2}{2}$, for $x, t \geq 0$

■ Find the inverse, $g(y)$, of the functions in Problems **69–71**.

69. $f(x) = \dfrac{x-2}{2x+3}$ **70.** $f(x) = \sqrt{\dfrac{4-7x}{4-x}}$

71. $f(x) = \dfrac{\sqrt{x}+3}{11-\sqrt{x}}$

■ In Problems **72–73**,
   (a) Write a function of x that performs the operations described.
   (b) Find the inverse and describe in words the sequence of operations in the inverse.

72. Multiply by 2, raise to the third power, and then add 5.

73. Divide by 7, add 4, and take the ninth root.

Chapter 9

CONTENTS

9.1   QUADRATIC FUNCTIONS

Unlike linear functions, quadratic functions can be used to describe situations where the speed or rate of change of an object changes as it moves.

| Example 1 | The height in feet of a ball thrown upward from the top of a building after t seconds is given by |

$$h(t) = -16t^2 + 32t + 128, \quad t \geq 0.$$

Find the ball's height after 0, 1, 2, 3, and 4 seconds and describe the path of the ball.

| Solution | We have |

$$h(0) = -16 \cdot 0^2 + 32 \cdot 0 + 128 = 128 \text{ ft.}$$
$$h(1) = -16 \cdot 1^2 + 32 \cdot 1 + 128 = 144 \text{ ft.}$$
$$h(2) = -16 \cdot 2^2 + 32 \cdot 2 + 128 = 128 \text{ ft.}$$

Continuing, we get Table 9.1. The ball rises from 128 ft when $t = 0$ to a height of 144 ft when $t = 1$, returns to 128 ft when $t = 2$ and reaches the ground, height 0 ft, when $t = 4$.

Table 9.1   *Height of a ball*

t (seconds)	$h(t)$ (ft)
0	128
1	144
2	128
3	80
4	0

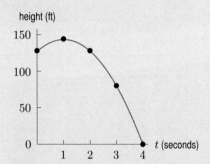

Figure 9.1: Height of a ball thrown from the top of a building

Interpreting Quadratic Functions Expressed in Standard Form

The expression defining the function in Example 1 has a term, $-16t^2$, involving the square of the independent variable. This term is called the *quadratic term*, the expression is called a *quadratic expression*, and the function it defines is called a *quadratic function*. In general,

> A **quadratic function** is one that can be written in the form
>
> $$y = f(x) = ax^2 + bx + c, \quad a, b, c \text{ constants}, a \neq 0.$$
>
> - The expression $ax^2 + bx + c$ is a **quadratic expression** in **standard form**.
> - The term ax^2 is called the **quadratic term** or **leading term**, and its coefficient a is the **leading coefficient**.
> - The term bx is called the **linear term**.
> - The term c is called the **constant term**.

Example 2   For the function $h(t) = -16t^2 + 32t + 128$ in Example 1, interpret in terms of the ball's motion

(a)   The constant term 128   (b)   The sign of the quadratic term $-16t^2$.

Solution   (a)   The constant term is the value of the function when $t = 0$, and so it represents the height of the building.

(b)   The negative quadratic term $-16t^2$ counteracts the positive terms $32t$ and 128 for $t > 0$, and eventually causes the values of $h(t)$ to decrease, which makes sense since the ball eventually starts to fall to the ground.

Figure 9.1 shows an important feature of quadratic functions. Unlike linear functions, quadratic functions have graphs that bend. This is the result of the presence of the quadratic term. See Figure 9.2, which illustrates this for the power functions $f(x) = x^2$ and $g(x) = -x^2$. The shape of the graph of a quadratic function is called a *parabola*. If the coefficient of the quadratic term is positive then the parabola opens upward, and if the coefficient is negative the parabola opens downward.

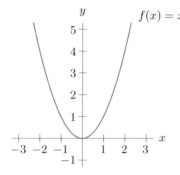 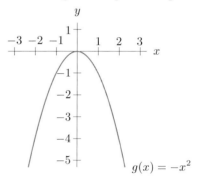

Figure 9.2: Graphs of $f(x) = x^2$ and $g(x) = -x^2$

Interpreting Quadratic Functions Expressed in Factored Form

Factoring a quadratic expression puts it in a form where we can easily see what values of the variable make it equal to zero.[1]

Example 3   The function $h(t) = -16t^2 + 32t + 128$ in Example 1 can be expressed in the form

$$h(t) = -16(t - 4)(t + 2), \quad t \geq 0.$$

What is the practical interpretation of the factors $(t - 4)$ and $(t + 2)$?

Solution   When $t = 4$, the factor $(t - 4)$ has the value $4 - 4 = 0$. So

$$h(4) = -16(0)(4 + 2) = 0.$$

In practical terms, this means that the ball hits the ground 4 seconds after it is thrown. When $t = -2$, the factor $(t + 2)$ has the value $-2 + 2 = 0$. However, there is no practical interpretation for this, since the domain of h is $t \geq 0$. See Figure 9.3.

---

[1] See Section 2.3 for a review of factoring.

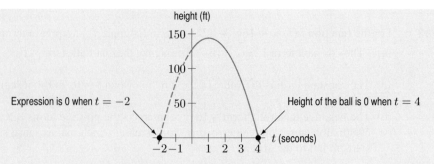

Figure 9.3: Interpretation of the factored form $h(t) = -16(t - 4)(t + 2)$

Values of the independent variable where a function has the value zero, such as the $t = 4$ and $t = -20$ in the previous example, are called **zeros** of the function. In general, we have the following definition.

A quadratic function in x is expressed in **factored form** if it is written as

$$y = f(x) = a(x - r)(x - s), \quad \text{where } a, r, \text{ and } s \text{ are constants and } a \neq 0.$$

- The constants r and s are zeros of the function $f(x) = a(x - r)(x - s)$.
- The constant a is the leading coefficient, the same as the constant a in the standard form.

The zeros of a function can be easily recognized from its graph as the points where the graph crosses the horizontal axis.

Example 4   Can each function graphed in Figure 9.4 be expressed in factored form $f(x) = a(x - r)(x - s)$? If so, is each of the parameters r and s positive, negative, or zero? (Assume $r \leq s$.)

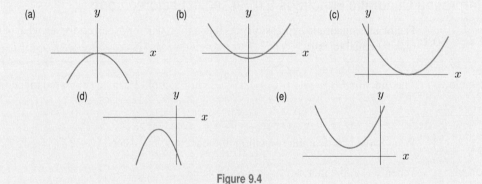

Figure 9.4

Solution   Since r and s are the x-intercepts, we look at whether the x-intercepts are positive, negative or zero.

(a) There is only one x-intercept at $x = 0$, so $r = s = 0$.
(b) There are two x-intercepts, one positive and one negative, so $r < 0$ and $s > 0$.
(c) There is only one x-intercept. It is positive and $r = s$.
(d) There are no x-intercepts, so it is not possible to write the quadratic in factored form.
(e) There are no x-intercepts, so it is not possible to write the quadratic in factored form.

Expressing a quadratic function in factored form allows us to see not only where it is zero, but also where it is positive and where it is negative.

Example 5   A college bookstore finds that if it charges p dollars for a T-shirt, it sells $1000 - 20p$ T-shirts. Its revenue is the product of the price and the number of T-shirts it sells.

(a) Express its revenue $R(p)$ as a quadratic function of the price p in factored form.
(b) For what prices is the revenue positive?

Solution   (a) The revenue $R(p)$ at price p is given by

$$\text{Revenue} = R(p) = (\text{Price})(\text{Number sold}) = p(1000 - 20p).$$

(b) We know the price, p, is positive, so to make R positive we need to make the factor $1000 - 20p$ positive as well. Writing it in the form

$$1000 - 20p = 20(50 - p)$$

we see that R is positive only when $p < 50$. Therefore the revenue is positive when $0 < p < 50$.

Interpreting Quadratic Functions Expressed in Vertex Form

The next example illustrates a form for expressing a quadratic function that shows conveniently where the function reaches its maximum value.

Example 6   The function $h(t) = -16t^2 + 32t + 128$ in Example 1 can be expressed in the form

$$h(t) = -16(t - 1)^2 + 144.$$

Use this form to show that the ball reaches its maximum height $h = 144$ when $t = 1$.

Solution   Looking at the right-hand side, we see that the term $-16(t-1)^2$ is a negative number times a square, so it is always negative or zero, and it is zero when $t = 1$. Therefore $h(t)$ is always less than or equal to 144 and is equal to 144 when $t = 1$. This means the maximum height the ball reaches is 144 feet, and it reaches that height after 1 second. See Table 9.2 and Figure 9.5.

Table 9.2   *Values of $h(t) = -16(t-1)^2 + 144$ are less than or equal to 144*

t (seconds)	$-16(t-1)^2$	$-16(t-1)^2 + 144$
0	-16	128
1	0	144
2	-16	128
3	-64	80
4	-144	0

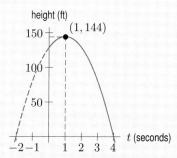

Figure 9.5: Ball reaches its greatest height of 144 ft at $t = 1$

The point $(1, 144)$ in Figure 9.5 is called the *vertex* of the graph. For quadratic functions the vertex shows where the function reaches either its largest value, called the *maximum*, or its smallest value, called the *minimum*. In Example 6 the function reaches its maximum value at the vertex because the coefficient is negative in the term $-16(t - 1)^2$.

In general:

A quadratic function in x is expressed in **vertex form** if it is written as

$$y = f(x) = a(x - h)^2 + k, \quad \text{where } a, h, \text{ and } k \text{ are constants and } a \neq 0.$$

For the function $f(x) = a(x - h)^2 + k$,
- $f(h) = k$, and the point (h, k) is the vertex of the graph.
- The coefficient a is the leading coefficient, the same a as in the standard form.
    - If $a > 0$ then k is the minimum value of the function, and the graph opens upward.
    - If $a < 0$ then k is the maximum value of the function, and the graph opens downward.

Example 7   For each function, find the maximum or minimum and sketch the graph, indicating the vertex.

(a) $g(x) = (x - 3)^2 + 2$ (b) $A(t) = 5 - (t + 2)^2$ (c) $h(x) = x^2 - 4x + 4$

Solution   (a) The expression for g is in vertex form. We have

$$g(x) = (x - 3)^2 + 2 = \text{Positive number (or zero)} + 2.$$

Thus $g(x) \geq 2$ for all values of x except $x = 3$, where it equals 2. The minimum value is 2, and the vertex is at $(3, 2)$ where the graph reaches its lowest point. See Figure 9.6(a).

(b) The expression for A is also in vertex form. We have

$$A(t) = 5 - (t + 2)^2 = 5 - \text{Positive number (or zero)}.$$

Thus $A(t) \leq 5$ for all t except $t = -2$, where it equals 5. The maximum value is 5, and the vertex is at $(-2, 5)$, where the graph reaches its highest point. See Figure 9.6(b).

(c) The expression for h is not in vertex form. However, recognizing that it is a perfect square,[2] we can write it as

$$h(x) = x^2 - 4x + 4 = (x - 2)^2,$$

which is in vertex form with $k = 0$. So $h(2) = 0$ and $h(x)$ is positive for all other values of x. Thus the minimum value is 0, and it occurs at $x = 2$. The vertex is at $(2, 0)$. See Figure 9.6(c).

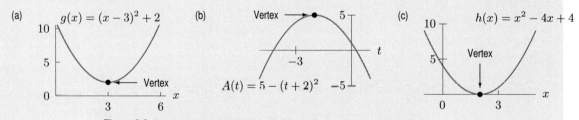

Figure 9.6: Interpretation of the vertex form of a quadratic expression

---

[2] See page 42 for a review of perfect squares.

Example 8   The functions in Example 7 can be thought of as resulting from shifts or scales of the basic function $f(x) = x^2$. Describe these operations in each case.

Solution   (a) The function $g(x)$ is formed from $f(x)$ by a horizontal shift 3 units to the right and a vertical shift up 2 units.
(b) The function $A(x)$ is formed from $f(x)$ by a horizontal shift 2 units to the left followed by a vertical shift up 5 units and a multiplication by -1.
(c) Using the form $h(x) = (x - 2)^2$ shows that $h(x)$ is formed from $f(x)$ by a horizontal shift 2 units to the right.

Exercises and Problems for Section 9.1

EXERCISES

1. A peanut, dropped at time $t = 0$ from an upper floor of the Empire State Building, has height in feet above the ground t seconds later given by

$$h(t) = -16t^2 + 1024.$$

What does the factored form

$$h(t) = -16(t - 8)(t + 8)$$

tell us about when the peanut hits the ground?

2. A coin, thrown upward at time $t = 0$ from an office in the Empire State Building, has height in feet above the ground t seconds later given by

$$h(t) = -16t^2 + 64t + 960 = -16(t - 10)(t + 6).$$

(a) From what height is the coin thrown?
(b) At what time does the coin reach the ground?

3. When a company charges a price p dollars for one of its products, its revenue is given by

$$\text{Revenue} = f(p) = 500p(30 - p).$$

(a) For what price(s) does the company have no revenue?
(b) What is a reasonable domain for $f(p)$?

In Exercises 4–7, for which values of x is the function positive and for which is it negative?

4. $f(x) = (x - 4)(x + 5)$
5. $g(x) = x^2 - x - 56$
6. $h(x) = x^2 - 12x + 36$
7. $k(x) = -(x - 1)(x - 2)$

For each function $f(x) = a(x - h)^2 + k$ graphed in 8–12, is each of the constants h and k positive, negative, or zero?

8.
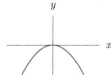

9.

10.

11.
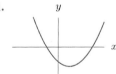

12.
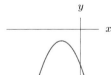

■ Match the graphs in Exercises 13–20 to the following equations, or state that there is no match.

(a) $y = (x - 2)^2 + 3$ (b) $y = -(x - 2)^2 + 3$
(c) $y = (x + 2)^2 + 3$ (d) $y = (x + 2)^2 - 3$
(e) $y = -(x + 2)^2 + 3$ (f) $y = 2(x - 2)^2 + 3$
(g) $y = (x - 3)^2 + 2$ (h) $y = (x + 3)^2 - 2$

13.
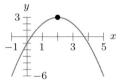

14.
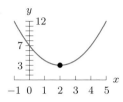

15.
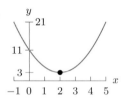

16.
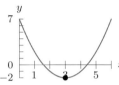

17.
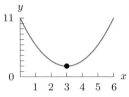

18.
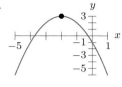

19.
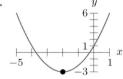

20.
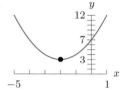

PROBLEMS

21. A ball is dropped from the top of a tower. Its height above the ground in feet t seconds after it is dropped is given by $100 - 16t^2$.

(a) Explain why the 16 tells you something about how fast the speed is changing.

(b) When dropped from the top of a tree, the height of the ball at time t is $120 - 16t^2$. Which is taller, the tower or the tree?

(c) When dropped from a building on another planet, the height of the ball is given by $100 - 20t^2$. How does the height of the building compare to the height of the tower? How does the motion of the ball on the other planet compare to its motion on the earth?

22. The average weight of a baby during the first year of life is roughly a quadratic function of time. At month m, its average weight, in pounds, is approximated by[3]

$$w(m) = -0.042m^2 + 1.75m + 8.$$

(a) What is the practical interpretation of the 8?

(b) What is the average weight of a one-year-old?

■ Each of the graphs in Problems 23–28 is the graph of a quadratic function.
(a) If the function is expressed in the form $y = ax^2 + bx + c$, say whether a and c are positive, negative, zero.
(b) If the function is expressed in the form $y = a(x-h)^2 + k$, say whether h and k are positive, negative, zero.

(c) Can the expression for the function be factored as $y = a(x - r)(x - s)$? If it can, are r and s equal to each other? Say whether they are positive, negative, or zero (assume $r \leq s$).

23.

24.

25.
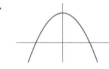

26.

27.

28.

29. A company finds that if it charges x dollars for a widget it can sell $1500 - 3x$ of them. It costs \$5 to produce a widget.

(a) Express the revenue, $R(x)$, as a function of price.

(b) Express the cost, $C(x)$, as a function of price.

(c) Express the profit, $P(x)$, which is revenue minus cost, as a function of price.

---
[3]http://www.cdc.gov/growthcharts/, accessed June 6, 2003.

■ What changes to the parameters a, h, k in the equation

$$y = a(x - h)^2 + k$$

produce the effects described in Problems 30–31?

30. The vertex of the graph is shifted down and to the right.

31. The graph changes from upward-opening to downward-opening, but the zeros (x-intercepts) do not change.

■ In Problems 32–33, use what you know about the graphs of quadratic functions.

32. Find the domain of g given that

$$g(x) = \sqrt{(x - 4)(x + 6)}.$$

33. Find the range of h given that

$$h(x) = \sqrt{25 + (x - 3)^2}.$$

9.2 WORKING WITH QUADRATIC EXPRESSIONS

In the previous section we saw how to interpret the form in which a quadratic function is expressed. In this section we see how to construct and manipulate expressions for quadratic functions.

Constructing Quadratic Expressions

In the previous section we saw that different forms give us different information about a quadratic function. The standard form tells us the vertical intercept, the factored form tells us the horizontal intercepts, and the vertex form tells us the maximum or minimum value of the function. We can also go the other way and use this information to construct an expression for a given quadratic function.

Example 1 Find a quadratic function whose graph could be

(a) 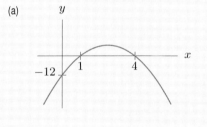   (b)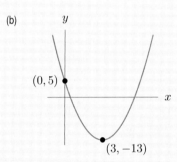

Figure 9.7: Find possible expressions for these functions

Solution (a) Since we know the zeros, we start with a function in factored form:

$$f(x) = a(x - r)(x - s).$$

The graph has x-intercepts at $x = 1$ and $x = 4$, so the function has zeros at those values, so we choose $r = 1$ and $s = 4$, which gives

$$f(x) = a(x - 1)(x - 4).$$

Since the y-intercept is -12, we know that $y = -12$ when $x = 0$. So

$$-12 = a(0 - 1)(0 - 4)$$

$$-12 = 4a$$
$$-3 = a.$$

So the function

$$f(x) = -3(x-1)(x-4)$$

has the right graph. Notice that the value of a is negative, which we expect because the graph opens downward.

(b) We are given the vertex of the parabola, so we try to write its equation using the vertex form $y = f(x) = a(x-h)^2 + k$. Since the coordinates of the vertex are $(3, -13)$, we let $h = 3$ and $k = -13$. This gives

$$f(x) = a(x-3)^2 - 13.$$

The y-intercept is $(0, 5)$, so we know that $y = 5$ when $x = 0$. Substituting, we get

$$5 = a(0-3)^2 - 13$$
$$5 = 9a - 13$$
$$18 = 9a$$
$$2 = a.$$

Therefore, the function

$$f(x) = 2(x-3)^2 - 13$$

has the correct graph. Notice that the value of a in positive, which we expect because the graph opens upward.

In Example 5 on page 259 about T-shirt sales at a bookstore, we saw how a quadratic function can arise in applications to economics that involve multiplying two linear functions together, one representing price and one representing sales. In the next example we see how a quadratic function can be used to represent profit.

Example 2   The revenue to a bookstore from selling $1000 - 20p$ T-shirts at p dollars each is

$$R(p) = p(1000 - 20p).$$

Suppose that each T-shirt costs the bookstore \$3 to make.

(a) Write an expression for the cost of making the T-shirts.
(b) Write an expression for the profit, which is the revenue minus the cost.
(c) For what values of p is the profit positive?

Solution   (a) Since each T-shirt costs \$3, we have

$$\text{Cost} = 3(\text{Number of T-shirts sold}) = 3(1000 - 2p).$$

(b) We have

$$\text{Profit} = \text{Revenue} - \text{Cost} = p(1000 - 20p) - 3(1000 - 20p).$$

(c) Factored form is the most useful for answering this question. Taking out a common factor of $(1000 - 20p)$, we get

$$\text{Profit} = p(1000 - 20p) - 3(1000 - 20p) = (p-3)(1000 - 20p) = 20(p-3)(50 - p).$$

This first factor is $p - 3$, which is positive when $p > 3$ and negative when $p < 3$. So

$$p > 50 : \quad g(p) = 20(p - 3)(50 - p) = \text{positive} \times \text{negative} = \text{negative}$$
$$3 < p < 50 : \quad g(p) = 20(p - 3)(50 - p) = \text{positive} \times \text{positive} = \text{positive}$$
$$p < 3 : \quad g(p) = 20(p - 3)(50 - p) = \text{negative} \times \text{positive} = \text{negative}.$$

So the profit is positive if the price is greater than \$3 but less than \$50.

In Example 2 we converted the expression for the function into factored form to see where it was positive. How do we convert between different forms in general?

Converting Quadratic Expressions to Standard and Factored Form

In the previous section we saw three forms for a function giving the height of a ball:

$$\begin{aligned} h(t) &= -16t^2 + 32t + 128 \quad \text{(standard form)} \\ &= -16(t - 4)(t + 2) \quad \text{(factored form)} \\ &= -16(t - 1)^2 + 144 \quad \text{(vertex form)}. \end{aligned}$$

One way to see that these forms are equivalent is to convert them all to standard form.

Converting to Standard Form

We convert an expression to standard form by expanding and collecting like terms, using the distributive law. For example, to check that $-16(t - 1)^2 + 144$ and $-16t^2 + 32t + 128$ are equivalent expressions, we expand the first term:

$$\begin{aligned} -16(t - 1)^2 + 144 &= -16(t^2 - 2t + 1) + 144 \\ &= -16t^2 + 32t - 16 + 144 \\ &= -16t^2 + 32t + 128. \end{aligned}$$

Similarly, expanding shows that $-16(t - 4)(t + 2)$ and $-16t^2 + 32t + 128$ are equivalent:

$$\begin{aligned} -16(t - 4)(t + 2) &= -16(t^2 - 4t + 2t - 8) \\ &= -16(t^2 - 2t - 8) \\ &= -16t^2 + 32t + 128. \end{aligned}$$

Converting to Factored Form

Factoring takes an expression from standard form to factored form, using the distributive law in reverse. Factoring is reviewed in Section 2.3.

Example 3   Write each of the following expressions in the indicated form.

(a) $2 - x^2 + 3x(2 - x)$ (standard)

(b) $\dfrac{(n - 1)(2 - n)}{2}$ (factored)

(c) $(z + 3)(z - 2) + z - 2$ (factored)

(d) $5(x^2 - 2x + 1) - 9$ (vertex)

Solution

(a) Expanding and collecting like terms, we have

$$2 - x^2 + 3x(2 - x) = 2 - x^2 + 6x - 3x^2 = -4x^2 + 6x + 2.$$

(b) This is already almost in factored form. All we need to do is express the division by 2 as multiplication by 1/2:

$$\frac{(n-1)(2-n)}{2} = \frac{1}{2}(n-1)(2-n) = \frac{1}{2}(n-1)(-(-2+n)) = -\frac{1}{2}(n-1)(n-2).$$

(c) We could expand this expression and then factor it, but notice that there is a common factor of $z - 2$, which enables us to get to the factored form more directly:

$$\begin{aligned}
(z+3)(z-2) + z - 2 &= (z-2)((z+3)+1) \quad \text{factor out } (z-2) \\
&= (z-2)(z+4) \\
&= (z-2)(z-(-4)).
\end{aligned}$$

(d) The key to putting this in vertex form is to recognize that the expression in parentheses is a perfect square:

$$5\underbrace{\left(x^2 - 2x + 1\right)}_{(x-1)^2} - 9 = 5(x-1)^2 - 9.$$

In Example 3(d) we had to rely on recognizing a perfect square to put the expression in vertex form. Now we give a more systematic method.

How Do We Put an Expression in Vertex Form?

Example 4   Find the vertex of the parabolas (a) $y = x^2 + 6x + 9$ (b) $y = x^2 + 6x + 8$.

Solution

(a) We recognize the expression on the right-hand side of the equal sign as a perfect square: $y = (x+3)^2$, so the vertex is at $(-3, 0)$.

(b) Unlike part (a), the expression on the right-hand side of the equal sign is not a perfect square. In order to have a perfect square, the $6x$ term should be followed by a 9. We can make this happen by adding a 9 and then subtracting it in order to keep the expressions equivalent:

$$\begin{aligned}
x^2 + 6x + 8 = \underbrace{x^2 + 6x + 9}_{\text{Perfect square}} -9 + 8 \quad &\text{add and subtract 9} \\
= (x+3)^2 - 1 \qquad &\text{since } -9 + 8 = -1.
\end{aligned}$$

Therefore, $y = x^2 + 6x + 8 = (x+3)^2 - 1$. This means the vertex is at the point $(-3, -1)$.

In the last example, we put the equation $y = x^2 + 6x + 8$ into a form where the right-hand side contains a perfect square, in a process that is called *completing the square*. To complete the square, we use the form of a perfect square

$$(x + p)^2 = x^2 + 2px + p^2.$$

Example 5   Put each expression in vertex form by completing the square.

(a) $x^2 - 8x$ (b) $x^2 + 4x - 7$

Solution   (a) We compare the expression with the form of a perfect square:

$$x^2 - 8x$$
$$x^2 + 2px + p^2.$$

To match the pattern, we must have $2p = -8$, so $p = -4$. Thus, if we add $(-4)^2 = 16$ to $x^2 - 8x$ we obtain a perfect square. Of course, adding a constant changes the value of the expression, so we must subtract the constant as well.

$$x^2 - 8x = \underbrace{x^2 - 8x + 16}_{\text{Perfect square}} - 16 \quad \text{add and subtract } 16$$

$$x^2 - 8x = (x - 4)^2 - 16.$$

Notice that the constant, 16, that was added and subtracted could have been obtained by taking half the coefficient of the x-term, $(-8/2)$, and squaring this result. This gives $(-8/2)^2 = 16$.

(b) We compare with the form of a perfect square:

$$x^2 + 4x - 7$$
$$x^2 + 2px + p^2.$$

Note that in each of the previous examples, we chose p to be half the coefficient of x. In this case, half the coefficient of x is 2, so we add and subtract $2^2 = 4$:

$$x^2 + 4x - 7 = \underbrace{x^2 + 4x + 4}_{\text{Perfect square}} - 4 - 7 \quad \text{add and subtract } 4$$

$$x^2 + 4x - 7 = (x + 2)^2 - 11.$$

An Alternative Method for Completing the Square

In Examples 4 and 5 we transform the quadratic expression for $f(x)$. Example 6 illustrates a variation on the method of completing the square that uses the equation $y = f(x)$. Since there are more operations available for transforming equations than for transforming expressions, this method is more flexible.

Example 6   Put the expression on the right-hand side in vertex form by completing the square.

(a) $y = x^2 + x + 1$ (b) $y = 3x^2 + 24x - 15$

Solution   (a) The overall strategy is to add or subtract constants from both sides to get the right-hand side in the form of a perfect square. We first subtract 1 from both sides so that the right-hand side has no constant term, then add a constant to complete the square:

$$y = x^2 + x + 1$$
$$y - 1 = x^2 + x \qquad \text{subtract 1 from both sides}$$
$$y - 1 + \frac{1}{4} = x^2 + x + \frac{1}{4} \qquad \text{add } 1/4 = (1/2)^2 \text{ to both sides}$$

$$y - \frac{3}{4} = \left(x + \frac{1}{2}\right)^2$$

$$y = \left(x + \frac{1}{2}\right)^2 + \frac{3}{4} \qquad \text{isolate } y \text{ on the left side.}$$

(b) This time we first divide both sides by 3 to make the coefficient of x^2 equal to 1, then proceed as before:

$$y = 3x^2 + 24x - 15$$

$$\frac{y}{3} = x^2 + 8x - 5 \qquad\qquad \text{divide both sides by 3}$$

$$\frac{y}{3} + 5 = x^2 + 8x \qquad\qquad \begin{array}{l}\text{add 5 to both sides so there is} \\ \text{no constant term on the right}\end{array}$$

$$\frac{y}{3} + 5 + 4^2 = x^2 + 8x + 4^2 \qquad \text{add } 4^2 = (8/2)^2 \text{ to both sides}$$

$$\frac{y}{3} + 21 = (x + 4)^2$$

$$\frac{y}{3} = (x + 4)^2 - 21$$

$$y = 3(x + 4)^2 - 63 \qquad \text{isolate } y \text{ on the left side.}$$

Example 7   A bookstore finds that if it charges $\$p$ for a T-shirt then its revenue from T-shirt sales is given by

$$R = f(p) = p(1000 - 20p)$$

What price should it charge in order to maximize the revenue?

Solution   We first expand the revenue function into standard form:

$$R = f(p) = p(1000 - 20p) = 1000p - 20p^2.$$

Since the coefficient of p^2 is -20, we know the graph of the quadratic opens downward. So the vertex form of the equation gives us its maximum value. The expression on the right is more complicated than the ones we have dealt with so far, because it has a coefficient -20 on the quadratic term. In order to deal with this, we work with the equation $R = 1000p - 20p^2$ rather than the expression $1000p - 20p^2$:

$$R = 1000p - 20p^2$$

$$-\frac{R}{20} = p^2 - 50p \qquad\qquad \text{divide by } -20$$

$$-\frac{R}{20} + 25^2 = p^2 - 50p + 25^2 \qquad \text{add } 25^2 \text{ to both sides}$$

$$-\frac{R}{20} + 25^2 = (p - 25)^2$$

$$R = -20(p - 25)^2 + 20 \cdot 25^2$$

$$R = -20(p - 25)^2 + 12{,}500.$$

So the vertex is $(25, 12{,}500)$, indicating that the price that maximizes the revenue is $p = \$25$. Figure 9.8 shows the graph of f. The graph reveals that revenue initially rises as the price increases, but eventually starts to fall again when the high price begins to deter customers.

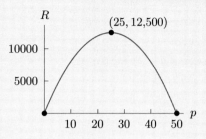

Figure 9.8: Revenue from the sale of T-shirts

Notice two differences between the method in Example 7 and the method in Examples 4 and 5: we can move the -20 out of the way temporarily by dividing both sides, and instead of adding and subtracting 25^2 to an expression, we add to both sides of an equation. Since in the end we subtract the 25^2 from both sides and multiply both sides by -20, we end up with an equivalent expression on the right-hand side.

Visualizing The Process of Completing The Square

We can visualize how to find the constant that needs to be added to $x^2 + bx$ in order to obtain a perfect square by thinking of $x^2 + bx$ as the area of a rectangle. For example, the rectangle in Figure 9.9 has area $x(x+8) = x^2 + 8x$. Now imagine cutting the rectangle into pieces as in Figure 9.10 and trying to rearrange them to make a square, as in Figure 9.11. The corner piece, whose area is $4^2 = 16$, is missing. By adding this piece to our expression, we "complete" the square: $x^2 + 8x + 16 = (x+4)^2$.

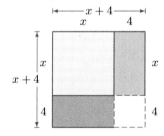

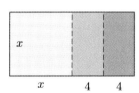

Figure 9.9: Rectangle with sides x and $x + 8$

Figure 9.10: Cutting off a strip of width 4

Figure 9.11: Rearranging the piece to make a square with a missing corner

Exercises and Problems for Section 9.2

EXERCISES

■ Exercises 1–4 show the graph of a quadratic function. Find a possible formula for the function.

1.
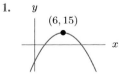

2.
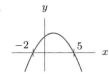

3.
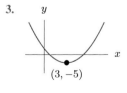

4.
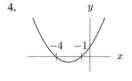

Find possible formulas for the quadratic functions described in Exercises 5–8.

5. Graph has vertex $(2, 3)$ and y-intercept -4.

6. Zeros $x = 3, -5$, and graph has y-intercept 12.

7. Graph has vertex $(3, 9)$ and passes through the origin.

8. Graph has x-intercepts 8 and 12 and y-intercept 50.

9. A rectangle is 6 feet narrower than it is long. Express its area as a function of its length l in feet.

10. The height of a triangle is 3 feet more than twice the length of its base. Express its area as a function of the length of its base, x, in feet.

In Exercises 11–12, a quadratic expression is written in vertex form.
(a) Write the expression in standard form and in factored form.
(b) Evaluate the expression at $x = 0$ and $x = 3$ using each of the three forms and compare the results.

11. $(x + 3)^2 - 1$ 12. $(x - 2)^2 - 25$

In Exercises 13–22, express the quadratic function in standard form, and identify a, b, and c.

13. $f(x) = x(x - 3)$ 14. $q(m) = (m - 7)^2$

15. $f(n) = (n - 4)(n + 7)$ 16. $g(p) = 1 - \sqrt{2}p^2$

17. $m(t) = 2(t - 1)^2 + 12$ 18. $p(q) = (q + 2)(3q - 4)$

19. $h(x) = (x - r)(x - s)$ 20. $p(x) = a(x - h)^2 + k$

21. $q(p) = (p - 1)(p - 6) + p(3p + 2)$

22. $h(t) = 3(2t - 1)(t + 5)$

In Exercises 23–30, write the expression in factored form.

23. $x^2 + 8x + 15$ 24. $x^2 + 2x - 24$

25. $(x - 4)^2 - 4$ 26. $x^2 + 7x + 10$

27. $4z^2 - 49$ 28. $9t^2 + 60t + 100$

29. $(x + 2)(x - 3) + 2(x - 3)$

30. $6w^2 + 31w + 40$

In Exercises 31–42, put the functions in vertex form $f(x) = a(x - h)^2 + k$ and state the values of a, h, k.

31. $y = -2(x - 5)^2 + 5$

32. $y = \dfrac{(x + 4)^2}{5} - 7$

33. $y = (2x - 4)^2 + 6$

34. $y - 5 = (2 - x)^2$

35. $y = x^2 + 12x + 20$

36. $y = 4x^2 + 24x + 17$

37. $y - 5 = x^2 + 2x + 1$

38. $y - 12 = 2x^2 + 8x + 8$

39. $y = x^2 + 6x + 4$

40. $y = 2x^2 + 20x + 12$

41. $y + 2 = x^2 - 3x$

42. $y = 3x^2 + 7x + 5$

In Exercises 43–46, find the x- and y-intercepts.

43. $y = 3x^2 + 15x + 12$

44. $y = x^2 - 3x - 10$

45. $y = x^2 + 5x + 2$

46. $y = 3x^2 + 18x - 30$

In Exercises 47–50, find the vertex of the parabola.

47. $y = x(x - 1)$ 48. $y = x^2 - 2x + 3$

49. $y = 2 - x^2$ 50. $y = 2x + 2 - x^2$

In Exercises 51–54, find the minimum value of the function, if it has one.

51. $f(x) = (x - 3)^2 + 10$

52. $g(x) = x^2 - 2x - 8$

53. $h(x) = (x - 5)(x - 1)$

54. $j(x) = -(x - 4)^2 + 7$

PROBLEMS

55. The profit (in thousands of dollars) a company makes from selling a certain item depends on the price of the item. The three different forms for the profit at a price of p dollars are:

Standard form: $-2p^2 + 24p - 54$

Factored form: $-2(p - 3)(p - 9)$

Vertex form: $-2(p - 6)^2 + 18$.

(a) Show that the three forms are equivalent.
(b) Which form is most useful for finding the prices that give a profit of zero dollars? (These are called the break-even prices.) Use it to find these prices.
(c) Which form is most useful for finding the profit when the price is zero? Use it to find that profit.
(d) The company would like to maximize profits. Which form is most useful for finding the price that gives the maximum profit? Use it to find the optimal price and the maximum profit.

■ In Problems 56–57, find a quadratic function with the given zeros and write it in standard form.

56. 3 and 4

57. $a + 1$ and $3a$, where a is a constant

58. Group expressions (a)–(f) together so that expressions in each group are equivalent. Note that some groups may contain only one expression.

(a) $(t + 3)^2$ **(b)** $t^2 + 9$
(c) $t^2 + 6t + 9$ **(d)** $t(t + 9) - 3(t - 3)$
(e) $\sqrt{t^4 + 81}$ **(f)** $\dfrac{2t^2 + 18}{2}$

■ Write the quadratic functions in Problems 59–60 in the following forms and state the values of all constants.
(a) Standard form $y = ax^2 + bx + c$.
(b) Factored form $y = a(x - r)(x - s)$.
(c) Vertex form $y = a(x - h)^2 + k$.

59. $y = x(x - 3) - 7(x - 3)$

60. $y = 21 - 23x + 6x^2$.

61. Find a quadratic function $F(x)$ that takes its largest value of 100 at $x = 3$, and express it in standard form.

62. A carpenter finds that if she charges p dollars for a chair, she sells $1200 - 3p$ of them each year.

(a) At what price will she price herself out of the market, that is, have no customers at all?
(b) How much should she charge to maximize her annual revenue?

63. The length of a rectangular swimming pool is twice its width. The pool is surrounded by a walk that is 2 feet wide. The area of the region consisting of the pool and the walk is 1056 square feet.

(a) Use the method of completing the square to determine the dimensions of the swimming pool.
(b) If the material for the walk costs $10 per square foot, how much would the material cost for the entire walk?

■ In Problems 64–65, put the quadratic function in factored form, and use the factored form to sketch a graph of the function without a calculator.

64. $y = x^2 + 8x + 12$ **65.** $y = x^2 - 6x - 7$

■ Write the quadratic function

$$y = 8x^2 - 2x - 15$$

in the forms indicated in Problems 66–68. Give the values of all constants.

66. $y = a(x - r)(x - s)$ **67.** $y = a(x - h)^2 + k$

68. $y = kx(vx + 1) + w$

69. Explain how you can determine the coefficient of x^2 in the standard form without expanding out:

$$x(2x + 3) - 5(x^2 + 2x + 1) - 5(10x + 2) + 3x + 25$$

What is the coefficient?

■ In Problems 70–71, write an expression $f(x)$ for the result of the given operations on x, and put it in standard form.

70. Add 5, multiply by x, subtract 2.

71. Subtract 3, multiply by x, add 2, multiply by 5.

9.3 SOLVING QUADRATIC EQUATIONS BY COMPLETING THE SQUARE

In Example 3 on page 257 we solved the equation $h(t) = 0$ to find when the ball hit the ground. When we find the zeros of a quadratic function $f(x) = ax^2 + bx + c$, we are solving the equation

$$ax^2 + bx + c = 0.$$

A **quadratic equation** in x is one which can be put into the standard form

$$ax^2 + bx + c = 0, \qquad \text{where } a, b, c \text{ are constants, with } a \neq 0.$$

Some quadratic equations can be solved by taking square roots.

Example 1 Solve (a) $x^2 - 4 = 0$ (b) $x^2 - 5 = 0$

Solution (a) Rewriting the equation as
$$x^2 = 4,$$
we see the solutions are
$$x = \pm\sqrt{4} = \pm 2.$$
(b) Similarly, the solutions to $x^2 - 5 = 0$ are $x = \pm\sqrt{5} = \pm 2.236$.

Example 2 Use the result of Example 1 to solve $(x - 2)^2 = 5$.

Solution Since the equation
$$x^2 = 5 \qquad \text{has solutions} \qquad x = \pm\sqrt{5},$$
the equation
$$(x - 2)^2 = 5 \qquad \text{has solutions} \qquad x - 2 = \pm\sqrt{5}.$$
Thus the solutions to $(x - 2)^2 = 5$ are
$$x = 2 + \sqrt{5} \qquad \text{and} \qquad x = 2 - \sqrt{5},$$
which can be combined as $x = 2 \pm \sqrt{5}$. Using a calculator, these solutions are approximately
$$x = 2 + 2.236 = 4.236 \qquad \text{and} \qquad x = 2 - 2.236 = -0.236.$$

Example 3 Solve (a) $2(y + 1)^2 = 0$ (b) $2(y - 3)^2 + 4 = 0$.

Solution (a) Since $2(y + 1)^2 = 0$, dividing by 2 gives
$$(y + 1)^2 = 0 \qquad \text{so} \qquad y + 1 = 0.$$
Thus the only solution is $y = -1$. It is possible for a quadratic equation to have only one solution.
(b) Since $2(y - 3)^2 + 4 = 0$, we have
$$2(y - 3)^2 = -4,$$
so dividing by 2 gives
$$(y - 3)^2 = -2.$$
But since no number squared is -2, this equation has no real number solutions.

In general, if we can put a quadratic equation in the form
$$(x - h)^2 = \text{Constant},$$
then we can solve it by taking square roots of both sides.

Example 4   For the function
$$h(t) = -16(t-1)^2 + 144$$
giving the height of a ball after t seconds, find the times where the ball reaches a height of 135 feet.

Solution   We want to find the values of t such that $h(t) = 135$, so we want to solve the equation
$$-16(t-1)^2 + 144 = 135.$$
Isolating the $(t-1)^2$ term we get
$$(t-1)^2 = \frac{135 - 144}{-16} = \frac{9}{16}$$
$$t - 1 = \pm\frac{3}{4}$$
$$t = 1 \pm \frac{3}{4}.$$

Therefore, the solutions are $t = 0.25$ and $t = 1.75$. So the ball reaches a height of 135 ft on its way up very soon after being thrown and again on its way down about 2 seconds after being thrown.

Next we develop a systematic method for solving quadratic equations by taking square roots. The solutions to quadratic equations are sometimes called *roots* of the equation.

The General Method of Completing the Square

In Section 9.2 we saw how to find the vertex of a parabola by completing the square. A similar method can be used to solve quadratic equations.

Example 5   Solve $x^2 + 6x + 8 = 1$.

Solution   First, we move the constant to the right by adding -8 to both sides:

$$x^2 + 6x + 8 - 8 = 1 - 8 \qquad \text{add } -8 \text{ to each side}$$
$$x^2 + 6x = -7$$
$$x^2 + 6x + 9 = -7 + 9 \qquad \text{complete the square by adding } (6/2)^2 = 9 \text{ to both sides}$$
$$(x+3)^2 = 2$$
$$x + 3 = \pm\sqrt{2} \qquad \text{take the square root of both sides}$$
$$x = -3 \pm \sqrt{2}.$$

If the coefficient of x^2 is not 1 we can divide through by it before completing the square.

Example 6   Solve $3x^2 + 6x - 2 = 0$ for x.

Solution   First, divide both sides of the equation by 3:

$$x^2 + 2x - \frac{2}{3} = 0.$$

Next, move the constant to the right by adding $2/3$ to both sides:

$$x^2 + 2x = \frac{2}{3}.$$

Now complete the square. The coefficient of x is 2, and half this is 1, so we add $1^2 = 1$ to each side:

$$x^2 + 2x + 1 = \frac{2}{3} + 1 \qquad \text{completing the square}$$
$$(x+1)^2 = \frac{5}{3}$$
$$x + 1 = \pm\sqrt{\frac{5}{3}}$$
$$x = -1 \pm \sqrt{\frac{5}{3}}.$$

The Quadratic Formula

Completing the square on the equation $ax^2 + bx + c = 0$ gives a formula for the solution of any quadratic equation. First we divide by a, getting

$$x^2 + \frac{b}{a}x + \frac{c}{a} = 0.$$

Now subtract the constant c/a from both sides:

$$x^2 + \frac{b}{a}x = -\frac{c}{a}.$$

We complete the square by adding a constant to both sides of the equation. The coefficient of x is b/a, and half this is

$$\frac{1}{2} \cdot \frac{b}{a} = \frac{b}{2a}.$$

We square this and add the result, $(b/2a)^2$, to each side:

$$x^2 + \frac{b}{a}x + \left(\frac{b}{2a}\right)^2 = -\frac{c}{a} + \left(\frac{b}{2a}\right)^2 \qquad \text{complete the square}$$
$$x^2 + \frac{b}{a}x + \frac{b^2}{4a^2} = -\frac{c}{a} + \frac{b^2}{4a^2} \qquad \text{expand parentheses}$$
$$= -\frac{c}{a} \cdot \frac{4a}{4a} + \frac{b^2}{4a^2} \qquad \text{find a common denominator}$$

$$= \frac{b^2 - 4ac}{4a^2} \qquad \text{simplify right-hand side.}$$

We now rewrite the left-hand side as a perfect square:

$$\left(x + \frac{b}{2a}\right)^2 = \frac{b^2 - 4ac}{4a^2}.$$

Taking square roots gives

$$x + \frac{b}{2a} = \frac{\sqrt{b^2 - 4ac}}{2a} \qquad \text{or} \qquad x + \frac{b}{2a} = \frac{-\sqrt{b^2 - 4ac}}{2a},$$

so

$$x = \frac{-b}{2a} + \frac{\sqrt{b^2 - 4ac}}{2a} \qquad \text{or} \qquad x = \frac{-b}{2a} - \frac{\sqrt{b^2 - 4ac}}{2a},$$

which gives the solutions

$$x = \frac{-b + \sqrt{b^2 - 4ac}}{2a} \qquad \text{and} \qquad x = \frac{-b - \sqrt{b^2 - 4ac}}{2a}.$$

The **quadratic formula** combines both solutions of $ax^2 + bx + c = 0$:

$$x = \frac{-b \pm \sqrt{b^2 - 4ac}}{2a}.$$

Example 7 Use the quadratic formula to solve for x: (a) $2x^2 - 2x - 7 = 0$ (b) $3x^2 + 3x = 10$

Solution (a) We have $a = 2$, $b = -2$, $c = -7$, so

$$x = \frac{-(-2) \pm \sqrt{(-2)^2 - 4 \cdot 2(-7)}}{2 \cdot 2} = \frac{2 \pm \sqrt{4 + 56}}{4} = \frac{2 \pm \sqrt{60}}{4}.$$

If we write $\sqrt{60} = \sqrt{4 \cdot 15} = 2\sqrt{15}$, we get

$$x = \frac{2 \pm 2\sqrt{15}}{4} = \frac{1}{2} \pm \frac{\sqrt{15}}{2}.$$

(b) We first put the equation in standard form by subtracting 10 from both sides:

$$3x^2 + 3x - 10 = 0.$$

Thus $a = 3$, $b = 3$, and $c = -10$, so

$$x = \frac{-3 \pm \sqrt{3^2 - 4 \cdot 3(-10)}}{2 \cdot 3} = \frac{-3 \pm \sqrt{9 + 120}}{6} = \frac{-3 \pm \sqrt{129}}{6} = -\frac{1}{2} \pm \frac{\sqrt{129}}{6}.$$

Example 8   A ball is thrown into the air, and its height, in feet, above the ground t seconds afterward is given by $y = -16t^2 + 32t + 8$. How long is the ball in the air?

Solution   The ball is in the air until it hits the ground, which is at a height of $y = 0$, so we want to know when

$$-16t^2 + 32t + 8 = 0.$$

Using the quadratic formula gives

$$t = \frac{-32 \pm \sqrt{32^2 - 4(-16)(8)}}{2(-16)} = \frac{-32 \pm \sqrt{1536}}{-32} = 1 \pm \frac{\sqrt{6}}{2} = 2.22, -0.22.$$

The negative root does not make sense in this context, so the time in the air is $t = 2.22$ seconds, a little more than 2 seconds.

Example 9   The distance it takes a driver to stop in an emergency is the sum of the reaction distance (the distance traveled while the driver is reacting to the emergency) and the braking distance (the distance traveled once the driver has put on the brakes). For a car traveling at v km/hr, the reaction distance is $0.42v$ meters and the stopping distance is $0.0085v^2$ meters. What is the speed of a car that stops in 100 meters?

Solution   The total stopping distance is $0.42v + 0.0085v^2$ meters, so we want to solve the equation

$$0.42v + 0.0085v^2 = 100, \quad \text{or, in standard form,} \quad 0.0085^2 + 0.42v - 100 = 0.$$

Using the quadratic formula, we get

$$v = \frac{-0.42 \pm \sqrt{0.42^2 - 4 \cdot 0.0085(-100)}}{2 \cdot 0.0085} = 86.537 \text{ or } -135.950.$$

The positive root is the only one that makes sense in this context, so the car was going about 87 km/hr.

The Discriminant

We can tell how many solutions a quadratic equation has without actually solving the equation. Look at the expression, $b^2 - 4ac$, under the square root sign in the quadratic formula.

For the equation $ax^2 + bx + c = 0$, we define the **discriminant** $D = b^2 - 4ac$.
- If $D = b^2 - 4ac$ is positive, then $\pm\sqrt{b^2 - 4ac}$ has two different values, so the quadratic equation has two distinct solutions (roots).
- If $D = b^2 - 4ac$ is negative, then $\sqrt{b^2 - 4ac}$ is the square root of a negative number, so the quadratic equation has no real solutions.
- If $D = b^2 - 4ac = 0$, then $\sqrt{b^2 - 4ac} = 0$, so the quadratic equation has only one solution. This is sometimes referred to as a repeated root.

Example 10   How many solutions does each equation have?

(a) $4x^2 - 10x + 7 = 0$ (b) $0.3w^2 + 1.5w + 1.8 = 0$ (c) $3t^2 - 18t + 27 = 0$

Solution   (a) We use the discriminant. In this case, $a = 4$, $b = -10$, and $c = 7$. The value of the discriminant is thus

$$b^2 - 4ac = (-10)^2 - 4 \cdot 4 \cdot 7 = 100 - 112 = -12.$$

Since the discriminant is negative, we know that the quadratic equation has no real solutions.

(b) We again use the discriminant. In this case, $a = 0.3$, $b = 1.5$, and $c = 1.8$. The value of the discriminant is thus

$$b^2 - 4ac = (1.5)^2 - 4 \cdot 0.3 \cdot 1.8 = 0.09.$$

Since the discriminant is positive, we know that the quadratic equation has two distinct solutions.

(c) Checking the discriminant, we find

$$b^2 - 4ac = (-18)^2 - 4 \cdot 3 \cdot 27 = 0.$$

Since the discriminant is zero, we know that the quadratic equation has only one solution.

Exercises and Problems for Section 9.3

EXERCISES

1. You wish to fence a circular garden of area 80 square meters. How much fence do you need?

■ Write the equations in Exercises **2–10** in the standard form $ax^2 + bx + c = 0$ and give possible values of a, b, c. Note that there may be more than one possible answer.

2. $2x^2 - 0.3x = 9$

3. $3x - 2x^2 = -7$

4. $4 - x^2 = 0$

5. $A = \pi \left(\dfrac{x}{2} \right)^2$

6. $-2(2x-3)(x-1) = 0$

7. $\dfrac{1}{1-x} - 4 = \dfrac{4-3x}{2-x}$

8. $7x(7 - x - 5(x - 7)) = (2x - 3)(3x - 2)$

9. $t^2 x - x^2 t^3 + t x^2 - t^3 - 4x^2 - 3x = 5$

10. $5x \left((x+1)^2 - 2 \right) = 5x \left((x+1)^2 - x \right)$

■ Solve the quadratic equations in Exercises **11–22** by taking square roots.

11. $x^2 = 9$

12. $x^2 - 7 = 0$

13. $x^2 + 3 = 17$

14. $(x+2)^2 - 4 = 0$

15. $(x - 3)^2 - 6 = 10$

16. $(x - 1)^2 + 5 = 0$

17. $(x - 5)^2 = 6$

18. $7(x - 3)^2 = 21$

19. $1 - 4(9 - x)^2 = 13$

20. $2(x - 1)^2 = 5$

21. $\left((x - 3)^2 + 1 \right)^2 = 16$

22. $\left(x^2 - 5 \right)^2 - 5 = 0$

■ Solve the quadratic equations in Exercises **23–28** or state that there are no solutions.

23. $x^2 + 6x + 9 = 4$

24. $x^2 - 12x - 5 = 0$

25. $2x^2 + 3x - 1 = 0$

26. $5x - 2x^2 - 5 = 0$

27. $x^2 + 5x - 7 = 0$

28. $(2x + 5)(x - 3) = 7$

■ In Exercises **29–45**, solve by

(a) Completing the square   (b) Using the quadratic formula

29. $x^2 + 8x + 12 = 0$

30. $x^2 - 10x - 15 = 0$

31. $2x^2 + 16x - 24 = 0$

32. $x^2 + 7x + 5 = 0$

33. $x^2 - 9x + 2 = 0$

34. $x^2 + 17x - 8 = 0$

35. $x^2 - 22x + 10 = 0$

36. $2x^2 - 32x + 7 = 0$

37. $3x^2 + 18x + 2 = 0$

38. $5x^2 + 17x + 1 = 0$

39. $6x^2 + 11x - 10 = 0$

40. $2x^2 = -3 - 7x$

41. $7x^2 = x + 8$

42. $4x^2 + 4x + 1 = 0$

43. $9x^2 - 6x + 1 = 0$

44. $4x^2 + 4x + 3 = 0$

45. $9x^2 - 6x + 2 = 0$

Find all zeros (if any) of the quadratic functions in Exercises 46–47.

46. $y = 3x^2 - 2x - 4$ **47.** $y = 5x^2 - 2x + 2$

In Exercises 48–53, use the discriminant to say whether the equation has two, one, or no solutions.

48. $2x^2 + 7x + 3 = 0$ **49.** $7x^2 - x - 8 = 0$

50. $9x^2 - 6x + 1 = 0$ **51.** $4x^2 + 4x + 1 = 0$

52. $9x^2 - 6x + 2 = 0$ **53.** $4x^2 + 4x + 3 = 0$

PROBLEMS

54. At time $t = 0$, in seconds, a pair of sunglasses is dropped from the Eiffel Tower in Paris. At time t, its height in feet above the ground is given by

$$h(t) = -16t^2 + 900.$$

  (a) What does this expression tell us about the height from which the sunglasses were dropped?
  (b) When do the sunglasses hit the ground?

55. The height of a ball t seconds after it is dropped from the top of a building is given by

$$h(t) = -16t^2 + 100.$$

  How long does it take the ball to fall k feet, where $0 \le k \le 100$?

In Problems 56–59, for what values of the constant A (if any) does the equation have no solution? Give a reason for your answer.

56. $3(x - 2)^2 = A$ **57.** $(x - A)^2 = 10$

58. $A(x - 2)^2 + 5 = 0$ **59.** $5(x - 3)^2 + A = 10$

60. Squaring both sides of the first equation below yields the second equation:

$$x = \sqrt{2x + 3}$$
$$x^2 = 2x + 3.$$

  Note that $x = 3$ is a solution to both equations. Are the equations are equivalent? Explain your reasoning.

61. A Norman window is composed of a rectangle surmounted by a semicircle whose diameter is equal to the width of the rectangle.

  (a) What is the area of a Norman window in which the rectangle is l feet long and w feet wide?
  (b) Find the dimensions of a Norman window with area 20 ft^2 and with rectangle twice as long as it is wide.

62. The New River Gorge Bridge in West Virginia is the second longest steel arch bridge in the world.[4] Its height above the ground, in feet, at a point x feet from the arch's center is $h(x) = -0.00121246x^2 + 876$.

  (a) What is the height of the top of the arch?
  (b) What is the span of the arch at a height of 575 feet above the ground?

63. A rectangle of paper is 2 inches longer than it is wide. A one inch square is cut from each corner, and the paper is folded up to make an open box with volume 80 cubic inches. Find the dimensions of the rectangle.

64. The stopping distance, in feet, of a car traveling at v miles per hour is given by[5]

$$d = 2.2v + \frac{v^2}{20}.$$

  (a) What is the stopping distance of a car going 30 mph? 60 mph? 90 mph?
  (b) If the stopping distance of a car is 500 feet, use a graph to determine how fast it was going when it braked, and check your answer using the quadratic formula.

65. If a and c have opposite signs, the equation $ax^2 + bx + c = 0$ has two solutions. Explain why this is true in two different ways:

  (a) Using what you know about the graph of $y = ax^2 + bx + c$.
  (b) Using what you know about the quadratic formula.

66. Graph each function on the same set of axes and count the number of intercepts. Explain your answer using the discriminant.

  (a) $f(x) = x^2 - 4x + 5$ **(b)** $f(x) = x^2 - 4x + 4$
  (c) $f(x) = x^2 - 4x + 3$ **(d)** $f(x) = x^2 - 4x + 2$

---

[4]www.nps.gov/neri/bridge.htm, accessed on February 19, 2005.
[5]http://www.arachnoid.com/lutusp/auto.html, accessed June 6, 2003.

67. Use what you know about the discriminant $b^2 - 4ac$ to decide what must be true about c in order for the quadratic equation $3x^2 + 2x + c = 0$ to have two different solutions.

68. Use what you know about the discriminant $b^2 - 4ac$ to decide what must be true about b in order for the quadratic equation $2x^2 + bx + 8 = 0$ to have two different solutions.

69. Show that $2ax^2 - 2(a-1)x - 1 = 0$ has two solutions for all values of the constant a, except for $a = 0$. What happens if $a = 0$?

70. Under what conditions on the constants b and c do the line $y = -x + b$ and the curve $y = c/x$ intersect in

(a) No points?

(b) Exactly one point?

(c) Exactly two points?

(d) Is it possible for the two graphs to intersect in more than two points?

71. If the equation $2x^2 - bx + 50 = 0$ has at least one real solution, what can you say about b?

72. What can you say about the constant c given that $x = 3$ is the largest solution to the equation $x^2 + 3x + c = 0$?

73. Use what you know about the quadratic formula to find a quadratic equation having

$$x = \frac{-2 \pm \sqrt{8}}{2}$$

as solutions. Your equation should be in standard form with integer (whole number) coefficients.

9.4 SOLVING QUADRATIC EQUATIONS BY FACTORING

In this section we explore an alternative method of solving quadratic equations. It does not always apply, but when it does it can be simpler than completing the square. In Section 9.1 we saw that r and s are solutions of the equation

$$a(x - r)(x - s) = 0.$$

The following principle shows that they are the only solutions.

The **zero-factor principle** states that

If $A \cdot B = 0$ then either $A = 0$ or $B = 0$ (or both).

The factors A and B can be any numbers, including those represented by algebraic expressions.

Example 1   Find the zeros of the quadratic function $f(x) = x^2 - 4x + 3$ by expressing it in factored form.

Solution   We have
$$f(x) = x^2 - 4x + 3 = (x - 1)(x - 3),$$

so $x = 1$ and $x = 3$ are zeros. To see that they are the only zeros, we let $A = x - 1$ and $B = x - 3$, and we apply the zero-factor principle. If x is a zero, then

$$\underbrace{(x - 1)}_{A} \underbrace{(x - 3)}_{B} = 0,$$

so either $A = 0$, which implies

$$x - 1 = 0$$

$$x = 1,$$

or $B = 0$, which implies

$$x - 3 = 0$$
$$x = 3.$$

In general, the zero-factor principle tells us that for a quadratic equation in the form

$$a(x - r)(x - s) = 0, \qquad \text{where } a, r, s \text{ are constants}, a \neq 0,$$

the only solutions are $x = r$ and $x = s$.

Example 2   Solve for x.

(a) $(x - 2)(x - 3) = 0$ (b) $(x - q)(x + 5) = 0$
(c) $x^2 + 4x = 21$ (d) $(2x - 2)(x - 4) = 20$

Solution   (a) This is a quadratic equation in factored form, with solutions $x = 2$ and $x = 3$.
(b) This is a quadratic equation in factored form, with solutions $x = q$ and $x = -5$.
(c) We put the equation into factored form:

$$x^2 + 4x = 21$$
$$x^2 + 4x - 21 = 0$$
$$(x + 7)(x - 3) = 0.$$

Thus the solutions are $x = -7$ and $x = 3$.
(d) The left side is in factored form, but the right side is not zero, so we cannot apply the zero-factor principle. We put the equation in a form where the right side is zero:

$$(2x - 2)(x - 4) = 20$$
$$2(x - 1)(x - 4) = 20 \quad \text{factoring out a 2}$$
$$(x - 1)(x - 4) = 10 \quad \text{dividing both sides by 2}$$
$$x^2 - 5x + 4 = 10 \quad \text{expanding left-hand side}$$
$$x^2 - 5x - 6 = 0$$
$$(x + 1)(x - 6) = 0.$$

Applying the zero-factor principle we have either

$$x + 1 = 0 \quad \text{or} \quad x - 6 = 0.$$

Solving these equations for x gives $x = -1$ and $x = 6$.

Example 3   Find the horizontal and vertical intercepts of the graph of $h(x) = 2x^2 - 16x + 30$.

Solution   The vertical intercept is the the constant term, $h(0) = 30$. The horizontal intercepts are the solutions to the equation $h(x) = 0$, which we find by factoring the expression for $h(x)$:

$$h(x) = 2(x^2 - 8x + 15)$$
$$0 = 2(x - 3)(x - 5) \quad \text{letting } h(x) = 0 \text{ and factoring}$$

The x-intercepts are at $x = 3$ and $x = 5$. See Figure 9.12.

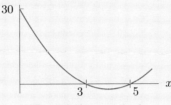

Figure 9.12: A graph of
$h(x) = 2x^2 - 16x + 30$

Example 4   Find the points where the graphs of

$$f(x) = 2x^2 - 4x + 7 \quad \text{and} \quad g(x) = x^2 + x + 1$$

intersect.

Solution   See Figure 9.13. At the points where the graphs intersect, their function values match so

$$2x^2 - 4x + 7 = x^2 + x + 1$$
$$2x^2 - x^2 - 4x - x + 7 - 1 = 0 \qquad \text{combine like terms}$$
$$x^2 - 5x + 6 = 0 \qquad \text{simplify.}$$

To apply the zero-factor principle, we factor the left-hand side:

$$(x - 2)(x - 3) = 0.$$

If x is a solution to this equation then either $x - 2 = 0$ or $x - 3 = 0$, by the zero-factor principle. So the possible solutions are $x = 2$ and $x = 3$. We check these values in the original equation:

$$f(2) = 2 \cdot 2^2 - 4 \cdot 2 + 7 = 7$$
$$g(2) = 2^2 + 2 + 1 \qquad = 7,$$

so the first intersection point is $(2, 7)$, and

$$f(3) = 2 \cdot 3^2 - 4 \cdot 3 + 7 = 13$$
$$g(3) = 3^2 + 3 + 1 \qquad = 13,$$

so the second intersection point is $(3, 13)$.

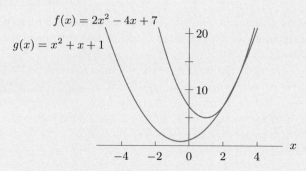

Figure 9.13: Where do the graphs of $f(x)$ and $g(x)$ intersect?

Example 5   Write a quadratic equation in t that has the given solutions.

(a) -1 and 2

(b) $2 + \sqrt{3}$ and $2 - \sqrt{3}$

(c) a and b, constants

Solution   (a) Using the factored form we get

$$(t - (-1))(t - 2) = (t + 1)(t - 2) = 0.$$

(b) If $t = 2 \pm \sqrt{3}$, then $t - 2 = \pm\sqrt{3}$, so

$$(t - 2)^2 = 3$$

is a quadratic equation with the required solutions.

(c) Again we use the factored form $(t - a)(t - b) = 0$.

Solving Other Equations Using Quadratic Equations

Factoring can be used to solve some equations that are not quadratic, as in the following examples.

Example 6   Solve $2x^3 - 5x^2 = 3x$.

Solution   We first put the equation in standard form:

$$2x^3 - 5x^2 - 3x = 0$$
$$x(2x^2 - 5x - 3) = 0 \quad \text{factor out an } x$$
$$x(2x + 1)(x - 3) = 0 \quad \text{factor the quadratic.}$$

If the product of three numbers is 0, then at least one of them must be 0. Thus

$$x = 0 \quad \text{or} \quad 2x + 1 = 0 \quad \text{or} \quad x - 3 = 0,$$

so

$$x = 0 \qquad \text{or} \qquad x = -\frac{1}{2} \qquad \text{or} \qquad x = 3.$$

We call $2x^3 - 5x^2 - 3x = 0$ a *cubic equation*, because the highest power of x is 3.

You might be tempted to divide through by x in the original equation, $2x^3 - 5x^2 = 3x$, giving $2x^2 - 5x = 3$. However this is not allowed, because x could be 0. In fact, the result of dividing by x is to lose the solution $x = 0$.

Example 7   Solve $y^4 - 10y^2 + 9 = 0$.

Solution   Since $y^4 = (y^2)^2$, the equation can be written as

$$y^4 - 10y^2 + 9 = (y^2)^2 - 10y^2 + 9 = 0.$$

We can think of this as a quadratic equation in y^2. Letting $z = y^2$, we get

$$
\begin{aligned}
z^2 - 10z + 9 &= 0 && \text{replacing } y^2 \text{ with } z \\
(z - 1)(z - 9) &= 0 && \text{factoring the left side} \\
(y^2 - 1)(y^2 - 9) &= 0 && \text{replacing } z \text{ with } y^2.
\end{aligned}
$$

Thus, the solutions are given by

$$y^2 - 1 = 0 \qquad \text{or} \qquad y^2 - 9 = 0.$$

Solving for y gives

$$y = \pm 1 \qquad \text{or} \qquad y = \pm 3.$$

The equation $y^4 - 10y^2 + 9 = 0$ is called a *quartic equation* because the highest power of y is 4.

Sometimes an equation that does not look like a quadratic equation can be transformed into one. We must be careful to check our answers after making a transformation, because sometimes the transformed equation has solutions that are not solutions to the original equation.

Example 8   Solve $\dfrac{a}{a - 1} = \dfrac{4}{a}$.

Solution   Multiplying both sides by $a(a - 1)$ gives

$$\frac{a}{a - 1} \cdot a(a - 1) = \frac{4}{a} \cdot a(a - 1).$$

Canceling $(a - 1)$ on the left and a on the right, we get

$$a^2 = 4(a - 1),$$

giving the quadratic equation

$$a^2 - 4a + 4 = 0.$$

Factoring gives

$$(a - 2)^2 = 0,$$

which has solution $a = 2$. Since we multiplied both sides by a factor that could be zero, we check the solution:

$$\frac{2}{2 - 1} = \frac{4}{2} = 2.$$

Example 9   Solve $\dfrac{x^2 - 1}{x + 3} = \dfrac{8}{x + 3}$.

Solution   Notice that the two equal fractions have the same denominator. Therefore, it will suffice to find the values of x that make the numerators equal, *provided that such values do not make the denominator zero*.

$$x^2 - 1 = 8$$

Rewriting this equation as

$$x^2 = 9$$

and taking the square root of both sides, we see that

$$x = 3 \quad \text{and} \quad x = -3.$$

The denominator of the original equation, $x + 3$, is not zero when $x = 3$. Therefore, it is a solution to the original equation. However, the denominator in the original equation is zero when $x = -3$. This means that $x = -3$ is not a solution to the original equation. So $x = -3$ is an *extraneous* solution that was introduced during the solving procedure.

Exercises and Problems for Section 9.4

EXERCISES

■ Solve the equations in Exercises **1–10** by factoring.

1. $(x - 2)(x - 3) = 0$ **2.** $x(5 - x) = 0$

3. $x^2 + 5x + 6 = 0$ **4.** $6x^2 + 13x + 6 = 0$

5. $(x - 1)(x - 3) = 8$ **6.** $(2x - 5)(x - 2)^2 = 0$

7. $x^4 + 2x^2 + 1 = 0$ **8.** $x^4 - 1 = 0$

9. $(x - 3)(x + 2)(x + 7) = 0$

10. $x(x^2 - 4)(x^2 + 1) = 0$

■ Solve the equations in Exercises **11–22** by any method.

11. $2(x - 3)(x + 5) = 0$ **12.** $x^2 - 4 = 0$

13. $5x(x + 2) = 0$ **14.** $x(x + 3) = 10$

15. $x^2 + 2x = 5x + 4$ **16.** $2x^2 + 5x = 0$

17. $x^2 - 8x + 12 = 0$ **18.** $x^2 + 3x + 7 = 0$

19. $x^2 + 6x - 4 = 0$ **20.** $2x^2 - 5x - 12 = 0$

21. $x^2 - 3x + 12 = 5x + 5$ **22.** $2x(x + 1) = 5(x - 4)$

■ Find the zeros (if any) of the quadratic functions in Exercises **23–27**.

23. $y = 3x^2 - 2x - 11$ **24.** $y = 5x^2 + 3x + 3$

25. $y = (2x - 3)(3x - 1)$ **26.** $y = 3x^2 - 5x - 1$

27. $y = x(6x - 10) - 7(3x - 5)$

■ In Exercises 28–36, write a quadratic equation in x with the given solutions.

28. $\sqrt{5}$ and $-\sqrt{5}$ **29.** 2 and -3

30. 0 and 3/2   **31.** $2 + \sqrt{3}$ and $2 - \sqrt{3}$

32. $-p$ and 0   **33.** $p + \sqrt{q}$ and $p - \sqrt{q}$

34. a and b **35.** a, no other solutions

36. With no solutions

■ Solve the equations in Exercises 37–45.

37. $x^2 - x^3 + 2x = 0$ **38.** $-2x^2 - 3 + x^4 = 0$

39. $x + \dfrac{1}{x} = 2.$ **40.** $\dfrac{2}{y} - \dfrac{2}{y-3} - 3 = 0$

41. $\dfrac{3}{z-2} - \dfrac{12}{z^2-4} = 1$ **42.** $t^4 - 13t^2 + 36 = 0.$

43. $t^4 - 3t^2 - 10 = 0.$ **44.** $t + 2\sqrt{t} - 15 = 0.$

45. $(t-3)^6 - 5(t-3)^3 + 6 = 0.$

46. Consider the equation $(x - 3)(x + 2) = 0$.

   (a) What are the solutions?
   (b) Use the quadratic formula as an alternative way to find the solutions. Compare your answers.

47. Consider the equation $x^2 + 7x + 12 = 0$.

   (a) Solve the equation by factoring.
   (b) Solve the equation using the quadratic formula. Compare your answers.

PROBLEMS

48. Which of the following equations have the same solution? Give reasons for your answers that do not depend on solving the equations.

   (a) $\dfrac{2x-1}{x-3} = \dfrac{1}{x+3}$
   (b) $\dfrac{2x-1}{x+3} = \dfrac{1}{x-3}$
   (c) $\dfrac{x-3}{2x-1} = \dfrac{1}{x+3}$
   (d) $\dfrac{x+3}{2x-1} = \dfrac{1}{x-3}$
   (e) $(2x-1)(x+3) = x - 3$
   (f) $(x-3)(x+3) = 2x - 1$
   (g) $x + 3 = (2x-1)(x-3)$

■ Without solving them, say whether the equations in Problems 49–56 have two solutions, one solution, or no solution. Give a reason for your answer.

49. $3(x-3)(x+2) = 0$ **50.** $(x-2)(x-2) = 0$

51. $(x+5)(x+5) = -10$ **52.** $(x+2)^2 = 17$

53. $(x-3)^2 = 0$ **54.** $3(x+2)^2 + 5 = 1$

55. $-2(x-1)^2 + 7 = 5$ **56.** $2(x-3)^2 + 10 = 10$

57. If a diver jumps off a diving board that is 6 ft above the water at a velocity of 20 ft/sec, his height, s, in feet, above the water can be modeled by $s(t) = -16t^2 + 20t + 6$, where $t \geq 0$ is in seconds.

   (a) How long is the diver in the air before he hits the water?
   (b) What is the maximum height achieved and when does it occur?

58. A ball is thrown straight upward from the ground. Its height above the ground in meters after t seconds is given by $-4.9t^2 + 30t + c$.

   (a) Find the constant c.
   (b) Find the values of t that make the height zero and give a practical interpretation of each value.

59. A gardener wishes to double the area of her 4 feet by 6 feet rectangular garden. She wishes to add a strip of uniform width to all of the sides of her garden. How wide should the strip be?

60. Does $x^{-1} + 2^{-1} = (x+2)^{-1}$ have solutions? If so, find them.

■ In Problems 61–64, solve **(a)** For p **(b)** For q. In each case, assume that the other quantity is nonzero and restricted so that solutions exist.

61. $p^2 + 2pq + 5q = 0$ **62.** $q^2 + 3pq = 10$

63. $p^2q^2 - p + 2 = 0$ **64.** $pq^2 + 2p^2q = 0$

65. Consider the equation $ax^2 + bx = 0$ with $a \neq 0$.

   (a) Use the discriminant to show that this equation has solutions.
   (b) Use factoring to find the solutions.
   (c) Use the quadratic formula to find the solutions.

66. We know that if $A \cdot B = 0$, then either $A = 0$ or $B = 0$. If $A \cdot B = 6$, does that imply that either $A = 6$ or $B = 6$? Explain your answer.

67. In response to the problem "Solve $x(x + 1) = 2 \cdot 6$," a student writes "We must have $x = 2$ or $x + 1 = 6$, which leads to $x = 2$ or $x = 5$ as the solutions." Is the student correct?

68. A flawed approach to solving the equation $z - 2\sqrt{z} = 8$ is shown below:

$$z - 2\sqrt{z} = 8 \qquad (1)$$
$$z - 8 = 2\sqrt{z} \qquad (2)$$
$$(z - 8)^2 = (2\sqrt{z})^2 \quad (3)$$

$$z^2 - 16z + 64 = 4z \qquad (4)$$
$$z^2 - 20z + 64 = 0 \qquad (5)$$
$$(z - 4)(z - 16) = 0 \qquad (6)$$
$$z = 4, 16. \qquad (7)$$

Identify and account for the flaw, specifying the step (1)–(7) where it is introduced.

69. The equation

$$10x^2 - 29x + 21 = 0$$

has solutions $x = 3/2$ and $x = 7/5$. Does this mean that the expressions $10x^2 - 29x + 21$ and $(x - 3/2)(x - 7/5)$ are equivalent? Explain your reasoning.

9.5 COMPLEX NUMBERS

Until now, we have been regarding expressions like $\sqrt{-4}$ as undefined, because there is no real number whose square is -4. In this section, we expand our idea of number to include *complex numbers*. In the system of complex numbers, there *is* a number whose square is -4.

Using Complex Numbers to Solve Equations

The general solution of cubic equations was discovered during the sixteenth century. The solution introduced square roots of negative numbers, called imaginary numbers. This discovery lead mathematicians to find the solutions of quadratic equations, such as

$$x^2 - 2x + 10 = 0,$$

which is not satisfied by any real number x. Applying the quadratic formula gives

$$x = \frac{2 \pm \sqrt{4 - 40}}{2} = 1 \pm \frac{\sqrt{-36}}{2}.$$

The number -36 does not have a square root which is a real number. To overcome this problem, we define the imaginary number $i = \sqrt{-1}$. Then

$$i^2 = -1.$$

Using i, we see that $(6i)^2 = 36i^2 = -36$, so

$$x = 1 \pm \frac{\sqrt{-36}}{2} = 1 \pm \frac{\sqrt{(6i)^2}}{2} = 1 \pm \frac{6i}{2} = 1 \pm 3i.$$

There are two solutions for this quadratic equation just as there were two solutions in the case of real numbers. The numbers $1 + 3i$ and $1 - 3i$ are examples of complex numbers.

A **complex number** is defined as any number that can be written in the form

$$z = a + bi,$$

where a and b are real numbers and $i = \sqrt{-1}$. The *real part* of z is the number a; the *imaginary part* is the number bi.

Calling the number i imaginary makes it sound as though i does not exist in the same way as real numbers exist. In practice, if we measure mass or position, we want our answers to be real. However, there are real-world phenomena, such as electromagnetic waves, which are described using complex numbers.

Example 1   Solve $x^2 - 6x + 15 = 4$.

Solution   To solve this equation, we put it in the form $ax^2 + bx + c = 0$.

$$x^2 - 6x + 15 = 4$$
$$x^2 - 6x + 11 = 0.$$

Applying the quadratic formula gives

$$\begin{aligned}
x &= \frac{6 \pm \sqrt{36 - 44}}{2} \\
&= \frac{6 \pm \sqrt{-8}}{2} \\
&= \frac{6 \pm \sqrt{-1} \cdot \sqrt{8}}{2} \\
&= \frac{6 \pm i\sqrt{8}}{2} \\
&= \frac{6}{2} \pm \frac{2i\sqrt{2}}{2} \\
&= 3 \pm i\sqrt{2}.
\end{aligned}$$

Algebra of Complex Numbers

We can perform operations on complex numbers much as we do on real numbers. Two complex numbers are equal if and only if their real parts are equal and their imaginary parts are equal. That is, $a + bi = c + di$ means that $a = c$ and $b = d$. In particular, the equality $a + bi = 0$ is equivalent to $a = 0$, $b = 0$. A complex number with an imaginary part equal to zero is a real number.

Two complex numbers are called *conjugates* if their real parts are equal and if their imaginary parts differ only in sign. The complex conjugate of the complex number $z = a + bi$ is denoted $\bar{z}$, so

$$\bar{z} = a - bi.$$

(Note that z is real if and only if $z = \bar{z}$.)[6]

Example 2   What is the conjugate of $5 - 7i$?

Solution   To find the conjugate, we simply change the subtraction sign to an addition sign. The conjugate of $5 - 7i$ is $5 + 7i$.

---

[6]When speaking, we say "z-bar" for $\bar{z}$.

Example 3   Find the real numbers a and b that will make the following equation true: $2 - 6i = 2a + 3bi$.

Solution   For the equation to be true, we must have $2a = 2$ and $3bi = -6i$. Thus, $a = 1$ and $3b = -6$, which gives us $b = -2$.

Addition and Subtraction of Complex Numbers

To add two complex numbers, we add the real and imaginary parts separately:

$$(a + bi) + (c + di) = (a + c) + (b + d)i.$$

Example 4   Compute the sum $(2 + 7i) + (3 - 5i)$.

Solution   Adding real and imaginary parts gives $(2 + 7i) + (3 - 5i) = (2 + 3) + (7i - 5i) = 5 + 2i$.

Subtracting one complex number from another is similar:

$$(a + bi) - (c + di) = (a - c) + (b - d)i.$$

Example 5   Compute the difference $(5 - 4i) - (8 - 3i)$.

Solution   Subtracting real and imaginary parts gives $(5 - 4i) - (8 - 3i) = (5 - 8) + (-4i - (-3i)) = (5 - 8) + (-4i + 3i) = -3 - i$.

Multiplication of Complex Numbers

Multiplication of complex numbers follows the distributive law. We use the identity $i^2 = -1$ and separate the real and imaginary parts to expand the product $(a + bi)(c + di)$:

$$
\begin{aligned}
(a + bi)(c + di) &= a(c + di) + bi(c + di) \\
&= ac + adi + bci + bdi^2 = ac + bd(-1) + adi + bci \\
&= ac - bd + adi + bci \\
&= (ac - bd) + (ad + bc)i.
\end{aligned}
$$

Example 6   Compute the product $(4 + 6i)(5 + 2i)$.

Solution   Multiplying out gives $(4 + 6i)(5 + 2i) = 20 + 8i + 30i + 12i^2 = 20 + 38i - 12 = 8 + 38i$.

Example 7   Simplify $(2 + 4i)(4 - i) + 6 + 10i$.

Solution   We wish to obtain a single complex number in the form $a + bi$. Multiplying out and adding real and imaginary parts separately:

$$(2 + 4i)(4 - i) + 6 + 10i = 8 - 2i + 16i - 4i^2 + 6 + 10i = 14 + 24i + 4 = 18 + 24i.$$

Multiplying a number by its complex conjugate gives a real, non-negative number. This property allows us to divide complex numbers easily.

Example 8   (a) Compute the product of $-5 + 4i$ and its conjugate.
(b) Compute $z \cdot \bar{z}$, where $z = a + bi$ and a, b are real.

Solution   (a) We have

$$(-5 + 4i)(-5 - 4i) = 25 + 20i - 20i - 16i^2 = 25 - 16(-1) = 25 + 16 = 41.$$

(b) We have

$$\begin{aligned} z \cdot \bar{z} &= (a + bi)(a - bi) \\ &= a^2 + abi - abi - b^2 i^2 \\ &= a^2 - b^2(-1) \\ &= a^2 + b^2. \end{aligned}$$

A special case of multiplication is the multiplication of i by itself, that is, powers of i. We know that $i^2 = -1$; then, $i^3 = i \cdot i^2 = -i$, and $i^4 = (i^2)^2 = (-1)^2 = 1$. Then $i^5 = i \cdot i^4 = i$, and so on. That is, for a nonnegative integer n, i^n takes on only four values. Thus we have

$$i^n = \begin{cases} i & \text{for } n = 1, 5, 9, 13, \ldots \\ -1 & \text{for } n = 2, 6, 10, 14, \ldots \\ -i & \text{for } n = 3, 7, 11, 15, \ldots \\ 1 & \text{for } n = 4, 8, 12, 16, \ldots \end{cases}$$

Example 9   Simplify each of the following.
(a) i^{34}
(b) $3i^8 + 2i^{21} - 3i^{43} - 4i^{26}$

Solution   Since $i^4 = 1$, we use this fact to help us simplify.
(a) $i^{34} = i^2 \cdot i^{32} = i^2(i^4)^8 = (-1)(1) = -1$
(b)

$$3i^8 + 2i^{21} - 3i^{43} - 4i^{26} = 3(i^8) + 2i \cdot i^{20} - 3i^3 \cdot i^{40} - 4i^2 \cdot i^{24}$$

$$= 3(i^4)^2 + 2i(i^4)^5 - 3i^3(i^4)^{10} - 4i^2(i^4)^6$$
$$= 3(1) + 2i(1) - 3i^3(1) - 4i^2(1)$$
$$= 3 + 2i - 3i^3 - 4i^2$$
$$= 3 + 2i - 3(-i) - 4(-1)$$
$$= 3 + 2i + 3i + 4$$
$$= 7 + 5i.$$

Division of Complex Numbers

How can we divide two complex numbers? Even a very simple case such as $(2 + i)/(1 - i)$ does not have an obvious solution. However, suppose that we divide $2 + i$ by a real number, such as 5. Then we have

$$\frac{2 + i}{5} = \frac{2}{5} + \frac{i}{5} = \frac{2}{5} + \frac{1}{5}i$$

In order to divide any two complex numbers, we use the complex conjugate to create a division by a real number, since the product of a number and its complex conjugate is always real.

Example 10   Compute $\dfrac{2 + 3i}{3 + 2i}$.

Solution   The conjugate of the denominator is $3 - 2i$, so we multiply by $(3 - 2i)/(3 - 2i)$.

$$\frac{2 + 3i}{3 + 2i} = \frac{2 + 3i}{3 + 2i} \cdot \frac{3 - 2i}{3 - 2i} = \frac{6 - 4i + 9i - 6i^2}{3^2 + 2^2} = \frac{12 + 5i}{13} = \frac{12}{13} + \frac{5}{13}i.$$

Exercises and Problems for Section 9.5

EXERCISES

Write the complex numbers in Exercises **1–18** in the form $a + bi$ where a and b are real numbers.

1. $(2 - 6i) + (23 - 14i)$

2. $(7 + i)(2 - 5i)$

3. $\dfrac{5}{4 + 3i}$

4. $\sqrt{-12}$

5. $\sqrt{-81} - \sqrt{-4}$

6. $3\sqrt{-25} - 4\sqrt{-64}$

7. $2\sqrt{-16} - 5\sqrt{-1}$

8. $(-4i^3)(-2i) + 6(5i^3)$

9. $(7 + 5i) - (12 - 11i)$

10. $6(3 + 4i) - 2i(i + 5)$

11. $(5 + 2i)^2$

12. $\dfrac{1}{4 - 5i}$

13. $(3 + 8i) + 2(4 - 7i)$

14. $(9 - 7i) - 3(5 + i)$

15. $3\sqrt{-5} + 5\sqrt{-45}$

16. $2i(3i^2 - 4i + 7)$

17. the conjugate of $11 + 13i$

18. the product of $6 - 7i$ and its conjugate

In Exercises **19–23**, find the real numbers a and b.

19. $8 + 4i = a + bi$

20. $28i = a + bi$

21. $6 = 3a + 5bi$

22. $15 - 25i = 3a + 5bi$

23. $36 + 12i = 9a - 3bi$

PROBLEMS

24. Which of the following statements is true? Explain your answer.

(a) $i^{14} = i^{24}$ (b) $i^{30} = i^{40}$

(c) $i^{32} = i^{48}$ (d) $i^{19} = i^{27}$

■ Solve the equations Problems 25–32.

25. $x^2 - 8x + 17 = 0$

26. $x^2 + 29 = 10x$

27. $x^2 - 6x + 14 = 0$

28. $x^2 - 12x + 45 = 6$

29. $2x^2 - 10x + 20 = 7$

30. $x^2 - 5x + 16 = 3$

31. $4x^2 + 2x + 1 = 4x$

32. $3x^2 + 3x + 10 = 7x + 6$

REVIEW EXERCISES AND PROBLEMS FOR CHAPTER 9

EXERCISES

■ In Exercises 1–8, sketch the graph without using a calculator.

1. $y = (x - 2)^2$

2. $y = 5 + 2(x + 1)^2$

3. $y = (3 - x)(x + 2)$

4. $y = (2x + 3)(x + 3)$

5. $y = (x - 1)(x - 5)$

6. $y = -2(x + 3)(x - 4)$

7. $y = 2(x - 3)^2 + 5$

8. $y = -(x + 1)^2 + 25$

■ In Exercises 9–14, write the expressions in standard form.

9. $-5x^2 - 2x + 3$

10. $7 - \dfrac{t^2}{2} - \dfrac{t}{3}$

11. $\dfrac{z^2 + 4z + 7}{5}$

12. $2(r - 2)(3 - 2r)$

13. $2\left(s - s(4 - s) - 1\right)$

14. $(z^2 + 3)(z^2 + 2) - (z^2 + 4)(z^2 - 1)$

■ In Exercises 15–18, find the minimum or maximum value of the quadratic expression in x.

15. $(x + 7)^2 - 8$

16. $a - (x + 2)^2$

17. $q - 7(x + a)^2$

18. $2(x^2 + 6x + 9) + 2$

■ In Exercises 19–26, write the expressions in vertex form and identify the constants a, h, and k.

19. $2(x^2 - 6x + 9) + 4$

20. $11 - 7(3 - x)^2$

21. $(2x + 4)^2 - 7$

22. $x^2 - 12x + 36$

23. $-x^2 + 2bx - b^2$

24. $6(x^2 - 8x + 16) + 2$

25. $\dfrac{(t - 6)^2 - 3}{4}$

26. $b + c(x^2 - 4dx + 4d^2) + 5$

■ In Exercises 27–29, rewrite the equation in a form that clearly shows its solutions and give the solutions.

27. $x^2 + 3x + 2 = 0$

28. $1 + x^2 + 2x = 0$

29. $5z + 6z^2 + 1 = 0$

■ Write the quadratic functions in Exercises 30–31 in the following forms and state the values of all constants.

(a) Standard form $y = ax^2 + bx + c$.

(b) Factored form $y = a(x - r)(x - s)$.

(c) Vertex form $y = a(x - h)^2 + k$.

30. $y - 8 = -2(x + 3)^2$

31. $y = 2x(3x - 7) + 5(7 - 3x)$

■ In Exercises 32–51, solve for x.

32. $x^2 - 16 = 0$

33. $x^2 + 5 = 9$

34. $(x + 4)^2 = 25$

35. $(x - 6)^2 + 7 = 7$

36. $(x - 5)^2 + 15 = 17$

37. $(x - 1)(2 - x) = 0$

38. $(x - 93)(x + 115) = 0$

39. $(x + 47)(x + 59) = 0$

40. $x^2 + 14x + 45 = 0$

41. $x^2 + 21x + 98 = 0$

42. $2x^2 + x - 55 = 0$

43. $x^2 + 16x + 64 = 0$

44. $x^2 + 26x + 169 = 0$

45. $x^2 + 12x + 7 = 0$

46. $x^2 + 24x - 56 = 0$

47. $3x^2 + 30x + 9 = 0$

48. $2x^3 - 4x^2 + 2x = 0$

49. $x^5 + 3x^3 + 2x = 0$

50. $x + \sqrt{x} = 6$

51. Solve $x^{-1} + 2^{-1} = -x - 2$.

■Solve Exercises **52–57** with the quadratic formula.

52. $x^2 - 4x - 12 = 0$ **53.** $2x^2 - 5x - 12 = 0$

54. $y^2 + 3y + 4 = 6$ **55.** $3y^2 + y - 2 = 7$

56. $7t^2 - 15t + 5 = 3$ **57.** $-2t^2 + 12t + 5 = 2t + 8$

■Solve Exercises **58–63** by completing the square.

58. $x^2 - 8x + 8 = 0$ **59.** $y^2 + 10y - 2 = 0$

60. $s^2 + 3s - 1 = 2$ **61.** $r^2 - r + 2 = 7$

62. $2t^2 - 4t + 4 = 6$ **63.** $3v^2 + 9v = 12$

64. Explain why the equation $(x - 3)^2 = -4$ has no real solution.

65. Explain why the equation $x^2 + x + 7 = 0$ has no positive solution.

■Find possible quadratic equations in standard form that have the solutions given in Exercises **66–69**.

66. $x = 2, -6$ **67.** $t = \dfrac{2}{3}, t = -\dfrac{1}{3}$

68. $x = 2 \pm \sqrt{5}$ **69.** $x = \sqrt{5}, -\sqrt{3}$

■Write the complex numbers in Exercises **70–83** in the form $a + bi$ where a and b are real numbers.

70. $(10 + 3i) - (18 - 4i)$ **71.** $\dfrac{4 - i}{3 + i}$

72. $\sqrt{-100}$ **73.** $\sqrt{-9} + 5\sqrt{-49}$

74. $\sqrt{-8} + 3\sqrt{-18}$ **75.** $2\sqrt{-3} - 4\sqrt{-27}$

76. $5\sqrt{-32} - 6\sqrt{-2}$ **77.** $(3 - i)(5 + 3i)$

78. $\dfrac{5i}{2 + i}$ **79.** $(1 + 4i) + (6 - 8i)$

80. $4i^3(6i^5) - 2i(4i^8) + 3i^2$

81. $(3 + i)(5 - i/2)$

82. $(3i)^3 + (5i)^2 - 4i + 30$

83. $\dfrac{5 + 10i}{2 - i}$

PROBLEMS

84. The three different forms for a quadratic expression are:

$$\text{Standard form: } x^2 - 10x + 16$$
$$\text{Factored form: } (x - 2)(x - 8)$$
$$\text{Vertex form: } (x - 5)^2 - 9.$$

(a) Show that the three forms are equivalent.

(b) Which form is most useful for finding the

(i) Smallest value of the expression? Use it to find that value.

(ii) Values of x when the expression is 0? Use it to find those values of x.

(iii) Value of the expression when $x = 0$? Use it to find that value.

■Problems **85–86** show the graph of a quadratic function.

(a) If the function is in standard form $y = ax^2 + bx + c$, is a positive or negative? What is c?

(b) If the function is in factored form $y = a(x - r)(x - s)$ with $r \le s$, what is r? What is s?

(c) If the function is in vertex form $y = a(x - h)^2 + k$, what is h? What is k?

85.

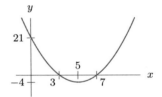

86.

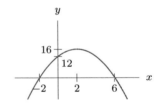

■Write the quadratic function

$$y = 4x - 30 + 2x^2$$

in the forms indicated in Problems **87–90**. Give the values of all constants.

87. $y = ax^2 + bx + c$ **88.** $y = a(x - r)(x - s)$

89. $y = a(x - h)^2 + k$ **90.** $y = ax(x - v) + w$

91. Explain why the smallest value of $f(x) = 3(x-6)^2 + 10$ occurs when $x = 6$ and give the value.

92. Explain why the largest value of $g(t) = -4(t+1)^2 + 21$ occurs when $t = -1$ and give the value.

93. Write the expression $x^2 - 6x + 9$ in a form that demonstrates that the value of the expression is always greater than or equal to zero, no matter what the value of x.

94. (a) Use the method of completing the square to write $y = x^2 - 6x + 20$ in vertex form.
(b) Use the vertex form to identify the smallest value of the function, and the x-value at which it occurs.

95. The height in feet of a stone, t seconds after it is dropped from the top of the Petronas Towers (one of the tallest buildings in the world), is given by

$$h = 1483 - 16t^2.$$

(a) How tall are the Petronas Towers?
(b) When does the stone hit the ground?

96. The height, in feet, of a rocket t seconds after it is launched is given by $h = -16t^2 + 160t$. How long does the rocket stay in the air?

97. A rectangle has one corner at $(0,0)$ and the opposite corner on the line $y = -x + b$, where b is a positive constant.

(a) Express the rectangle's area, A, in terms of x.
(b) What value of x gives the largest area?
(c) With x as in part (b), what is the shape of the rectangle?

98. A farmer encloses a rectangular paddock with 200 feet of fencing.

(a) Express the area A in square feet of the paddock as a function of the width w in feet.
(b) Find the maximum area that can be enclosed.

99. A farmer makes two adjacent rectangular paddocks with 200 feet of fencing.

(a) Express the total area A in square feet as a function of the length y in feet of the shared side.
(b) Find the maximum total area that can be enclosed.

100. A farmer has a square plot of size x feet on each side. He wants to put hedges, which cost $10 for every foot, around the plot, and on the inside of the square he wants to sow seed, which costs $0.01 for every square foot. Write an expression for the total cost.

■ In Problems **101–104**, find the number of x-intercepts, and give an explanation of your answer that does not involve changing the form of the right-hand side.

101. $y = -3(x+3)(x-5)$ **102.** $y = 1.5(x-10)^2$

103. $y = -(x-1)^2 + 5$ **104.** $y = 2(x+3)^2 + 4$

■ In Problems **105–107**, use a calculator or computer to sketch the graphs of the equations. What do you observe? Use algebra to confirm your observation.

105. $y = (x+2)(x+1) + (x-2)(x-3) - 2x(x-1)$
106. $y = (x-1)(x-2) + 3x$ and $y = x^2$
107. $y = (x-a)^2 + 2a(x-a) + a^2$ for $a = 0$, $a = 1$, and $a = 2$

108. When the square of a certain number is added to the number, the result is the same as when 48 is added to three times the number. Use the method of completing the square to determine the number.

■ In Problems **109–112**, decide for what values of the constant A (if any) the equation has a positive solution. Give a reason for your answer.

109. $x^2 + x + A = 0$ **110.** $3(x+4)(x+A) = 0$

111. $A(x-3)^2 - 6 = 0$ **112.** $5(x-3)^2 - A = 0$

■ In Problems **113–115**, the equation has the solution $x = 0$. What does this tell you about the parameters? Assume $a \neq 0$.

113. $ax^2 + bx + c = 0$

114. $a(x-r)(x-s) = 0$

115. $a(x-h)^2 + k = 0$

■ Problems **116–119** refer to the equation $(x-a)(x-1) = 0$.

116. Solve the equation for x, assuming a is a constant.

117. Solve the equation for a, assuming x is a constant not equal to 1.

118. For what value(s) of the constant a does the equation have only one solution in x?

119. For what value(s) of x does the equation have infinitely many solutions in a?

In Problems 120–123, give the values of a for which the equation has a solution in x. You do not need to solve the equation.

120. $(3x - 2)^{10} + a = 0$

121. $ax - 5 + 6x = 0$

122. $6x - 5 = 8x + a - 2x$

123. $4x^2 - (1 - 2x)^2 + a = 0$

124. For what values of a does the equation $ax^2 = x^2 - x$ have exactly one solution in x?

125. **(a)** Show that if $(x - h)^2 + k = 0$ has two distinct solutions in x, then k must be negative.

(b) Use your answer to part (a) to explain why k must be negative if $(x - h)^2 + k = (x - r)(x - s)$, with $r \neq s$.

(c) What can you conclude about k if $-2(x - h)^2 + k = -2(x - r)(x - s)$, with $r \neq s$?

126. **(a)** Explain why $2(x - h)^2 + k$ is greater than or equal to k for all values of x.

(b) If $2(x - h)^2 + k = ax^2 + bx + c$, explain why it must be true that $c \geq k$.

(c) What can you conclude about c and k if $-2(x - h)^2 + k = ax^2 + bx + c$?

127. If a is a constant, rewrite $x^2 + (a - 1)x - a = 0$ in a form that clearly shows its solutions. What are the solutions?

128. Rewrite $2a + 2 + a^2 = 0$ in a form that clearly shows it has no solutions.

SOLVING DRILL

In Problems 1–30, solve the equation for the indicated variable.

1. $x^2 - 4x + 3 = 0$; for x
2. $10 + 3x = 18 + 5x - x^2$; for x
3. $6 - 3t = 2t - 10 + 5t$; for t
4. $t^2 + 7t + 3 = 0$; for t
5. $8r + 5 = r^2 + 5r + 4$; for r
6. $5s^3 + 33 = 0$; for s
7. $2w^2 + 16w + 15 = 5w + 3$; for w
8. $5x + 13 - 2x = 10x - 9$; for x
9. $3p^2 + 3p - 8 = p^2 + 5p - 14$; for p
10. $3t^2 + 5t - 4 = 2t^2 + 8t + 6$; for t
11. $2x^5 = 12x^2$; for x
12. $5.2w - 2.3w^2 + 7.2 = 0$; for w
13. $2.1q^2 - 4.2q + 9.8 = 14.9 + 1.7q - 0.4q^2$; for q
14. $4.2t + 5.9 - 0.7t = 12.2$; for t
15. $5x^2 = 20$; for x
16. $2x(x + 5) - 4x + 9 = 0$; for x
17. $p(p - 3) = 2(p + 4)$; for p
18. $x(x + 5) - 2(4x + 1) = 2$; for x
19. $5(2s + 1) - 3(7s + 10) = 0$; for s
20. $2t(1.3t - 4.8) + 4.6(t^2 - 8) = 5t(0.8t + 3.9) - 5(3t + 4)$; for t
21. $p^2 + 5pq + 8qr^3 = 0$; for p
22. $p^2 + 5pq + 8qr^3 = 0$; for q
23. $p^2 + 5pq + 8qr^3 = 0$; for r
24. $as - bs^2 + ab = 2bs - a^2$; for s
25. $as - bs^2 + ab = 2bs - a^2$; for a
26. $as - bs^2 + ab = 2bs - a^2$; for b
27. $p(2q + p) = 3q(p + 4)$; for p
28. $p(2q + p) = 3q(p + 4)$; for q
29. $2[T]V_0 = 5V(H^5 - 3V_0)$; for H
30. $2[T]V_0 = 5V(H^5 - 3V_0)$; for V_0

Chapter 10

Exponential Functions, Expressions, and Equations

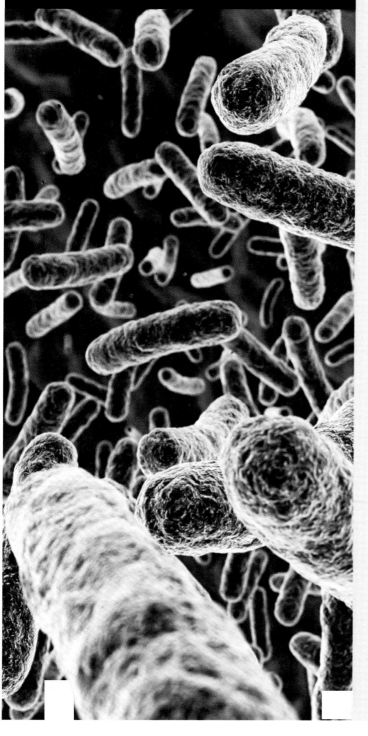

CONTENTS

10.1 EXPONENTIAL FUNCTIONS

We often describe the change in a quantity in terms of factors instead of absolute amounts. For instance, if ticket prices rise from $7 to $14, instead of saying "Ticket prices went up by $7" we might say "Ticket prices doubled." Likewise, we might say theater attendance has dropped by a third, or that online ordering of tickets has increased tenfold.

In Chapter 5 we saw that linear functions describe quantities that grow at a constant rate. Now we consider functions describing quantities that grow by a constant factor.

Example 1	In a population of bacteria, there are initially 5 million bacteria, and the number doubles every hour. Find a formula for $f(t)$, the number of bacteria (in millions) after t hours.
Solution	At $t = 0$, we have

$$f(0) = 5.$$

One hour later, at $t = 1$, the population has doubled, so we have

$$f(1) = 5 \cdot 2 = 10.$$

After 2 hours, at $t = 2$, the population has doubled again, so we have

$$f(2) = 5 \cdot 2 \cdot 2 = 5 \cdot 2^2 = 20.$$

After each hour, the population grows by another factor of 2 (that is, it gets multiplied by 2), so after 3 hours the population is $5 \cdot 2^3$, after 4 hours the population is $5 \cdot 2^4$, and so on. Thus, after t hours the population is

$$f(t) = 5 \cdot 2^t.$$

Exponents Are Used to Group Repeated Growth Factors

We can express repeated addition as multiplication. For example, in an investment that starts with $500 and grows by $50 per year for three years, the final balance is given by

$$\text{Final balance} = 500 + \underbrace{50 + 50 + 50}_{\text{3 terms of 50}} = 500 + 50 \cdot 3.$$

Now we express repeated multiplication using exponents. For example, after 3 hours, the bacteria population has doubled 3 times:

$$f(3) = 5 \cdot \underbrace{2 \cdot 2 \cdot 2}_{\text{3 factors of 2}} = 5 \cdot 2^3.$$

Example 2	Write an expression that represents the final value of a property that is initially worth $180,000 and

(a) Increases tenfold   (b) Quadruples twice
(c) Increases seven times by a factor of 1.12

Solution

(a) The final value is $180,000 \cdot 10$ dollars.

(b) Since the initial value quadruples two times, the final value is $180,000 \cdot 4 \cdot 4$ dollars, which can be written $180,000 \cdot 4^2$ dollars.

(c) Since the initial value increases by a factor of 1.12 seven times, the final value is

$$180,000 \underbrace{(1.12)(1.12)\cdots(1.12)}_{7 \text{ times}},$$

which can be written $180,000(1.12)^7$ dollars.

Exponential Functions Change by a Constant Factor

Functions describing quantities that grow by a constant positive factor, called the *growth factor*, are called *exponential functions*. For instance, the function $Q = 5 \cdot 2^t$ in Example 1 has the form

$$Q = (\text{Initial value}) \cdot (\text{Growth factor})^t,$$

where the initial value is 5 and the growth factor is 2. In general:

Exponential Functions

A quantity Q is an **exponential function** of t if it can be written as

$$Q = f(t) = a \cdot b^t, \quad a \text{ and } b \text{ constants}, b > 0.$$

Here, a is the **initial value** and b is the **growth factor**. We sometimes call b the **base**.

The base b is restricted to positive values because if $b \le 0$, then b^t is not defined for some exponents t. For example, $(-2)^{1/2}$ and 0^{-1} are not defined.

Example 3

Are the following functions exponential? If so, identify the initial value and growth factor.

(a) $Q = 275 \cdot 3^t$ (b) $Q = 80(1.05)^t$ (c) $Q = 120 \cdot t^3$

Solution

(a) This is exponential with an initial value of 275 and a growth factor of 3.

(b) This is exponential with an initial value of 80 and a growth factor of 1.05.

(c) This is not exponential because the variable t appears in the base, not the exponent. It is a power function.

Example 4

Model each situation with an exponential function.

(a) The number of animals N in a population is initially 880, and the size of the population increases by a factor of 1.14 each year.

(b) The price P of an item is initially \$55, and it increases by a factor of 1.031 each year.

Solution

(a) The initial size is 880 and the growth factor is 1.14, so $N = 880(1.14)^t$ after t years.

(b) The initial price is \$55 and the growth factor is 1.031, so $P = 55(1.031)^t$ after t years.

Example 5   The following exponential functions describe different quantities. What do they tell you about how the quantities grow?

(a) The number of bacteria (in millions) in a sample after t hours is given by $N = 25 \cdot 2^t$.
(b) The balance (in dollars) of a bank account after t years is given by $B = 1200(1.033)^t$.
(c) The value of a commercial property (in millions of dollars) after t years is given by $V = 22.1(1.041)^t$.

Solution   (a) Initially there are 25 million bacteria, and the number doubles every hour.
(b) The initial balance is \$1200, and the balance increases by a factor of 1.033 every year.
(c) The initial value is \$22.1 million, and the value increases by a factor of 1.041 every year.

Decreasing by a Constant Factor

So far, we have only considered increasing exponential functions. However, a function described by a quantity that decreases by a constant factor is also exponential.

Example 6   After being treated with bleach, a population of bacteria begins to decrease. The number of bacteria remaining after t minutes is given by

$$g(t) = 800 \left(\frac{3}{4}\right)^t.$$

Find the number of bacteria at $t = 0, 1, 2$ minutes.

Solution   We have

$$g(0) = 800 \left(\frac{3}{4}\right)^0 = 800$$

$$g(1) = 800 \left(\frac{3}{4}\right)^1 = 600$$

$$g(2) = 800 \left(\frac{3}{4}\right)^2 = 450.$$

Here the constant 800 tells us the initial size of the bacteria population, and the constant $3/4$ tells us that each minute the population is three-fourths as large as it was a minute earlier.

Graphs of Exponential Functions

Figure 10.1 shows a graph of the bacteria population from Example 1, and Figure 10.2 shows a graph of the population from Example 6. These graphs are typical of exponential functions. Notice in particular that:

- Both graphs cross the vertical axis at the initial value, $f(0) = 5$ in the case of Figure 10.1, and $g(0) = 800$ in the case of Figure 10.2.
- Neither graph crosses the horizontal axis.

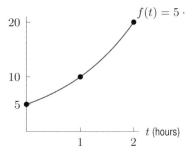

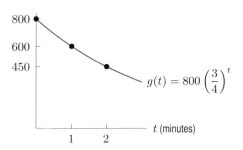

Figure 10.1: The bacteria population
$f(t) = 5 \cdot 2^t$ from Example 1

Figure 10.2: The bacteria population
$g(t) = 800(3/4)^t$ from Example 6

- The graph of $f(t) = 5 \cdot 2^t$ rises as we move from left to right, indicating that the number of bacteria increases over time. This is because the growth factor, 2, is greater than 1.

- The graph of $g(t) = 800(3/4)^t$ falls as we move from left to right, indicating that the number of bacteria decreases over time. This is because the growth factor, $3/4$, is less than 1.

Notice that exponential functions for which $b \neq 1$ are either increasing or decreasing—in other words, their graphs do not change directions. This is because an exponential function represents a quantity that changes by a constant factor. If the factor is greater than 1, the function's value increases, and the graph rises (when read from left to right); if the factor is between 0 and 1, the function's value decreases, and the graph falls.

The Domain of an Exponential Function

In Example 1 about a population of bacteria, we considered positive integer values of the exponent. But other values are possible. For example, $5 \cdot 2^{1.5}$ tells you how many bacteria there are 1.5 hours after the beginning.

Example 7   Let $P = f(t) = 1500(1.025)^t$ give the population of a town t years after 2005. What do the following expressions tell you about the town?

(a) $f(0)$ (b) $f(2.9)$ (c) $f(-4)$

Solution   (a) We have
$$f(0) = 1500(1.025)^0 = 1500.$$

This tells us that in year 2005, the population is 1500.

(b) We have
$$f(2.9) = 1500(1.025)^{2.9} = 1611.352.$$

This tells us that 2.9 years after 2005, close to the end of 2007, the population is about 1610.

(c) We have
$$f(-4) = 1500(1.025)^{-4} = 1359.926.$$

This tells us that 4 years *before* 2005, or in year 2001, the population was about 1360.

In Example 7, negative input values are interpreted as times in the past, and positive input values are interpreted as times in the future. As the example suggests, any number can serve as an input for an exponential function. Thus, the domain for an exponential function is the set of all real numbers.

Recognizing Exponential Functions

Sometimes we need to use the exponent laws to find out whether an expression can be put in a form that defines an exponential function.

Example 8   Are the following functions exponential? If so, identify the values of a and b.

(a) $Q = \dfrac{2^{-t}}{3}$ (b) $Q = \sqrt{5 \cdot 4^t}$ (c) $Q = \sqrt{5t^4}$

Solution   (a) Writing this as

$$\frac{2^{-t}}{3} = \frac{1}{3}\left(2^{-1}\right)^t = \frac{1}{3}\left(\frac{1}{2}\right)^t,$$

we see that the function is exponential. We have $a = 1/3$ and $b = 1/2$.

(b) Writing this as

$$\sqrt{5 \cdot 4^t} = (5 \cdot 4^t)^{1/2} = 5^{1/2}(4^{1/2})^t = \sqrt{5} \cdot 2^t,$$

we see that the function is exponential. We have $a = \sqrt{5}$ and $b = 2$.

(c) Writing this as

$$\sqrt{5t^4} = \sqrt{5}t^2,$$

we see that it is a power function with exponent 2 and not an exponential function.

Exercises and Problems for Section 10.1

EXERCISES

■Are the functions in Exercises **1–4** exponential? If so, identify the initial value and the growth factor.

1. $Q = 12t^4$ **2.** $Q = t \cdot 12^4$

3. $Q = 0.75(0.2)^t$ **4.** $Q = 0.2(3)^{0.75t}$

■Write the exponential functions in Exercises **5–8** in the form $Q = ab^t$ and identify the initial value and growth factor.

5. $Q = 300 \cdot 3^t$ **6.** $Q = \dfrac{190}{3^t}$

7. $Q = 200 \cdot 3^{2t}$ **8.** $Q = 50 \cdot 2^{-t}$

■Can the quantities in Exercises **9–12** be represented by exponential functions? Explain.

9. The price of gas if it grows by $0.02 a week.

10. The quantity of a prescribed drug in the bloodstream if it shrinks by a factor of 0.915 every 4 hours.

11. The speed of personal computers if it doubles every 3 years.

12. The height of a baseball thrown straight up into the air and then caught.

13. The population of a town is M in 2005 and is growing with a yearly growth factor of Z. Write a formula that gives the population, P, at time t years after 2005.

■Write expressions representing the quantities described in Exercises **14–19**.

14. The balance B increases n times in a row by a factor of 1.11.

15. The population is initially 800 and grows by a factor of d a total of k times in a row.

16. The population starts at 2000 and increases by a factor of k five times in a row.

17. The investment value V_0 drops by a third n times in a row.

18. The area covered by marsh starts at R acres and decreases by a factor of s nine times in a row.

19. The number infected N increases by a factor of z for t times in a row.

20. A lump of uranium is decaying exponentially with time, t. Which of graphs (I)–(IV) could represent this?

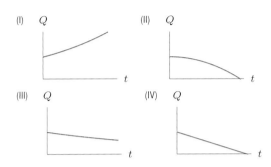

21. Which of graphs (I)–(IV) might be exponential?

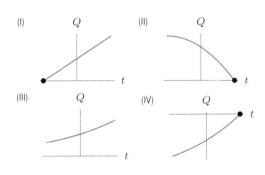

22. An investment initially worth $3000 grows by a factor of 1.062 every year. Complete Table 10.1, which shows the value of the investment over time.

Table 10.1

t	1	3	5	10	25	50
V						

23. Match the exponential functions (a)–(d) with their graphs (I)–(IV).

(a) $Q = 4(1.2)^t$ **(b)** $Q = 4(0.7)^t$

(c) $Q = 8(1.2)^t$ **(d)** $Q = 8(1.4)^t$

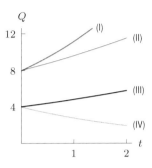

Sketch graphs of the exponential functions in Exercises 24–25. Label your axes.

24. $y = 100(1.04)^t$ **25.** $y = 800(0.9)^t$

PROBLEMS

26. In 2005, Bhutan had a population[1] of about 2,200,000 and an annual growth factor of 1.0211. Let $f(t)$ be the population t years after 2005 assuming growth continues at this rate.

(a) Evaluate the following expressions and explain what they tell you about Bhutan.

(i) $f(1)$ (ii) $f(2)$
(iii) $f(3)$ (iv) $f(5)$

(b) Write a formula for $f(t)$.

27. In 2005, the Czech Republic had a population of about 10,200,000 and a growth factor of 0.9995 (that is, the population was shrinking).[2] Let $f(t)$ be the population t years after 2005 assuming growth continues at this rate.

(a) Evaluate the following expressions and explain what they tell you about the Czech Republic.

(i) $f(1)$ (ii) $f(2)$
(iii) $f(3)$ (iv) $f(5)$

(b) Write a formula for $f(t)$.

28. Write an expression that represents the population of a bacteria colony that starts with 10,000 members and

(a) Halves twice
(b) Is multiplied 8 times by 0.7
(c) Is multiplied 4 times by 1.5

[1] www.odci.gov/cia/publications/factbook/geos/bt.html#People, accessed August 9, 2005.
[2] www.odci.gov/cia/publications/factbook/geos/ez.html, accessed August 9, 2005.

29. The population of a city after t years is given by $220{,}000(1.016)^t$. Identify the initial value and the growth factor and explain what they mean in terms of the city.

30. The number of grams of palladium-98 remaining after t minutes is given by

$$Q = 600(0.962)^t.$$

Find the quantity remaining after

(a) 1 minute   (b) 2 minutes   (c) 1 hour

31. An investment grows by 1.2 times in the first year, by 1.3 times in the second year, then drops by a factor of 0.88 in the third year. By what overall factor does the value change over the three-year period?

32. The value V, in dollars, of an investment after t years is given by the function

$$V = g(t) = 3000(1.08)^t.$$

Plot the value of the investment at 5-year intervals over a 30-year period beginning with $t = 0$. By how much does the investment grow during the first ten years? The second ten years? The third ten years?

33. A mill in a town with a population of 400 closes and people begin to move away. Find a possible formula for the number of people P in year t if each year one-fifth of the remaining population leaves.

■ Write the exponential functions in Problems 34–39 in the form $Q = ab^t$, and identify the initial value and the growth factor.

34. $Q = \dfrac{50}{2^{t/12}}$

35. $Q = 250 \cdot 5^{-2t-1}$

36. $Q = 2^t \sqrt{300}$

37. $Q = 40 \cdot 2^{2t}$

38. $Q = 45 \cdot 3^{t-2}$

39. $Q = \dfrac{\sqrt{100^t}}{5}$

40. Do the following expressions define exponential functions? If so, identify the values of a and b.

(a) $3 \cdot 0.5^x$ (b) 8^t (c) 8^{1-2x}

■ For Problems 41–44, put the function in the required form and state the values of all constants.

41. $y = 3x + 8$ in the form $y = 5 + m(x - x_0)$.

42. $y = 3\left(x\sqrt{7}\right)^3$ in the form $y = kx^p$.

43. $y = 3x^3 - 2x^2 + 4x + 5$ in the form

$$y = a + x\left(b + x(c + dx)\right).$$

44. $y = \dfrac{\left(\sqrt{3}\right)^{4t}}{5}$ in the form $y = ab^t$.

10.2 WORKING WITH THE BASE OF AN EXPONENTIAL EXPRESSION

It is sometimes convenient to describe growth in terms of a percentage rate. In Example 1, we see that growth at a constant percent rate is the same as growth by a constant factor. In other words, a quantity that grows at a constant percent rate defines an exponential function.

Example 1   A person invests \$500 in year $t = 0$ and earns 10% interest each year. Find the balance after:

(a)   10 years   (b) t years.

Solution   (a) After 1 year, the balance in dollars is given by

$$\text{Balance after 1 year} = \text{Original deposit} + \text{Interest}$$
$$= 500 + 0.10 \cdot 500$$
$$= 500(1 + 0.10) \qquad \text{factor out 500}$$
$$= 500 \cdot 1.10 = 550 \qquad \text{growth factor is } 1.10.$$

At the end of 2 years, we add another 10%, which is the same as multiplying by another growth factor of 1.10:

$$\text{Balance at end of 2 years} = \underbrace{(500 \cdot 1.10)}_{550} 1.10 = 500 \cdot 1.10^2 = 605 \text{ dollars.}$$

Since the growth factor is 1.10 every year, we get

$$\text{Balance at end of 10 years} = 500(1.10)^{10} = 1296.87 \text{ dollars.}$$

(b) Extending the pattern from part (a), we see

$$\text{Balance at end of } t \text{ years} = 500(1.10)^t \text{ dollars.}$$

Converting Between Growth Rates and Growth Factors

Example 1 shows that if a balance grows by 10% a year, the growth factor is

$$b = 1 + 0.10 = 1.10.$$

In general,

> In the exponential function $f(t) = ab^t$, the growth factor b is given by
>
> $$b = 1 + r,$$
>
> where r is the percent rate of change, called the **growth rate**. For example, if $r = 9\%$, $b = 1 + r = 1 + 9\% = 1 + 0.09 = 1.09$.

Example 2 State the growth rate r and the growth factor b for the following quantities.

(a) The size of a population grows by 0.22% each year.
(b) The value of an investment grows by 11.3% each year.
(c) The area of a lake decreases by 7.4% each year.

Solution (a) We have $r = 0.22\% = 0.0022$ and $b = 1 + r = 1.0022$.
(b) We have $r = 11.3\% = 0.113$ and $b = 1 + r = 1.113$.
(c) We have $r = -7.4\% = -0.074$ and $b = 1 + r = 1 + (-0.074) = 0.926$.

When the growth rate is negative, we can also express it as a *decay rate*. For example, the area of the lake in part (c) of Example 2 has a decay rate of 7.4%.

Example 3 What is the growth or decay rate r of the following exponential functions?

(a) $f(t) = 200(1.041)^t$ (b) $g(t) = 50(0.992)^t$ (c) $h(t) = 75\left(\sqrt{5}\right)^t$

Solution (a) Here, $b = 1.041$, so the growth rate is

$$1 + r = 1.041$$
$$r = 0.041 = 4.1\%.$$

(b) Here, $b = 0.992$, so the growth rate is

$$1 + r = 0.992$$

$$r = -0.008 = -0.8\%.$$

We could also say that the decay rate is 0.8% per year.

(c) Here, $b = \sqrt{5} = 2.2361$, so the growth rate is

$$1 + r = 2.2361$$
$$r = 1.2361 = 123.61\%.$$

It is easy to confuse growth rates and growth factors. Keep in mind that:

$$\text{Change in value} = \text{Starting value} \times \text{Growth rate}$$
$$\text{Ending value} = \text{Starting value} \times \text{Growth factor.}$$

Example 4   The starting value of a bank account is $1000. What does an annual growth rate of 4% tell us? What does a growth factor of 1.04 tell us?

Solution   A growth rate of 4% tells us that over the course of the year,

$$\text{Change in value} = \text{Starting value} \times \text{Growth rate}$$
$$= 1000(4\%) = \$40.$$

A growth factor of 1.04 tells us that after 1 year,

$$\text{Ending value} = \text{Starting value} \times \text{Growth factor}$$
$$= 1000(1.04) = \$1040.$$

Of course, both statements have the same consequence for the bank balance.

Example 5   Find the growth rate r and annual growth factor b for a quantity that

(a)   Doubles each year   (b)   Grows by 200% each year (c)   Halves each year.

Solution   (a) The quantity doubles each year, so $b = 2$. We have

$$b = 1 + r$$
$$r = 2 - 1 = 1 = 100\%.$$

In other words, doubling each year is the same as growing by 100% each year.

(b) We have $r = 200\% = 2$, so

$$b = 1 + 200\%$$
$$= 1 + 2 = 3.$$

In other words, growing by 200% each year is the same as tripling each year.

(c) The quantity halves each year, so $b = 1/2$. We have

$$b = 1 + r$$

$$r = \frac{1}{2} - 1 = -\frac{1}{2} = -50\%.$$

In other words, halving each year is the same as decreasing by 50% each year.

Recognizing Exponential Growth from a Table of Data

We can sometimes use the fact that exponential functions describe growth by a constant factor to determine if a table of data describes exponential growth.

Example 6   Could the table give points on the graph of an exponential function $y = ab^x$? If so, find a and b.

Table 10.2

x	0	1	2	3
y	4	4.4	4.84	5.324

Solution   Taking ratios, we see that

$$\frac{4.4}{4} = 1.1, \quad \frac{4.84}{4.4} = 1.1, \quad \frac{5.324}{4.84} = 1.1.$$

Thus, the value of y increases by a factor of 1.1 each time x increases by 1. This is an exponential function with growth factor $b = 1.1$. From the table, we see that the starting value is $a = 4$, so $y = 4(1.1)^x$.

The next example compares growth at a constant percent rate with linear growth.

Example 7   A person invests \$500 in year $t = 0$. Find the annual balance over a five-year period assuming:
(a)   It grows by \$50 each year.   (b)   It grows by 10% each year.

Solution   (a) See Table 10.3, which shows that the balance is a linear function of the year.

Table 10.3   *An investment earning \$50 per year*

Year	0	1	2	3	4	5
Balance	500.00	550.00	600.00	650.00	700.00	750.00

(b) See Table 10.4. Notice that the balance goes up by a larger amount each year, because the 10% is taken of an ever-larger balance.

Balance in year 1 is \$500 + 10\%(500) = \$550.00

Balance in year 2 is \$550 + 10\%(550) = \$605.00

Balance in year 3 is \$605 + 10\%(605) = \$665.50,

and so on. So the balance is *not* a linear function of the year since it does not grow by the same dollar amount each year. Rather, it is an exponential function that grows by a constant factor.

Table 10.4   *An investment growing by 10% each year*

Year	0	1	2	3	4	5
Balance	500.00	550.00	605.00	665.50	732.05	805.26

Exercises and Problems for Section 10.2

EXERCISES

■ Do the exponential expressions in Exercises **1–6** represent growth or decay?

1. $203(1.03)^t$

2. $7.04(1.372)^t$

3. $42.7(0.92)^t$

4. $0.98(1.003)^t$

5. $109(0.81)^t$

6. $0.22(0.04)^t$

■ In Exercises **7–10** give the growth factor that corresponds to the given growth rate.

7. 8.5% growth

8. 215% growth

9. 46% decay

10. 99.99% decay

■ In Exercises **11–14** give the growth rate that corresponds to the given growth factor.

11. 1.7   **12.** 5   **13.** 0.27   **14.** 0.639

■ State the starting value a, the growth factor b, and the percentage growth rate r for the exponential functions in Exercises **15–19**.

15. $Q = 200(1.031)^t$

16. $Q = 700(0.988)^t$

17. $Q = \sqrt{3}\left(\sqrt{2}\right)^t$

18. $Q = 50\left(\dfrac{3}{4}\right)^t$

19. $Q = 5 \cdot 2^t$

■ The functions in Exercises **20–23** describe the value of different investments in year t. What do the functions tell you about the investments?

20. $V = 800(1.073)^t$

21. $V = 2200(1.211)^t$

22. $V = 4000 + 100t$

23. $V = 8800(0.954)^t$

■ Write an expression representing the quantities in Exercises **24–27**.

24. A population at time t years if it is initially 2 million and growing at 3% per year.

25. The value of an investment which starts at $5 million and grows at 30% per year for t years.

26. The quantity of pollutant remaining in a lake if it is removed at 2% a year for 5 years.

27. The cost of doing a project, initially priced at $2 million, if the cost increases exponentially for 10 years.

■ In words, give a possible interpretation in terms of percentage growth of the expressions in Exercises **28–31**.

28. $\$10,000(1.06)^t$

29. $(47 \text{ grams})(0.97)^x$

30. $(400 \text{ people})(1.006)^y$

31. $\$100,000(0.98)^a$

32. An investment initially worth $3000 grows by 6.2% per year. Complete Table 10.5, which shows the value of the investment over time.

Table 10.5

t	0	1	3	5	10	25	50
V ($)	3000						

33. A commercial property initially worth $200,000 decreases in value by 8.3% per year. Complete Table 10.6, which shows the value of the property (in $1000s) over time.

Table 10.6

t	0	3	5	10	25	50
V, $1000s	200					

34. With time t, in years, on the horizontal axis and the balance on the vertical axis, plot the balance at $t = 0, 1, 2, 3$ of an account that starts in year 0 with $5000 and increases by 12% per year.

PROBLEMS

35. After t years, an initial population P_0 has grown to $P_0(1 + r)^t$. If the population at least doubles during the first year, which of the following are possible values of r?

    (a) $r = 2\%$ **(b)** $r = 50\%$

    (c) $r = 100\%$ **(d)** $r = 200\%$

■ State the starting value a, the growth factor b, and the percentage growth rate r for the exponential functions in Problems **36–40**.

36. $Q = 90 \cdot 10^{-t}$ **37.** $Q = \dfrac{10^t}{40}$

38. $Q = \dfrac{210}{3 \cdot 2^t}$ **39.** $Q = 50 \left(\dfrac{1}{2}\right)^{t/25}$

40. $Q = 2000 \left(1 + \dfrac{0.06}{12}\right)^{12t}$

41. In Example 6 on page 300, describe the change in population per minute as a percentage.

42. With t in years since 2000, the population[3] of Plano, Texas, in thousands is given by $222(1.056)^t$.

    (a) What was the population in 2000? What is its growth rate?

    (b) What population is expected in 2010?

43. A quinine tablet is taken to prevent malaria; t days later, $50(0.23)^t$ mg remain in the body.

    (a) How much quinine is in the tablet? At what rate is it decaying?

    (b) How much quinine remains 12 hours after a tablet is taken?

44. In 2005, the population of Burkina Faso was about 13,900,000 and growing at 2.53% per year.[4] Estimate the population in

    (a) 2006   **(b)** 2007

    (c) 2008   **(d)** 2010

45. The mass of tritium in a 2000 mg sample decreases by 5.47% per year. Find the amount remaining after

    (a) 1 year   **(b)** 2 years

■ The population of Austin, Texas, is increasing by 2% a year, that of Bismark, North Dakota, is shrinking by 1% a year, and that of Phoenix, Arizona, is increasing by 3% a year. Interpret the quantities in Problems **46–48** in terms of one of these cities.

46. $(1.02)^3$ **47.** $(0.99)^1$ **48.** $(1.03)^2 - 1$

■ In Problems **49–51**, could the table give points on the graph of a function $y = ab^x$, for constants a and b? If so, find the function.

49.

x	0	1	2	3
y	1	7	49	343

50.

x	0	2	4	6
y	0	1	2	4

51.

x	4	9	14	24
y	5	4.5	4.05	3.2805

52. The percent change (increase or decrease) in the value of an investment each year over a five-year period is shown in Table 10.7. By what percent does the investment's value change over this five-year period?

Table 10.7

t	1	2	3	4	5
r	12.9%	9.2%	11.3%	−4.5%	−13.6%

10.3 WORKING WITH THE EXPONENT OF AN EXPONENTIAL EXPRESSION

In Example 1 on page 298 we saw that the function $f(t) = 5 \cdot 2^t$ describes a population of bacteria that doubles every hour. What if we wanted to describe a population that doubles every 5 hours, or every 1.6 hours?

---

[3] E. Glaeser and J. Shapiro, "City Growth and the 2000 Census: Which Places Grew, and Why," at www.brookings.edu, accessed on October 24, 2004.

[4] www.odci.gov/cia/publications/factbook/geos/uv.html#People, accessed August 9, 2005.

Doubling Time and Half Life

Example 1   The value V, in dollars, of an investment t years after 2004 is given by

$$V = 4000 \cdot 2^{t/12}.$$

(a) How much is it worth in 2016? 2028? 2040?

(b) Use the form of the expression for V to explain why the value doubles every 12 years.

(c) Use exponent laws to put the expression for V in a form that shows it is an exponential function.

Solution   (a) In 2016, twelve years have passed, and $t = 12$. In 2028, twelve more years have passed, so $t = 24$, and in 2040 we have $t = 36$.

$$\text{At } t = 12: \quad V = 4000 \cdot 2^{12/12} = 4000 \cdot 2^1 = 8000$$
$$\text{At } t = 24: \quad V = 4000 \cdot 2^{24/12} = 4000 \cdot 2^2 = 16{,}000$$
$$\text{At } t = 36: \quad V = 4000 \cdot 2^{36/12} = 4000 \cdot 2^3 = 32{,}000.$$

Notice that the value seems to double every 12 years.

(b) As we see in part (a), the $t/12$ in the exponent of the power $2^{t/12}$ goes up by 1 every time t increases by 12, that is, every 12 years. If the exponent goes up by one, then the power doubles. So the value doubles every 12 years.

(c) Using exponent laws, we write

$$V = 4000 \cdot 2^{t/12} = 4000(2^{1/12})^t = 4000 \cdot 1.0595^t.$$

Thus, the value is growing exponentially, at a rate of 5.95% per year.

The *doubling time* of an exponentially growing quantity is the amount of time it takes for the quantity to double. In the previous example, if you start with an investment worth $4000, then after 12 years you have $8000, and after another 12 years you have $16,000, and so on. The doubling time is 12 years.

For exponentially decaying quantities, we consider the half-life, the time it takes for the quantity to be halved.

Example 2   Let

$$P(t) = 1000 \left(\frac{1}{2}\right)^{t/7}$$

be the population of a town t years after it was founded.

(a) What is the initial population? What is the population after 7 years? After 14 years? When is the population 125?

(b) Rewrite $P(t)$ in the form $P(t) = ab^t$ for constants a and b, and give the annual growth rate of the population.

Solution   (a) Since $P(0) = 1000(1/2)^{0/7} = 1000$, the initial population is 1000. After 7 years the population is

$$P(7) = 1000 \left(\frac{1}{2}\right)^{7/7} = 1000 \left(\frac{1}{2}\right)^1 = 1000 \cdot \frac{1}{2} = 500,$$

and after 14 years it is

$$P(14) = 1000 \left(\frac{1}{2}\right)^{14/7} = 1000 \left(\frac{1}{2}\right)^2 = 1000 \cdot \frac{1}{4} = 250.$$

Notice that the population halves in the first 7 years and then halves again in the next 7 years. To reach 125 it must halve yet again, which happens after the next 7 years, when $t = 21$:

$$P(21) = 1000 \left(\frac{1}{2}\right)^{21/7} = 1000 \left(\frac{1}{2}\right)^3 = 1000 \cdot \frac{1}{8} = 125.$$

(b) We have

$$P(t) = 1000 \left(\frac{1}{2}\right)^{t/7} = 1000 \left(\left(\frac{1}{2}\right)^{1/7}\right)^t = 1000(0.906)^t.$$

So $a = 1000$ and $b = 0.906$. Since $b = 1 + r$, the growth rate is $r = -0.094 = -9.4\%$. Alternatively, we can say the population is decaying at a rate of 9.4% a year.

So far, we have considered quantities that double or halve in a fixed time period (12 years, 7 years, etc.). Example 3 uses the base 3 to represent a quantity that *triples* in a fixed time period.

Example 3   The number of people owning the latest phone after t months is given by

$$N(t) = 450 \cdot 3^{t/6}.$$

How many people own the phone in 6 months? In 20 months?

Solution   In 6 months

$$N(6) = 450 \cdot 3^{6/6}$$
$$= 450 \cdot 3$$
$$= 1350 \text{ people.}$$

In 20 months, there are

$$N(20) = 450 \cdot 3^{20/6}$$
$$= 17,523.332,$$

or approximately 17,523 people.

Comparing the functions
$$V = 4000 \cdot 2^{t/12},$$
which describes a quantity that doubles every 12 years, and
$$N = 450 \cdot 3^{t/6},$$
which describes a quantity that triples every 6 months, we can generalize as follows: The function
$$Q = ab^{t/T}$$

describes a quantity that increases by a factor of b every T years.

Example 4   Describe what the following functions tell you about the value of the investments they describe, where t is measured in years.

(a) $V = 2000 \cdot 3^{t/15}$ (b) $V = 3000 \cdot 4^{t/40}$ (c) $V = 2500 \cdot 10^{t/100}$

Solution   (a) This investment begins with $2000 and triples (goes up by a factor of 3) every 15 years.
(b) This investment begins with $3000 and quadruples (goes up by a factor of 4) every 40 years.
(c) This investment begins with $2500 goes up by a factor of 10 every 100 years.

Negative Exponents

We have seen that exponential decay is represented by ab^t with $0 < b < 1$. An alternative way to represent it is by using a negative sign in the exponent.

Example 5   Let $Q(t) = 4^{-t}$. Show that Q represents an exponentially decaying quantity that decreases to $1/4$ of its previous amount for every unit increase in t.

Solution   We rewrite $Q(t)$ as follows:

$$Q(t) = 4^{-t} = (4^{-1})^t = \left(\frac{1}{4}\right)^t.$$

This represents exponential decay with a growth factor of $1/4$, so the quantity gets multiplied by $1/4$ every unit of time.

Example 6   Express the population in Example 2 using a negative exponent.

Solution   Since $1/2 = 2^{-1}$, we can write

$$P(t) = 1000 \left(\frac{1}{2}\right)^{t/7} = 1000(2^{-1})^{t/7} = 1000 \cdot 2^{-t/7}.$$

Converting Between Different Time Scales

If a bank account starts at $500 and grows by 1% each month, its value after t months is given by $B(t) = 500 \cdot 1.01^t$. What if we want to know how much the account grows each year? Since a year is 12 months, we have

$$\text{Amount after one year} = 500 \cdot 1.01^{12} = \$563.41$$
$$\text{Amount after two years} = 500 \cdot 1.01^{24} = \$634.87$$
$$\text{Amount after } n \text{ years} = 500 \cdot 1.01^{12n}.$$

Example 7   Write a formula for the balance $C(n)$ after n years that shows the annual growth rate of the account. Which is better, an account earning 1% per month or one earning 12% per year?

Solution   Using the exponent rules, we get

$$C(n) = 500 \cdot 1.01^{12n} = 500 \left(1.01^{12}\right)^n$$
$$= 500 \cdot 1.127^n,$$

because $1.01^{12} = 1.127$. This represents the account balance after n years, so the annual growth factor is 1.127. Since $1.127 = 1 + 0.127 = 1 + r$, the annual growth rate r is 0.127, or 12.7%. This is more than 12%, so an account earning 1% per month is better than one earning 12% per year.

The next example shows how to go the other way, converting from an annual growth rate to a monthly growth rate.

Example 8   The population, in thousands, of a town after t years is given by

$$P = 50 \cdot 1.035^t.$$

(a) What is the annual growth rate?
(b) Write an expression that gives the population after m months.
(c) Use the expression in part (b) to find the town's monthly growth rate.

Solution   (a) From the given expression, we see that the growth factor is 1.035, so each year the population is 3.5% larger than the year before, and it is growing at 3.5% per year.
(b) We know that

$$\underbrace{\text{Number of months}}_{m} = 12 \times \underbrace{\text{Number of years}}_{t}$$
$$m = 12t,$$

so $t = m/12$. This means

$$50 \cdot 1.035^t = 50 \cdot 1.035^{m/12},$$

so the population after m months is given by

$$50 \cdot 1.035^{m/12}.$$

(c) Using the exponent laws, we rewrite this as

$$50 \left(1.035^{1/12}\right)^m = 50 \cdot 1.0029^m,$$

because $1.035^{1/12} = 1.0029$. So the monthly growth rate 0.0029, or 0.29%.

Shifting the Exponent

Sometimes it is useful to express an exponential function in a form that has a constant subtracted from the exponent. We call this shifting the exponent.

Example 9   The population of a town was 1500 people in 2005 and increasing at rate of 2.5% per year. Express the population in terms of (a) the number of years, t, since 2005, and   (b) the year, y. Use each expression to calculate the population in the year 2014.

Solution   (a) Here the starting value, when $t = 0$, is 1500, so

$$P = f(t) = 1500 \cdot 1.025^t.$$

Since 2014 is 9 years since 2005, we have $t = 9$. Substituting this into our equation, we get:

$$P = f(9) = 1500 \cdot 1.025^9$$
$$= 1873.294.$$

Thus, in the year 2014, we would have approximately 1873 people.
(b) If y is the year, then the relationship between y and t is given by

$$y = 2005 + t.$$

For instance, at $t = 9$ we have $y = 2005 + 9 = 2014$. Solving for t,

$$t = y - 2005,$$

we can rewrite our equation for P by substituting:

$$P = g(y) = 1500 \cdot 1.025^{y-2005} \quad \text{because } t = y - 2005.$$

So, in 2014, we would have

$$P = g(2014) = 1500 \cdot 1.025^{2014-2005}$$
$$= 1500 \cdot 1.025^9$$
$$= 1873.294,$$

which is the same answer as before.

In Example 9, we shifted the exponent by 2005. We can convert the resulting formula to standard form by using exponent rules:

$$P = 1500 \cdot 1.025^{y-2005}$$
$$= 1500 \cdot 1.025^y \cdot 1.025^{-2005}$$
$$= \frac{1500}{1.025^{2005}} \cdot 1.205^y.$$

However, in this form of the expression, the starting value, $a = 1500/1.025^{2005}$, does not have a reasonable interpretation as the size of the population in the year 0, because it is a very small number, and the town probably did not exist in the year 0.[5] It is more convenient to use the form $P = 1500 \cdot 1.025^{y-2005}$, because we can see at a glance that the population is 1500 in the year 2005.
In general, a shift of t_0 years gives us the formula

$$P = P_0 b^{t-t_0}.$$

---

[5] In fact, there is no year 0 in the Gregorian calendar. The year before year 1 is the year 1 B.C.E.

Here, P_0 is the value of P in year t_0. This form is like the point-slope form $y = y_0 + m(x - x_0)$ for linear functions, where y_0 is the value of y at $x = x_0$.

Example 10   Let $B = 5000(1.071)^{t-1998}$ be the value of an investment in year t. Here, we have $B_0 = 5000$ and $t_0 = 1998$, which tells us that the investment's value in 1998 was \$5000.

Exercises and Problems for Section 10.3

EXERCISES

1. The average rainfall in Hong Kong in January and February is about 1 inch each month. From March to June, however, average rainfall in each month is double the average rainfall of the previous month.

   (a) Make a table showing average rainfall for each month from January to June.
   (b) Write a formula for the average rainfall in month n, where $2 \leq n \leq 6$ and January is month 1.
   (c) What is the total average rainfall in the first six months of the year?

2. Find a formula for the value of an investment initially worth \$12,000 that grows by 12% every 5 years.

3. Find the starting value a, the growth factor b, and the growth rate r for the exponential function $Q = 500 \cdot 2^{t/7}$.

4. Which is better, an account earning 2% per month or one earning 7% every 3 months?

5. A sunflower grows at a rate of 1% a day; another grows at a rate of 7% per week. Which is growing faster? Assuming they start at the same height, compare their heights at the end of 1 and 5 weeks.

Write the functions in Exercises **6–9** in the form $Q = ab^t$. Give the values of the constants a and b.

6. $Q = \dfrac{1}{3} \cdot 2^{t/3}$

7. $Q = -\dfrac{5}{3^t}$

8. $Q = 7 \cdot 2^t \cdot 4^t$

9. $Q = 4(2 \cdot 3^t)^3$

What do the functions in Exercises **10–11** tell you about the quantities they describe?

10. The size P of a population of animals in year t is $P = 1200(0.985)^{12t}$.

11. The value V of an investment in year t is $V = 3500 \cdot 2^{t/7}$.

Find values for the constants a, b, and T so that the quantities described in Exercises **12–15** are represented by the function
$$Q = ab^{t/T}.$$

12. A population begins with 1000 members and doubles every twelve years.

13. An investment is initially worth 400 and triples in value every four years.

14. A community that began with only fifty households having cable television service has seen the number increase fivefold every fifteen years.

15. The number of people in a village of 6000 susceptible to a certain strain of virus has gone down by one-third every two months since the virus first appeared.

Find values for the constants a, b, and T so that the quantities described in Exercises **16–19** are represented by the function
$$Q = ab^{-t/T}.$$

16. A lake begins with 250 fish of a certain species, and one-half disappear every six years.

17. A commercial property initially worth \$120,000 loses 25% of its value every two years.

18. The amount of a 200 mg injection of a therapeutic drug remaining in a patient's blood stream goes down by one-fifth every 90 minutes.

19. The number of people who have not heard about a new movie is initially N and decreases by 10% every five days.

Find values for the constants a, b, and t_0 so that the quantities described in Exercises **20–23** are represented by the function
$$Q = ab^{t-t_0}.$$

20. The size of a population in year $t = 1995$ is 800 and it grows by 2.1% per year.

21. After $t = 3$ hours there are 5 million bacteria in a culture and the number doubles every hour.

22. A house whose value increases by 8.2% per year is worth $220,000 in year $t = 7$.

23. A 200 mg sample of radioactive material is isolated at time $t = 8$ hours, and it decays at a rate of 17.5% per hour.

PROBLEMS

24. Formulas I–III all describe the growth of the same population, with time, t, in years: I. $P = 15(2)^{t/6}$ II. $P = 15(4)^{t/12}$ III. $P = 15(16)^{t/24}$

    (a) Show that the three formulas are equivalent.
    (b) What does formula I tell you about the doubling time of the population?
    (c) What do formulas II and III tell you about the growth of the population? Give answers similar to the statement which is the answer to part (b).

25. Find the half-lives of the following quantities:

    (a) A decays by 50% in 1 week.
    (b) B decays to one quarter of the original in 10 days.
    (c) C decays by 7/8 in 6 weeks.

26. An exponentially growing population quadruples in 22 years. How long does it take to double?

27. A radioactive compound decays to 25% of its original quantity in 90 minutes. What is its half-life?

28. Arrange the following expressions in order of increasing half-lives (from smallest to largest):

    $$20(0.92)^t; \quad 120(0.98)^t; \quad 0.27(0.9)^t; \quad 90(0.09)^t.$$

29. The population of bacteria m doubles every 24 hours. The population of bacteria q grows by 3% per hour. Which has the larger population in the long term?

30. The half-life of substance A is 17 years, and substance B decays at a rate of 30% per decade. Of which substance is there less in the long term?

For the functions in the form $P = ab^{t/T}$ describing population growth in Problems **31–36**:
(a) Give the values of the constants a, b, and T. What do these constants tell you about population growth?
(b) Give the annual growth rate.

31. $P = 400 \cdot 2^{t/4}$ 32. $P = 800 \cdot 2^{t/15}$

33. $P = 80 \cdot 3^{t/5}$ 34. $P = 75 \cdot 10^{t/30}$

35. $P = 50 \left(\frac{1}{2}\right)^{t/6}$ 36. $P = 400 \left(\frac{2}{3}\right)^{t/14}$

State the starting value a, the growth factor b, and the growth rate r as a percent correct to 2 decimals for the exponential functions in Problems **37–40**.

37. $V = 2000(1.0058)^t$

38. $V = 500 \left(3 \cdot 2^{t/5}\right)$

39. $V = 5000(0.95)^{t/4}$

40. $V = 3000 \cdot 2^{t/9} + 5000 \cdot 2^{t/9}$

41. The value of an investment grows by a factor of 1.011 each month. By what percent does it grow each year?

42. The value of an investment grows by 0.06% every day. By what percent does it increase in a year?

43. A property decreases in value by 0.5% each week. By what percent does it decrease after one year (52 weeks)?

44. The area of a wetland drops by a third every five years. What percent of its total area disappears after twenty years?

Prices are increasing at 5% per year. What is wrong with the statements in Problems **45–53**? Correct the formula in the statement.

45. A $6 item costs $\$(6 \cdot 1.05)^7$ in 7 years' time.

46. A $3 item costs $\$3(0.05)^{10}$ in ten years' time.

47. The percent increase in prices over a 25-year period is $(1.05)^{25}$.

48. If time t is measured in months, then the price of a $100 item at the end of one year is $\$100(1.05)^{12t}$.

49. If the rate at which prices increase is doubled, then the price of a $20 object in 7 years' time is $\$20(2.10)^7$.

50. If time t is measured in decades (10 years), then the price of a $45 item in t decades is $\$45(1.05)^{0.1t}$.

51. Prices change by $10 \cdot 5\% = 50\%$ over a decade.

52. Prices change by $(5/12)\%$ in one month.

53. A $250 million town budget is trimmed by 1% but then increases with inflation as prices go up. Ten years later, the budget is $\$250(1.04)^{10}$ million.

For the investments described in Problems **54–59**, assume that t is the elapsed number of years and that T is the elapsed number of months.
(a) Describe in words how the value of the investment changes over time.
(b) Give the annual growth rate.

54. $V = 3000(1.0088)^{12t}$ 55. $V = 6000(1.021)^{4t}$

56. $V = 250(1.0011)^{365t}$ 57. $V = 400(1.007)^T$

58. $V = 625(1.03)^{T/3}$ 59. $V = 500(1.2)^{t/2}$

10.4   SOLVING EXPONENTIAL EQUATIONS

Suppose that a sample of 80 million bacteria is needed from the population in Example 1 on page 298, given in millions by $Q = 5 \cdot 2^t$ after t hours. The population is large enough when

$$80 = 5 \cdot 2^t.$$

This is an example of an exponential equation. The first step in solving it is to divide by the initial value and get

$$\frac{80}{5} = \frac{5 \cdot 2^t}{5}$$
$$16 = 2^t.$$

Since $16 = 2^4$, we see that $t = 4$ is a solution to the equation, so there are 80 million bacteria after 4 hours.

Example 1   A population is given by $P = 149(2^{1/9})^t$, where t is in years. When does the population reach 596?

Solution   We solve the equation

$$149(2^{1/9})^t = 596.$$

Dividing by 149, we get

$$(2^{1/9})^t = 4$$
$$2^{t/9} = 2^2.$$

Since the two exponents must be equal, we can see that

$$\frac{t}{9} = 2$$
$$t = 18 \text{ years.}$$

Notice that solving the equation $80 = 5 \cdot 2^4$ depended on 16 being a whole number power of 2, and solving the equation in Example 1 depended on 4 being the square of 2. What do we do if we encounter an exponential equation that does not work out so neatly?

Using Tables and Graphs to Approximate Solutions to Equations

Suppose we want 70 million bacteria in Example 1 on page 298, so we want to solve

$$5 \cdot 2^t = 70.$$

This equation cannot be solved so easily. Since there are 40 million bacteria after 3 hours and 80 million bacteria after 4 hours, we know that the solution must be between 3 and 4 hours, but can we find a more precise value?

In Chapter 11 we use logarithms to solve exponential equations algebraically. For now, we consider numerical and graphical methods. We try values of t between 3 and 4. Table 10.8 shows that $t \approx 3.81$ hours is approximately the solution to $5 \cdot 2^t = 70$. Alternatively, we can trace along a graph of $y = 5 \cdot 2^t$ until we find the point with y-coordinate 70, as shown in Figure 10.3.

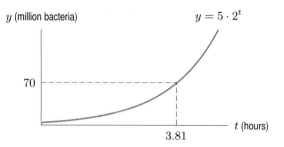

Figure 10.3: Graph of bacteria after t hours

Table 10.8   *Million bacteria after t hours*

t (hours)	3	3.5	3.6	3.7	3.8	3.81	3.9	4
$y = 5 \cdot 2^t$	40	56.6	60.6	65.0	69.6	70.1	74.6	80

Example 2   A bank account begins at $22,000 and earns 3% interest per year. When does the balance reach $62,000?

Solution   After t years the balance is given by

$$B = 22{,}000(1.03)^t.$$

Figure 10.4 shows that when $B = 62{,}000$ we have $t \approx 35$. So it takes about 35 years for the balance to reach $62,000.

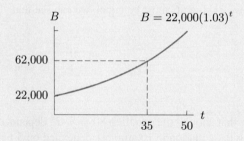

Figure 10.4: Graph of balance after t years

Describing Solutions to Exponential Equations

Even without solving an exponential equation exactly, it is often useful to be able to identify qualitative characteristics of the solution, such as whether it is positive or negative. For example, the equation

$$5 \cdot 2^t = 3$$

must have a negative solution because $5 > 3$, so 2^t must be less than 1. Positive powers of 2, like $2^2 = 4$ or $2^{1/2} = 1.4142$, are greater than 1, while negative powers, like $2^{-1} = 0.5$, are less than 1. In this case the solution is approximately $t = -0.7370$. You can check this on a calculator by evaluating $5 \cdot 2^{-0.7370}$, which is very close to 3.

Example 3   Is the solution to each of the following equations a positive or negative number?

(a) $2^x = 12$ (b) $2^x = 0.3$ (c) $(0.5)^x = 0.3$ (d) $3(0.5)^x = 12$

Solution   (a) The solution must be positive, since $12 > 2$. In fact, $2^{3.585} = 12.000$, so the solution is approximately 3.585.

(b) This time, since $0.3 < 2$, we need a negative power of 2, so the solution is negative. In fact, $2^{-1.737} = 0.3000$, so the solution is approximately -1.737.

(c) Here we want 0.5^x to be less than 1. Since $0.5 < 1$, positive powers of 0.5 are also less than 1. So the solution must be positive. In fact, the solution is about 1.737, since $0.5^{1.737} = 0.3000$.

(d) Since $12 > 3$, we 0.5^x to be greater than 1. Since $0.5 < 1$, we must use a negative exponent, so the solution is negative. In fact, the solution is -2, since

$$3(0.5)^{-2} = 3\left(\frac{1}{2}\right)^{-2} = 3\frac{1}{2^{-2}} = 3\frac{1}{1/4} = 3 \cdot 4 = 12.$$

See Figure 10.5.

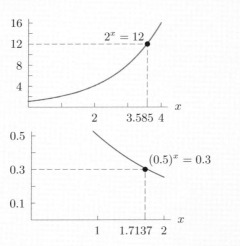
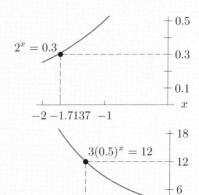

Figure 10.5

On page 317 we reasoned that

$$5 \cdot 2^t = 70$$

has a solution between 3 and 4, because 70 is between $40 - 5 \cdot 2^3$ and $80 = 5 \cdot 2^4$. This sort of reasoning is often useful in getting a rough estimate of the solution to an exponential equation.

Example 4   Say whether the solution is greater than 1, between 0 and 1, between -1 and 0, or less than -1.

(a) $4^x = 3$ (b) $5 \cdot 4^x = 26$ (c) $\left(\frac{1}{4}\right)^x = 3$

Solution   (a) Since 3 is between $4^0 = 1$ and $4^1 = 4$, the value of x that makes $4^x = 3$ is between 0 and 1.

     (b) Since 26 is between $5 \cdot 4^1 = 20$ and $5 \cdot 4^2 = 80$, the value of x that makes $5 \cdot 4^x = 26$ is between 1 and 2 and thus greater than 1.

     (c) Here x must be negative, since $1/4$ raised to a positive power is less than $1/4$, so it can never equal 3. Moreover, since 3 is between $(1/4)^0 = 1$ and $(1/4)^{-1} = 4$, then the value of x that makes $(1/4)^x = 3$ is between -1 and 0.

The Range of an Exponential Function

Not every exponential equation has a solution, as Example 5 illustrates.

Example 5   Describe the solution to

$$2^x = -4.$$

Solution   When 2 is raised to any power, positive or negative, the result is always a positive number. So, $2^x = -4$ has no solutions.

The observation in Example 5 that 2^x is positive for all values of x applies to any positive base: Provided $b > 0$,

$$b^x > 0 \quad \text{for all } x.$$

This means that the output value of the exponential function $f(x) = ab^x$ has the same sign as the starting value a, regardless of the value of x:

$$\text{Sign of output value} = (\text{Sign of } a) \cdot \underbrace{(\text{Sign of } b^x)}_{\text{Always positive}}$$

$$= \text{Sign of } a.$$

Thus, if a is positive, $f(x) = ab^x$ is positive for all x, and if a is negative, $f(x)$ is negative for all x. In fact, we have

- The range of an exponential function with $a > 0$ is the set of all positive numbers.
- The range of an exponential function with $a < 0$ is the set of all negative numbers.

Exercises and Problems for Section 10.4

EXERCISES

Solve the equations in Exercises 1–8 given that

$$f(t) = 2^t, \quad g(t) = 3^t, \quad h(t) = 4^t.$$

1. $f(t) = 4$

2. $h(t) = 4$

3. $g(t) = 1$

4. $2g(t) = 162$

5. $2 + h(t) = \dfrac{33}{16}$

6. $2(1 - f(t)) - 1$

7. $f(t) = h(t)$

8. $h(t) = f(6)$

Answer Exercises 9–12 based on Table 10.9, which gives values of the exponential function $Q = 12(1.32)^t$. Your answers may be approximate.

Table 10.9

t	2.0	2.2	2.4	2.6	2.8	3.0
Q	20.9	22.1	23.4	24.7	26.1	27.6

9. Solve $12(1.32)^t = 24.7$.

10. Solve $6(1.32)^t = 10.45$.

11. Solve $12(1.032)^t = 25$.

12. Solve $25 - 12(1.32)^t = 2.9$.

Without solving them, say whether the equations in Exercises **13–16** have a positive solution, a negative solution, a zero solution, or no solution. Give a reason for your answer.

13. $9^x = 250$

14. $2.5 = 5^t$

15. $7^x = 0.3$

16. $6^t = -1$

PROBLEMS

17. **(a)** What are the domain and range of $Q = 200(0.97)^t$?

(b) If Q represents the quantity of a 200-gram sample of a radioactive substance remaining after t days in a lab, what are the domain and range?

18. Match each statement (a)–(b) with the solutions to one or more of the equations (I)–(VI).

I. $10(1.2)^t = 5$ II. $10 = 5(1.2)^t$
III. $10 + 5(1.2)^t = 0$ IV. $5 + 10(1.2)^t = 0$
V. $10(0.8)^t = 5$ VI. $5(0.8)^t = 10$

(a) The time an exponentially growing quantity takes to grow from 5 to 10 grams.

(b) The time an exponentially decaying quantity takes to drop from 10 to 5 grams.

19. Match the statements (a)–(c) with one or more of the equations (I)–(VIII). The solution of the equation for $t > 0$ should be the quantity described in the statement. An equation may be used more than once. (Do not solve the equations.)

I. $4(1.1)^t = 2$ II. $(1.1)^{2t} = 4$
III. $2^t = 4$ IV. $2(1.1)^t = 4$
V. $4(0.9)^t = 2$ VI. $(0.9)^{2t} = 4$
VII. $2^{-t} = 4$ VIII. $2(0.9)^t = 4$

(a) The doubling time for a bank balance.

(b) The half-life of a radioactive compound.

(c) The time for a quantity growing exponentially to quadruple if its doubling time is 1.

20. The balance in dollars in a bank account after t years is given by the function $f(t) = 4622(1.04)^t$ rounded to the nearest dollar. For which values of t in Table 10.10 is

(a) $f(t) < 5199$?

(b) $f(t) > 5199$?

(c) $f(t) = 5199$?

What do your answers tell you about the bank account?

Table 10.10

t	0	1	2	3	4
$4622(1.04)^t$	4622	4807	4999	5199	5407

21. **(a)** Construct a table of values for the function $g(x) = 28(1.1)^x$ for $x = 0, 1, 2, 3, 4$.

(b) For which values of x in the table is

(i) $g(x) < 33.88$ (ii) $g(x) > 30.8$
(iii) $g(x) = 37.268$

22. **(a)** Construct a table of values for the function $y = 526(0.87)^x$ for $x = -2, -1, 0, 1, 2$.

(b) Use your table to solve $604.598 = 526(0.87)^x$ for x.

23. **(a)** Construct a table of values for the function $y = 253(2.65)^x$ for $x = -3.5, -1.5, 0.5, 2.5$.

(b) At which x-values in your table is $253(2.65)^x \geq 58.648$?

24. The value, V, of a \$5000 investment at time t in years is $V = 5000(1.012)^t$.

(a) Complete the following table:

t	55	56	57	58	59	60
V						

(b) What does your answer to part (a) tell you about the doubling time of your investment?

25. The quantity, Q, of caffeine in the body t hours after drinking a cup of coffee containing 100 mg is $Q = 100(0.83)^t$.

(a) Complete the following table:

t	3	3.5	3.6	3.7	3.8	3.9	4	5
Q								

(b) What does your answer to part (a) tell you about the half-life of caffeine?

26. A lab receives a 1000 grams of an unknown radioactive substance that decays at a rate of 7% per day.

(a) Write an expression for Q, the quantity of substance remaining after t days.
(b) Make a table showing the quantity of the substance remaining at the end of $8, 9, 10, 11, 12$ days.
(c) For what values of t in the table is the quantity left

(i) Less than 500 gm?
(ii) More than 500 gm?

(d) A lab worker says that the half-life of the substance is between 11 and 12 days. Is this consistent with your table? If not, how would you correct the estimate?

27. In year t, the population, L, of a colony of large ants is $L = 2000(1.05)^t$, and the population of a colony of small ants is $S = 1000(1.1)^t$.

(a) Construct a table showing each colony's population in years $t = 5, 10, 15, 20, 25, 30, 35, 40$.
(b) The small ants go to war against the large ants; they destroy the large ant colony when there are twice as many small ants as large ants. Use your table to determine in which year this happens.
(c) As long as the large ant population is greater than the small ant population, the large ants harvest fruit that falls on the ground between the two colonies. In which years in your table do the large ants harvest the fruit?

28. From Figure 10.6, when is

(a) $100(1.05)^t > 121.55$?
(b) $100(1.05)^t < 121.55$?

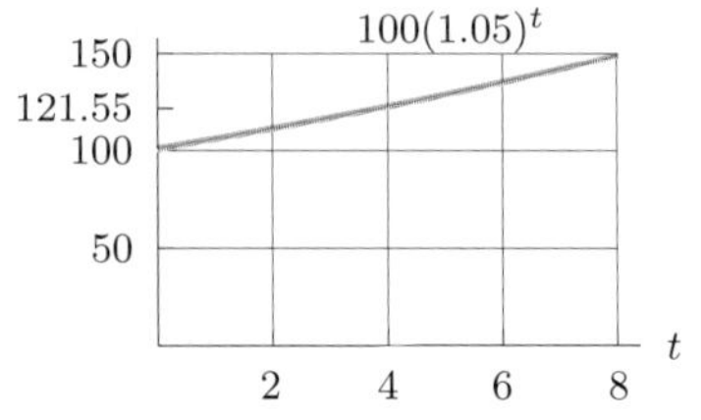

Figure 10.6

29. Figure 10.7 shows the resale value of a car given by $\$10{,}000(0.5)^t$, where t is in years.

(a) In which years in the figure is the car's value greater than $\$1250$?
(b) What value of t is a solution to $5000 = 10{,}000(0.5)^t$?

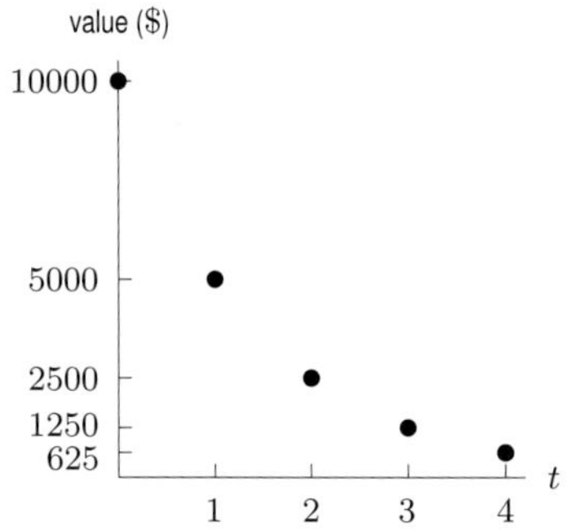

Figure 10.7

30. **(a)** Graph $y = 500(0.8)^x$ using the points $x = -2, -1.5, -1, -0.5, 0, 0.5, 1, 1.5, 2$.
(b) Use your graph to solve

(i) $500(0.8)^x = 698.71$
(ii) $500(0.8)^x \geq 400$

31. **(a)** Graph $y = 250(1.1)^x$ and $y = 200(1.2)^x$ using the points $x = 1, 2, 3, 4, 5$.
(b) Using the points in your graph, for what x-values is

(i) $250(1.1)^x > 200(1.2)^x$
(ii) $250(1.1)^x < 200(1.2)^x$

(c) How might you make your answers to part (b) more precise?

32. An employee signs a contract for a salary t years in the future given in thousands of dollars by $45(1.041)^t$.

(a) What do the numbers 45 and 1.041 represent in terms of the salary?
(b) What is the salary in 15 years? 20 years?
(c) By trial and error, find to the nearest year how long it takes for the salary to double.

Without solving them, say whether the equations in Problems 33–44 have a positive solution, a negative solution, a zero solution, or no solution. Give a reason for your answer.

33. $7 + 2^y = 5$

34. $25 \cdot 3^z = 15$

35. $13 \cdot 5^{t+1} = 5^{2t}$

36. $(0.1)^x = 2$

37. $5(0.5)^y = 1$

38. $5 = -(0.7)^t$

39. $28 = 7(0.4)^z$

40. $7 = 28(0.4)^z$

41. $0.01(0.3)^t = 0.1$

42. $10^t = 7 \cdot 5^t$

43. $4^t \cdot 3^t = 5$

44. $(3.2)^{2y+1}(1 + 3.2) = (3.2)^y$

45. Assume $0 < r < 1$ and x is positive. Without solving equations (I)–(IV) for x, decide which one has

(a) The largest solution
(b) The smallest solution
(c) No solution

I. $3(1 + r)^x = 7$ II. $3(1 + 2r)^x = 7$
III. $3(1 + 0.01r)^x = 7$ IV. $3(1 - r)^x = 7$

46. Assume that a, b, and r are positive and that $a < b$. Consider the solution for x to the equation $a(1+r)^x = b$. Without solving the equation, what is the effect of increasing each of a, r, and b, while keeping each of the other two fixed? Does the solution increase or decrease?

(a) a (b) r (c) b

■ In Problems **47–54**, decide for what values of the constant A the equation has

(a) A solution   (b) The solution $t = 0$
(c) A positive solution

47. $5^t = A$ **48.** $3^{-t} = A$

49. $(0.2)^t = A$ **50.** $A - 2^{-t} = 0$

51. $6.3A - 3 \cdot 7^t = 0$ **52.** $2 \cdot 3^t + A = 0$

53. $A5^{-t} + 1 = 0$ **54.** $2(0.7)^t + 0.2A = 0$

■ Solve the equations in Problems **55–60** for x. Your solutions will involve u.

55. $10^x = 1000^u$ **56.** $8^x = 2^u$

57. $5^{2x+1} = 125^u$ **58.** $\left(\dfrac{1}{2}\right)^x = \left(\dfrac{1}{16}\right)^u$

59. $\left(\dfrac{1}{3}\right)^x = 9^u$ **60.** $5^{-x} = \dfrac{1}{25^u}$

■ Solve the equations in Problems **61–64** using the following approximations:

$$10^{0.301} = 2, \qquad 10^{0.477} = 3, \qquad 10^{0.699} = 5.$$

Example. Solve $10^x = 6$. *Solution.* We have

$$
\begin{aligned}
10^x &= 6 \\
&= 2 \cdot 3 \\
&= 10^{0.301} \cdot 10^{0.477} \\
&= 10^{0.301+0.477} \\
&= 10^{0.778},
\end{aligned}
$$

so $x = 0.778$.

61. $10^x = 15$ **62.** $10^x = 9$

63. $10^x = 32$ **64.** $10^x = 180$

10.5   MODELING WITH EXPONENTIAL FUNCTIONS

Many real-world situations involve growth factors and growth rates, so they lend themselves to being modeled using exponential functions.

Example 1

The quantity, Q, in grams, of a radioactive sample after t days is given by

$$Q = 150(0.94)^t.$$

Identify the initial value and the growth factor and explain what they mean in terms of the sample.

Solution

The initial value, when $t = 0$, is $Q = 150(0.94)^0 = 150$. This means we start with 150 g of material. The growth factor is 0.94, which gives a growth rate of -6% per day, since $0.94 - 1 = -0.06$. As we mentioned in the follow-up to Example 2 on page 305, we call a negative rate a *decay rate*, so we say that the substance is decaying by 6% per day. Thus, there is 6% less of the substance remaining after each day passes.

Finding the Formula for an Exponential Function

If we are given the initial value a and the growth factor $b = 1 + r$ of an exponentially growing quantity Q, we can write a formula for Q,

$$Q = ab^t.$$

However, we are not always given a and b directly. Often, we must find them from the available information.

Example 2   A youth soccer league initially has $N = 68$ members. After two years, the league grows to 96 members. Assuming a constant percent growth rate, find a formula for $N = f(t)$ where t the number of years after the league was formed.

Solution   Since the number of players is increasing at a constant percent rate, we use the exponential form $N = f(t) = ab^t$. Initially, there are 68 members, so $a = 68$, and we can write $f(t) = 68b^t$. We now use the fact that $f(2) = 96$ to find the value of b:

$$f(t) = 68b^t$$
$$96 = 68b^2$$
$$\frac{96}{68} = b^2$$
$$\pm\sqrt{\frac{96}{68}} = b$$
$$b = \pm 1.188.$$

Since $b > 0$, we have $b = 1.188$. Thus, the function is $N = f(t) = 68(1.188)^t$.

Example 3   Find a formula for the exponential function whose graph is in Figure 10.8.

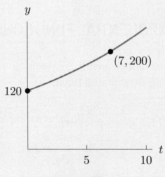

Figure 10.8

Solution   We know that the formula is of the form $y = ab^t$, and we want to solve for a and b. The initial value, a, is the y-intercept, so $a = 120$. Thus

$$y = 120b^t.$$

Because $y = 200$ when $t = 7$, we have

$$200 = 120b^7$$
$$\frac{200}{120} = b^7.$$

Raising both sides to the $1/7$ power,

$$(b^7)^{1/7} = \left(\frac{200}{120}\right)^{1/7}$$
$$b = 1.076.$$

Thus, $y = 120(1.076)^t$ is a formula for the function.

In Section 10.3 we saw how to use exponent laws to rewrite an expression like $4000 \cdot 2^{t/12}$, which represents a quantity that doubles every 12 years, in a form that shows the annual growth rate. The next example shows another method for finding the growth rate from the doubling time.

Example 4 The population of Malaysia is doubling every 38 years.[6]

(a) If the population is G in 2008, what will it be 2046?
(b) When will the population reach $4G$?
(c) What is the annual growth rate?

Solution (a) Assuming that the population continues to double every 38 years, it will be twice as large in 2046 as it is in 2008, or $2G$. See Figure 10.9.
(b) The population will double yet again, rising from $2G$ to $4G$, after another 38 years, or $2046 + 38 = 2084$. See Figure 10.9.
(c) Let $s(t)$ give the population Malaysia in year t. We have

$$s(t) = ab^t,$$

where a is the initial size in year $t = 0$. The population after 38 years, written $s(38)$, is twice the initial population. This means

$$s(38) = 2a$$
$$ab^{38} = 2a$$
$$b^{38} = 2$$
$$b = 2^{1/38} = 1.018.$$

The population increases by a factor of 1.018 every year, and its annual growth rate is 1.8%.

---

[6]CIA World Factbook at www.cia.gov.

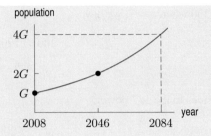

Figure 10.9: The population of Malaysia

Example 5   Hafnium has a half-life of 12.2 hours.

(a) If you begin with 1000 g, when will you have 500 g? 250 g? 125 g?
(b) What is the hourly decay rate?

Solution   (a) The half-life is 12.2 hours, so the level will drop from 1000 g to 500 g in this time. After 12.2 more hours, or 24.4 hours total, the level will drop to half of 500 g, or 250 g. After another 12.2 hours, or 36.6 hours total, the level will drop to half of 250 g, or 125 g. See Figure 10.10.

(b) Let $v(t)$ be the quantity remaining after t hours. Then

$$v(t) = ab^t,$$

where a is the initial amount of hafnium and b is the hourly decay factor. After 12.2 hours, only half the initial amount remains, so

$$v(12.2) = 0.5a$$
$$ab^{12.2} = 0.5a$$
$$b^{12.2} = 0.5$$
$$\left(b^{12.2}\right)^{1/12.2} = 0.5^{1/12.2}$$
$$b = 0.9448.$$

If r is the hourly growth rate then $r = b - 1 = 0.9448 - 1 = -0.0552 = -5.52\%$. Thus, hafnium decays at a rate of 5.52% per hour.

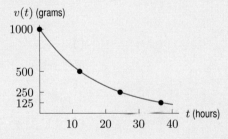

Figure 10.10: Quantity of hafnium remaining

Finding an Exponential Function When We Don't Know the Starting Value

In Examples 2 and 3, we are given the starting value a, so all we need to do is find the growth factor b. However, even if we are not given the starting value, we can sometimes use what we know about exponents to find a formula for an exponential function.

Example 6 An account growing at a constant percent rate contains \$5000 in 2004 and \$8000 in 2014. What is its annual growth rate? Express the account balance as a function of the number, t, of years since 2000.

Solution Since the account balance grows at a constant percent rate, it is an exponential function of t. Thus, if $f(t)$ is the account balance t years after 2000, then $f(t) = ab^t$ for positive constants a and b. Since the balance is \$5000 in 2004, we know that

$$f(4) = ab^4 = 5000.$$

Since the balance is \$8000 in 2014, when $t = 14$, we know that

$$f(14) = ab^{14} = 8000.$$

We can eliminate a by taking ratios

$$\frac{ab^{14}}{ab^4} = \frac{8000}{5000} = 1.6$$
$$b^{10} = 1.6$$
$$b = 1.6^{1/10} = 1.04812.$$

This tells us the annual growth rate is 4.812%. To solve for a, we use $f(4) = 5000$:

$$ab^4 = 5000$$
$$a = \frac{5000}{1.04812^4} = 4143.$$

This tells us that the initial account balance in 2000 was \$4143. Thus, $f(t) = 4143(1.04812)^t$.

Example 7 A lab receives a sample of germanium. The sample decays radioactively over time at a constant daily percent rate. After 15 days, 200 grams of germanium remain. After 37 days, 51.253 grams remain. What is the daily percent decay rate of germanium? How much did the lab receive initially?

Solution If $g(t)$ is the amount of germanium remaining t days after the sample is received, then $g(t) = ab^t$ for positive constants a and b. Since 200 g of germanium remain after 15 days we have

$$g(15) = ab^{15} = 200,$$

and since 51.253 g of germanium remain after 37 days we have

$$g(37) = ab^{37} = 51.253.$$

We can eliminate a by taking ratios:

$$\frac{\text{Amount after 37 days}}{\text{Amount after 15 days}} = \frac{ab^{37}}{ab^{15}} = \frac{51.253}{200}$$

$$\frac{b^{37}}{b^{15}} = 0.2563 \quad a \text{ is eliminated}$$
$$b^{22} = 0.2563.$$

So

$$b = (0.2563)^{1/22} = 0.93999.$$

Since $b = 1 + r$, we have $r = 0.93999 - 1 = -0.06001$, so the decay rate is about 6% per day. To find a, the amount of germanium initially received by the lab, we substitute for b in either of our two original equations, say, $ab^{15} = 200$:

$$a(0.93999)^{15} = 200$$
$$a = \frac{200}{(0.93999)^{15}}$$
$$a = 506.0.$$

Thus the lab received 506 grams of germanium. You can check that we obtain the same value for a by using the other equation, $ab^{37} = 51.253$. See Figure 10.11.

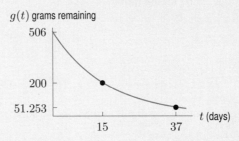

Figure 10.11: Graph showing amount of germanium remaining after t days

Exercises and Problems for Section 10.5

EXERCISES

In Exercises 1–3, write a formula for the quantity described.

1. An exponentially growing population that doubles every 9 years.

2. The amount of a radioactive substance whose half-life is 5 days.

3. The balance in an interest-bearing bank account, if the balance triples in 20 years.

Exercises 4–6 give a population's doubling time in years. Find its growth rate per year.

4. 50   5. 143   6. 14

7. The doubling time of an investment is 7 years. What is its yearly growth rate?

8. The half-life of nicotine in the body is 2 hours. What is the hourly decay rate?

9. Bismuth-210 has a half-life of 5 days. What is the decay rate per day?

Find the annual growth rate of the quantities described in Exercises 10–15.

10. A population doubles in size after 8 years.

11. The value of a house triples over a 14-year period.

12. A population goes down by half after 7 years.

13. A stock portfolio drops to one-fifth its former value over a 4-year period.

14. The amount of water used in a community increases by 25% over a 5-year period.

15. The amount of arsenic in a well drops by 24% over 8 years.

In Exercises 16–21, a pair of points on the graph of an exponential function is given. Find a formula for the function.

16. $f(t)$: $(0, 40)$, $(8, 100)$ 17. $g(t)$: $(0, 80)$, $(25, 10)$

18. $p(x)$: $(5, 20)$, $(35, 60)$ 19. $q(x)$: $(-12, 72)$, $(6, 8)$

20. $v(x)$: $(0.1, 2)$, $(0.9, 8)$

21. $w(x)$: $(5, 20)$, $(20, 5)$

22. Table 10.11 shows values for an exponential function. Find a formula for the function.

Table 10.11

t	2	3	4	5
$f(t)$	96	76.8	61.44	49.152

PROBLEMS

Find possible formulas for the exponential functions described in Problems 23–33.

23. $P = g(t)$ is the size of an animal population that numbers 8000 in year $t = 0$ and 3000 in year $t = 12$.

24. The graph of f contains the points $(24, 120)$ and $(72, 30)$.

25. The value V of an investment in year t if the investment is initially worth $3450 and grows by 4.75% annually.

26. An investment is worth $V = \$400$ in year $t = 5$ and $V = 1200$ in year $t = 12$.

27. A cohort of fruit flies (that is, a group of flies all the same age) initially numbers $P = 800$ and decreases by half every 19 days as the flies age and die.

28. $V = u(t)$ is the value in year t of an investment initially worth $3500 that doubles every eight years.

29. $f(12) = 290$, $f(23) = 175$

30. $f(10) = 65$, $f(65) = 20$.

31. An investment initially worth $3000 grows by 30% over a 5-year period.

32. A population of 10,000 declines by 0.42% per year.

33. The graph of g is shown in Figure 10.12.

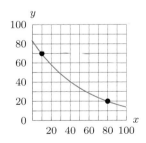

Figure 10.12

34. The balance in a bank account grows by a factor of a each year and doubles every 7 years.

   (a) By what factor does the balance change in 14 years? In 21 years?
   (b) Find a.

35. A radioactive substance has a 62 day half-life. Initially there are Q_0 grams of the substance.

   (a) How much remains after 62 days? 124 days?
   (b) When will only 12.5% of the original amount remain?
   (c) How much remains after 1 day?

36. Between 1994 and 1999, the national health expenditures in the United States were rising at an average of 5.3% per year. The U.S. health expenditures in 1994 were 936.7 billion dollars.

   (a) Express the national health expenditures, P, in billions of dollars, as a function of the year, t, with $t = 0$ corresponding to the year 1994.
   (b) Use this model to estimate the national health expenditures in the year 1999. Compare this number to the actual 1999 expenditures, which were 1210.7 billion dollars.

37. Chicago's population grew from 2.8 million in 1990 to 2.9 million in 2000.

   (a) What was the yearly percent growth rate?
   (b) Assuming the growth rate continues unchanged, what population is predicted for 2010?

38. Let $P(t)$ be the population[7] of Charlotte, NC, in thousands, t years after 1990.

     (a) Interpret $P(0) = 396$ and $P(10) = 541$.
     (b) Find a formula for $P(t)$. Assume the population grows at a constant percentage rate and give your answer to 5 decimal places.
     (c) Find and interpret $P(20)$.

39. Nicotine leaves the body at a constant percent rate. Two hours after smoking a cigarette, 58 mg remain in a person's body; three hours later (5 hours after the cigarette), there are 20 mg. Let $n(t)$ represent the amount of nicotine in the body t hours after smoking the cigarette.

     (a) What is the hourly percent decay rate of nicotine?
     (b) What is the initial quantity of nicotine in the body?
     (c) How much nicotine remains 6 hours after smoking the cigarette?

10.6   EXPONENTIAL FUNCTIONS AND BASE e

The mathematician Leonhard Euler[8] first used the letter e to stand for the important constant $2.71828\ldots$. Many functions in science, economics, medicine, and other disciplines involve this curious number. Just as π often appears in geometrical contexts, e often appears in situations involving exponential growth and probability. For example, the standard normal distribution in statistics, whose graph is the well-known bell-shaped curve, is given by the function

$$f(x) = \frac{1}{\sqrt{2\pi}} \cdot e^{-x^2/2}.$$

Biologists use the function

$$P(t) = P_0 e^{kt}$$

to describe population growth, and physicists use the function

$$V = Ae^{-t/(RC)}$$

to describe the voltage across the resistor in a circuit.

Writing Exponential Functions Using Base e

Example 1 on page 310 gives two equivalent ways of writing the value of an investment as an exponential function of time:

$$V = 4000 \cdot 1.0595^t \quad \text{This form emphasizes the investment's annual growth rate.}$$
$$V = 4000 \cdot 2^{t/12}. \quad \text{This form emphasizes the investment's doubling time.}$$

An important feature of an exponentially growing quantity, in addition to annual growth rate or its doubling time, is its *continuous growth rate*. For the investment described by Example 1 on page 310, it can be shown that the continuous growth rate is $k = 5.78\% = 0.0578$, and we can write

$$V = 4000e^{0.0578t}. \quad \text{This form emphasizes the continuous growth rate.}$$

As different as these three formulas appear, they all describe the same underlying function, as we can see by writing

$$4000 \cdot 2^{t/12} = 4000\left(2^{1/12}\right)^t = 4000 \cdot 1.0595^t \quad \text{so the first and second formulas are the same}$$
$$4000e^{0.0578t} = 4000\left(e^{0.0578}\right)^t = 4000 \cdot 1.0595^t. \quad \text{so the first and third formulas are the same}$$

---

[7]J. Glaeser and N. Shapiro, "City Growth and the 2000 Census: Which Places Grew and Why," www.brookings.edu, accessed on October 24, 2004.

[8]http://en.wikipedia.org/wiki/E_%28mathematical_constant%29, page last accessed June 11, 2007.

In the last step, we relied on the fact that $e \approx 2.718$: Using a calculator,[9] this gives

$$e^{0.0578} = 2.718^{0.0578} = 1.0595.$$

Depending on the point we wish to make, we might prefer one of these three equations to the other. The first one, having base $b = 1.0595$, emphasizes the investment's 5.95% annual growth rate, and the second, having base 2, emphasizes its doubling time of 12 years. The third, having base e, emphasizes the continuous growth rate of 5.78%. In general:

Any exponential function can be written in the form

$$Q = a \cdot e^{kt}, \quad a \text{ and } k \text{ constants.}$$

Here,
- a is the **initial value**.
- k is the **continuous growth rate**.
- Letting $b = e^k$, we can convert from base-e form to standard form $y = a \cdot b^t$ where b is the growth factor.

In Example 1, we see how to convert an exponential function given in base-e form to standard form.

Example 1   Equations (a)–(c) describe the value of investments after t years. For each investment, give the initial value, the continuous growth rate, the annual growth factor, and the annual growth rate.

(a) $V = 2500e^{0.035t}$ (b) $V = 5000e^{0.147t}$ (c) $V = 6250e^{-0.095t}$

Solution   (a) We have

$$a = 2500 \quad \text{starting value is \$2500}$$
$$k = 0.035 \quad \text{continuous growth rate is 3.5\%.}$$

Using the exponent laws we rewrite $V = 2500e^{0.035t}$ as

$$V = 2500 \left(e^{0.035}\right)^t = 2500 \cdot 1.0356^t,$$

where $b = e^k = e^{0.035} = 1.0356$ is the annual growth factor. Thus $r = b - 1 = 0.0356 = 3.56\%$ is the annual growth rate.

(b) We have

$$a = 5000 \quad \text{starting value is \$5000}$$
$$k = 0.147 \quad \text{continuous growth rate is 14.7.\%}$$

Again, we write

$$V = 5000e^{0.0147t} = 5000(e^{0.0147})^t = 5000 \cdot 1.158^t.$$

---
[9] Most scientific and graphing calculators have an $e^{\wedge}$ button.

So the annual growth factor is $b = e^k = e^{0.0147} = 1.158$ and the annual growth rate is $r = b - 1 = 0.158 = 15.8\%$.

(c) We have

$$a = 6250 \qquad \text{starting value is \$6250}$$
$$k = -0.095 \quad \text{continuous growth rate is } -9.5\%.$$

In this case we have an exponentially decaying quantity, but the conversion works the same way as before. We write
$$V = 6250\left(e^{-0.095}\right)^t = 6250 \cdot 0.9094^t.$$

So the annual growth factor is $b = e^k = e^{-0.095} = 0.9094$ and the annual growth rate is $r = b - 1 = -0.0906 = -9.06\%$.

Comparing Continuous vs. Annual Growth Rates

The value of k is close to the value of r provided k is reasonably small. In Example 1(a) we see that if $k = 0.035$ then $r = 0.0356$, and in Example 1(b) we see that if $k = 0.147$, then $r = 0.158$. Notice that in each case, the value of r is close to the value of k. This does not hold true for larger values of k. To see this, we try letting $k = 3$:

$$b = e^k = e^3 \quad = 20.086$$
$$r = b - 1 = 19.086.$$

In this case, k and r are not at all close in value.

Another important property illustrated by Example 1 is that sign of k is the same as the sign of r. In other words, if the growth rate r is positive, the continuous growth rate k is also positive, and if r is negative, so is k. It is also true that $k = 0$ if $r = 0$.

Manipulating Expressions Involving e

Sometimes we change the form of an expression to emphasize a certain point, as illustrated by Example 2.

| Example 2 | In a certain type of electrical circuit, the voltage V, in volts, is a function of the resistance R, in ohms, given by $$V = Ae^{-t/(RC)}, \quad \text{where } A \text{ and } C \text{ are constants.}$$ Suppose $R = 200{,}000$ and $V = 12e^{-0.1t}$. Find A and C. |

| Solution | We are told that $V = 12e^{-0.1t}$. Matching $12e^{-0.1t}$ with the form $Ae^{-t/(RC)}$, we see that $A = 12$ and $$-0.1t = -\frac{t}{RC} = \left(\frac{-1}{RC}\right)t.$$ Matching the coefficients of t, we get $$-0.1 = \frac{-1}{RC}$$ |

$$\frac{-1}{10} = \frac{-1}{RC}$$

$$RC = 10$$

$$C = \frac{10}{R} = \frac{10}{200{,}000} = 0.00005.$$

Simplifying Expressions Involving e^{kt}

Often, scientists and engineers write expressions involving e in different ways. In Example 3, we rewrite a complicated expression in a simpler form using the fact that

$$\frac{1}{e^{kt}} = e^{-kt}.$$

Example 3 The function

$$P = \frac{1600e^{kt}}{60 + 20e^{kt}}$$

describes a population that starts at 20 and levels out at 80. See Figure 10.13.

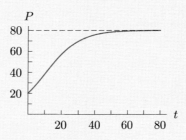

Figure 10.13: Growth of a population

Biologists sometimes write the expression for P in the form

$$P = \frac{L}{P_0 + Ae^{-kt}}.$$

Put P in this form and give the values of L, P_0, and A.

Solution To eliminate the e^{kt} in the numerator, we divide the numerator and denominator by e^{kt}:

$$P = \frac{1600e^{kt}}{60 + 20e^{kt}} = \frac{1600}{\frac{60}{e^{kt}} + 20} = \frac{1600}{60e^{-kt} + 20} = \frac{1600}{20 + 60e^{-kt}}.$$

Comparing with the form

$$P = \frac{L}{P_0 + Ae^{-kt}}$$

we have $L = 1600$, $P_0 = 20$, and $A = 60$.

Exercises and Problems for Section 10.6

EXERCISES

■The equations in Exercises **1–6** describe the value of investments after t years. For each investment, give the initial value, the continuous growth rate, the annual growth factor, and the annual growth rate.

1. $V = 1200e^{0.041t}$ **2.** $V = 3500e^{0.173t}$

3. $V = 7500e^{-0.059t}$ **4.** $V = 17{,}000e^{0.322t}$

5. $V = 20{,}000e^{-0.44t}$ **6.** $V = 1800e^{1.21t}$

■Write the exponential functions in Exercises **7–12** in the form $Q = ae^{kt}$ and state the values of a and k.

7. $Q = 20e^{-t/5}$ **8.** $Q = 200e^{0.5t-3}$

9. $Q = \dfrac{1}{40e^{0.4t}}$ **10.** $Q = 37.5 \left(e^{1-3t}\right)^2$

11. $Q = \dfrac{e^{\pi}e^{2t}}{e^{7t}}$ **12.** $Q = 90\sqrt{e^{-0.4t}}$

PROBLEMS

13. Investment A grows at an annual rate of $r = 10\%$, and investment B grows at a continuous rate of $k = 10\%$. Compare these two investments, both initially worth \$1000, over a 50-year period by completing Table 10.12.

Table 10.12

t (yr)	Inv A, \$	Inv B, \$
0		
10		
20		
30		
40		
50		

14. A population is initially 15,000 and grows at a continuous rate of 2% a year. Find the population after 20 years.

15. You buy a house for \$350,000, and its value declines at a continuous rate of -9% a year. What is it worth after 5 years?

16. You invest \$3500 in a bank that gives a continuous annual growth rate of 15%. What is your balance $B(t)$ after t years?

17. The voltage across a resistor at time t is $V = Ae^{-t/(RC)}$, where $R = 150{,}000$ and $C = 0.0004$. Is the voltage growing or decaying? What is the continuous rate?

■The functions in Problems **18–21** describe the growth of a population. Give the starting population at time $t = 0$.

18. $P(t) = 4000e^{rt}$ **19.** $P(t) = P_0e^{0.37t}$

20. $P(t) = \dfrac{1500e^{kt}}{30 + 20e^{kt}}$

21. $P(t) = \dfrac{L}{P_0 + Ae^{-kt}}$

22. Write
$$P(t) = \frac{1500e^{kt}}{30 + 20e^{kt}}$$
in the form
$$P = \frac{L}{P_0 + Ae^{-kt}}$$
and identify the constants L, P_0, and A.

■In Problems **23–26**, rewrite the given expression as indicated, and state the values of all constants.

23. Rewrite $5e^{2+3t}$ in the form ae^{kt}.

24. Rewrite $50 - \dfrac{50}{e^{0.25t}}$ in the form $A\left(1 - e^{-rt}\right)$.

25. Rewrite $\left(e^{3x} + 2\right)^2$ in the form $e^{rx} + ae^{sx} + b$.

26. Rewrite $\dfrac{e^{0.1t}}{2e^{0.1t} + 3}$ in the form $\dfrac{1}{a + be^{-kt}}$.

27. Two important functions in engineering are

$$\sinh x = \frac{e^x - e^{-x}}{2} \quad \text{hyperbolic sine function}$$

$$\cosh x = \frac{e^x + e^{-x}}{2} \quad \text{hyperbolic cosine function.}$$

Show that the following expression can be greatly simplified:
$$\left(\cosh x\right)^2 - \left(\sinh x\right)^2 .$$

REVIEW EXERCISES AND PROBLEMS FOR CHAPTER 10

EXERCISES

In Exercises 1–5, say whether the quantity is changing in an exponential or linear fashion.

1. An account receives a deposit of $723 per month.

2. A machine depreciates by 17% per year.

3. Every week, 9/10 of a radioactive substance remains from the beginning of the week.

4. One liter of water is added to a trough every day.

5. After 124 minutes, 1/2 of a drug remains in the body.

The functions in Exercises 6–9 describe different investments. For each investment, what are the starting value and the percent growth rate?

6. $V = 6000(1.052)^t$ 7. $V = 500(1.232)^t$

8. $V = 14{,}000(1.0088)^t$ 9. $V = 900(0.989)^t$

10. A young sunflower grows in height by 5% every day. If the sunflower is 6 inches tall on the first day it was measured, write an expression that describes the height h of the sunflower in inches t days later.

11. A radioactive metal weighs 10 grams and loses 1% of its mass every hour. Write an expression that describes its weight w in grams after t hours.

Can the expressions in Exercises 12–17 be put in the form ab^t? If so, identify the values of a and b.

12. $2 \cdot 5^t$ 13. 10^{-t} 14. $1.3t^{4.5}$

15. $10 \cdot 2^{3+2t}$ 16. $3^{-2t} \cdot 2^t$ 17. $\dfrac{3}{5^{2+t}}$

Give the initial value a, the growth factor b, and the percentage growth rate r for the exponential functions in Exercises 18–21.

18. $Q = 1700(1.117)^t$ 19. $Q = 1250(0.923)^t$

20. $Q = 120 \cdot (3.2)^t$ 21. $Q = 80(0.113)^t$

In Exercises 22–24, write the expression either as a constant times a power of x, or a constant times an exponential in x. Identify the constant. If a power, identify the exponent, and if an exponential, identify the base.

22. $\dfrac{-21}{2^x}$ 23. $\dfrac{2^x}{3^x}$ 24. $\dfrac{x^2}{x^3}$

In Exercises 25–28, give the growth factor that corresponds to the given growth rate.

25. 60% growth   26. 18% shrinkage

27. 100% growth   28. 99% shrinkage

In Exercises 29–32, give the rate of growth that corresponds to the given growth factor.

29. 1.095   30. 0.91   31. 2.16   32. 0.95

Find the percent rate of change for the quantities described in Exercises 33–37.

33. An investment doubles in value every 8 years.

34. A radioactive substance has a half-life of 11 days.

35. After treatment with antibiotic, the blood count of a particular strain of bacteria has a half-life of 3 days.

36. The value of an investment triples every seven years.

37. A radioactive substance has a half-life of 25 years.

Write expressions representing the quantities described in Exercises 38–43.

38. The balance B doubles n times in a row.

39. The value V increases n times in a row by a factor of 1.04.

40. The investment value begins at V_0 and grows by a factor of k every four years over a twelve-year period.

41. The investment value begins at V_0 and grows by a factor of k every h years over a twenty-year period.

42. The investment value begins at V_0 and grows by a factor of k every h years over an N-year period.

43. The number of people living in a city increases from P_0 by a factor of r for 4 years, and then by a factor of s for 7 years.

■The values in the tables in Exercises **44–46** are values of an exponential function $y = ab^x$. Find the function.

44.

x	0	1	2	3
y	200	194	188.18	182.5346

45.

x	0	2	4	6
y	1	2	4	8

46.

x	0	3	12	15
y	10	9	6.561	5.9049

■In Exercises **47–54**, find the formula for the exponential function whose graph goes through the two points.

47. $(3, 4), (6, 10)$ **48.** $(2, 6), (7, 1)$

49. $(-6, 2), (3, 6)$ **50.** $(-5, 8), (-2, 1)$

51. $(0, 2), (3, 7)$ **52.** $(1, 7), (5, 9)$

53. $(3, 6.2), (6, 5.1)$ **54.** $(2.5, 3.7), (5.1, 9.3)$

■Identify the expressions in Exercises **55–60** as exponential, linear, or quadratic in t.

55. $7 + 2t^2$ **56.** $6m + 7t^2 + 8t$

57. $5q^t$ **58.** $7t + 8 - 4w$

59. $5h^n + t$ **60.** $a(7.08)^t$

PROBLEMS

■Write an expression representing the quantities in Problems **61–65**.

61. The amount of caffeine at time t hours if there are 90 mg at the start and the quantity decays by 17% per hour.

62. The US population t years after 2000, when it was 281 million. The growth rate is 1% per year.

63. The price of a $30 item in t years if prices increase by 2.2% a year.

64. The difference in value in t years between $1000 invested today at 5% per year and $800 invested at 4% per year.

65. The difference in value in t years between a $1500 investment earning 3% a year and a $1500 investment earning 2% a year.

■As x gets larger, are the expressions in Exercises **66–70** increasing or decreasing? By what percent per unit increase or decrease of x?

66. $10(1.07)^x$ **67.** $5(0.96)^x$

68. $5(1.13)^{2x}$ **69.** $6\left(\dfrac{1.05}{2}\right)^x$

70. $\dfrac{5(0.97)^x}{3}$

■Find formulas for the exponential functions described in Problems **71–74**.

71. $f(3) = 12, f(20) = 80$

72. An investment is worth $3000 in year $t = 6$ and $7000 in year $t = 14$.

73. The graph of $y = v(x)$ contains the points $(-4, 8)$ and $(20, 40)$.

74. $w(30) = 40, w(80) = 30$

■Find possible formulas for the exponential functions in Problems **75–76**.

75. $V = f(t)$ is the value of an investment worth $2000 in year $t = 0$ and $5000 in year $t = 5$.

76. The value of the expression $g(8)$ is 80. The solution to the equation $g(t) = 60$ is 11.

■In Problems **77–79**, correct the mistake in the formula.

77. A population which doubles every 5 years is given by $P = 7 \cdot 2^{5t}$.

78. The quantity of pollutant remaining after t minutes if 10% is removed each minute is given by $Q = Q_0(0.1)^t$.

79. The quantity, Q, which doubles every T years, has a yearly growth factor of $a = 2^T$.

■The US population is growing by about 1% a year. In 2000, it was 282 million. What is wrong with the statements in Problems **80–82**? Correct the equation in the statement.

80. The population will be 300 million t years after 2000, where $282(0.01)^t = 300$.

81. The population, P, in 2020 is the solution to $P - (282 \cdot 1.01)^{20} = 0$.

82. The solution to $282(1.01)^t = 2$ is the number of years it takes for the population to double.

83. Which grows faster, an investment that doubles every 10 years or one that triples every 15 years?

84. Which disappears faster, a population whose half-life is 12 years or a population that loses $1/3$ of its members every 8 years?

85. For each description (a)–(c), select the expressions (I)–(VI) that could represent it. Assume P_0, $0 < r < 1$, and t are positive.

I. $P_0(1 + r)^t$ II. $P_0(1 - r)^t$

III. $P_0(1 + r)^{-t}$ IV. $P_0(1 - r)^{-t}$

V. $P_0 t/(1 + r)$ VI. $P_0 t/(1 - r)$

(a) A population increasing with time, t.
(b) A population decreasing with time, t.
(c) A population growing linearly with time, t.

86. Without calculating them, put the following quantities in increasing order.

(a) The solution to $2^t = 0.2$
(b) The solution to $3 \cdot 4^{-x} = 1$
(c) The solution to $49 = 7 \cdot 5^z$
(d) The number 0
(e) The number 1

87. A quantity grows or decays exponentially according to the formula $Q = ab^t$. Match the statements (a)–(b) with the solutions to one or more of the equations (I)–(VI).

I. $b^{2t} = 1$ II. $2b^t = 1$

III. $b^t = 2$ IV. $b^{2t} + 1 = 0$

V. $2b^t + 1 = 0$ VI. $b^{2t} + 2 = 0$

(a) The doubling time for an exponentially growing quantity.
(b) The half-life of an exponentially decaying quantity.

■ Without solving them, indicate the statement that best describes the solution x (if any) to the equations in Problems 88–97. *Example.* The equation $3 \cdot 4^x = 5$ has a positive solution because $4^x > 1$ for $x > 0$, and so the best-fitting statement is (a).

(a) Positive, because $b^x > 1$ for $b > 1$ and $x > 0$.
(b) Negative, because $b^x < 1$ for $b > 1$ and $x < 0$.
(c) Positive, because $b^x < 1$ for $0 < b < 1$ and $x > 0$.
(d) Negative, because $b^x > 1$ for $0 < b < 1$ and $x < 0$.
(e) There is no solution because $b^x > 0$ for $b > 0$.

88. $2 \cdot 5^x = 2070$

89. $3^x = 0.62$

90. $7^x = -3$

91. $6 + 4.6^x = 2$

92. $17 \cdot 1.8^x = 8$

93. $24 \cdot 0.31^x = 85$

94. $0.07 \cdot 0.02^x = 0.13$

95. $240 \cdot 0.55^x = 170$

96. $\dfrac{8}{2^x} = 9$

97. $\dfrac{12}{0.52^x} = 40$

98. A container of ice cream is taken from the freezer and sits in a room for t minutes. Its temperature in degrees Fahrenheit is $a - b \cdot 2^{-t} + b$, where a and b are positive constants.

(a) Write this expression in a form that shows that the temperature is always

(i) Less than $a + b$ (ii) Greater than a

(b) What are reasonable values for a and b?

■ In Problems 99–101, is 10^a is greater than 10^b?

99. $0 < a < b$

100. $0 < a$ and $b < 0$

101. $a < b < 0$

■ In Problems 102–104, is x^a is greater than x^b?

102. $a < b$ and $0 < x < 1$

103. $a < b$ and $x > 1$

104. $x < 0$, a is an even integer, b is an odd integer

■ In Problems 105–107, is $\dfrac{1}{10^a}$ greater than $\dfrac{1}{10^b}$?

105. $0 < a < b$

106. $0 < a$ and $b < 0$

107. $a < b < 0$

■ The expressions in Problems 108–111 involve several letters. Think of one letter at a time representing a variable and the rest as nonzero constants. In which cases is the expression linear? In which cases is it exponential? In this case, what is the base?

108. $2^n a^n$

109. $a^n b^n + c^n$

110. $AB^q C^q$

111. Ab^{2t}

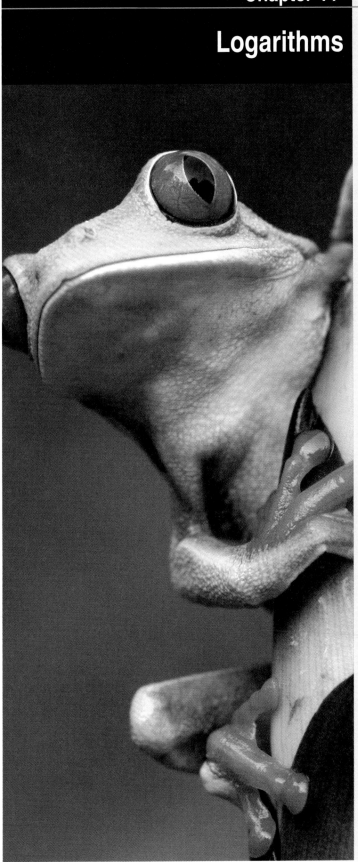

Chapter 11

Logarithms

CONTENTS

11.1   INTRODUCTION TO LOGARITHMS

Suppose that a population of bacteria is given by $P = 10^t$ after t hours. When will the population reach 4000? To find out, we must solve the following equation for the exponent t:

$$10^t = 4000.$$

Evaluating the left side at $t = 3$ gives $10^3 = 1000$, which is less than 4000, and evaluating at $t = 4$ gives $10^4 = 10,000$, which is greater than 4000. So we estimate that the solution is between 3 and 4. We can narrow down the solution further by calculating 10^t for values of t between 3 and 4.

Example 1   Estimate the solution to $10^t = 4000$ to within one decimal place.

Solution   Table 11.1 shows values of 10^t for t between 3.2 and 3.8.

Table 11.1   *Solve* $10^t = 4000$

t	3.2	3.3	3.4	3.5	3.6	3.7	3.8
10^t	1584.89	1995.26	2511.89	3162.28	3981.07	5011.87	6309.57

Since 4000 falls between $10^{3.6}$ and $10^{3.7}$, we estimate that the solution to $10^t = 4000$ is between 3.6 and 3.7. So it takes about 3.6 hours for the population to reach 4000.

What Is a Logarithm?

The solution in Example 1 is an example of a logarithm. In general, we define the *common logarithm*, written $\log_{10} x$, or $\log x$, as follows.

If x is a positive number,

$$\log x \text{ is the exponent to which we raise 10 to get } x.$$

In other words,

$$\text{if} \quad y = \log x \quad \text{then} \quad 10^y = x$$

and

$$\text{if} \quad 10^y = x \quad \text{then} \quad y = \log x.$$

A Logarithm Is an Exponent

This is the most important thing to keep in mind about logarithms. When you are looking for the logarithm of a number, you are looking for the exponent to which you need to raise 10 to get that number.

Example 2   Use the definition to find $\log 1000$, $\log 100$, $\log 10$.

Solution   Since $\log x$ is the power to which you raise 10 to get x, we have

$$\log 1000 = 3 \quad \text{because} \quad 10^3 = 1000$$
$$\log 100 = 2 \quad \text{because} \quad 10^2 = 100$$
$$\log 10 = 1 \quad \text{because} \quad 10^1 = 10.$$

In the previous example we recognized the number inside the logarithm as a power of 10. Sometimes we need to use exponent laws to put a number in the right form so that we can see what power of 10 it is.

Example 3   Without a calculator, evaluate the following, if possible:

(a) $\log 1$

(b) $\log \sqrt{10}$

(c) $\log \dfrac{1}{100,000}$

(d) $\log 0.01$

(e) $\log \dfrac{1}{\sqrt{1000}}$

(f) $\log(-10)$

Solution   (a) We have $\log 1 = 0$ because $10^0 = 1$.
(b) We have $\log \sqrt{10} = 1/2$ because $10^{1/2} = \sqrt{10}$.
(c) We have $\log \dfrac{1}{100,000} = -5$ because $10^{-5} = \dfrac{1}{100,000}$.
(d) We have $\log 0.01 = -2$ because $10^{-2} = 0.01$.
(e) We have

$$\log \frac{1}{\sqrt{1000}} = -\frac{3}{2} \quad \text{because} \quad 10^{-3/2} = \frac{1}{(10^3)^{1/2}} = \frac{1}{\sqrt{1000}}.$$

(f) Since any power of 10 is positive, -10 cannot be written as a power of 10. Thus, $\log(-10)$ is undefined.

In Example 3(f) we see that we cannot take the logarithm of a negative number. Also, $\log 0$ is not defined since there is no power of 10 that equals zero. However, the value of a logarithm itself can be negative, as in Example 3(c), (d), and (e), and it can be zero, as in Example 3(a).

For a number that is not an easy-to-see power of 10, you can estimate the logarithm by finding two powers of 10 on either side of it.

Example 4   Estimate $\log 63$.

Solution   We use the fact that $10 < 63 < 100$. Since $10^1 = 10$ and $10^2 = 100$, we can say that $1 < \log 63 < 2$. In fact, using a calculator, we have $\log 63 = 1.799$.

The definition of a logarithm as an exponent means that we can rewrite any statement about logarithms as a statement about powers of 10.

Example 5

Rewrite the following statements using exponents instead of logarithms.

(a) $\log 100 = 2$ (b) $\log 0.1 = -1$ (c) $\log 40 = 1.602$

Solution

For each statement, we use the fact that if $y = \log x$ then $10^y = x$.

(a) $\log 100 = 2$ means that $10^2 = 100$.
(b) $\log 0.1 = -1$ means that $10^{-1} = 0.1$.
(c) $\log 40 = 1.602$ means that $10^{1.602} = 40$.

Conversely, a statement about powers of 10 can be rewritten as a statement about logarithms.

Example 6

Rewrite the following statements using logarithms instead of exponents.

(a) $10^4 = 10,000$ (b) $10^{-3} = 0.001$ (c) $10^{0.6} = 3.981$

Solution

For each statement, we use the fact that if $10^y = x$, then $y = \log x$.

(a) $10^4 = 10,000$ means that $\log 10,000 = 4$.
(b) $10^{-3} = 0.001$ means that $\log 0.001 = -3$.
(c) $10^{0.6} = 3.981$ means that $\log 3.981 = 0.6$.

A Logarithm Is a Solution to an Exponential Equation

From the definition of $\log a$, we can solve the exponential equation

$$10^x = a.$$

Example 7

Solve the equation $10^t = 4000$.

Solution

In Example 1 we estimated the solution by using a calculator to find powers of 10 that are close to 4000. But the calculator will also calculate the solution directly. The solution to $10^t = 4000$ is $t = \log 4000$. Using a calculator, we get $t = \log 4000 = 3.602$.[1]

An equation like $10^t = 4000$ can be solved in one step by logarithms. More complicated equations can sometimes be simplified into the form $10^x = a$, and then solved in the same way.

Example 8

Solve the equation $3 \cdot 10^t - 8 = 13$.

Solution

We first simplify algebraically, then use the definition of the logarithm:

$$3 \cdot 10^t - 8 = 13$$
$$3 \cdot 10^t = 21 \qquad \text{add 8 to both sides}$$
$$10^t = 7 \qquad \text{divide both sides by 3}$$
$$t = \log 7 = 0.845. \quad \text{use the definition}$$

---

[1] This leaves unanswered the question of *how* the calculator finds this value, which requires more advanced mathematics.

Properties of Logarithms

We can think of taking logarithms as an operation, just as taking square roots is an operation. In fact, the logarithm is the inverse operation to the operation of raising 10 to a power.

The Logarithm Operation Undoes Raising 10 to a Power

We know that the operation of subtracting a constant undoes the operation of adding that constant, and the operation of dividing by a constant undoes the operation of multiplying by that constant. Likewise, the operation of taking a logarithm undoes the operation of raising 10 to a power. For example,

$$\log 10^5 = \log 100,\!000 = 5 \quad \longrightarrow \quad \text{the logarithm undoes raising 10 to a power, leaving 5}$$

$$10^{\log 5} = 10^{0.69897} = 5 \quad \longrightarrow \quad \text{raising 10 to the power of a logarithm undoes the logarithm, leaving 5.}$$

Example 9   Evaluate without using a calculator:

(a) $\log\left(10^{6.7}\right)$

(b) $10^{\log 2.5}$

(c) $10^{\log(-3.2)}$

(d) $10^{\log(x+1)}, \quad x > -1$

Solution

(a) Since the logarithm undoes raising 10 to a power, we have $\log\left(10^{6.7}\right) = 6.7$.

(b) Since raising 10 to a power undoes the logarithm, we have $10^{\log 2.5} = 2.5$.

(c) For an operation to be reversed it needs to be defined in the first place. In this case $\log(-3.2)$ is not defined, so $10^{\log(-3.2)}$ is also not defined.

(d) Since raising 10 to a power undoes the logarithm, we have $10^{\log(x+1)} = x + 1$. Notice that $\log(x + 1)$ is defined in this case since $x > -1$, making $x + 1 > 0$.

In general,

Inverse Property of Logarithms

The operation of raising 10 to a power and the logarithm operation undo each other:

$$\log(10^N) = N \qquad \text{for all } N$$
$$10^{\log N} = N \qquad \text{for } N > 0$$

Properties of Logarithms Coming from the Exponent Laws

When we multiply 10^2 and 10^3, we get $10^{2+3} = 10^5$, by the laws of exponents. Since the logarithms of these numbers are just their exponents, the equation $5 = 2 + 3$ can be interpreted as

$$\log(10^5) = \log(10^2) + \log(10^3).$$

In general, when we multiply two numbers together, the logarithm of their product is the sum of their logarithms. The following properties are derived from the exponent laws.

Properties of the Common Logarithm
- For a and b both positive and any value of t,

$$\log(ab) = \log a + \log b$$
$$\log\left(\frac{a}{b}\right) = \log a - \log b$$
$$\log(b^t) = t \cdot \log b$$

- Since $10^0 = 1$,

$$\log 1 = 0.$$

We justify these properties on page 346.

The properties of logarithms allow us to write expressions involving logarithms in many different ways.

Example 10   Show that the expressions $\log(1/b)$ and $-\log b$ are equivalent.

Solution   There are at least two ways to see this using the logarithm rules we know. First, we know that $\log(a/b) = \log a - \log b$, so $\log(1/b) = \log 1 - \log b$. Since $\log 1 = 0$, we have

$$\log \frac{1}{b} = \log 1 - \log b = 0 - \log b = -\log b.$$

Alternatively, we know that

$$\log \frac{1}{b} = \log\left(b^{-1}\right)$$
$$= -1 \cdot \log b \quad \text{using } \log\left(b^t\right) = t \cdot \log b.$$

When keeping track of manipulations involving logarithms, it is sometimes useful to think of $\log x$ and $\log y$ as simple quantities by using a substitution, as in the next example.

Example 11   Let $u = \log x$ and $v = \log y$. Rewrite the following expressions in terms of u and v, or state that this is not possible.

(a) $\log(4x/y)$ (b) $\log(4xy^3)$ (c) $\log(4x - y)$

Solution   (a) We have

$$\log(4x/y) = \log(4x) - \log y \qquad \text{using } \log\left(\frac{a}{b}\right) = \log a - \log b$$
$$= \log 4 + \log x - \log y \quad \text{using } \log(ab) = \log a + \log b$$
$$= \log 4 + u - v.$$

(b) We have

$$\log(4xy^3) = \log(4x) + \log(y^3) \qquad \text{using } \log(ab) = \log a + \log b$$

$$\begin{aligned}
&= \log 4 + \log x + \log(y^3) && \text{again using } \log(ab) = \log a + \log b \\
&= \log 4 + \log x + 3 \log y && \text{using } \log(b^t) = t \cdot \log b \\
&= \log 4 + u + 3v.
\end{aligned}$$

Note that it would be incorrect to write

$$\log(4xy^3) = 3\log(4xy).$$

This is because the exponent of 3 applies only to the y and not to the expression $4xy$.

(c) It is not possible to rewrite $\log(4x - y)$ in terms of u and v. Although there are properties for the logarithm of a product or a quotient, there are no rules for the logarithm of a sum or a difference.

The Logarithm Function

Just as we can use the square root operation to define the function $y = \sqrt{x}$, we can use logarithms to define the *logarithm function*, $y = \log x$.

The Domain of the Logarithm Function

Since the logarithm of a number is the exponent to which you raise 10 to get that number, and since any power of 10 is a positive number, we can only take logarithms of positive numbers. In fact, the domain of the logarithm function is all positive numbers.

The Graph of the Logarithm Function

Table 11.2 gives values of $y = \log x$ and Figure 11.1 shows the graph of the logarithm function.

Table 11.2 *Values of* $\log x$

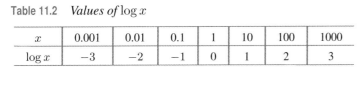

x	0.001	0.01	0.1	1	10	100	1000
$\log x$	-3	-2	-1	0	1	2	3

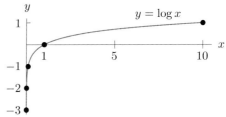

Figure 11.1: Graph $\log x$

The Range of the Logarithm Function

Note that the graph in Figure 11.1 increases as x increases and has an x-intercept at $x = 1$. The values of $\log x$ are positive for x greater than one and are negative for x less than one. The graph climbs to $y = 1$ at $x = 10$ and to $y = 2$ at $x = 100$. To reach the modest height of $y = 10$ requires x to equal 10^{10}, or 10 billion! This suggests that the logarithm function increases very slowly. Nonetheless, the graph of $y = \log x$ eventually climbs to any value we choose. Although x cannot equal zero in the logarithm function, we can choose $x > 0$ to be as small as we like. As x decreases toward zero, the values of $\log x$ become more and more negative. In summary, the range of $\log x$ is all real numbers.

Justification of $\log(a \cdot b) = \log a + \log b$ and $\log(a/b) = \log a - \log b$

Since a and b are both positive, we can write $a = 10^m$ and $b = 10^n$, so $\log a = m$ and $\log b = n$. Then the product $a \cdot b$ can be written

$$a \cdot b = 10^m \cdot 10^n = 10^{m+n}.$$

Therefore $m + n$ is the power of 10 needed to give $a \cdot b$, so

$$\log(a \cdot b) = m + n,$$

which gives

$$\boxed{\log(a \cdot b) = \log a + \log b.}$$

Similarly, the quotient a/b can be written as

$$\frac{a}{b} = \frac{10^m}{10^n} = 10^{m-n}.$$

Therefore $m - n$ is the power of 10 needed to give a/b, so

$$\log\left(\frac{a}{b}\right) = m - n,$$

and thus

$$\boxed{\log\left(\frac{a}{b}\right) = \log a - \log b.}$$

Justification of $\log(b^t) = t \cdot \log b$

Suppose that b is positive, so we can write $b = 10^k$ for some value of k. Then

$$b^t = (10^k)^t.$$

We have rewritten the expression b^t so that the base is a power of 10. Using a property of exponents, we can write $(10^k)^t$ as 10^{kt}, so

$$b^t = (10^k)^t = 10^{kt}.$$

Therefore kt is the power of 10 that gives b^t, so

$$\log(b^t) = kt.$$

But since $b = 10^k$, we know $k = \log b$. This means

$$\log(b^t) = (\log b)t = t \cdot \log b.$$

Thus, for $b > 0$ we have

$$\boxed{\log\left(b^t\right) = t \cdot \log b.}$$

Exercises and Problems for Section 11.1

EXERCISES

In Exercises 1–8, rewrite the equation using exponents instead of logarithms.

1. $\log 0.01 = -2$

2. $\log 1000 = 3$

3. $\log 20 = 1.301$

4. $\log A^2 = B$

5. $\log 5000 = 3.699$

6. $\log \frac{1}{10^3} = -3$

7. $\log(\alpha\beta) = 3x^2 + 2y^2$

8. $\log\left(\frac{a}{b}\right) = 9$

In Exercises 9–16, rewrite the equation using logarithms instead of exponents.

9. $10^5 = 100,000$

10. $10^{-4} = 0.0001$

11. $10^{2.301} = 200$

12. $10^m = n$

13. $100^{2.301} = 39,994$

14. $10^{-0.08} = 0.832$

15. $10^{a^2 b} = 97$

16. $10^{qp} = nR$

Solve the equations in Exercises 17–24.

17. $10^x = 10,000$

18. $10^x = 421$

19. $10^x = 0.7162$

20. $10^x = \frac{23}{37}$

21. $10^x = 17.717$

22. $10^x = \frac{3}{\sqrt{17}}$

23. $10^x = 57$

24. $15.5 = 10^y$

In Exercises 25–34, evaluate without a calculator, or say if the expression is undefined.

25. $\log 0.001$

26. $\log \frac{1}{10}$

27. $\log \sqrt{1000}$

28. $\log \frac{1}{\sqrt{100,000}}$

29. $\log(-100)$

30. $10^{\log(-1)}$

31. $10^{\log 1}$

32. $\log 10^{-5.4}$

33. $\log 10^7$

34. $10^{\log 100}$

35. A trillion is one million million. What is the logarithm of a trillion?

36. As of June 1, 2008, the US national debt[2] was $9,391,288,825,656.43, or about $9.391 trillion. What is the logarithm of this figure, correct to three decimals? **Hint**: $\log 9.391 = 0.9727$.

PROBLEMS

37. (a) Calculate $\log 2$, $\log 20$, $\log 200$ and $\log 2000$ and describe the pattern.
   (b) Using the pattern in part (a) make a guess about the values of $\log 20,000$ and $\log 0.2$.
   (c) Justify the guess you made in part (b) using the properties of logarithms.

In Problems 38–43, without using a calculator, find two consecutive integers, one lying above and the other lying below the logarithm of the number.

38. 205

39. 8991

40. $1.22 \cdot 10^4$

41. $0.99 \cdot 10^5$

42. 0.6

43. 0.012

44. Complete the following table. Based on your answers, which table entry is closest to $\log 8$?

x	0.901	0.902	0.903	0.904	0.905
10^x					

45. What does the table tell you about $\log 27$?

x	1.40	1.42	1.44	1.46
10^x	25.1	26.3	27.5	28.8

46. Complete the following table. Use your answer to estimate $\log 0.5$.

x	−0.296	−0.298	−0.300	−0.302
10^x				

---

[2]The US Bureau of the Public Debt posts the value of the debt to the nearest penny on the website www.treasurydirect.gov/govt/govt.htm.

■ Solve the equations in Problems **47–52**, first approximately, as in Example 1, by filling in the given table, and then to four decimal places by using logarithms.

47.

Table 11.3   *Solve $10^x = 500$*

x	2.6	2.7	2.8	2.9
10^x				

48.

Table 11.4   *Solve $10^x = 3200$*

x	3.4	3.5	3.6	3.7
10^x				

49.

Table 11.5   *Solve $10^x = 0.03$*

x	−1.6	−1.5	−1.4	−1.3
10^x				

50.

Table 11.6   *Solve $2^x = 20$*

x	4.1	4.2	4.3	4.4
2^x				

51.

Table 11.7   *Solve $5^x = 130$*

x	2.9	3	3.1	3.2
5^x				

52.

Table 11.8   *Solve $0.5^x = 0.1$*

x	3.1	3.2	3.3	3.4
0.5^x				

■ Are the expressions in Problems **53–55** equivalent for positive a and b? If so, explain why. If not, give values for a and b that lead to different values for the two expressions.

53. $\log(10b)$ and $1 + \log b$

54. $\log(10^{a+b})$ and $\log(a + b)$

55. $10^{\log(a+b)}$ and $a + b$

■ In Problems **56–63**, rewrite the expression in terms of $\log A$ and $\log B$, or state that this is not possible.

56. $\log(AB^2)$

57. $\log(2A/B)$

58. $\log(A + B/A)$

59. $\log(1/(AB))$

60. $\log(A + B)$

61. $\log(AB^2 + B)$

62. $\log\left(A\sqrt{B}\right) + \log\left(A^2\right)$

63. $\log(A(A + B)) - \log(A + B)$

■ If possible, use logarithm properties to rewrite the expressions in Problems **64–69** in terms of u, v, w given that

$$u = \log x, v = \log y, w = \log z.$$

Your answers should not involve logs.

64. $\log xy$

65. $\log \dfrac{x}{z}$

66. $\log x^2 y^3 \sqrt{z}$

67. $\dfrac{\log \sqrt{x^3}}{\log \frac{y}{z^2}}$

68. $\left(\log \dfrac{1}{y^3}\right)^2$

69. $\log\left(x^2 + y^2\right)$

■ In Problems **70–75**, find possible formulas for the functions using logs or exponentials.

70.
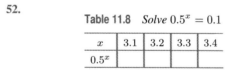

71.

72.
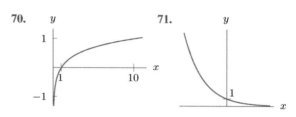

73.
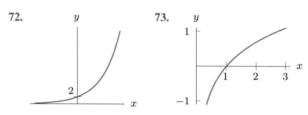

74.
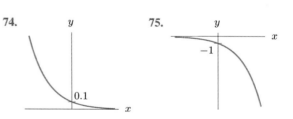

75.

76. Suppose

$$\log A = 2 + \log B.$$

How many times as large as B is A?

11.2   SOLVING EXPONENTIAL EQUATIONS USING LOGARITHMS

In the last section, we saw that $t = \log 4000$ is the solution to the equation $10^t = 4000$. But how do we solve an equation like $3^t = 4000$?

Example 1   Estimate the solution to $3^t = 4000$.

Solution   We have:

$$3^7 = 2187 \qquad \longrightarrow \qquad \text{7 is too small}$$
$$3^8 = 6561 \qquad \longrightarrow \qquad \text{8 is too big}$$
$$3^{7.5} = 3787.995 \qquad \longrightarrow \qquad \text{7.5 is too small}$$
$$3^{7.6} = 4227.869 \qquad \longrightarrow \qquad \text{7.6 is too big.}$$

Thus, the exponent of 3 that gives 4000 is between 7.5 and 7.6.

It turns out that we can use logarithms to solve the equation in Example 1, even though the base is 3. This is because we can use the rule $10^{\log N} = N$ to convert the equation to one with base 10.

Example 2   By writing 3 as a power of 10, find the exact solution to $3^t = 4000$.

Solution   Using the fact that $3 = 10^{\log 3}$, we have

$$3^t = 4000 \quad \text{original equation}$$
$$(10^{\log 3})^t = 4000 \quad \text{because } 3 = 10^{\log 3}.$$

Now we rewrite $(10^{\log 3})^t$ as $10^{t \log 3}$ using the exponent laws. This gives

$$10^{t \log 3} = 4000.$$

We now use the definition to write:

$$t \log 3 = \log 4000 \qquad \qquad \text{the variable } t \text{ is no longer in the exponent}$$
$$t = \frac{\log 4000}{\log 3} = 7.5496 \quad \text{dividing both sides by } \log 3.$$

The solution, $t = \log 4000 / \log 3 = 7.5496$, agrees with our earlier result.

Taking the Logarithm of Both Sides of an Equation

The logarithm provides us with an important new operation for solving equations. Like the operation of multiplying by a nonzero constant, the operation of taking a logarithm does not change whether or not two quantities are equal, and so it is a valid operation to be performed on both sides of an equation (as long is it is defined).

Example 3   Solve $3^t = 4000$ by taking the logarithm of both sides and using the properties of logarithms.

Solution   We have

$$3^t = 4000$$
$$\log(3^t) = \log 4000 \qquad \text{taking the logarithm of both sides}$$
$$t \log 3 = \log 4000 \qquad \text{using } \log(b^t) = t \cdot \log b$$
$$t = \frac{\log 4000}{\log 3} = 7.5496. \quad \text{dividing by } \log 3$$

This is the same answer that we got in Example 2.

Example 4   In Example 1 on page 310, the value of an investment t years after 2004 is given by $V = 4000 \cdot 2^{t/12}$. Solve the equation

$$4000 \cdot 2^{t/12} = 7000.$$

What does the solution tell you about the investment?

Solution   We have

$$4000 \cdot 2^{t/12} = 7000$$
$$2^{t/12} = \frac{7000}{4000} = 1.75$$
$$\log\left(2^{t/12}\right) = \log 1.75$$
$$\frac{t}{12} \cdot \log 2 = \log 1.75 \qquad \text{using the property } \log(b^t) = t \cdot \log b$$
$$t = \frac{12 \log 1.75}{\log 2}$$
$$= 9.688.$$

This tells us that the investment's value will reach \$7000 after about 9.7 years, or midway into 2013.

In the solution to the last example, we first divide through by 4000 and then take the logarithm of both sides. However, we can also solve equations like this by taking the logarithm right away:

Example 5   Solve $5 \cdot 7^x = 20$ by

(a)   Dividing through by 5 first   (b)   Taking the logarithm first.

Solution   (a) We have

$$5 \cdot 7^x = 20$$
$$7^x = 4 \qquad \text{dividing through by 5}$$
$$\log(7^x) = \log 4 \qquad \text{taking the logarithm of both sides}$$
$$x \log 7 = \log 4 \qquad \text{using } \log(b^t) = t \cdot \log b$$
$$x = \frac{\log 4}{\log 7} = 0.712. \quad \text{dividing by } \log 7$$

(b) Taking the logarithm first, we have

$$5 \cdot 7^x = 20$$

$$\log(5 \cdot 7^x) = \log 20 \qquad \text{taking the logarithm of both sides}$$

$$\log 5 + \log(7^x) = \log 20 \qquad \text{using } \log(ab) = \log a + \log b$$

$$\log(7^x) = \log 20 - \log 5 \qquad \text{subtracting } \log 5 \text{ from both sides}$$

$$x \log 7 = \log 20 - \log 5 \qquad \text{using } \log(b^t) = t \cdot \log b$$

$$x = \frac{\log 20 - \log 5}{\log 7} = 0.712. \qquad \text{dividing by } \log 7$$

Although the expressions we get for the answers look different, they have the same value, 0.712. We can also see that the expressions have the same value using logarithm properties:

$$\log 20 - \log 5 = \log\left(\frac{20}{5}\right) = \log 4 \quad \text{using } \log a - \log b = \log\left(\frac{a}{b}\right).$$

Notice that the solution to Example 5 is positive. In the next example, the solution is negative.

Example 6   Solve the equation $300 \cdot 7^x = 45$.

Solution   We have:

$$300 \cdot 7^x = 45$$

$$7^x = \frac{45}{300} = 0.15$$

$$\log(7^x) = \log 0.15 \qquad \text{take the logarithm of both sides}$$

$$x \log 7 = \log 0.15 \qquad \text{using } \log b^t = t \cdot \log b$$

$$x = \frac{\log 0.15}{\log 7} = -0.975.$$

In Example 5, we have

$$(\text{Smaller number}) \cdot 7^x = \text{Larger number}.$$

This means that 7^x must be greater than 1, so $x > 0$. In contrast, in Example 6 we have

$$(\text{Larger number}) \cdot 7^x = \text{Smaller number}.$$

This means that 7^x must be less than 1, so $x < 0$. See Figures 11.2 and 11.3 for a comparison.

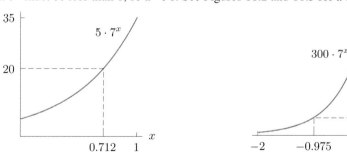

Figure 11.2: The graphical solution to Example 5   Figure 11.3: The graphical solution to Example 6

In Examples 5 and 6, we use logarithms to solve an equation that contains a variable in the exponent on only one side of the equation. In Example 7, we solve an equation that has a variable in the exponent on both sides.

Example 7   Katie invests $1200 at 5.1% in a retirement fund, and at the same time John invests $2500 at 4.4%. When do their investments have the same value? If they retire at the same time, whose investment is worth more at retirement?

Solution   The solution to the equation $1200(1.051)^x = 2500(1.044)^x$ tells us how many years it takes after the initial investment for the two investments to reach the same value. We take logarithms of both sides of the equation to convert it to a linear equation in $\log x$.

$$1200(1.051)^x = 2500(1.044)^x$$
$$\log 1200 + x \log 1.051 = \log 2500 + x \log 1.044$$
$$x(\log 1.051 - \log 1.044) = \log 2500 - \log 1200$$
$$x = \frac{\log 2500 - \log 1200}{\log 1.051 - \log 1.044} = 109.833.$$

Thus, we see that the two investments have equal value after about 110 years. Since John's investment starts at a greater value, and since they will no doubt retire before 110 years have passed, John's investment will be worth more when they retire.

Equations Containing Logarithms

Just as we can use the operation of taking logarithms to solve equations involving exponentials, we can use the operation of raising to a power to solve equations involving logarithms.

Example 8   Solve the equation $\log N = 1.9294$.

Solution   Using the definition of the logarithm, we know that

$$\log N = 1.9294 \quad \text{means that} \quad N = 10^{1.9294}.$$

Using a calculator, $N = 10^{1.9294} = 84.996$.

In Example 8, we use the definition of the logarithm directly. In the next example we take the same approach, but we need to do some algebraic simplification as well.

Example 9   Solve $4 \log(2x - 6) = 8$.

Solution   We have

$$\begin{aligned}
4 \log(2x - 6) &= 8 \\
\log(2x - 6) &= 2 \quad &&\text{divide both sides by 4} \\
2x - 6 &= 10^2 \quad &&\text{by the definition} \\
2x - 6 &= 100 \quad &&\text{because } 10^2 = 100 \\
2x &= 106 \quad &&\text{add 6 to both sides} \\
x &= 53 \quad &&\text{divide both sides by 2.}
\end{aligned}$$

Sometimes we transform the equation before using the rules of logarithms.

Example 10   Solve $\log(x + 198) - 2 = \log x$ for x.

Solution   We start by putting all the logarithms on one side:

$$\log(x + 198) - \log x = 2$$
$$\log \frac{(x + 198)}{x} = 2 \qquad \text{using } \log\left(\frac{a}{b}\right) = \log a - \log b$$
$$\frac{(x + 198)}{x} = 100 \qquad \text{using the definition}$$
$$x + 198 = 100x \qquad \text{multiplying both sides by } x$$
$$198 = 99x \qquad \text{gathering terms}$$
$$x = 2.$$

Substituting $x = 2$ into the original equation gives $\log(2 + 198) - \log(2) = 2$, which is valid, because

$$\log(2 + 198) - \log 2 = \log 200 - \log 2 = \log \frac{200}{2} = \log 100 = 2.$$

Since the logarithm is not defined for all real numbers, we must be careful to make sure our solutions are valid.

Example 11   Solve $\log x + \log(x - 3) = 1$ for x.

Solution   Using the properties of logarithms, we have

$$\log x + \log(x - 3) = 1$$
$$\log(x(x - 3)) = 1 \qquad \text{using } \log(ab) = \log a + \log b$$
$$x(x - 3) = 10 \qquad \text{using the definition}$$
$$x^2 - 3x - 10 = 0 \qquad \text{subtracting 10 from both sides}$$
$$(x - 5)(x + 2) = 0 \qquad \text{factoring.}$$

So, solving for x gives $x = 5$ or $x = -2$.

Although $x = 5$ and $x = -2$ satisfy the quadratic equation $x^2 - 3x - 10 = 0$, we must check both values of x to see if they satisfy our original equation.

For $x = 5$, we have $\log(5) + \log(2) = 1$, which is a valid statement because $\log(5) + \log(2) = \log(10) = 1$ by the properties of logarithms. For $x = -2$, we have $\log(-2) + \log(-5) = 1$. Note that this is not valid, since log is not defined for negative input values. Thus, $x = 5$ is the only solution. The extraneous solution, $x = -2$, is introduced when we combine the two logarithm expressions into one.

Exercises and Problems for Section 11.2

EXERCISES

Solve the exponential equations in Exercises **1–5** without using logarithms, then use logarithms to confirm your answer.

1. $2^4 = 4^x$

2. $\dfrac{1}{81} = 3^x$

3. $1024 = 2^x$

4. $5^3 = 25^x$

5. $64^3 = 4^x$

Solve the equations in Exercises **6–15**.

6. $2^t = 19$

7. $1.071^x = 3.25$

8. $17^z = 12$

9. $\left(\dfrac{2}{3}\right)^y = \dfrac{5}{7}$

10. $80^w = 100$

11. $0.088^a = 0.54$

12. $(1.041)^t = 520$

13. $2^p = 90$

14. $(1.033)^q = 600$

15. $(0.988)^r = 55$

Solve the equations in Exercises **16–39**.

16. $3 \cdot 10^t = 99$

17. $5 \cdot 10^x - 35 = 0$

18. $2.6 = 50 \cdot 10^t$

19. $10^t + 5 = 0$

20. $10^t - 5.4 = 7.2 - 10^t$

21. $100(1.041)^t = 520$

22. $5 \cdot 2^p = 90$

23. $40(1.033)^q = 600$

24. $100(0.988)^r = 55$

25. $8 \cdot \left(\frac{2}{3}\right)^t = 20$

26. $500 \cdot 1.31^b = 3200$

27. $700 \cdot 0.882^t = 80$

28. $50 \cdot 2^{x/12} = 320$

29. $120 \left(\dfrac{2}{3}\right)^{z/17} = 15$

30. $2(10^y + 3) = 4 - 5(1 - 10^y)$

31. $130 \cdot 1.031^w = 220 \cdot 1.022^w$

32. $700 \cdot 0.923^p = 300 \cdot 0.891^p$

33. $820(1.031)^t = 1140(1.029)^t$

34. $84.2(0.982)^y = 97(0.891)^y$

35. $1320(1.045)^q = 1700(1.067)^q$

36. $0.315(0.782)^x = 0.877(0.916)^x$

37. $500(1.032)^t = 750$

38. $2300(1.0417)^t = 8400$

39. $2215(0.944)^t = 800$

Solve the equations in Exercises **40–47**.

40. $\log x = \dfrac{1}{2}$

41. $\log x = -3$

42. $\log x = 5$

43. $\log x = 1.172$

44. $3 \log x = 6$

45. $7 - 2 \log x = 10$

46. $2(\log x - 1) = 9$

47. $\log(x - 3) = 1$

PROBLEMS

In Problems **48–55**, assume a and b are positive constants. Imagine solving for x (but do not actually do so). Will your answer involve logarithms? Explain how you can tell.

48. $10^x = a$

49. $10^2 x = 10^3 + 10^2$

50. $x^{10} - 10a = 0$

51. $2^{x+1} - 3 = 0$

52. $Q = b^x$

53. $a = \log x$

54. $3(\log x) + a = a^2 + \log x$

55. $Pa^{-kx} = Q$

56. A travel mug of $90°C$ coffee is left on the roof of a parked car on a $0°C$ winter day. The temperature of the coffee after t minutes is given by $H = 90(0.5)^{t/10}$. When will the coffee be only lukewarm ($30°C$)?

57. The population of a city t years after 1990 is given by $f(t) = 1800(1.055)^t$. Solve the equation $f(t) = 2500$. What does your answer tell you about the city?

58. The dollar value of two investments after t years is given by $f(t) = 5000(1.062)^t$ and $g(t) = 9500(1.041)^t$. Solve the equation $f(t) = g(t)$. What does your solution tell you about the investments?

59. The number of working lightbulbs in a large office building after t months is given by $g(t) = 4000(0.8)^t$. Solve the equation $g(t) = 1000$. What does your answer tell you about the lightbulbs?

60. The number of acres of wetland infested by an invasive plant species after t years is given by $u(t) = 12 \cdot 2^{t/3}$. Solve the equation $u(t) = 50$. What does your answer tell you about the wetland?

61. The amount of natural gas in thousands of cubic feet per day delivered by a well after t months is given by $v(t) = 80 \cdot 2^{-t/5}$. Solve the equation $v(t) = 5$. What does your answer tell you about the well?

62. The amount of contamination remaining in two different wells t days after a chemical spill, in parts per billion (ppb), is given by $r(t) = 320(0.94)^t$ and $s(t) = 540(0.91)^t$. Solve the equation $r(t) = s(t)$. What does the solution tell you about the wells?

63. The size of two towns t years after 2000 is given by

$$u(t) = 1200(1.019)^t \text{ and } v(t) = 1550(1.038)^t.$$

Solve the equation $u(t) = v(t)$. What does the solution tell you about the towns?

64. What properties are required of a and b if $\log(a/b) = r$ has a solution for

(a) $r > 0$?
(b) $r < 0$?
(c) $r = 0$?

11.3 APPLICATIONS OF LOGARITHMS TO MODELING

In the previous section we saw how to solve exponential equations using logarithms. Exponential equations often arise when we ask questions about situations that are modeled by exponential functions.

Example 1   The balance of a bank account begins at \$32,000 and earns 3.5% interest per year. When does it reach \$70,000?

Solution   The balance in year t is given by $32,000(1.035)^t$ dollars. We solve:

$$32,000(1.035)^t = 70,000$$
$$1.035^t = \frac{70,000}{32,000}$$
$$\log 1.035^t = \log\left(\frac{70}{32}\right) \qquad \text{taking the logarithm of both sides}$$
$$t \cdot \log 1.035 = \log\left(\frac{70}{32}\right) \qquad \text{using } \log(b^t) = t \cdot \log b$$
$$t = \frac{\log(70/32)}{\log 1.035} = 22.754.$$

The balance reaches \$70,000 in 22.754 years.

Example 2   Between the years 1991 and 2001, the population of Naples, Italy decreased, on average, by 0.605% per year.[3] In 2001, the population was 1,004,500. If the population continues to shrink by 0.605% per year:

(a) What will the population be in 2010?
(b) When will the population reach half its 2001 level?

Solution   The population t years after 2001 is given by $1,004,500(0.99395)^t$.

(a) The year 2010 corresponds to $t = 9$. We have

$$\text{Population in year } 9 = 1,004,500(0.99395)^9 = 951,110.$$

---

[3] http://www.citypopulation.de/Italien.html, accessed May 30, 2003.

(b) Half the 2001 level is $1{,}004{,}500/2 = 502{,}250$. Solving, we have:

$$1{,}004{,}500(0.99395)^t = 502{,}250$$

$$0.99395^t = \frac{502{,}250}{1{,}004{,}500} = \frac{1}{2}$$

$$\log 0.99395^t = \log\left(\frac{1}{2}\right) \qquad \text{taking the logarithm of both sides}$$

$$t \log 0.99395 = \log\left(\frac{1}{2}\right) \qquad \text{using } \log b^t = t \log b$$

$$t = \frac{\log(1/2)}{\log 0.99395} = 114.223,$$

so Naples' population will halve in about 114 years, in 2115.

The models in Examples 1 and 2 lead to equations of the form

$$f(x) = N, \qquad \text{where } f \text{ is an exponential function.}$$

The next example gives rise to an equation of the form

$$f(x) = g(x), \qquad \text{where } f \text{ and } g \text{ are exponential functions.}$$

Example 3   The population of City A begins at 25,000 people and grows at 2.5% per year. City B begins with a larger population of 200,000 people but grows at the slower rate of 1.2% per year. Assuming that these growth rates hold constant, will the population of City A ever be as large as the population of City B? If so, when?

Solution   In year t, the population (in 1000s) of City A is $25(1.025)^t$ and the population of City B is $200(1.012)^t$. To find out when the population of City A grows to be as large as in City B, we solve the equation

$$\underbrace{25(1.025)^t}_{\text{population of } A} = \underbrace{200(1.012)^t}_{\text{population of } B}.$$

$$25(1.025)^t = 200(1.012)^t$$

$$\frac{(1.025)^t}{(1.012)^t} = \frac{200}{25}$$

$$\left(\frac{1.025}{1.012}\right)^t = 8 \qquad \text{using } \frac{a^t}{b^t} = \left(\frac{a}{b}\right)^t$$

$$\log\left(\frac{1.025}{1.012}\right)^t = \log 8 \qquad \text{taking the logarithm of both sides}$$

$$t \log\left(\frac{1.025}{1.012}\right) = \log 8 \qquad \text{using } \log b^t = t \log b$$

$$t = \frac{\log 8}{\log(1.025/1.012)} = 162.914.$$

The cities' populations will be the same in about 163 years. We can verify our solution by finding the populations of Cities A and B in year $t = 162.914$:

$$\text{Population of City } A = 25(1.025)^{162.914} = 1{,}396.393 \text{ thousand}$$
$$\text{Population of City } B = 200(1.012)^{162.914} = 1{,}396.394 \text{ thousand}.$$

The answers are not exactly equal because we rounded the value of t.

Doubling Times and Half-Lives

We can use logarithms to calculate doubling times and half-lives.

Example 4   The size of a population of frogs is given by $185(1.135)^t$ where t is in years.

(a) What is the initial size of the population?
(b) How long does it take the population to double in size?
(c) How long does this population take to quadruple in size? To increase by a factor of 8?

Solution   (a) The initial size is $185(1.135)^0 = 185$ frogs.
(b) In order to double, the population must reach 370 frogs. We solve the following equation for t:

$$185(1.135)^t = 370$$
$$1.135^t = 2$$
$$\log\left(1.135^t\right) = \log 2$$
$$t \cdot \log 1.135 = \log 2$$
$$t = \frac{\log 2}{\log 1.135} = 5.474 \text{ years.}$$

(c) We have seen that the population increases by 100% during the first 5.474 years. But since the population is growing at a constant percent rate, it increases by 100% every 5.474 years, not just the first 5.474 years. Thus it doubles its initial size in the first 5.474 years, quadruples its initial size in two 5.474-year periods, or 10.948 years, and increases by a factor of 8 in three 5.474-year periods, or 16.422 years. We can also find the time periods directly with logarithms.

We can verify the results of the last example as follows:

$$185(1.135)^{5.474} = 370 \qquad \text{twice the initial size}$$
$$185(1.135)^{10.948} = 740 \qquad \text{4 times the initial size}$$
$$185(1.135)^{16.422} = 1480 \qquad \text{8 times the initial size.}$$

Example 5   Polonium-218 decays at a rate of 16.578% per day. What is the half-life if you start with:

(a)   100 grams? $\qquad\qquad\qquad\qquad$ (b) a grams?

Solution   Decaying at a rate of 16.578% per day means that there is 83.422% remaining after each day.
$\qquad$ In part (a), we start with 100 g of polonium-218 and need to find how long it takes for there to be 50 g, which means we must solve the equation
$$100(0.83422)^t = 50.$$

In part (b), we start with a grams and need to find how long it takes for there to be $a/2$ grams, which means we must solve the equation

$$a(0.83422)^t = \frac{a}{2}.$$

We solve these equations side-by-side to highlight the equivalent steps:

$$100(0.83422)^t = 50 \qquad\qquad a(0.83422)^t = \frac{a}{2}$$

$$0.83422^t = \frac{50}{100} \qquad\qquad 0.83422^t = \frac{\left(\frac{a}{2}\right)}{a}$$

$$0.83422^t = \frac{1}{2} \qquad\qquad 0.83422^t = \frac{1}{2}$$

$$t\log 0.83422 = \log\frac{1}{2} \qquad\qquad t\log 0.83422 = \log\frac{1}{2}$$

$$t = \frac{\log\frac{1}{2}}{\log 0.83422} = 3.824 \qquad\qquad t = \frac{\log\frac{1}{2}}{\log 0.83422} = 3.824,$$

so polonium-218 has a half-life of about 3.824 days. Notice that from the third step on, the calculations are exactly the same, so we arrive at the same answer. The point of using a letter like a instead of a number like 100 is to show that the half-life does not depend on how much polonium-218 we start with.

For another way to think about Example 5, notice that since any exponential function is of the form

$$\text{(Initial value)} \cdot \text{(Growth factor)}^t,$$

when we reach half the initial value, we know that the (Growth factor)^t component must equal $1/2$:

$$\text{(Initial value)} \cdot \underbrace{\text{(Growth factor)}^t}_{\frac{1}{2}} = \frac{1}{2} \cdot \text{Initial value}.$$

This means that in order to find the half-life, we must solve an equation of the form

$$\text{(Growth factor)}^t = \frac{1}{2}.$$

This is why the third step of both equations in Example 5 is

$$0.83422^t = \frac{1}{2}.$$

Exercises and Problems for Section 11.3

EXERCISES

In Exercises 1–4, find the doubling time (for increasing quantities) or half-life (for decreasing quantities).

1. $Q = 200 \cdot 1.21^t$

2. $Q = 450 \cdot 0.81^t$

3. $Q = 55 \cdot 5^t$

4. $Q = 80 \cdot 0.22^t$

5. A 10 kg radioactive metal loses 5% of its mass every 14 days. How many days will it take until the metal weighs 1 kg?

6. An investment begins with $500 and earns 10.1% per year. When will it be worth $1200?

7. An abandoned building's value is $250(0.95)^t$ thousand dollars. When will Etienne be able to buy the building with his savings account, which has \$180,000 at year $t = 0$ and is growing by 12% per year?

8. City A, with population 60,000, is growing at a rate of 3% per year, while city B, with population 100,000, is losing its population at a rate of 4% per year. After how many years are their populations equal?

9. In 2003 the number of sport utility vehicles sold in China was predicted to grow by 30% a year for the next 5 years.[4] What is the doubling time?

10. A substance decays at 4% per hour. Suppose we start with 100 grams of the substance. What is the half-life?

11. A population grows at a rate of 11% per day. Find its doubling time. How long does it take for the population to quadruple?

12. Cesium-137, which is found in spent nuclear fuel, decays at 2.27% per year. What is its half-life?

13. An investment initially worth \$1900 grows at an annual rate of 6.1%. In how many years will the investment be worth \$5000?

PROBLEMS

You deposit \$10,000 into a bank account. In Problems 14–17, use the annual interest rate given to find the number of years before the account holds

(a) \$15,000   (b) \$20,000

14. 2%   **15.** 3.5%   **16.** 7%   **17.** 14%

18. A population of prairie dogs grows exponentially. The colony begins with 35 prairie dogs; three years later there are 200 prairie dogs.

   (a) Give a formula for the population as a function of time.

   (b) Use logarithms to find, to the nearest year, when the population reaches 1000 prairie dogs.

19. The population in millions of a bacteria culture after t hours is given by $y = 20 \cdot 3^t$.

   (a) What is the initial population?

   (b) What is the population after 2 hours?

   (c) How long does it take for the population to reach 1000 million bacteria?

   (d) What is the doubling time of the population?

20. The population in millions of bacteria after t hours is given by $y = 30 \cdot 4^t$.

   (a) What is the initial population?

   (b) What is the population after 2 hours?

   (c) How long does it take for the population to reach 1000 million bacteria?

   (d) What is the doubling time of the population?

21. A balloon is inflated in such a way that each minute its volume doubles in size. After 10 minutes, the volume of the balloon is 512 cm^3.

   (a) What is the volume of the balloon initially?

   (b) When is the volume of the balloon 256 cm^3, half of what it is at 10 minutes?

   (c) When is the volume of the balloon 100 cm^3?

22. The population of Nevada, the fastest growing state, increased by 12.2% between April 2000 and July 2003.[5]

   (a) What was the yearly percent growth rate?

   (b) What is the doubling time?

23. The population of Florida is 17 million and growing at 1.96% per year.[6]

   (a) What is the doubling time?

   (b) If the population continues to grow at the same rate, in how many years will the population reach 100 million?

24. When the minimum wage was enacted in 1938, it was \$0.25 per hour.[7]

   (a) If the minimum wage increased at the average rate of inflation, 4.3% per year, how much would it be in 2004?

   (b) The minimum wage in 2004 was \$5.15. Has the minimum wage been increasing faster or slower than inflation?

   (c) If the minimum wage grows at 4.3% per year from 1938, when does it reach \$10 per hour?

25. Iodine-131, used in medicine, has a half-life of 8 days.

   (a) If 5 mg are stored for a week, how much is left?

   (b) How many days does it take before only 1 mg remains?

---

[4] "They're Huge, Heavy, and Loveable", *Business Week*, p. 58 (December 29, 2003).

[5] http://quickfacts.census.gov/qfd/states, accessed October 18, 2004.

[6] http://quickfacts.census.gov/qfd/states, accessed October 18, 2004.

[7] www.dol.gov, accessed October 18, 2004.

26. Three cities have the populations and annual growth rates in the table.

   (a) Without any calculation, how can you tell which pairs of cities will have the same population at some time in the future?

   (b) Find the time(s) at which each pair of cities has the same population.

City	A	B	C
Population (millions)	1.2	3.1	1.5
Growth rate (%/yr)	2.1	1.1	2.9

27. Water is passed through a pipe containing porous material to filter out pollutants. Each inch of pipe removes 50% of the pollutant entering.

   (a) True or false: The first inch of the pipe removes 50% of the pollutants and the second inch removes the remaining 50%.

   (b) What length of pipe is needed to remove all but 10% of the pollutants?

28. A radioactive substance decays at a constant percentage rate per year.

   (a) Find the half-life if it decays at a rate of

       (i) 10% per year.   **(ii)** 19% per year.

   (b) Compare your answers in parts (i) and (ii). Why is one exactly half the other?

29. What is the doubling time of the population of a bacteria culture whose growth is modeled by $y = ab^t$, with t in hours?

11.4 NATURAL LOGARITHMS AND LOGARITHMS TO OTHER BASES

The exponential function $V = 12e^{-0.1t}$ from Example 2 on page 332 tells us the voltage across a resistor at time t. We can find the voltage for a few values of t using a calculator:

$$\text{At } t = 0: \quad 12e^{-0.1(0)} = 12e^0 \quad = 12 \qquad \text{12 volts initially}$$
$$\text{At } t = 5: \quad 12e^{-0.1(5)} = 12e^{-0.5} = 7.278 \quad \text{7.278 volts after 5 seconds}$$
$$\text{At } t = 10: \quad 12e^{-0.1(10)} = 12e^{-1} \quad = 4.415 \quad \text{4.415 volts after 10 seconds}$$
$$\text{At } t = 15: \quad 12e^{-0.1(15)} = 12e^{-1.5} = 2.678 \quad \text{2.678 volts after 15 seconds}$$

Notice that the voltage drops rapidly from its initial level of 12V. Suppose we need to know when the voltage reaches a certain value, for instance, 3V. This must happen sometime between $t = 10$ seconds (when it reaches 4.415V) and $t = 15$ seconds (when it reaches 2.678V). In Example 1, we use logarithms to find the exact value of t.

Example 1   Using logarithms, solve the equation $12e^{-0.1t} = 3$.

Solution   We have

$$12e^{-0.1t} = 3$$
$$e^{-0.1t} = 0.25 \qquad\qquad \text{divide both sides by 12}$$
$$\log\left(e^{-0.1t}\right) = \log 0.25 \qquad\qquad \text{take the logarithm}$$
$$-0.1t \log e = \log 0.25 \qquad\qquad \text{use a logarithm property}$$
$$-0.1t = \frac{\log 0.25}{\log e} \qquad\qquad \text{divide by } \log e$$
$$t = -10 \cdot \frac{\log 0.25}{\log e} = 13.863,$$

so the voltage should drop to 3V after about 13.863 seconds. This is what we expected—we predicted the voltage should drop to 3V sometime between 10 and 15 seconds. To confirm our answer, we see that

$$12e^{-0.1(13.863)} = 12e^{-1.3863} = 3.$$

Solving Equations Using e as a Logarithm Base

When we solve an equation like $10^x = 0.25$, we seek a power 10 that is equal to 0.25:

$$10^? = 0.25.$$

The function $\log x$ is defined to give us the exponent we need: Here, $\log 0.25 = -0.602$ because $10^{-0.602} = 0.25$. Likewise, when we solve an equation like $e^x = 0.25$, we seek a power of the number e that is equal to 0.25:

$$e^? = 0.25.$$

We define a new function, $\log_e x$ or "log base e of x," to give us the power of the number e that is equal to x. This function is often written $\ln x$, for *natural logarithm.*

If x is a positive number, we define the **natural logarithm** of x, written $\ln x$ or $\log_e x$, so that

$$\ln x \text{ is the exponent to which we raise } e \text{ to get } x.$$

In other words,

$$\text{if} \quad y = \ln x \quad \text{then} \quad e^y = x$$

and

$$\text{if} \quad e^y = x \quad \text{then} \quad y = \ln x.$$

Example 2   Solve $12e^{-0.1t} = 3$ using the natural logarithm function.

Solution   Since $12e^{-0.1t} = 3$, we have $e^{-0.1t} = 0.25$. So we need to find the power of e that is equal to 0.25. By definition of the natural logarithm function, the exponent we need is $\ln 0.25$, so

$$-0.1t = \ln 0.25 \qquad \text{by definition of ln function}$$
$$t = -10 \ln 0.25 \qquad \text{multiply by } -10$$
$$= 13.863 \qquad \text{using a calculator.}$$

Nearly all scientific and graphing calculators have an $\boxed{\ln}$ button to find the natural log. The answer agrees with our solution in Example 1, where we solved the same equation to obtain

$$t = -10 \cdot \frac{\log 0.25}{\log e} = 13.863.$$

Properties of Natural Logarithms

The natural logarithm function $\ln x$ has similar properties to the common log function $\log x$:

Properties of the Natural Logarithm

- The operation of raising e to a power and the ln operation undo each other:

$$\ln(e^N) = N \qquad \text{for all } N$$
$$e^{\ln N} = N \qquad \text{for } N > 0$$

- For a and b both positive and any value of t,

$$\ln(ab) = \ln a + \ln b$$
$$\ln\left(\frac{a}{b}\right) = \ln a - \ln b$$
$$\ln(b^t) = t \cdot \ln b$$

- Since $e^0 = 1$,

$$\ln 1 = 0.$$

Using Natural Logarithms to Find Continuous Growth Rates

In Example 1 on page 331, we saw how to determine the growth factor b and the percent growth rate r of an exponential function like $V = 5000e^{0.147t}$. Using the natural logarithm, we can determine the continuous growth rate if we know the growth factor.

Example 3

The value of an investment in year t is given by $V = 1000(1.082)^t$. Give the initial value, the annual growth factor, the percent growth rate, and the continuous growth rate.

Solution

The initial value is $a = \$1000$, the growth factor is $b = 1.082$, and the percent growth rate is $r = b - 1 = 0.082 = 8.2\%$. To find the continuous growth rate k, we write the function in the form ae^{kt}:

$$
\begin{array}{ll}
1000e^{kt} = 1000(1.082)^t & \text{two different ways to write the same function} \\
e^{kt} = 1.082^t & \text{divide by 1000} \\
\left(e^k\right)^t = 1.082^t & \text{rewrite left-hand side using exponent laws} \\
e^k = 1.082 & \text{compare bases} \\
k = \ln 1.082 & \text{definition of ln} \\
 = 0.0788. &
\end{array}
$$

This means the continuous growth rate is 7.88%, and that an alternative form of this function is $V = 1000e^{0.0788t}$. We can check our answer by writing

$$
\begin{aligned}
V &= 1000e^{0.0788t} \\
&= 1000\left(e^{0.0788}\right)^t \\
&= 1000(1.082)^t,
\end{aligned}
$$

as required.

In general,

Converting an Exponential Function to and from Base e
- To convert a function of the form $Q = ae^{kt}$ to a function of the form $Q = ab^t$, use the fact that $b = e^k$.
- To convert a function of the form $Q = ab^t$ to a function of the form $Q = ae^{kt}$, use the fact that $k = \ln b$.

When Should We Use ln x instead of log x?

In Example 1, we solved $e^{-0.1x} = 0.25$ using the log function, and in Example 2 we solved the same equation using the ln function:

$$\text{Using } \log x: \quad x = -10 \cdot \frac{\log 0.25}{\log e} \qquad\qquad \text{Using } \ln x: \quad x = -10 \ln 0.25$$
$$= 13.863 \qquad\qquad\qquad\qquad\qquad = 13.863.$$

Even though we get the right numerical answer in both cases, our solution to Example 1 involves an extra factor of $1/\log e$. This is because we used log base 10 to solve an equation involving base e.

Example 4   Find when the investment from Example 3 will be worth \$5000 using the:

(a)   log function   (b)   ln function.

Solution   We have:

(a) $1000(1.082)^t = 5000$ (b) $1000(1.082)^t = 5000$

$1.082^t = 5$ $1.082^t = 5$ dividing both sides by 1000

$\log\left(1.082^t\right) = \log 5$ $\ln\left(1.082^t\right) = \ln 5$ applying (a) log, (b) ln

$t \log 1.082 = \log 5$ $t \ln 1.082 = \ln 5$

$t = \dfrac{\log 5}{\log 1.082}$ $t = \dfrac{\ln 5}{\ln 1.082}$

$= 20.421.$ $= 20.421.$

In Example 4, we obtain the same numerical answer using either log or ln. In summary,

---

In all cases, you can get the right numerical answer using either ln or log. However:
- If an equation involves base 10, using log might be easier than ln.
- If an equation involves base e, using ln might be easier than log.

Otherwise, there is often no strong reason to prefer one approach to the other.[8]

---

[8]This changes in calculus, where it is often easier to use the ln function than the log function regardless of the base.

Logarithms to Other Bases

In addition to base 10 and base e, other bases are sometimes used for logarithms. Perhaps the most frequent of these is $\log_2 x$, the logarithm with base 2. For instance, since computers represent data internally using the base 2 number system, computer scientists often use logarithms to base 2.

To see how these different bases work, suppose we want to solve the equation $b^y = x$ where b and x are positive numbers. Taking the logarithm of both sides gives

$$\log b^y = \log x$$
$$y \cdot \log b = \log x$$
$$y = \frac{\log x}{\log b}.$$

This means y is the exponent of b that yields x, and we say that y is the "log base b" of x, written $y = \log_b x$. As you can check for yourself, a similar derivation works using $\ln x$ instead of $\log x$. By analogy with common and natural logarithms, we have:

If x and b are positive numbers,

$$\log_b x \text{ is the exponent to which we raise } b \text{ to get } x.$$

In other words,

$$\text{if} \quad y = \log_b x \quad \text{then} \quad b^y = x$$

and

$$\text{if} \quad b^y = x \quad \text{then} \quad y = \log_b x.$$

- We can find $\log_b x$ using either common or natural logarithms:

$$\log_b x = \frac{\log x}{\log b} \quad \text{or} \quad \log_b x = \frac{\ln x}{\ln b}.$$

- The $\log_b x$ function has properties similar to the $\log x$ and $\ln x$ function.

Though we can always use the formula $\log_b x = \log x / \log b$, sometimes we do not need it.

Example 5   Evaluate the following expressions.

(a) $\log_2 16$ (b) $\log_3 81$ (c) $\log_{100} 1{,}000{,}000$ (d) $\log_5 20$

Solution   (a) By definition, $\log_2 16$ is the exponent of 2 that gives 16. We know that $2^4 = 16$, so $\log_2 16 = 4$.
(b) By definition, $\log_3 81$ is the exponent of 3 that gives 81. We know that $3^4 = 81$, so $\log_3 81 = 4$.
(c) By definition, $\log_{100} 1{,}000{,}000$ is the exponent of 100 that gives 1,000,000. We know that $100^3 = 1{,}000{,}000$, so $\log_{100} 1{,}000{,}000 = 3$.
(d) By definition, $\log_5 20$ is the exponent of 5 that gives 20. We know that $5^1 = 5$ and $5^2 = 25$, so this exponent must be between 1 and 2. Using either of our formulas for $\log_b$, we have

$$\log_5 20 = \frac{\log 20}{\log 5} = \frac{1.3010}{0.6990} = 1.8614 \quad \text{using common logarithms}$$

$$\log_5 20 = \frac{\ln 20}{\ln 5} = \frac{2.996}{1.609} = 1.8614 \quad \text{using natural logarithms.}$$

Note that we get the same answer using either log or ln. To check our answer, we see that

$$5^{\log_5 20} = 5^{1.8614} = 20,$$

as required.

Using logarithms to other bases allows us to write expressions in a more compact form.

Example 6   Write the following expressions in the form $\log_b x$ and state the values of b and x. Verify your answers using a calculator.

(a) $\dfrac{\ln 8}{\ln 3}$ (b) $\dfrac{\log 50}{\log 5}$ (c) $\dfrac{\ln 0.2}{\ln 4}$

Solution   (a) We have

$$\frac{\ln 8}{\ln 3} = \log_3 8 \quad \text{so } b = 3, x = 8$$
$$= 1.893. \quad \text{using a calculator}$$

To verify our answer, we need to show that 1.893 is the exponent of 3 that gives 8:

$$3^{1.893} = 8. \quad \text{as required}$$

(b) We have

$$\frac{\log 50}{\log 5} = \log_5 50 \quad \text{so } b = 5, x = 50$$
$$= 2.431. \quad \text{using a calculator}$$

To verify our answer, we need to show that 2.431 is the exponent of 5 that gives 50:

$$5^{2.431} = 50. \quad \text{as required}$$

(c) We have

$$\frac{\ln 0.2}{\ln 4} = \log_4 0.2 \quad \text{so } b = 4, x = 0.2$$
$$= -1.161. \quad \text{using a calculator}$$

To verify our answer, we need to show that -1.161 is the exponent of 4 that gives 0.2:

$$4^{-1.161} = 0.2. \quad \text{as required}$$

An Application of $\log_2$ to Music

In music, the pitch of a note depends on its frequency. For instance, the A above middle C on a piano has, by international agreement, a frequency of 440 Hz, and is known as *concert pitch*.[9] If the frequency is doubled, the resulting note is an octave higher, so the A above concert pitch has frequency 880 Hz, and the A below has frequency 220 Hz. So, for a note with frequency f_1,

$$\text{Frequency of a note 1 octave higher} = 2f_1$$
$$\text{Frequency of a note 2 octaves higher} = 2^2 f_1$$
$$\text{Frequency of a note 3 octaves higher} = 2^3 f_1$$
$$\text{Frequency of a note } n \text{ octaves higher} = 2^n f_1.$$

So if f_1 and f_2 are the frequencies of two notes (with $f_1 \geq f_2$) the number of octaves between them is the exponent n such that $f_2 = 2^n f_1$. Rewriting this as

$$2^n = \frac{f_1}{f_2},$$

we take logarithms to base 2 and get

$$n = \text{Number of octaves} = \log_2\left(\frac{f_1}{f_2}\right)$$

Example 7   Humans can hear a pitch as high as 20,000 Hz. How many octaves above concert pitch is this?

Solution   We have

$$
\begin{aligned}
\text{Number of octaves} &= \log_2\left(\frac{\text{Highest audible pitch}}{\text{Concert pitch}}\right) \\
&= \log_2\left(\frac{20{,}000}{440}\right) \\
&= \log_2 45.455 \\
&= \frac{\log 45.455}{\log 2} \qquad\qquad \text{using formula for } \log_2 x \\
&= 5.506,
\end{aligned}
$$

so the highest audible pitch is between 5 and 6 octaves above concert pitch.

There are twelve piano keys in an octave: A, $A\sharp$, B, C, $C\sharp$, D, $D\sharp$, E, F, $F\sharp$, G, $G\sharp$. Another unit of measurement in pitch is the cent. The interval between two adjacent keys on a piano is defined to be 100 cents, so each octave measures 1200 cents.[10] Thus the number of cents between two notes with frequencies f_1 and f_2 is

$$\text{Number of cents} = 1200 \log_2\left(\frac{f_1}{f_2}\right).$$

---

[9] http://en.wikipedia.org/wiki/Middle_C, accessed June 14, 2007.
[10] http://en.wikipedia.org/wiki/Cent_%28music%29, accessed June 14, 2007.

Example 8   How many piano keys above concert pitch is the 20,000 Hz note in Example 7?

Solution   We know from Example 7 that this note is 5.506 octave above concert pitch. Multiplying by 1200 gives

$$\text{Number of cents} = 1200 \times \text{Number of octaves} = 1200 \cdot 5.506 = 6607.2,$$

so the highest audible pitch is 6607.2 cents above concert pitch. There are 100 cents per piano key, so this amounts to just over

$$\frac{6607.2 \text{ cents}}{100 \text{ cents per key}} \approx 66 \text{ keys higher than concert pitch.}$$

We know every 12 keys is an octave, so if we start at the A above middle C and count up $5 \cdot 12 = 60$ keys, we come to another (barely audible) A key. There are $66 - 60 = 6$ keys to go—$A\sharp$, B, C, $C\sharp$, D, bringing us to $D\sharp$, the 66th key above concert pitch. At 6600 cents, this $D\sharp$ key is 7.2 cents lower (that is, very slightly flatter) than a pitch of 6607.2 cents or 20,000 Hz.[11]

Example 9   Find the frequency of middle C.

Solution   Counting down keys from concert pitch (A) to middle C, we have $G\sharp$, G, $F\sharp$, F, E, $D\sharp$, D, $C\sharp$, C. This means middle C is nine keys or $9 \cdot 100$ cents below middle A. Letting f_C be the pitch of middle C, we see that

$$\text{Number of cents} = 1200 \log_2 \left(\frac{\text{Concert pitch}}{f_C} \right)$$
$$900 = 1200 \log_2 \left(\frac{440}{f_C} \right)$$
$$\log_2 \left(\frac{440}{f_C} \right) = \frac{900}{1200}$$
$$= 0.75.$$

We know that $\log_2 x = y$ means $x = 2^y$, so here,

$$\frac{440}{f_C} = 2^{0.75}$$
$$f_C = \frac{440}{2^{0.75}}$$
$$= 261.626 \text{ Hz.}$$

Thus, middle C has a pitch of 261.626 Hz.[12]

---

[11]Modern pianos have 88 keys, with 39 of them above concert pitch, so this would add 27 keys, making the piano about 1/3 wider.

[12]Another pitch standard known as *scientific pitch* tunes middle C to the nearby, slightly flatter 256 Hz. Since 256 Hz is a power of 2, and since every octave above (or below) is twice (or half) the pitch, this means every C note is a power of 2. But this convention has not caught on. See http://en.wikipedia.org/wiki/Pitch_(music) for more details (accessed June 18, 2007).

Exercises and Problems for Section 11.4

EXERCISES

■ Solve the equations in Exercises **1–6** using natural logs.

1. $8e^{-0.5t} = 3$

2. $200e^{0.315t} = 750$

3. $0.03e^{-0.117t} = 0.12$

4. $7000e^{t/45} = 1200$

5. $23e^{t/90} = 1700$

6. $90 - 5e^{-0.776t} = 40$

■ Solve the equations in Exercises **7–10** using

(a) the log function   (b) the ln function.

Verify that you obtain the same numerical value either way.

7. $250e^{0.231t} = 750$

8. $40 \cdot 10^{t/5} = 360$

9. $3000(0.926)^t = 600$

10. $700 \cdot 2^{t/9} = 42175$

■ In Exercises **11–16**, give the starting value a, the growth factor b, the percent growth rate r, and the continuous growth rate k of the exponential function.

11. $Q = 210e^{-0.211t}$

12. $Q = 395(1.382)^t$

13. $Q = 350(1.318)^t$

14. $Q = 2000 \cdot 2^{-t/25}$

15. $Q = 27.2(1.399)^t$

16. $Q = 0.071(1.308)^t$

17. If \$2500 is invested at 8.2% annual interest, compounded annually, when is it worth \$12,000?

18. If \$1200 is invested at 7.5% annual interest, compounded continuously, when is it worth \$15,000?

■ Evaluate the expressions in Exercises **19–30** without using a calculator. Verify your answers. *Example.* We have $\log_2 32 = 5$ because $2^5 = 32$.

19. $\log_2 64$

20. $\log_4 64$

21. $\log_2 0.25$

22. $\log_3 9$

23. $\log_9 3$

24. $\log_9 9$

25. $\log_{20} 400$

26. $\log_{20} 1$

27. $\log_{20} 0.05$

28. $\log_{100} 10$

29. $\log_{100} 10,000$

30. $\log_{100} 1000$

■ Write the expressions in Exercises **31–33** in the form $\log_b x$ and state the values of b and x. Verify your answers using a calculator as in Example 6.

31. $\dfrac{\log 12}{\log 2}$

32. $\dfrac{\ln 20}{\ln 7}$

33. $\dfrac{\log 0.75}{\log 5}$

PROBLEMS

■ Languages diverge over time, and as part of this process, old words are replaced with new ones.[13] Using methods of *glottochronology*, linguists have estimated that the number of words on a standardized list of 100 words that remain unchanged after t millennia is given by

$$f(t) = 100e^{-Lt}, \quad L = 0.14.$$

Refer to this formula to answer Problems **34–36**. What do your answers tell you about word replacement?

34. Evaluate $f(3)$.

35. Solve $f(t) = 10$.

36. Find the average rate of change of $f(t)$ between $t = 3$ and $t = 5$.

■ Wikipedia is an online, collaboratively edited encyclopedia. The number of articles N (in 1000s) for the English-language site can be approximated by the exponential function[14]

$$N = f(t) = 90e^{t/\tau}, \quad \tau = 499.7 \text{ days},$$

where t is the number of days since October 1, 2002. Answer Problems **37–39**. What do your answers tell you about Wikipedia?

37. Evaluate $f(200)$.

38. Solve $N = 250$.

39. Find the average rate of change of f between $t = 500$ and $t = 600$.

---

[13] http://en.wikipedia.org/wiki/Glottochronology, accessed May 6, 2007.

[14] http://en.wikipedia.org/wiki/Wikipedia:Modelling_Wikipedia%27s_growth, accessed May 8, 2007. Prior to 2002 the growth was more complicated, and more recently the growth has slowed somewhat.

Problems 40–43 concern the *Krumbein phi (φ)* scale of particle size, which geologists use to classify soil and rocks, defined by the formula[15]

$$\phi = -\log_2 D,$$

where D is the diameter of the particle in mm.

40. Some particles of clay have diameter 0.0035 mm. What do they measure on the ϕ scale?

41. A cobblestone has diameter 3 inches. Given that 1 inch is 25.4 mm, what does it measure on the ϕ scale?

42. Geologists define a boulder as a rock measuring -8 or less on the phi scale. What does this imply about the diameter of a boulder?

43. On the ϕ scale, two particles measure $\phi_1 = 3$ and $\phi_2 = -1$, respectively. Which particle is larger in diameter? How many times larger?

Write the expressions in Problems 44–49 in the form $\log_b x$ for the given value of b. State the value of x, and verify your answer using a calculator.

44. $\dfrac{\log 17}{2}$, $b = 100$

45. $\dfrac{\log 17}{2}$, $b = 10$

46. $\dfrac{\log 90}{4 \log 5}$, $b = 5$

47. $\dfrac{\log 90}{4 \log 5}$, $b = 25$

48. $\dfrac{4}{\log_2 5}$, $b = 5$

49. $\dfrac{4}{\log_3 5}$, $b = 5$

50. In 1936, researchers at the Wright-Patterson Air Force Base found that each new aircraft took less time to produce than the one before, owing to what is now known as the *learning curve effect*. Specifically, researchers found that[16]

$$f(n) = t_0 n^{\log_2 k}$$

where $f(n) = $ Time required to produce n^{th} unit

$t_0 = $ Time required to produce the first unit.

The value of k will vary from industry to industry, but for a particular industry it is often a constant.

(a) Suppose for a particular industry, the first unit takes 10,000 hours of labor to produce, so $n_0 = 10{,}000$, and that $k = 0.8$. The learning-curve effect states that the additional hours of labor goes down by a fixed percentage each time production doubles. Show that this is the case by completing the table. By what percent does the required time drop when production is doubled?

Table 11.9

n	1	2	4	8	16
$f(n)$					

(b) What kind of function (exponential, power, logarithmic, other) is $f(n)$? Discuss.

(c) Suppose a different industry has a lower value of t_0, say, $t_0 = 7000$, but a higher value of b, say, $b = 0.9$. Explain what this tells you about the difference in the learning-curve effect between these two industries.

(d) What would it mean for $k > 1$ in terms of production time? Explain why this is unlikely to be the case.

REVIEW EXERCISES AND PROBLEMS FOR CHAPTER 11

EXERCISES

In Exercises 1–16, evaluate without a calculator, or say if the expression is undefined.

1. $\log 100$

2. $\log(1/100)$

3. $\log 1$

4. $\log 0$

5. $\log(-1)$

6. $\log \sqrt{10}$

7. $\log \sqrt[3]{10}$

8. $\log \dfrac{1}{\sqrt[4]{10}}$

9. $\log 10^{3.68}$

10. $\log 10^{-0.584}$

11. $\log 10^{2n+1}$

12. $10^{\log 10}$

13. $10^{\log(-10)}$

14. $10^{\log(1/100)}$

15. $10^{\log(5.9)}$

16. $10^{\log(3a-b)}$

---

[15]http://en.wikipedia.org/wiki/Krumbein_scale, page last accessed June 24, 2007.

[16]http://en.wikipedia.org/wiki/Experience_curve_effects, page last accessed June 23, 2007.

■ In Exercises 17–22, rewrite the equation using exponents instead of logarithms.

17. $\log 10 = 1$ **18.** $\log 0.001 = -3$

19. $\log 54.1 = 1.733$ **20.** $\log 0.328 = -0.484$

21. $\log w = r$ **22.** $\log 384 = n$

■ In Exercises 23–28, rewrite the equation using logarithms instead of exponents.

23. $10^6 = 1,000,000$ **24.** $10^0 = 1$

25. $10^{-1} = 0.1$ **26.** $10^{1.952} = 89.536$

27. $10^{-0.253} = 0.558$ **28.** $10^a = b$

29. (a) Rewrite the statement $10^2 = 100$ using logs.
    (b) Rewrite the statement $10^2 = 100$ using the square root operation.

30. (a) Rewrite the statement $10^3 = 1000$ using logs.
    (b) Rewrite the statement $10^3 = 1000$ using the cube root operation.

■ Simplify the expressions in Exercises 31–37.

31. $\log 10^{5x-3}$ **32.** $\log \sqrt{10^x}$

33. $\log \left(1000^{2x^3} \right)$ **34.** $(0.1)^{-\log(1+x^3)}$

35. $\log \dfrac{1}{\sqrt[3]{10^x}}$ **36.** $10^{2\log(3x+1)}$

37. $\log \sqrt{1000^x}$

■ Solve the equations in Exercises 38–75.

38. $5^t = 20$ **39.** $\pi^x = 100$

40. $(\sqrt{2})^x = 30$ **41.** $8(1.5)^t = 60.75$

42. $127(16)^t = 26$ **43.** $40(15)^t = 6000$

44. $3(64)^t = 12$ **45.** $16 \left(\dfrac{1}{2} \right)^t = 4$

46. $100 \cdot 4^t - 254$ **47.** $(0.03)^{t/3} = 0.0027$

48. $155/3^t = 15$ **49.** $5 \cdot 6^x = 2$

50. $(0.9)^{2t} = 3/2$ **51.** $15^{t-2} = 75$

52. $2^{2t-5} = 1024$ **53.** $4(3^{t^2-7t+19}) = 4 \cdot 3^9$

54. $2(17)^t - 5 = 25$ **55.** $\dfrac{1}{4}(5)^t + 86.75 = 3^5$

56. $80 - (1.02)^t = 20$ **57.** $4(0.6)^t = 10(0.6)^{3t}$

58. $AB^t = C, A \neq 0$ **59.** $10^x = 1,000,000,000$

60. $10^x = 0.00000001$ **61.** $10^x = \dfrac{1}{\sqrt[3]{100}}$

62. $10^x = 100^3$ **63.** $8 \cdot 10^x - 5 = 4$

64. $10 + 2(1 - 10^x) = 8$ **65.** $10^x = 3 \cdot 10^x - 3$

66. $10^x \sqrt{5} = 2$ **67.** $12 - 4(1.117)^t = 10$

68. $50 \cdot 2^t = 1500$ **69.** $30 \left(1 - 0.5^t \right) = 25$

70. $800 \left(\dfrac{9}{8} \right)^t = 1000$ **71.** $r + b^t = z$

72. $v + wb^t = L$

73. $250(0.8)^t = 800(1.1)^t$

74. $40(1.118)^t = 90(1.007)^t$

75. $0.0315(0.988)^t = 0.0422(0.976)^t$

■ Solve the equations in Exercises 76–85.

76. $\log x = 2 \log \sqrt{2}$ **77.** $2 \log x = \log 9$

78. $\log(x - 1) = 3$ **79.** $\log(x^2) = 6$

80. $\log x = -5$ **81.** $\log x = \dfrac{1}{3}$

82. $\log x = 4.41$ **83.** $2 \log x = 5$

84. $15 - 3 \log x = 10$ **85.** $2 \log x + 1 = 5 - \log x$

■ A population grows at the constant percent growth rate given in Exercises 86–88. Find the doubling time.

86. 5%   **87.** 10%   **88.** 25%

■ In Exercises 89–94, rewrite the equations using exponents instead of natural logarithms.

89. $\ln 1 = 0$ **90.** $\ln x = -2$

91. $\ln 10 = y$ **92.** $\ln(2p^2) = 3$

93. $4 = \ln 7r$ **94.** $2q + 1 = \ln 0.5$

In Exercises **95–100**, rewrite the equation using the natural logarithm instead of exponents.

95. $e^0 = 1$

96. $e^5 = 148.413$

97. $e^{-4} = 0.018$

98. $e^x = 12$

99. $1.221 = e^{0.2}$

100. $7 = e^{x^2}$

In Exercises **101–108**, evaluate without a calculator, or say if the expression is undefined.

101. $\ln(\ln e)$

102. $3\ln e^2$

103. $\ln \dfrac{1}{e^3}$

104. $e^{\ln 5} - \ln e^5$

105. $\ln \sqrt{e} - \sqrt{\ln e}$

106. $\dfrac{2\ln e}{\ln 1}$

107. $\ln e + \ln(-e)$

108. $\dfrac{2}{3}\ln \dfrac{1}{\sqrt{e^3}}$

In Exercises **109–114**, simplify the expressions if possible.

109. $3\ln x^{-2} + \ln x^6$

110. $e^{-4\ln t}$

111. $\dfrac{\ln e^{t/2}}{t}$

112. $\dfrac{1}{3}\ln \sqrt{e^x}$

113. $\ln(x + e)$

114. $\dfrac{4\ln x - \ln x^3}{\ln x}$

In Exercises **115–118**, evaluate without a calculator.

115. $\log_2 16$

116. $\log_3 27$

117. $\log_5 \dfrac{1}{25}$

118. $\log_7 \sqrt[3]{49}$

In Exercises **119–122**, rewrite the expression using (a) base 10 logarithms and (b) natural logarithms. Use a calculator to verify that you obtain the same numerical value in either case.

119. $\log_2 100$

120. $\log_5 3$

121. $\log_3 5$

122. $\log_{0.01} 10$

In Exercises **123–126**, evaluate the expression to the nearest thousandth.

123. $\log_2 9$

124. $\log_2 6$

125. $\log_2 3$

126. $\log_2 0.75$

PROBLEMS

Without calculating them, explain why the two numbers in Problems **127–130** are equal.

127. $\log(1/2)$ and $-\log 2$

128. $\log(0.2)$ and $-\log 5$

129. $10^{\frac{1}{2}\log 5}$ and $\sqrt{5}$

130. $\log 303$ and $2 + \log 3.03$

131. Complete the following table. Based on your answers, which table entry is closest to $\log 15$?

x	1.175	1.176	1.177	1.178	1.179
10^x					

132. What does the table tell you about $\log 45$?

x	1.64	1.66	1.68	1.70
10^x	43.7	45.7	47.9	50.1

In Problems **133–138**, assume x and b are positive. Which of statements (I)–(VI) is equivalent to the given statement?

I. $x = b$

II. $\log b = x$

III. $\log x = b$

IV. $x = \dfrac{1}{10}\log b$

V. $x = b/10$

VI. $x = b^{1/10}$

133. $10^b = x$

134. $x^{10} = b$

135. $x^3 - b^3 = 0$

136. $b^3 = 10^{3x}$

137. $10^x = b^{1/10}$

138. $10^{b/10} = 10^x$

If possible, use logarithm properties to rewrite the expressions in Problems 139–144 in terms of u, v, w given that

$$u = \log x, v = \log y, w = \log z.$$

Your answers may involve constants such as $\log 2$ but nonconstant logs and exponents are not allowed.

139. $\log xy$

140. $\log \dfrac{y}{x}$

141. $\log\left(2xz^3\sqrt{y}\right)$

142. $\dfrac{\log\left(x^2\sqrt{y}\right)}{\log\left(z^3\sqrt[3]{x}\right)}$

143. $\log\left(\log\left(3^x\right)\right)$

144. $\left(\log\dfrac{1}{x}\right)\left(\log\left(2y\right)\right)\left(\log\left(z^5\right)\right)$

Simplify the expressions in Problems 145–148.

145. $10^{\log(2x+1)}$

146. $10^{\log\sqrt{x^2+1}}$

147. $100^{\log(x+1)}$

148. $10^{\log x + \log(1/x)}$

Solve the equations in Problems 149–155.

149. $50 \cdot 3^{t/4} = 175$.

150. $0.35 \cdot 2^{-12t} = 0.05$

151. $3\log(x/2) = 6$

152. $10^{x+5} = 17$

153. $25 = 2 \cdot 10^{2x+1}$

154. $3 - 10^{x-0.5} = 10^{x-0.5}$

155. $2(10^{5x+1} + 2) = 3(1 + 10^{5x+1})$

Solve the equations in Problems 156–161.

156. $\log(2x) - 1 = 3$

157. $2\log x - 1 = 5$

158. $2\log(x-1) = 6$

159. $\dfrac{1}{2}\log(2x-1) + 2 = 5$

160. $2(\log(x-1) - 1) = 4$

161. $\log(\log x) = 1$

162. An investment initially worth \$1200 grows by 11.2% per year. When will the investment be worth \$5000?

163. In 2005, the population of Turkey was about 70 million and growing at 1.09% per year, and the population of the European Union (EU) was about 457 million and growing at 0.15% per year.[17] If current growth rates continue, when will Turkey's population equal that of the EU?

164. A predatory species of fish is introduced into a lake. The number of these fish after t years is given by $N = 20 \cdot 2^{t/3}$.

    (a) What is the initial size of the fish population?

    (b) How long does it take the population to double in size? To triple in size?

    (c) The population will stop growing when there are 200 fish. When will this occur?

165. Percy Weasley's first job with the Ministry of International Magical Cooperation was to write a report trying to standardize the thickness of cauldrons, because "leakages have been increasing at a rate of almost three percent a year."[18] If this continues, when will there be twice as many leakages as there are now?

166. The population of a town decreases from an initial level of 800 by 6.2% each year. How long until the town is half its original size? 10% of its original size?

167. The number of regular radio listeners who have not heard a new song begins at 600 and drops by half every two weeks. How long will it be until there are only 50 people who have not heard the new song?

168. One investment begins with \$5000 and earns 6.2% per year. A second investment begins with less money, \$2000, but it earns 9.2% per year. Will the second investment ever be worth as much as the first? If so, when?

---

[17]http://www.cia.gov/cia/publications/factbook, accessed October 14, 2005.

[18]J.K. Rowling, *Harry Potter and the Goblet of Fire* (Scholastic Press, 2000).

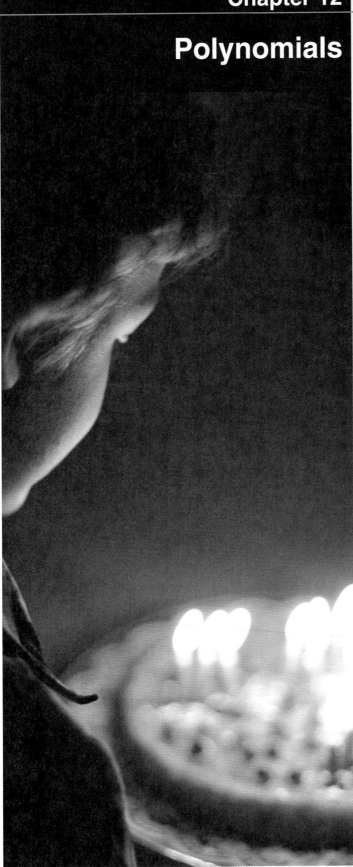

Chapter 12

Polynomials

CONTENTS

12.1 POLYNOMIAL FUNCTIONS

Have you ever wondered how your calculator works? The electronic circuits inside a calculator can perform only the basic operations of arithmetic. But what about the square-root button, or the e^x button? How can a calculator evaluate expressions like $\sqrt[9]{7}$ or $e^{-0.6}$?

The answer is that some calculators use special built-in formulas that return excellent approximations for $\sqrt{x}$ or e^x. These formulas require only simple arithmetic, so they can be written using only the low-level instructions understood by computer chips. So what can we do using only the simple arithmetic operations of addition and multiplication?

Example 1 The expression

$$0.042x^4 + 0.167x^3 + 0.5x^2 + x + 1$$

can be used to approximate the values of e^x.

(a) Show that this expression gives a good estimate for the value of $e^{0.5}$ and $e^{-0.6}$.
(b) What operations are used in evaluating these two estimates?

Solution (a) We have

$$0.042(0.5)^4 + 0.167(0.5)^3 + 0.5(0.5)^2 + (0.5) + 1 = 1.6485,$$

whereas

$$e^{0.5} = 1.6487,$$

so the expression $0.042x^4 + 0.167x^3 + 0.5x^2 + x + 1$ gives an excellent approximation of e^x at $x = 0.5$. Likewise,

$$0.042(-0.6)^4 + 0.167(-0.6)^3 + 0.5(-0.6)^2 + (-0.6) + 1 = 0.5494,$$

whereas

$$e^{-0.6} = 0.5488.$$

Once again, the approximation is good.

(b) Although the expression $0.042x^4 + 0.167x^3 + 0.5x^2 + x + 1$ looks more complicated than e^x, it is simpler in one important way. You can calculate the values of the expression using only multiplication and addition. For instance, at $x = 0.5$, the first two terms are

$$0.042x^4 = 0.042(0.5)^4 = (0.042)(0.5)(0.5)(0.5)(0.5) = 0.002625$$
$$0.167x^3 = 0.167(0.5)^3 = (0.167)(0.5)(0.5)(0.5) = 0.020875.$$

The other terms are calculated in a similar way, then added together. The same comments apply at $x = -0.6$.

Approximation Formulas Used by Calculators

The techniques used by a calculator to approximate the value of expressions like e^x are more complicated than the one we have shown here. However, the basic principle is the same: the calculator uses only basic operations.

How Good Is Our Approximation Formula?

Although our formula for approximating e^x works reasonably well for values of $x = 0.5$ and $x = -0.6$, it works less well for other values. Figure 12.1 shows graphs of $y = e^x$ and the expression

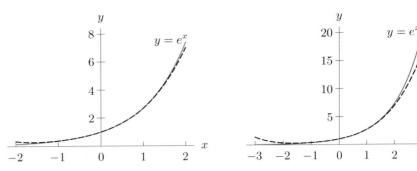

Figure 12.1: The graphs are close near $x = 0.5$ and $x = -0.6$ but draw farther apart as we zoom out

we used to approximate it. Notice that the two graphs are quite close in the general vicinity of $x = 0.5$ and $x = -0.6$. However, for other values of x, the two graphs draw farther apart. This means that we cannot rely on our formula for all values of x.

What Is a Polynomial?

The expression in Example 1 is formed by adding together the terms $0.042x^4$, $0.167x^3$, $0.5x^2$, x, and 1. They are all of the form kx^n (even 1, which is equal to kx^n with $k = 1$ and $n = 0$).

> A **monomial** in x is an expression of the form kx^n, where k is a constant (nonzero) and n is a whole number (positive or zero).

Example 2　Which of the following are monomials in x?

(a)　$-\dfrac{x^2}{9}$

(b)　$2\sqrt{x^3}$

(c)　$\sqrt{2}$

(d)　$2 \cdot 7^x$

(e)　$x^2 - 2x$

(f)　$\dfrac{7}{x^3}$

Solution

(a) This is a monomial with $k = -1/9$ and $n = 2$.

(b) This is not a monomial because $2\sqrt{x^3} = 2x^{3/2}$, and $3/2$ is not a whole number.

(c) This is a monomial with $k = \sqrt{2}$ and $n = 0$.

(d) A monomial cannot have expressions like 7^x, where the variable appears in the exponent.

(e) This is not a monomial because it is not of the form kx^n.

(f) This is not a monomial because x appears in the denominator. If we write this as

$$7x^{-3},$$

we see that $k = 7$ and $n = -3$, and n is a negative integer.

The expression in Example 1, which is the sum of five monomials, is an example of a *polynomial*. In general,

> A **polynomial** in x is an expression that is equivalent to a sum of monomials in x.

For example, the polynomial $2x^2 + 3x$ is a sum of two monomials in x, and the polynomial $1 + x + x^4$ is a sum of three monomials in x.

Example 3   Which of the following are polynomials in x?

(a) $x^2 + 3x^4 - \dfrac{x}{3}$ (b) 5 (c) $5\sqrt{x} - 25x^3$

(d) $9 \cdot 2^x + 4 \cdot 3^x + 15x$ (e) $(x-2)(x-4)$ (f) $\dfrac{3}{x} - \dfrac{x}{7}$

Solution   (a) This is a polynomial. Writing it as

$$x^2 + 3x^4 - \frac{x}{3} = x^2 + 3x^4 - \frac{1}{3}x,$$

we see that it is the sum of three monomials, namely, x^2, $3x^4$ and $-x/3$.

(b) This is a polynomial because it can be written as $5x^0$.

(c) This is not a polynomial because it has a term in $\sqrt{x} = x^{1/2}$, and $1/2$ is not a whole number.

(d) While a polynomial can have whole number powers of x (like x^2 or x^3), it cannot have expressions like 2^x, where the variable appears in the exponent.

(e) This is a polynomial because it is equivalent to a sum of monomials:

$$(x-2)(x-4) = x^2 - 4x - 2x + 8 = x^2 - 6x + 8.$$

(f) This is not a polynomial because x appears in the denominator. If we write this as

$$3x^{-1} - \frac{x}{7},$$

we see that one of the operations is raising x to a negative power.

Polynomial Functions

Just as a linear function is one defined by a linear expression, and a quadratic function is one defined by a quadratic expression, a polynomial function is a function of the form

$$y = p(x) = \text{Polynomial in } x.$$

Example 4   A box is 4 ft longer than it is wide and twice as high as it is wide. Express the volume of the box as a polynomial function of the width.

Solution   Let w be the width of the box in feet, and let V be its volume in cubic feet. Then its length is $w + 4$, and its height is $2w$, so

$$V = w(w + 4)(2w) = 2w^2(w + 4) = 2w^3 + 8w^2.$$

Example 5   A deposit of $800 is made into a bank account with an annual interest rate of r. Express B, the balance after 4 years, as a polynomial function of r.

Solution   Each year the balance is multiplied by $1 + r$, so at the end of four years we have

$$B = 800(1 + r)^4.$$

This is probably the most convenient form for the expression describing this function. However, by expanding and collecting like terms, we could express it as a sum of terms involving powers up to r^4, namely,

$$B = 800 + 3200r + 4800r^2 + 3200r^3 + 800r^4.$$

We do this in Problem 27 on page 378.

Example 6   Which of the following describe polynomial functions?

(a) The surface area of a cylinder of radius R and fixed height h is $f(R) = 2\pi Rh + 2\pi R^2$.

(b) The balance on a deposit P after t years in a bank account with an annual interest rate of r is $B(t) = P(1 + r)^t$.

(c) The effective potential energy of a sun-planet system is a function of the distance r between the planet and the sun given by $g(r) = -3r^{-1} + r^{-2}$.

Solution   (a) This is a polynomial function because it is expressed as a sum of a constant times R plus another constant times R^2.

(b) Although this function looks similar to the one in Example 5, the situation is different because here r is a constant, not the independent variable. This is not a polynomial function because the independent variable occurs in the exponent.

(c) This is not a polynomial because r appears to a negative power.

Exercises and Problems for Section 12.1

EXERCISES

Which of the expressions in Exercises 1–6 are equivalent to monomials in x?

1. $-\dfrac{x^3}{5}$

2. $-\dfrac{5}{x^3}$

3. $-x \cdot x^2$

4. $x - x^2$

5. $3 + 17$

6. 5^x

Which of the expressions in Exercises 7–12 are polynomials in x? If an expression is not a polynomial in x, what rules it out?

7. $\dfrac{2x^3}{5} - 2x^7$

8. $x(x-1) - x^2(1 - x^3)$

9. $\sqrt{2}x - x^2 + x^4$

10. $\sqrt{2x} - x^3 + x^5$

11. $\dfrac{1}{x} + \dfrac{x^2}{3} + 4x^3$

12. $14x^4 + 7x^3 - 3 \cdot 2^x + 2x + 15$

In Exercises 13–18, is the expression a polynomial in the given variable?

13. $(4 - 2p^2)p + 3p - (p+2)^2$, in p

14. $(x-1)(x-2)(x-3)(x-4) + 29$, in x

15. $\left(a + \dfrac{1}{x}\right)^2 - \left(a - \dfrac{1}{x}\right)^2$, in x

16. $\left(a + \dfrac{1}{x}\right)^2 - \left(a - \dfrac{1}{x}\right)^2$, in a

17. $\dfrac{n(n+1)(n+2)}{6}$, in n

18. $P\left(1 + \dfrac{r}{12}\right)^{10}$, in r

In Exercises 19–25, $p(z) = 4z^3 - z$. Find the given values and simplify if possible.

19. $p(0)$ 20. $p(\sqrt{5})$ 21. $p(-1)$

22. $p(4^t)$ 23. $p(t+1)$ 24. $p(3x)$

25. The values of z such that $p(z) = 0$

26. If $p(x) = x^4 - 2x^2 + 1$, find

(a) $p(0)$ (b) $p(2)$ (c) $p(t^2)$.

(d) The values of x such that $p(x) = 0$

PROBLEMS

27. Expand the polynomial in Example 5.

▪ Problems **28–31** refer to the functions $f(x)$ and $g(x)$, where the function

$$g(x) = 1 + \frac{1}{2}x + \frac{3}{8}x^2 + \frac{5}{16}x^3$$

is used to approximate the values of

$$f(x) = \frac{1}{\sqrt{1-x}}.$$

28. Evaluate f and g at $x = 0$. What does this tell you about the graphs of these two functions?

29. Evaluate $f(x)$ and $g(x)$ at $x = 0.1, 0.2, 0.3$ and record your answers to three decimal places in a table. Does

your table support the claim that $g(x)$ is a good approximation to $f(x)$ for these values of x?

30. Show that $f(x)$ is undefined at $x = 1$ and $x = 2$, but that $g(x)$ is defined at these values. Explain why the algebraic operations used to define f may lead to undefined values, whereas the operations used to define g will not.

31. Given that

$$f(1/2) = \frac{1}{\sqrt{1 - \frac{1}{2}}} = \frac{1}{\sqrt{\frac{1}{2}}} = \frac{1}{\frac{1}{\sqrt{2}}} = \sqrt{2},$$

use $g(x)$ to find a rational number (a fraction) that approximately equals $\sqrt{2}$.

12.2   WORKING WITH POLYNOMIALS

Polynomials are classified by the highest power of x in them. Linear expressions $b + mx$ are polynomials whose highest power, 1, occurs in the term $x^1 = x$, and quadratic expressions $ax^2 + bx + c$ are polynomials whose highest power, 2, occurs in x^2. We sometimes call these *linear polynomials* and *quadratic polynomials* respectively. *Cubic* polynomials are next in line, and can be written $ax^3 + bx^2 + cx + d$. Then come *quartic* polynomials, with a term in x^4, and *quintic* polynomials, with a term in x^5.

The Degree of a Polynomial

The highest power of the variable occurring in the polynomial is called the *degree* of the polynomial. Thus, a linear polynomial has degree 1 and is a first-degree polynomial, a quadratic polynomial has degree 2 and is a second-degree polynomial, and so on.

Example 1   Give the degree of

(a) $4x^5 - x^3 + 3x^2 + x + 1$ (b) $9 - x^3 - x^2 + 3(x^3 - 1)$
(c) $(1 - x)(1 + x^2)^2$ (d) -1

Solution   (a) The polynomial $4x^5 - x^3 + 3x^2 + x + 1$ is of degree 5 because the term with the highest power is $4x^5$.

(b) Expanding and collecting like terms, we have

$$9 - x^3 - x^2 + 3(x^3 - 1) = 9 - x^3 - x^2 + 3x^3 - 3 = 2x^3 - x^2 + 6,$$

so this is a polynomial of degree 3.

(c) Expanding and collecting like terms, we have

$$(1 - x)(1 + x^2)^2 = (1 - x)(1 + 2x^2 + x^4)$$
$$= 1 + 2x^2 + x^4 - x - 2x^3 - x^5$$
$$= -x^5 + x^4 - 2x^3 + 2x^2 - x + 1,$$

so this is a polynomial of degree 5.

(d) Because -1 can be written as $(-1)x^0$, it is a polynomial of degree 0.

Example 2   Every year on his birthday Elliot's parents invest money for him at an annual interest rate of r, starting with \$500 on his 15th birthday, \$600 on his 16th, and going up by \$100 each year after that.

(a) If $r = 4\%$, what is his investment worth on his 21st birthday?

(b) If $x = 1 + r$, then x is the annual growth factor. Write a polynomial in x for the value of his investment on his 21st birthday.

(c) Find the degree of the polynomial and interpret it in terms of the investment.

Solution   (a) On his 21st birthday the investment is worth the sum of all of the birthday presents plus interest. The first gift of \$500 has been earning interest for 6 years, and so is worth $\$500(1.04)^6 = \$632.66$. The second gift of \$600 has been earning interest for 5 years, so it is worth $\$600(1.04)^5 = \$729.99$, and so on. Adding up all the gifts, we get

$$\text{Total value} = 500(1.04)^6 + 600(1.04)^5 + 700(1.04)^4 + 800(1.04)^3$$
$$+ 900(1.04)^2 + 1000(1.04) + 1100$$
$$= \$6194.88.$$

(b) Replacing 1.04 by x, we get

$$\text{Total value} = 500x^6 + 600x^5 + 700x^4 + 800x^3 + 900x^2 + 1000x + 1100.$$

(c) Since the term with the highest power is $500x^6$, the degree is 6. This makes sense because the first present has been earning interest for 6 years, so it has been multiplied by x six times.

The Standard Form of a Polynomial

We get the standard form of a polynomial by expanding and collecting like terms. Since a polynomial can have a large number of coefficients, we number them using subscripts, like this:

Degree of polynomial	Standard form
1	$a_1 x + a_0$
2	$a_2 x^2 + a_1 x + a_0$
3	$a_3 x^3 + a_2 x^2 + a_1 x + a_0$
4	$a_4 x^4 + a_3 x^3 + a_2 x^2 + a_1 x + a_0$
5	$a_5 x^5 + a_4 x^4 + a_3 x^3 + a_2 x^2 + a_1 x + a_0$

The coefficient of x^4 is a_4, the coefficient of x^3 is a_3, and so on. Notice that a_0 is the coefficient of $x^0 = 1$, and therefore a_0 is the constant term.

Example 3   Identify the coefficients of (a) $2x^3 + 3x^2 + 4x - 1$ (b) $4x^5 - x^3 + 3x^2 + x + 1$ (c) x^2.

Solution   (a) The subscript of the coefficient should match up with the power of x, so

$$2x^3 + 3x^2 + 4x - 1 = a_3 x^3 + a_2 x^2 + a_1 x + a_0.$$

Thus, $a_3 = 2$, $a_2 = 3$, $a_1 = 4$, and $a_0 = -1$.

(b) Writing this as

$$4x^5 - x^3 + 3x^2 + x + 1 = 4x^5 + 0 \cdot x^4 + (-1)x^3 + 3x^2 + 1 \cdot x^1 + 1 \cdot x^0,$$

we see that it has coefficients

$$a_5 = 4, \quad a_4 = 0, \quad a_3 = -1, \quad a_2 = 3, \quad a_1 = 1, \quad a_0 = 1.$$

(c) Writing this as

$$x^2 = 1 \cdot x^2 + 0 \cdot x + 0,$$

we get $a_2 = 1$, $a_1 = 0$, and $a_0 = 0$.

Polynomials of Degree n

We sometimes use a letter, such as n, to represent the degree of a polynomial. So n is the highest power in the polynomial, and the coefficient of the corresponding term is written a_n.

Example 4   Consider the polynomial
$$13x^{17} - 9x^{16} + 4x^{14} + 3x^2 + 7.$$

(a)   What is the degree, n?   (b)   Write out the terms $a_n x^n$ and $a_{n-1}x^{n-1}$.
(c)   What is the value of a_{n-3}?   (d)   What is the value of a_1?

Solution   (a) The degree is $n = 17$.
(b) Since $n = 17$, then a_n is the coefficient of x^{17}, so $a_n = 13$. Likewise, a_{n-1} is the coefficient of $x^{n-1} = x^{16}$, so $a_{n-1} = -9$. Therefore, $a_n x^n = 13x^{17}$ and $a_{n-1}x^{n-1} = -9x^{16}$.
(c) The degree is $n = 17$, so $a_{n-3} = a_{14}$ is the coefficient of x^{14}, which means that $a_{n-3} = 4$.
(d) Since a_1 is the coefficient of x, and there is no x-term, $a_1 = 0$.

In general, we have:

The Standard Form of a Polynomial

The standard form of a polynomial in x is

$$p(x) = a_n x^n + a_{n-1}x^{n-1} + \cdots + a_1 x + a_0,$$

where n is a non-negative integer and $a_0, a_1, \ldots, a_n$ are constants.
- The constants $a_0, a_1, \ldots, a_n$ are the *coefficients*.
- If $a_n \neq 0$, then n is the *degree* of the polynomial, and a_n is the *leading coefficient*.

A polynomial with degree zero is equal to its constant term, and is called a *constant polynomial*. If that constant is zero, then we have the *zero polynomial* (whose degree is undefined).

The Constant Term

The *constant term* a_0 gives the value of a polynomial at $x = 0$:
$$p(0) = a_n \cdot 0^n + a_{n-1} \cdot 0^{n-1} + \cdots + a_1 \cdot 0 + a_0 = a_0.$$

Example 5   Give the constant term of (a) $4 - x^3 - x^2$ (b) $z^3 + z^2$ (c) $(q-1)(q-2)(q-3)$.

Solution

(a) Writing this as $-x^3 - x^2 + 4$, we see that its constant term is $a_0 = 4$.

(b) Since the only two terms in this polynomial involve positive powers of z, it has constant term $a_0 = 0$.

(c) We could calculate the constant term of this polynomial by expanding it out. However, since we know that the constant term is the value of the polynomial when $q = 0$, we can say that

$$a_0 = (0-1)(0-2)(0-3) = (-1)(-2)(-3) = -6.$$

The constant term tells us the vertical intercept of the graph of p.

Example 6   Which of Figures 12.2 and 12.3 is the graph of $p(x) = x^4 - 4x^3 + 16x - 16$?

Solution   Since $p(0) = -16$, we choose the graph that appears to intercept the y-axis at around $y = -16$, Figure 12.3.

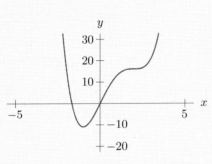

Figure 12.2: A quartic polynomial

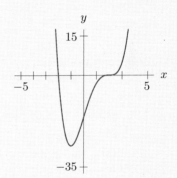

Figure 12.3: A quartic polynomial

Example 7   Find the constant term of the polynomial $800(1 + r)^4$ in Example 5 on page 376, and interpret it in terms of interest rates.

Solution   The polynomial gives the bank balance 4 years after an initial deposit of $800. Although we have not expanded this polynomial, we know that the constant term is the value when $r = 0$, so

$$\text{Constant term} = 800(1 + 0)^4 = 800.$$

This is the value of the bank balance after 4 years if the interest rate is 0. It makes sense that this is the same as the initial deposit, since if the interest rate is 0 the balance does not grow or shrink.

The Leading Term and Leading Coefficient

We already know that all first-degree polynomials have graphs that are lines, and all second-degree polynomials have graphs that are parabolas. This suggests that polynomials of the same degree have certain features in common and that the term with the highest power plays an important role in the overall behavior of a polynomial function. In a polynomial $a_n x^n + a_{n-1} x^{n-1} + \cdots + a_1 x + a_0$, with $a_n \neq 0$, the highest-power term $a_n x^n$ is called the *leading term* of the polynomial. The coefficient, a_n, of the leading term is called the *leading coefficient* of the polynomial.

Example 8   Give the degree and leading term of

(a) $s - 3s^3 + s^2$

(b) $\dfrac{u(u+1)(2u+1)}{6}$

Solution

(a) The degree is 3, and the leading term is $-3s^3$.

(b) We could expand this out to find the leading term, but it is not necessary. The highest degree term comes from multiplying the leading terms from each factor, giving

$$\frac{u \cdot u \cdot (2u)}{6} = \frac{u^3}{3}.$$

Thus, the degree is 3 and the leading term is $\dfrac{u^3}{3}$.

As we saw in Example 8(b), we can sometimes deduce information about a polynomial without having all the terms.

Example 9   If $p(x)$ has degree 4 and constant term -2, and $q(x)$ has degree 5 and constant term 7, what can we say about

(a)   The degree of $p(x)q(x)$?

(b)   The coefficient of x in $p(x)q(x)$?

(c)   The degree of $p(x) + q(x)$?

(d)   The constant term of $3p(x) - 2q(x)$?

(e)   The degree of $p(x)^2$, that is, $(p(x))(p(x))$?

Solution

(a) When we multiply two polynomials together, the leading term in the product comes from multiplying together the leading terms in each factor. The leading term in $p(x)$ is a constant times x^4 and the leading term in $q(x)$ is a constant times x^5, so the leading term in $p(x)q(x)$ is a constant times $x^4 \cdot x^5 = x^9$. Thus the degree of $p(x)q(x)$ is 9.

(b) We do not have enough information: to know the coefficient of x in $p(x)q(x)$ we would need to know more about the coefficients in $p(x)$ and $q(x)$.

(c) Since $p(x)$ has terms only up to degree 4, the leading term in $p(x) + q(x)$ is the term in x^5 coming from $q(x)$, so the degree is 5.

(d) The constant terms in $p(x)$ and $q(x)$ are the only terms contributing to the constant term in $3p(x) - 2q(x)$, so that term is $3(-2) - 2 \cdot 7 = -20$.

(e) When we multiply $p(x)$ by itself, the highest-degree term comes from the highest degree of $p(x)$, which is a constant times x^4, times itself, giving a constant times x^8. Thus the degree of $p(x)^2$ is 8.

Example 10 Consider the polynomial $(ax^3 + 1)^2 - x^6$, where a is a constant.

(a) What is the leading term?
(b) For what values of a is the degree less than 6?

Solution

(a) One approach would be to place this polynomial in standard form. However, we can determine the leading term with fewer steps as follows:

$$\begin{aligned}(ax^3 + 1)^2 - x^6 &= (ax^3)^2 + \text{lower-order terms} - x^6 \\ &= a^2 x^6 - x^6 + \text{lower-order terms} \\ &= (a^2 - 1)x^6 + \text{lower-order terms}.\end{aligned}$$

So the leading term is $(a^2 - 1)x^6$.

(b) Notice that for most values of a, the degree of the polynomial is 6. However, the term $(a^2 - 1)x^6$ disappears if $a^2 - 1 = 0$, that is, if $a = \pm 1$. Therefore, for these values of a, the degree is less than 6.

Summary

- The standard form for a polynomial in x is

$$a_n x^n + a_{n-1} x^{n-1} + \cdots + a_1 x + a_0, \quad a_n \neq 0.$$

- Here, n is called the *degree* of the polynomial and is the power of the highest-power term or *leading term*.

- The *coefficients* are $a_0, a_1, \ldots, a_{n-1}$, and a_n.

- The coefficient a_0 is also called the *constant term*. It determines the value of the expression at $x = 0$.

- Special categories of polynomials include (in order of increasing degree) *linear, quadratic, cubic, quartic,* and *quintic*.

Exercises and Problems for Section 12.2

EXERCISES

In Exercises 1–4, write the polynomials in standard form.

1. $3x - 2x^2 + 5x^7 + 4x^5$
2. $3x^2 + 2x + 2x^7 - 5x^2 - 3x^7$
3. $x(x - 2) + x^2(3 - x)$ 4. $\dfrac{x^4 - 2x - 14x^3}{7}$

In Exercises 5–10, give the constant term, a_0.

5. $4t^3 - 2t^2 + 17$ 6. $12t - 2t^3 + 6$
7. $15 - 11t^9 - 8t^4$ 8. $7t^3 + 2t^2 + 5t$
9. $(3t + 1)(2t - 1)$ 10. $t(t - 1)(t - 2)$

In Exercises 11–13, find the degree.

11. $2s^6 - 3s^5 - 6s^4 - 4s + 1$
12. $2s^3 - s^2 + 1 - s^3 + 2s^2 - s + 3s^3$
13. $3s^2 + 2s^4 + s - s^4 + 2s^3 - 1 - s^4 + 3s$

In Exercises 14–19, give the leading term.

14. $3x^5 - 2x^3 + 4$ 15. $2x^7 - 4x^{11} + 6$
16. $12 - 3x^5 - 15x^3$ 17. $12x^{13} + 4x^5 - 11x^{13}$
18. $13x^4(2x^2 + 1)$ 19. x^8

■ In Exercises 20–24, give the leading coefficient.

20. $5x^6 - 4x^5 + 3x^4 - 2x^3 + x^2 + 1$

21. $1 - 6r^2 + 40r - \frac{1}{2}r^3 + 16r$

22. $100 - \sqrt{6}s + 15s^2$

23. $\sqrt{7}u^3 + 12u - 4 + 6u^2$

24. $t^3 - 2t^2 - \sqrt[3]{9}t^3 + 1$

■ List the nonzero coefficients of the polynomials in Exercises 25–28.

25. $3u^4 + 6u^3 - 3u^2 + 8u + 1$

26. $2x^5 - 3x^3 + x^7 + 1$

27. $\dfrac{s^{13}}{3}$

28. πx

■ Write the polynomials in Exercises 29–38 in standard form

$$a_n x^n + a_{n-1} x^{n-1} + \cdots + a_1 x + a_0.$$

What are the values of the coefficients $a_0, a_1, \ldots, a_n$? Give the degree of the polynomial.

29. $2x^3 + x - 2$ 30. $5 - 3x^7$

31. $20 - 5x$ 32. $\sqrt{7}$

33. $\dfrac{x^2 \sqrt[3]{5}}{7}$ 34. $3 - 2(x - 5)^2$

35. $15x - 4x^3 + 12x - 5x^4 + 9x - 6x^5$

36. $(x - 3)(2x - 1)(x - 2)$ 37. $(x + 1)^3$

38. $1 + x + \dfrac{x^2}{2} + \dfrac{x^3}{6} + \dfrac{x^4}{24} + \dfrac{x^5}{120}$

PROBLEMS

39. Without expanding, what is the constant term of

$$(x + 2)(x + 3)(x + 4)(x + 5)(x + 6)?$$

40. Without expanding, what is the leading term of

$$(2s + 5)(3s + 1)(s - 10)?$$

41. What is the degree and leading coefficient of the polynomial $r(x) = 4$?

■ Problems 42–45 refer to Example 2 on page 379 about the value of annual gifts to Elliot growing at an annual growth factor of $x = 1 + r$, where r is the annual interest rate.

42. Suppose that the first three gifts were $1000, $500, and $750.

(a) If $r = 5\%$, what is the total value of the investments on his 17th birthday? On his 18th birthday?

(b) Write polynomial expression in x for the value on his 17th and 18th birthday.

43. The value of his investments on his 20th birthday is

$$800x^5 + 900x^4 + 300x^2 + 500x + 1200.$$

How much money was invested on each birthday?

44. The total value of his investments on his 20th birthday is

$$1000x^5 + 500x^4 + 750x^3 + 1200x + 650.$$

(a) What were the gifts on his 18th, 19th and 20th birthdays?

(b) Evaluate the polynomial in part (a) for $x = 1.05, 1.06, 1.07$. What do these values tell you about the investment?

45. Graph the polynomial

$$500x^6 + 600x^5 + 700x^4 + 800x^3 + 900x^2 + 1000x + 1100$$

in part (c) of Example 2 and estimate the interest rate if Elliot has $6000 on his 21st birthday.

■ In Problems 46–48, what is the degree of the resulting polynomial?

46. The product of a quadratic and a linear polynomial.

47. The product of two linear polynomials.

48. The sum of a degree 8 polynomial and a degree 4 polynomial.

49. What is the value of

$$5(x - 1)(x - 2) + 2(x - 1)(x - 3) - 4(x - 2)(x - 3)$$

when $x = 3$?

50. What values of the constants A, B, and C, will make

$$A(x - 1)(x - 2) + B(x - 1)(x - 3) - C(x - 2)(x - 3)$$

have the value 7 when $x = 3$?

■ Evaluate the expressions in Problems 51–54 given that

$$f(x) = 2x^3 + 3x - 3, \quad g(x) = 3x^2 - 2x - 4,$$

$$h(x) = f(x)g(x) = a_n x^n + a_{n-1} x^{n-1} + \cdots + a_0.$$

51. n

52. a_{n-1}

53. a_0

54. $h(1)$

55. If the following product of two polynomials,

$$(3t^2 - 7t - 2)(4t^3 - 3t^2 + 5),$$

is written in standard form, what are the constant and leading terms?

In Problems **56–60**, state the given quantities if $p(x)$ is a polynomial of degree 5 with constant term 3, and $q(x)$ is a polynomial of degree 8 with constant term -2.

56. The constant term of $p(x)q(x)$.

57. The degree of $p(x) + q(x)$.

58. The degree of $p(x)q(x)$.

59. The constant term of $p(x) - 2q(x)$.

60. The degree of $p(x)^3 q(x)^2$.

61. Given that $(3x^3 + a)(2x^b + 3) = 6x^7 + cx^4 + dx^3 + 3$, find possible values for the constants a, b, c, and d.

62. Find the constants r, s, p, and q if multiplying out the polynomial $(rx^5 + 2x^4 + 3)(2x^3 - sx^2 + p)$ gives

$$6x^8 - 11x^7 - 10x^6 - 12x^5 - 8x^4 + qx^3 - 15x^2 - 12.$$

In Problems **63–67**, give polynomials satisfying the given conditions if possible, or say why it is impossible to do so.

63. Two polynomials of degree 5 whose sum has degree 4.

64. Two polynomials of degree 5 whose sum has degree 8.

65. Two polynomials of degree 5 whose product has degree 8.

66. Two polynomials whose product has degree 9.

67. A polynomial whose product with itself has degree 9.

In Problems **68–71**, give the value of a that makes the statement true.

68. The degree of $(t - 1)^3 + a(t + 1)^3$ is less than 3.

69. The constant term of $(t - 1)^5 - (t - a)$ is zero.

70. The coefficient of t in $t(a + (t + 1)^{10})$ is zero.

71. The constant term of $(t + 2)^2(t - a)^2$ is 9.

72. Suppose that two polynomials $p(x)$ and $q(x)$ have constant term 1, the coefficient of x in $p(x)$ is a and the coefficient of x in $q(x)$ is b. What is the coefficient of x in $p(x)q(x)$?

73. Find the product of $5x^2 - 3x + 1$ and $10x^3 - 3x^2 - 1$.

74. Find the leading term and constant term of

(a) $(x - 1)^2$ (b) $(x - 1)^3$ (c) $(x - 1)^4$

75. What is the coefficient of x^{n-1} in $(x+1)^n$ for $n = 2, 3$ and 4?

12.3 SOLVING POLYNOMIAL EQUATIONS

A *polynomial equation* is an equation in which the expressions on both sides are polynomials. As with quadratic equations, we can sometimes solve polynomial equations by factoring, using the zero-factor principle on page 279.

Example 1   Solve $x^2 = x^3$.

Solution   We subtract x^3 from both sides and factor:

$$x^2 - x^3 = 0$$
$$x^2(1 - x) = 0.$$

For the expression on the left to be zero, one of the factors x^2 or $1 - x$ must be zero, so

$$x = 0 \quad \text{or} \quad x = 1.$$

Notice that we cannot solve this equation by dividing both sides by x^2, because x^2 could be zero. Doing this would leave only the solution $x = 1$, missing the solution $x = 0$.

Example 2   Solve $w^3 - 2w - 1 = (w + 1)^3$.

Solution   Expanding the right-hand side,

$$w^3 - 2w - 1 = w^3 + 3w^2 + 3w + 1$$

$w^3 - 2w - 1 - w^3 - 3w^2 - 3w - 1 = 0$ subtract right side from both sides

$-3w^2 - 5w - 2 = 0$ collect like terms (w^3 terms cancel)

$3w^2 + 5w + 2 = 0$ multiply both sides by -1

$(3w + 2)(w + 1) = 0,$ factor the left side,

so the solutions are $w = -2/3, w = -1$.

Example 3   Show that 2 and 3 are the only consecutive positive integers such that the cube of the first is 1 less than the square of the second.

Solution   If we have two consecutive numbers, say n and $n + 1$, such that the cube of the first, n^3, is one less than the square of the second, $(n + 1)^2$, then we have the polynomial equation

$$n^3 = (n + 1)^2 - 1.$$

Solving this equation gives

$$n^3 = n^2 + 2n + 1 - 1$$
$$n^3 - n^2 - 2n = 0$$
$$n(n + 1)(n - 2) = 0.$$

The solutions are $n = 0$, $n = -1$, and $n = 2$. The only positive integer solution is $n = 2$, so the only pair of consecutive positive integers with this property is $n = 2, n + 1 = 3$.

In the previous examples we converted the equation to the form $p(x) = 0$. For a polynomial $p(x)$, the solutions to $p(x) = 0$ are called the *zeros* of the polynomial.

Finding Zeros by Graphing

It is not always easy to factor a polynomial of degree greater than 2. Graphing the polynomial to see where it is zero can help us find approximate zeros, or to guess an exact zero.

Example 4   Solve $2x^3 + x^2 - 12x + 9 = 0$ by graphing the polynomial $p(x) = 2x^3 + x^2 - 12x + 9$.

Solution   In Figure 12.4, the graph of $p(x)$ crosses the x-axis at approximately $x = -3$, $x = 1$, and $x = 3/2$, so these are approximate solutions to $p(x) = 0$. To test whether they are exact solutions, we evaluate

$$p(-3) = 2(-3)^3 + (-3)^2 - 12(-3) + 9 \quad = -54 + 9 + 36 + 9 = 0$$
$$p(1) = 2(1)^3 + (1)^2 - 12(1) + 9 \quad\quad = 2 + 1 - 12 + 9 = 0$$

$$p\left(\frac{3}{2}\right) = 2\left(\frac{3}{2}\right)^3 + \left(\frac{3}{2}\right)^2 - 12\left(\frac{3}{2}\right) + 9 = \frac{27}{4} + \frac{9}{4} - 18 + 9 = 0,$$

so all three are exact solutions. In the next subsection we see why they are the only solutions.

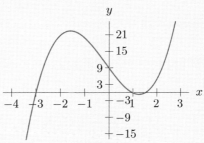

Figure 12.4: Graph of
$p(x) = 2x^3 + x^2 - 12x + 9$

The Relation Between Factors and Zeros

We have seen how to find zeros from factors. For example, if $p(x) = (x - 5)(x^2 + 2x + 4)$ then $p(5) = 0(5^2 + 2 \cdot 5 + 4) = 0$. This works in general:

> ### Finding Zeros from Factors
> If
> $$p(x) = (x - k) \cdot \text{(Other factors)},$$
> then
> $$p(k) = 0 \cdot \text{(Another number)} = 0,$$
> so $x = k$ is a zero of $p(x)$.

Example 5 Find a polynomial that has degree 4 and has zeros at $t = -1$, $t = 0$, $t = 1$ and $t = 2$.

Solution We can make the polynomial have the zeros we want by choosing the right factors. A factor of $t - (-1) = t + 1$ will make it have a zero at $t = -1$, a factor of $t - 0 = t$ will make it have a zero at $t = 0$, and factors $t - 1$ and $t - 2$ will make it have zeros at $t = 1$ and $t = 2$. Multiplying these factors we get the degree 4 polynomial

$$p(t) = t(t + 1)(t - 1)(t - 2).$$

It turns out that not only do factors give zeros, but zeros give factors:

Finding Factors from Zeros

If a polynomial has a zero at $x = k$, then it has a factor $(x - k)$.

We prove this fact on page 427. We can use it to factor polynomials if we already know their zeros.

Example 6   Factor the polynomial $p(x)$ in Example 4.

Solution   We know $p(x)$ has zeros $x = -3$, 1, and 3/2, so it must have factors $x - k$, where $k = -3$, 1, or 3/2, that is,

$$x - (-3) = x + 3, \quad x - 1, \quad \text{and} \quad x - 3/2.$$

Multiplying these together we get

$$(x + 3)(x - 1)\left(x - \frac{3}{2}\right) = (x^2 + 2x - 3)\left(x - \frac{3}{2}\right) = x^3 + \frac{1}{2}x^2 - 6x + \frac{9}{2}.$$

Multiplying all the coefficients on the right-hand side by 2 gives the coefficients of $p(x)$, so

$$p(x) = 2(x + 3)(x - 1)\left(x - \frac{3}{2}\right) = (x + 3)(x - 1)(2x - 3).$$

How Many Zeros Can a Polynomial Have?

A polynomial cannot have more factors than its degree. For example, a fourth-degree polynomial cannot have more than four factors, because if it did then the highest power of x would be greater than four when we multiplied the factors out. Since each zero of a polynomial corresponds to a factor, the number of zeros cannot be more than the degree of the polynomial. It is possible, however, for the number of zeros to be less than the degree.

Example 7   Without solving, say how many solutions each equation could have.
 (a) $x^4 + x^3 - 3x^2 - 2x + 2 = 0$ (b) $3(x^3 - x) = 1 + 3x^3$
 (c) $(s^2 + 5)(s - 3) = 0$

Solution   (a) Since the degree of the polynomial on the left is 4, the equation can have at most 4 solutions.
 (b) This looks as if it could simplify to an equation of the form $p(x) = 0$, where $p(x)$ is a cubic polynomial, so there would be up to 3 solutions. However, the leading term on both sides is $3x^3$, so when we simplify, the leading terms cancel, leaving only the linear term. Thus the equation can only have one solution.
 (c) The polynomial on the left is a cubic, so there could be up to 3 solutions. However, the quadratic factor, $s^2 + 5$, cannot have any zeros because it is 5 plus a square, so it is always positive. Thus, there is only one solution, coming from the linear factor.

Multiple Zeros

Sometimes the factor of a polynomial is repeated, as in

$$(x-4)^2 = \underbrace{(x-4)(x-4)}_{\text{Occurs twice}}$$

$$\text{or} \qquad (x+1)^3 = \underbrace{(x+1)(x+1)(x+1)}_{\text{Occurs three times}}.$$

In this case we say the zero is a *multiple zero*. For instance, we say that $x = 4$ is a *double zero* of $(x-4)^2$, and $x = -1$ is a *triple zero* of $(x+1)^3$. If a zero is not repeated, we call it a *simple zero*.

Visualizing the Zeros of a Polynomial

The polynomial $p(x) = x^3 - x^2 - 6x$ factors as

$$p(x) = x(x-3)(x+2),$$

and so has zeros at $x = -2$, $x = 0$, and $x = 3$. Since there are no repeated factors, these are all simple zeros. Figure 12.5 shows the graph of $y = p(x)$ crossing the x-axis at these zeros. In general, the graph of a polynomial crosses the axis at a simple zero.

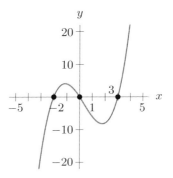

Figure 12.5: Graph of $y = x^3 - x^2 - 6x$

What Does the Graph Look Like at Multiple Zeros?

The graph of $(x-4)^2$ in Figure 12.6 show typical behavior near double zeros. Here the graph touches, but does not cross, the horizontal axis at the double zero $x = 4$. This behavior is typical of the graph of $(x-k)^n$ for even n: the graph of the polynomial does not cross the x-axis at $x = k$, but "bounces" off the x-axis at $x = k$.

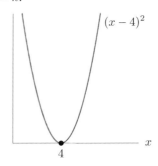

Figure 12.6: Double zero at $x = 4$

The graph of $(x + 1)^3$ in Figure 12.7 show typical behavior near triple zeros. Here the graph crosses the horizontal axis at the triple zero $x = -1$ but is flattened there compared to a simple zero (see Figure 12.5). This behavior is typical of the graph of $(x - k)^n$ for odd n: the graph of the polynomial crosses the x-axis at $x = k$, but it looks flattened there.

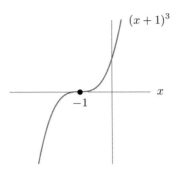

Figure 12.7: Triple zero at $x = -1$

In general:

The Appearance of the Graph at a Multiple Zero

Suppose that the polynomial $p(x)$ has a multiple zero at $x = k$ that occurs exactly n times, so

$$p(x) = (x - k)^n \cdot (\text{Other factors}), \quad n \geq 2.$$

- If n is even, the graph of the polynomial does not cross the x-axis at $x = k$, but "bounces" off the x-axis at $x = k$. (See Figure 12.6.)

- If n is odd, the graph of the polynomial crosses the x-axis at $x = k$, but it looks flattened there. (See Figure 12.7.)

Example 8   Describe in words the zeros of the polynomials of degree 4 graphed in Figure 12.8.

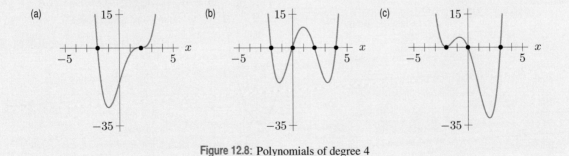

Figure 12.8: Polynomials of degree 4

Solution   (a) The first graph has two x-intercepts, one at $x = -2$ and one at $x = 2$. The flattened appearance near $x = 2$ suggests that the polynomial has a multiple zero there. Since the graph crosses the x-axis at $x = 2$ (instead of bouncing off it), the corresponding factor must occur an odd number of times. Since the polynomial is of degree 4, we conclude there is a triple zero at $x = 2$ and a simple zero at $x = -2$.

(b) The second graph has four x-intercepts, corresponding to four simple zeros. We know none of the zeros are repeated because the polynomial is degree 4.

(c) The third graph has three x-intercepts at $x = 0$, $x = 3$, and $x = -2$. The graph crosses the axis at $x = 0$ and $x = 3$, so they are simple zeros. The graph bounces at $x = -2$, so this is a double zero of the polynomial. It is not a higher-order zero (such as a quadruple zero) because the polynomial is only of degree 4.

A Connection between Polynomials and Decimal Notation

We can think of an integer written in decimal notation as a polynomial expression involving a sum of powers of ten (as opposed to powers of a variable like x).

Example 9   (a) Use the fact that
$$3000 + 700 + 20 = 3720$$
to find a solution to the equation
$$3x^3 + 7x^2 + 2x = 3720.$$

(b) What does your answer tell you about the factors of the polynomial $3x^3 + 7x^2 + 2x - 3720$?

Solution   (a) We can rewrite $3000 + 700 + 20 = 3720$ as
$$3 \cdot 10^3 + 7 \cdot 10^2 + 2 \cdot 10^1 + 0 \cdot 10^0 = 3720.$$

This means that the equation
$$3x^3 + 7x^2 + 2x = 3720$$
has a solution at $x = 10$. Consequently, the polynomial
$$3x^3 + 7x^2 + 2x - 3720$$
has a zero at $x = 10$.

(b) According to the box on page 388, we know $(x - 10)$ must be a factor. In fact,
$$3x^3 + 7x^2 + 2x - 3720 = (x - 10)(3x^2 + 37x + 372).$$

How did we find the factor $3x^2 + 37x + 372$ on the right-hand side? We used polynomial long division, which is analogous to ordinary long division of integers. We review polynomial long division in Section 13.3.

Exercises and Problems for Section 12.3

EXERCISES

■Find the zeros of the polynomials in Exercises 1–6.

1. $(x-3)(x-4)(x+2)$ 2. $x(x+5)(x-7)^2$

3. $x^2 - x - 6$ 4. $x^4 - 4x^3 + 4x^2$

5. $x^2 + 1$ 6. $x^4 - 1$

■Give all the solutions of the equations in Exercises 7–18.

7. $(x-1)(x+2)(x-3) = 0$
8. $(x+3)\left(1 - x^2\right) = 0$
9. $x^3 + 3x^2 + 2x = 0$
10. $x^3 - 2x^2 + 2^2 x - 2^3 = 0$
11. $(x-1)x(x+3) = 0$
12. $x^4 + x^2 - 2 = 0$
13. $s(s^2 + 1) = s(2s^2 - 3)$
14. $(t+3)^3 + 4(t+3)^2 = 0$

15. $(u+3)^3 = (u+3)^3$
16. $(u+3)^3 = -(u+3)^3$
17. $(u+3)^3 = -(u+3)^2$
18. $(s+10)^2 - 6(s+10) - 16 = 0$

■Find possible formulas for the polynomials described in Exercises 19–24.

19. The degree is $n = 2$ and the zeros are $x = 2, -3$.
20. The degree is 5 and the zeros are $x = -4, -1, 0, 3, 9$.
21. The degree is $n = 3$ and there is one zero at $x = 5$ and one double zero at $x = -13$.
22. The degree is $n = 3$ and the only zeros are $x = 2, -3$.
23. The degree is $n = 5$ and there is one zero at $x = -6$.
24. The degree is $n = 6$ and there is one simple zero at $x = -1$, one double zero at $x = 3$, and one multiple zero at $x = 5$.

PROBLEMS

25. The profit from selling q items of a certain product is $P(q) = 36q - 0.0001q^3$ dollars. Find the values of q such that $P(q) = 0$. Which of these values make sense in the context of the problem? Interpret the values that make sense.

26. American Airlines limits the size of carry-on baggage to 45 linear inches (length + width + height), with a weight of no more than 40 pounds.[1]

    (a) If the length and width of a piece of luggage both measure x inches, express the maximum height of the luggage in terms of x.
    (b) Express the volume of the piece of luggage in part (a) in terms of x.
    (c) Find the zeros of your equation from part (b). What does this tell you about the dimensions of the piece of luggage?

27. Find a polynomial of degree 4 that has zeros at $x = -2, x = -1, x = 2$, and $x = 3$ and whose graph contains the point $(0, 6)$.

28. Find two different polynomials of degree 3 with zeros 1, 2, and 3.

29. (a) Find two different polynomials with zeros $x = -1$ and $x = 5/2$.

    _____

    [1] www.aa.com, accessed October 16, 2007.

    (b) Find a polynomial with zeros $x = -1$ and $x = 5/2$ and leading coefficient 4.

■In Problems 30–37, without solving the equation, decide how many solutions it has.

30. $(x-1)(x-2) = 0$
31. $(x^2 + 1)(x-2) = 0$
32. $(x^2 + 2x)(x-3) = 0$
33. $(x^2 - 4)(x+5) = 0$
34. $(x-2)x = 3(x-2)$
35. $(2 - x^2)(x-4)(5-x) = 0$
36. $(2 + x^2)(x-4)(5-x) = 0$
37. $(x^4 + 2)(3 + x^2) = 0$

38. Find the solutions of

$$(x^2 - a^2)(x+1) = 0, \quad a \text{ a constant.}$$

39. For what value(s) of the constant a does $\left(x^2 - a^2\right)(x + 1) = 0$ have exactly two solutions?

In Problems 40–48, for what values of a does the equation have a solution in x?

40. $x^2 - a = 0$

41. $2x^2 + a = 0$

42. $ax^2 - 5 = 0$

43. $(ax^2 + 1)(x - a) = 0$

44. $a^2 + ax^2 = 0$

45. $(x^2 + a)(x^2 - a) = 0$

46. $x^3 + a = 0$

47. $x^4 + 5a = 0$

48. $a - x^5 = 0$

49. Find approximate solutions to

$$3x^3 - 2x^2 \ 6x + 4 = 0$$

by graphing the polynomial.

50. Consider the polynomial $x^5 - 3x^4 + 4x^3 - 2x + 1$.

    (a) What is the value of the polynomial when $x = 4$?

    (b) If a is the answer you found in part (a), show that $x - 4$ is a factor of $x^5 - 3x^4 + 4x^3 - 2x + 1 - a$.

51. Use the identity $(x-1)(x^4+x^3+x^2+x+1) = x^5 - 1$ to show that $8^5 - 1$ is divisible by 7.

52. Consider the polynomial $p(x) = (x - k)^n$, where k is a constant and n is a positive integer.

    (a) If n is even explain why the graph of $p(x)$ is never below the x-axis.

    (b) If n is odd explain why the graph of $p(x)$ is below the x axis for $x < k$, and above it for $x > k$.

12.4 LONG-RUN BEHAVIOR OF POLYNOMIAL FUNCTIONS

The graph of a polynomial function

$$y = p(x)$$

can have numerous x-intercepts and bumps. However, if we zoom out and consider the appearance of the graph from a distance and ignore fine details, the picture becomes much simpler.

Example 1   Describe how the features of the graph of $f(x) = x^3 + x^2$ change when zooming out.

Solution   In Figure 12.9, we see that the graph of f has two x-intercepts and two bumps. In Figure 12.10, we zoom out to show the same function on a larger viewing window. Here, the bumps are no longer apparent, nor is the fact that there are two x-intercepts.

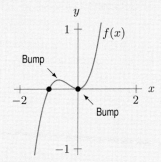

Figure 12.9: On this scale, we see that the graph of $y = x^3 + x^2$ has two bumps and two x-intercepts

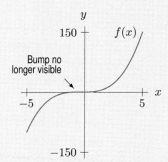

Figure 12.10: On this scale, some features of the graph of $y = x^3 + x^2$ are no longer apparent

The Importance of the Leading Term

When we zoom out from a graph, as in Example 1, we turn our attention to values of the polynomial for large (positive or negative) values of x. Although every term in a polynomial contributes to its value, we see in Example 2 that for large enough values of x, the leading term's contribution can be more important than the contributions of all the other terms combined.

Example 2   Show that, provided the value of x is large enough, the value of the term $4x^5$ in the polynomial $p(x) = 4x^5 - x^3 + 3x^2 + x + 1$ is far larger than the value of the other terms combined.

Solution   To see this, we let $x = 100$. Then

$$4x^5 = 4(100)^5 = 40{,}000{,}000{,}000.$$

On the other hand, the value of the other terms combined is

$$-x^3 + 3x^2 + x + 1 = -(100)^3 + 3(100)^2 + 100 + 1$$
$$= -1{,}000{,}000 + 30{,}000 + 100 + 1 = -969{,}899.$$

Therefore, at $x = 100$, the combined value is

$$p(100) = \underbrace{40{,}000{,}000{,}000}_{\text{largest contribution}} + \underbrace{-969{,}899}_{\text{relatively small correction}}$$
$$= 39{,}999{,}030{,}101.$$

Thinking in terms of money, the value of $4x^5$ by itself is \$40 billion, whereas if we include the other terms the combined value is \$39.999 billion. Comparing these two values, the difference is negligible, even though (by itself) \$969,899 is a lot of money.

Hence, for large values of x,

$$p(x) = 4x^5 + \underbrace{-x^3 + 3x^2 + x + 1}_{\text{relatively small correction}}$$
$$p(x) \approx 4x^5.$$

In general, if the value of x is large enough, we can give a reasonable approximation for a polynomial's value by ignoring or neglecting the lower-degree terms. Consequently, when viewed on a large enough scale, the graph of the polynomial function

$$y = a_n x^n + a_{n-1} x^{n-1} + \cdots + a_1 x + a_0$$

strongly resembles the graph of the power function

$$y = a_n x^n,$$

and we say that the *long-run behavior* of the polynomial is given by its leading term.

The Long-Run Behavior of a Polynomial

In a polynomial $a_n x^n + a_{n-1} x^{n-1} + \cdots + a_1 x + a_0$, with $a_n \neq 0$, the leading term $a_n x^n$ determines the polynomial's long-run behavior.

Example 3   Find a window in which the graph of $y = x^3 + x^2$ resembles the graph of its leading term $y = x^3$.

Solution   Figure 12.11 gives the graphs of $y = x^3 + x^2$ and $y = x^3$. On this scale, $y = x^3 + x^2$ does not look like a power. On the larger scale in Figure 12.12, the graph of $y = x^3 + x^2$ resembles the graph of $y = x^3$. On this larger scale, the "bumps" in the graph of $y = x^3 + x^2$ are too small to be seen. On an even larger scale, as in Figure 12.13, the graph of $y = x^3 + x^2$ is nearly indistinguishable from the graph of $y = x^3$.

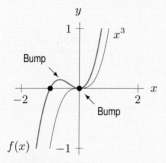

Figure 12.11: On this scale, $y = x^3 + x^2$ does not look like a power function

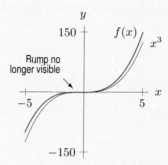

Figure 12.12: On this scale, $y = x^3 + x^2$ resembles the power function $y = x^3$

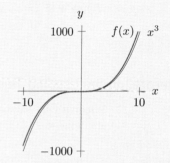

Figure 12.13: On this scale, $y = x^3 + x^2$ is nearly indistinguishable from $y = x^3$

Since the long-run behavior of a polynomial is given by its leading term, polynomials with the same leading term have similar graphs on a large scale. However, they can look quite different on a small scale.

Example 4   Compare the graphs of the polynomial functions

$$f(x) = x^4 - 4x^3 + 16x - 16, \quad g(x) = x^4 - 4x^3 - 4x^2 + 16x, \quad h(x) = x^4 + x^3 - 8x^2 - 12x.$$

Solution   Since each of these polynomials has x^4 as its leading term, all their graphs resemble the graph of x^4 on a large scale. See Figure 12.14. On a smaller scale, the polynomials look different. See Figure 12.15. Two of the graphs go through the origin while the third does not. The graphs also differ from one another in the number of bumps each one has and in the number of times each one crosses the x-axis.

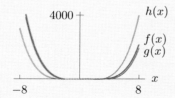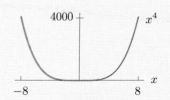

Figure 12.14: On a large scale, the three polynomials resemble the power $y = x^4$

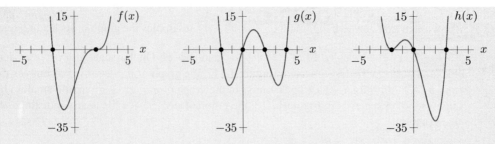

Figure 12.15: On a smaller scale, the three polynomials look quite different from each other

Interpreting the Leading Term in Mathematical Models

When using a polynomial for a mathematical model, the leading term can give us important insights about the real-world situation being described.

Example 5   The box described in Example 4 on page 376 has width w, length $w + 4$, height $2w$, and volume

$$V = w(w + 4)(2w) = 2w^3 + 8w^2.$$

Find the leading term of this polynomial and give a practical interpretation of it.

Solution   The leading term is $2w^3$. This represents the approximate volume of the box when w is large. To see this, suppose the box is relatively small, having width $w = 2$ ft. Then

$$\text{Volume} = 2 \cdot 6 \cdot 4 = 48 \text{ ft}^3$$
$$\text{Leading term} = \quad 2 \cdot 2^3 = 16 \text{ ft}^3.$$

Here, the leading term does not give a good approximation to the volume. But now imagine an enormous box—a giant warehouse, say—having width $w = 1000$ ft. Then

$$\text{Volume} = 1000 \cdot 1004 \cdot 2000 = 2{,}008{,}000{,}000 \text{ ft}^3$$
$$\text{Leading term} = \quad\quad 2 \cdot 1000^3 = 2{,}000{,}000{,}000 \text{ ft}^3.$$

We see that the true volume, 2.008 billion ft^3, is only slightly larger than the approximation of 2 billion ft^3 provided by the leading term. Unlike the small box, whose width is just a fraction of its length (2 ft versus 6 ft), the width of the large box is almost the same as its length (1000 ft versus 1004 ft). This means

$$\text{Volume} = \underbrace{\text{Width}}_{w} \times \underbrace{\text{Length}}_{w+4} \times \underbrace{\text{Height}}_{2w}$$
$$\approx \underbrace{\text{Width}}_{w} \times \underbrace{\text{Width}}_{w} \times \underbrace{\text{Height}}_{2w} \quad \text{because width and height are close}$$
$$= w \cdot w \cdot 2w$$
$$= 2w^3,$$
$$\text{so} \quad \text{Volume} \approx \text{Leading term}.$$

Finding a Possible Formula for a Polynomial Function from Its Graph

Often, we can determine a possible expression for a polynomial function from its graph by looking at its large-scale appearance, by finding zeros, and by using specific points on the graph.

Example 6
Find a possible formula for the polynomial $f(x)$ graphed in Figure 12.16.

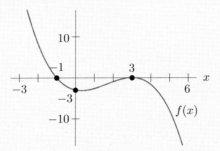

Figure 12.16: Find a formula for this polynomial

Solution
Based on its large-scale shape, which is similar in appearance to $y = -x^3$, the degree of $f(x)$ is odd and is larger than 1. Note that, judging from the shape alone, we cannot be sure the degree is 3; a fifth-degree polynomial might have the same general long-run appearance. But we do know the degree is not 1 (otherwise the graph would be a straight line).

The graph has x-intercepts at $x = -1$ and $x = 3$. We see that $x = 3$ is a multiple zero of even power, because the graph bounces off the x-axis here instead of crossing it. (See page 390.) Therefore, a possible formula is

$$f(x) = k(x + 1)(x - 3)^2, \quad k \text{ a non-zero constant.}$$

which has degree 3 and agrees with our previous observation.

To find k, we note that the y-intercept of the graph tells us $f(0) = -3$, so

$$f(0) = k(0 + 1)(0 - 3)^2 = -3,$$

which gives

$$9k = -3 \quad \text{so} \quad k = -\frac{1}{3}.$$

Thus, $f(x) = -\frac{1}{3}(x + 1)(x - 3)^2$ is a possible formula for this polynomial function.

The Constant k

If we had not accounted for the constant k in Example 6, we would have obtained the formula

$$y = (x + 1)(x - 3)^2.$$

Although this function has the right zeros (at $x = -1$ and a double zero at $x = 3$), it has the wrong long-run behavior: On large scales, its graph more closely resembles $y = x^3$, not $y = -x^3$.

Although other choices of k lead to graphs similar to Figure 12.16, only one value of k yields the correct y-intercept. For instance, the function

$$y = -0.5(x + 1)(x - 3)^2$$

has the right zeros and in the long run resembles $y = -0.5x^3$. But its y-intercept is $y = -4.5$, not $y = -3$.

Example 7   Consider the polynomial

$$q(x) = 3x^6 - 2x^5 + 4x^2 - 1.$$

(a) What does the leading term tell us about the graph?

(b) What does the constant term tell us about the graph?

(c) What do the answers to (a) and (b) tell us about the number of solutions to $q(x) = 0$?

Solution   (a) The leading term is $3x^6$, so on a large scale, the graph of $y = 3x^6 - 2x^5 + 4x^2 - 1$ looks like $y = 3x^6$. (See Figure 12.17.) This means that this function takes on large positive values as x grows large (either positive or negative).

(b) The constant term is -1. This tells us that the graph crosses the y-axis at $y = -1$, which is below the x-axis.

(c) The equation $3x^6 - 2x^5 + 4x^2 - 1 = 0$ must have at least two solutions. We know this because the graph of $y = 3x^6 - 2x^5 + 4x^2 - 1$ is smooth and unbroken, so it must cross the x-axis at least twice to get from its y-intercept of $y = -1$ to the positive values it attains as x grows larger in both the positive and negative directions.

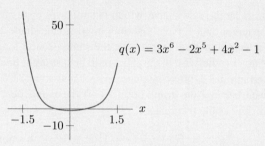

Figure 12.17: Graph must cross x-axis at least twice since $3x^6 - 2x^5 + 4x^2 - 1$ looks like $3x^6$ for large x, and $3x^6 - 2x^5 + 4x^2 - 1 = -1$ when $x = 0$

Note that a sixth-degree polynomial can have as many as six real zeros, or none at all. For instance, the sixth-degree polynomial $x^6 + 1$ has no zeros, because its value is always greater than or equal to 1.

Exercises and Problems for Section 12.4

EXERCISES

Find possible formulas for the polynomials shown in Problems **1–6**.

1.

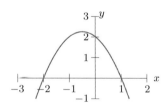

2.

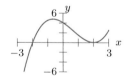

3.

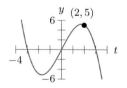

4.

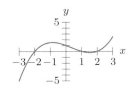

5.

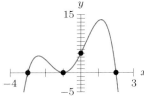

6.

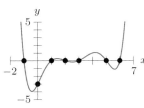

Find possible formulas for the polynomial functions described in Exercises **7–10.**

7. The graph crosses the x-axis at $x = -2$ and $x = 3$ and its long-run behavior is like $y = x^2$.

8. The graph crosses the x-axis at $x = -2$ and $x = 3$ and its long-run behavior is like $y = -2x^2$.

9. The graph bounces off the x-axis at $x = -2$, crosses the x-axis at $x = 3$, and has long-run behavior like $y = x^3$.

10. The graph bounces off the x-axis at $x = -2$, crosses the x-axis at $x = 3$, and has long-run behavior like $y = x^5$.

PROBLEMS

11. The polynomial $p(x)$ can be written in two forms:

   I. $p(x) = 2x^3 - 3x^2 - 11x + 6$

  II. $p(x) = (x - 3)(x + 2)(2x - 1)$

     Which form most readily shows

  (a) The zeros of $p(x)$? What are they?

  (b) The vertical intercept? What is it?

  (c) The sign of $p(x)$ as x gets large, either positive or negative? What are the signs?

  (d) The number of times $p(x)$ changes sign as x increases from large negative to large positive x? How many times is this?

12. A polynomial $p(x)$ can be written in two forms:

   I. $p(x) = (x^2 + 4)(4 - x^2)$

  II. $p(x) = 16 - x^4$

     Which form most readily shows

  (a) The number of zeros of $p(x)$? Find them.

  (b) The vertical intercept? What is it?

  (c) The sign of $p(x)$ as x gets large, either positive or negative. What are the signs?

13. Graph $y = 2(x - 3)^2(x + 1)$. Label all axis intercepts.

14. A cubic polynomial, $p(x)$, has $p(1) = -4$ and $p(2) = 7$. What can you say about the zeros of $p(x)$?

15. Explain why $p(x) = x^5 + 3x^3 + 2$ must have at least one zero.

16. Find two different formulas for a polynomial $p(x)$ of degree 3 with $p(0) = 6$ and $p(-2) = 0$, and graph them.

17. By calculating $p(0)$ and $p(1)$ show that

$$p(x) = -x^5 + 3x^2 - 1$$

has at least three zeros.

Answer Problems **18–19** based on Figure 12.18, which shows the same polynomial in 4 different viewing windows.

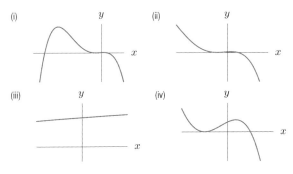

Figure 12.18

18. Rank the windows in order from smallest to largest.

19. One of the following equations describes the polynomial shown in Figure 12.18. Which is it?

  (a) $y = -(x + 1)^2(x - 1)(x + 10)$

  (b) $y = (x + 1)^2(x - 1)(x + 10)$

  (c) $y = (x + 1)(x - 1)(x + 10)$

  (d) $y = -(x - 1)^2(x + 1)(x - 10)$

REVIEW EXERCISES AND PROBLEMS FOR CHAPTER 12

EXERCISES

■ In Exercises **1–5**, give the degree.

1. $-x + 2x^2 + 1$
2. $x(x - 1)(x + 3)$
3. $(x + 1)^5 - x^5$
4. $(x^2 - 1)(x^2 - 2)(x^2 - x)$
5. $p(x^2)$, where $p(x) = x^4 - x^3 + 3x + 1$.

■ In Exercises **6–10**, give the constant term.

6. $(x - 1)(x - 2)(x - 3)(x - 4)(x - 5)$
7. $(x - a)^3(x + 1/a)^3$, $a \neq 0$
8. $(x - 1)^5 - (x + 1)^5$
9. $(x - 1)^6 - (x + 1)^6$
10. $(x - x_0)^3(x^4 + 3x^2 - 2x + 10)$

■ In Exercises **11–16**, is the expression a polynomial in x? If so, give the degree and the leading coefficient. (Assume a is a positive integer.)

11. $8x^5 - x^3 + 11x^2 + 50$ 12. $5 - \sqrt{2}x^2 + x$
13. $ax^{-1} + 2$ 14. a^{-1}
15. $x^a + a^x$, $a \neq 1$ 16. $\dfrac{1 - x(x + 1)}{3}$

■ Give the degree, leading term, leading coefficient, and constant term of the polynomials in Exercises **17–20**.

17. $-3x^2 + x + 5$ 18. $x(x + 1)^2$
19. $(2x^3 - 3x^2)^2$
20. $(2x^3 - 3x^2)^2 - (x^2 + 1)^2$

■ In Exercises **21–23**, for the given polynomials $p(x)$, find:
(a) $p(0)$ (b) $p(1)$ (c) $p(-1)$ (d) $p(-t)$

21. $p(x) = x^3 + x^2 + x + 1$
22. $p(x) = x^5 + 2x^3 - x$
23. $p(x) = x^4 + 3x^2 + 1$

■ In Exercises **24–27**, write polynomials in standard form in x based on the coefficients given assuming all other coefficients equal zero.

24. $a_3 = 2, a_2 = 4, a_1 = -3, a_0 = 5$
25. $a_6 = 17, a_5 = 4, a_2 = 9$
26. $a_{20} = 33, a_{14} = 20, a_{33} = 7, a_7 = 14$

27. $a_1 = 2$

■ Give all the solutions of the equations in Exercises **28–32**.

28. $(x + 3)(1 - x)^2 = 0$
29. $(x + 3^2)(1 - x^2) = 0$
30. $x^3 - x^2 + x - 1 = 0$
31. $x^4 = 2x^2 - 1$
32. $4x^2 - 3x = x^3$

■ Find possible formulas for the polynomials shown in Exercises **33–36**.

33.
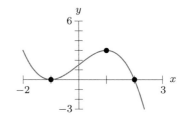

34.
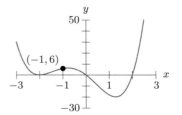

35.
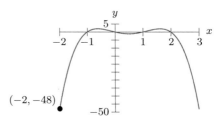

36.
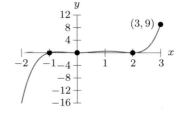

PROBLEMS

In Problems **37–39**, let $p(x)$ be a polynomial in x with degree m not equal to zero and leading coefficient a. For each polynomial given, find the degree and leading coefficient.

37. $5p(x)$ **38.** $xp(x)$ **39.** $p(x) + 5$

In Problems **40–41**, the polynomial $p(x)$ has leading coefficient a and degree m, and the polynomial $q(x)$ has leading coefficient b and degree n. Suppose that $m > n$.

40. What is the leading coefficient and degree of $p(x)q(x)$?

41. What is the leading coefficient and degree of $p(x) + q(x)$?

Without solving them, decide how many solutions there are to the equations in Problems **42–47**.

42. $x^2 - 10^3 = 0$

43. $x^3 - 7^4 = 0$

44. $x^2 + 10^{2.4} = 0$

45. $x^3 - 10^{-4.52} = 0$

46. $x^2 - \log 3 = 0$

47. $\log(0.2) - x^2 = 0$

48. If the cubic $x^3 + ax^2 + bx + c$ has zeros at $x = 1$, $x = 2$, and $x = 3$, what are the values, of a, b, and c?

49. **(a)** Show that $(1 + x + x^2)(1 - x + x^2) = 1 + x^2 + x^4$.
     (b) Use part (a) to show that neither $1 + x + x^2$ nor $1 - x + x^2$ have zeros.

50. If the roots of $x^3 + 2x^2 - x - 2 = 0$ are -1, 1, and 2, what are the roots of $(x-3)^3 + 2(x-3)^2 - (x-3) - 2 = 0$? [Hint: Consider what values of $x - 3$ satisfy the second equation.]

51. If a is a constant, rewrite the equation $x^3 - ax^2 + a^2x - a^3 = 0$ in a form that clearly shows its solutions. What are the solutions?

52. Find viewing windows on which the graph of $f(x) = x^3 + x^2$ resembles the plots in (a)–(d).

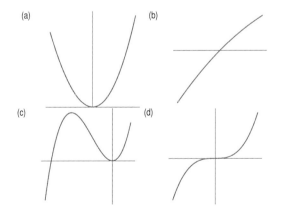

53. Find a cubic polynomial with all of the following properties:

    (i) The graph crosses the x-axis at -7, -2, and 3.
   (ii) From left to right, the graph rises, falls, and rises.
  (iii) The graph crosses the y-axis at -3.

54. Without using a calculator or computer, sketch $p(x) = x^2(x + 2)^3(x - 1)$.

55. Someone tells you that $x^6 + x^2 + 1$ has no zeros. How do you know the person is right?

56. Someone tells you that $-x^6 + x^2 + 1$ has at least two zeros. How do you know the person is right?

57. Someone tells you that $x^5 + x^2 + 1$ has at least one zero. How do you know the person is right?

Match each description in Problems **58–59** to the polynomial function below whose graph it best describes:

(a) $y = (x^2 - 6x + 8)(x^2 - 5x + 4)(4x - 3 - x^2)$

(b) $y = (x^2 - 5x + 6)(x^2 - 7x + 12)(x^2 - 6x + 8)$

(c) $y = (x^2 - 3x + 2)(x^2 - 7x + 12)(x^2 + 1)$

(d) $y = (x^4 + 2x^2 + 1)(x^2 - 3x + 2)$

58. This graph crosses the x-axis the most times (not counting bounces).

59. In the long run, this graph most resembles the power function $y = -x^6$.

60. The polynomial $f(x)$ has three zeros, and the polynomial $g(x)$ has four zeros. What can you say about the zeros of the polynomial

$$h(x) = f(x) + f(x)g(x)?$$

61. State intervals of x on which the graph of

$$f(x) = (x + 4)(x + 1)^2(x - 3)$$

lies below the x-axis.

62. Give intervals of x on which the graph of the following equation lies above the x-axis:

$$y = -2(x - 3)^2(x + 5)(x + 8)$$

63. Consider the polynomial $g(t) = (t^2 + 5)(t - a)$, where a is a constant. For which value(s) of t is $g(t)$ positive?

64. If a is a nonzero constant, for what values of x is $x^3 - ax^2 + a^2x - a^3$ positive?

65. Does the polynomial function $y = ax^4 + 2x^3 - 9x^2 + 3x - 18$, with $a > 0$, have any zeros? If so, at least how many, and how do you know? If not, why not?

66. Using a calculator or computer, sketch the graph of

$$y = (x - a)^3 + 3a(x - a)^2 + 3a^2(x - a) + a^3$$

for $a = 0$, $a = 1$, and $a = 2$. What do you observe? Use algebra to confirm your observation.

Pascal's triangle[2] is an arrangement of integers that begins

$$
\begin{array}{ccccccccc}
 & & & & 1 & & & & \\
 & & & 1 & & 1 & & & \\
 & & 1 & & 2 & & 1 & & \\
 & 1 & & \mathbf{3} & & 3 & & 1 & \\
1 & & 4 & & 6 & & \mathbf{4} & & 1
\end{array}
$$

with n entries in each row formed by adding the two entries above it. For instance, the boldface 3 above is formed by adding the preceding entries 1 and 2, and the boldface 4 is formed by adding the preceding entries 3 and 1. Given this information, answer Problems **67–70**.

67. Find the next 3 rows of Pascal's triangle.

68. Write the following polynomials in standard form:

$$(x + 1)^0, \quad (x + 1)^1, \quad (x + 1)^2, \quad (x + 1)^3.$$

What do you notice?

69. Without multiplying it out, write $(x + 1)^7$ in standard form. [Hint: Make an educated guess based on your answers to Problems 67 and 68.]

70. Consider the following pattern:

$$(2x + 3)^0 = 1$$
$$(2x + 3)^1 = 2x + 3$$
$$(2x + 3)^2 = (2x)^2 + 2(2x)3 + 3$$
$$(2x + 3)^3 = (2x)^3 + 3(2x)^2 3 + 3(2x)3^2 + 3^3.$$

Based on this pattern, write $(2x + 3)^4$ in standard form.

In Problems **71–76**, find the given term or coefficient for $n = 2, 3$ and 4, and describe the pattern.

71. The coefficient of x in $(x + 2)^n$

72. The coefficient of x^{n-1} in $(2x + 1)^n$

73. The leading term of $(3x - 2)^n$

74. The coefficient of x^{n-1} in $(x - 1)^n$

75. The coefficient of x in $(1 - x)^n$

76. The constant term of $(x + a)^n$

In Problems **77–79**, Gloria has a credit card balance and decides to pay it off by making monthly payments. For example, suppose her balance is $b_0 = \$2000$ in month $t = 0$ and she makes no more purchases using the card. Each month thereafter, she is charged 1% interest on the debt, and then she makes a \$25 payment. To find her balance in month $t = 1$, we multiply \$2000 by 1.01 (because the balance goes up by 1%) and then subtract the \$25 payment. This gives $b_1 = 2000.00(1.01) - 25 = 1995.00$. Likewise, her balance in months $t = 2, 3$ is given by

$$b_2 = 1995.00(1.01) - 25 = \$1989.95$$
$$b_3 = 1989.95(1.01) - 25 = \$1984.85.$$

Let $x = 1 + r$ where $r = 1\%$ is the monthly interest rate.

77. **(a)** What do b_4 and b_5 represent in terms of Gloria's credit card balance? Evaluate b_4 and b_5.

(b) We can write an expression for b_1 in terms of $x = 1.01$:

$$b_1 = 2000 \underbrace{(1.01)}_{x} - 25 = 2000x - 25.$$

Notice that $b_2 = b_1(1.01) - 25$, that $b_3 = b_2(1.01) - 25$, and so on. Given this, write simplified polynomial expressions for b_2, b_3, b_4, and b_5 in terms of x.

(c) Evaluate your expression for b_5 for $r = 1.5\%$, that is, for $x = 1.015$. What does you answer tell you about interest rates and credit card debt?

78. **(a)** Gloria's first two monthly payments (in months $t = 1$ and 2) are \$25 and \$60, respectively. With $x = 1.01$, show that b_2, her balance in month $t = 2$, is given by the polynomial

$$2000x^2 - 25x - 60.$$

(b) Her next three monthly payments (in months $t = 3, 4$, and 5) are \$45, \$80, and \$25, respectively. Write a simplified polynomial expression for b_5.

---

[2]Though named for Blaise Pascal, who described it in 1655, it was known to medieval Chinese and Islamic scholars and may have been known as early as 450 BC. See http://en.wikipedia.org/wiki/Pascal%27s_triangle.

79. In some months, Gloria makes expenditures on her credit card, rather than payments. If $x = 1.01$, after six months her balance is given by the polynomial

$$3000x^6 - 120x^5 - 90x^4 + 50x^3 + 100x^2 - 200x - 250.$$

Describe in words her credit card debt, giving her initial balance and each of her monthly payments or expenditures.

■ Find formulas for the functions described in Problems **80–83.** Some problems have more than one possible answer.

80. The graph of h is a parabola with vertex $(h, k) = (3, 4)$ and with y-intercept 12.

81. The graph of w is given in Figure 12.19.

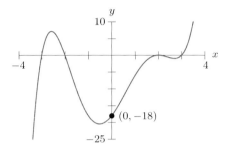

Figure 12.19

82. The graph of h is a parabola with vertex $(-2, 5)$ and y-intercept -1.

83. The graph of w is given in Figure 12.20.

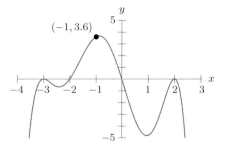

Figure 12.20

Chapter 13

Rational Functions

CONTENTS

13.1   RATIONAL FUNCTIONS

If we add, subtract, or multiply two polynomials, the result is always a polynomial. However, when we divide one polynomial by another, the result is not necessarily a polynomial, just as when we divide one integer by another, the result is not necessarily an integer.

Example 1   Which of the following is a polynomial?

(a) $\dfrac{x^4 - 1}{x^2 + 1}$

(b) $\dfrac{x^4 + 1}{x^2 + 1}$

Solution   (a) We have (because $x^4 - 1$ is a difference of squares)

$$\frac{x^4 - 1}{x^2 + 1} = \frac{\left(x^2 + 1\right)\left(x^2 - 1\right)}{x^2 + 1} = x^2 - 1.$$

This is a quadratic polynomial.

(b) In this case, since $x^4 + 1$ cannot be factored, the denominator cannot be divided into the numerator. Thus, the expression cannot be simplified and written as a polynomial.

Rational Expressions

We define a *rational expression* to be any expression equivalent to the ratio of two polynomials (except where the expressions are undefined).

Example 2   Write each expression as the ratio of two polynomials.

(a) $\dfrac{1}{x^2} + 6$

(b) $\dfrac{x^2 + 2}{x - 3} + 2$

(c) $\dfrac{1 + \frac{1}{x}}{x - \frac{1}{x}}$

(d) $5x^3 - 2x^2 + 7$

Solution   (a) We put each term over a common denominator and add:

$$\frac{1}{x^2} + 6 = \frac{1}{x^2} + \frac{6x^2}{x^2} = \frac{1 + 6x^2}{x^2}.$$

(b) We put each term over a common denominator and add:

$$\frac{x^2 + 2}{x - 3} + 2 = \frac{x^2 + 2}{x - 3} + \frac{2(x - 3)}{x - 3} = \frac{x^2 + 2 + 2x - 6}{x - 3} = \frac{x^2 + 2x - 4}{x - 3}.$$

(c) Since both the numerator and denominator involve algebraic fractions with denominator x, we first multiply them by x to simplify them:

$$\frac{1 + \frac{1}{x}}{x - \frac{1}{x}} = \frac{x\left(1 + \frac{1}{x}\right)}{x\left(x - \frac{1}{x}\right)} = \frac{x + 1}{x^2 - 1} = \frac{x + 1}{(x + 1)(x - 1)} = \frac{1}{x - 1}.$$

(d) We have

$$5x^3 - 2x^2 + 7 = \frac{5x^3 - 2x^2 + 7}{1}.$$

Here, the denominator is 1 (a constant polynomial). Any polynomial is a rational expression, because we can think of it as a ratio with denominator equal to the constant polynomial 1.

Rational Functions

Just as we define polynomial functions using polynomial expressions, we define rational functions using rational expressions:

A **rational function** is a function that can be put in the form

$$f(x) = \frac{a(x)}{b(x)}, \quad \text{where } a(x) \text{ and } b(x) \text{ are polynomials, and } b(x) \text{ is not the zero polynomial.}$$

Example 3   Let $f(t) = 50 + 0.1t$ represent the population, in millions, of a country in year t, and let $g(t) = 200 + 0.05t + 0.03t^2$ represent the gross domestic product (GDP), in billions of dollars.[1] The *per capita GDP*, $h(t)$, is the GDP divided by the population.

(a) Show that $h(t)$ is a rational function.

(b) Evaluate $h(0)$ and $h(20)$. What does this tell us about the country's economy?

Solution   (a) We have

$$\text{Per capita GDP} = h(t) = \frac{g(t)}{f(t)} = \underbrace{\frac{200 + 0.05t + 0.03t^2}{50 + 0.1t}}_{\text{A rational expression}}.$$

Since $h(t)$ can be defined using a rational expression in t, then $h(t)$ is a rational function of t.

(b) We have

$$h(0) = \frac{g(0)}{f(0)} = \frac{200 + 0.05(0) + 0.03\left(0^2\right)}{50 + 0.1(0)} = \frac{\$200 \text{ billion}}{50 \text{ million people}} = \$4,000 \text{ per person}$$

$$h(20) = \frac{g(20)}{f(20)} = \frac{200 + 0.05(20) + 0.03\left(20^2\right)}{50 + 0.1(20)} = \frac{\$213 \text{ billion}}{52 \text{ million people}} = \$4,096 \text{ per person.}$$

We see that after 20 years, the GDP, when divided by the population, rises from \$4,000 per person to \$4,096 per person. This suggests that on average, the people living in the country have become slightly more prosperous over time.[2]

The Graph of a Rational Function

Figure 13.1 shows the graph of the rational function $f(x) = \dfrac{x^2 - x - 6}{x^2 - x - 2}$. This graph has many features that graphs of polynomials do not have. It levels off to the horizontal line $y = 1$ as the value of x gets large (either in a positive or negative direction). This line is called a *horizontal asymptote*. Furthermore, on either side of the two vertical lines, $x = -1$ and $x = 2$, the y-values get very large (either in a positive or negative direction). These lines are called *vertical asymptotes*.

---

[1] The GDP of a country is a measure of the overall strength of its economy.

[2] At least, if there has been no inflation or the measure of GDP is in constant year 0 dollars.

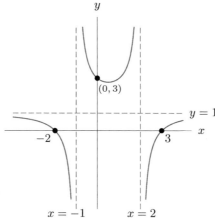

Figure 13.1: Graph of $f(x) = \dfrac{x^2 - x - 6}{x^2 - x - 2}$

Using the Factored Form of a Rational Function

Just as with polynomials, many aspects of the behavior of rational functions are determined by the factors of the numerator and denominator.

Finding the Zeros of a Rational Function: Factoring the Numerator

In Figure 13.1, the two x-intercepts, at $x = -2$ and $x = 3$, correspond to the zeros of f. For polynomials, x-intercepts can be found by factoring. For rational functions, they can be found by factoring the numerator.

Example 4   Find the zeros of $f(x) = \dfrac{x^2 - x - 6}{x^2 - x - 2}$ algebraically.

Solution   A fraction is zero provided its numerator is zero. Putting the numerator $a(x) = x^2 - x - 6$ in factored form $(x - 3)(x + 2)$, we see that it has zeros at $x = 3, -2$.

Finding the Domain of a Rational Function: Factoring the Denominator

The dashed vertical lines in Figure 13.1 tell us that the graph has vertical asymptotes at $x = -1$ and $x = 2$. The function f is undefined at these values, so they are not in its domain. Since fractions are undefined only when the denominator is zero, we look for such values by factoring the denominator.

Example 5   Show that the domain of $f(x) = \dfrac{x^2 - x - 6}{x^2 - x - 2}$ is all values of x except $x = -1$ and $x = 2$.

Solution   The numerator, $a(x) = x^2 - x - 6$, and denominator, $b(x) = x^2 - x - 2$, are defined everywhere. However, their ratio is undefined if $b(x) = 0$. Putting $b(x) = x^2 - x - 2$ into factored form $(x - 2)(x + 1)$, we see that its zeros are 2 and -1, so $f(x)$ is undefined when $x = 2$ and $x = -1$.

What Causes Vertical Asymptotes?

The graph in Figure 13.1 is very steep near the vertical asymptotes. The rapid rise (or fall) of the graph of a rational function near a vertical asymptote is because the denominator becomes small but the numerator does not. Table 13.1 shows this for the asymptote at $x = 2$. As x gets close to 2, two things happen: The value of the numerator gets close to -4, and the value of the denominator gets close to 0. Thus, the value of y grows larger and larger (either positive or negative), since we are dividing a number nearly equal to -4 by a number nearly equal to 0.

Table 13.1   *Values of* $f(x) = \dfrac{x^2 - x - 6}{x^2 - x - 2}$

x	Numerator	Denominator	$f(x)$
1.900	-4.290	-0.290	14.793
1.990	-4.030	-0.030	134.779
1.999	-4.003	-0.003	1334.778
2.000	-4.000	0.000	undefined
2.001	-3.997	0.003	-1331.889
2.010	-3.970	0.030	-131.890
2.100	-3.690	0.310	-11.903

Holes in the Graph of a Rational Function

Sometimes the numerator and denominator of a rational function are both zero at the same time. In this case the zero in the denominator need not cause the graph to have a vertical asymptote.

Example 6   Cancel the common factor in the expression for

$$f(x) = \frac{x^2 - 5x + 6}{x^2 - 4}.$$

Is the resulting expression equivalent to the original expression? What does the answer say about the graph of f?

Solution   We have

$$\frac{x^2 - 5x + 6}{x^2 - 4} = \frac{(x - 2)(x - 3)}{(x - 2)(x + 2)} \qquad \text{common factor of } x - 2$$

$$= \frac{\cancel{(x - 2)}(x - 3)}{\cancel{(x - 2)}(x + 2)} \qquad \text{cancel}$$

$$= \frac{x - 3}{x + 2} \qquad \text{provided } x \neq 2.$$

The expression $(x^2 - 5x + 6)/(x^2 - 4)$ has the same value as the expression $(x - 3)/(x + 2)$, except at $x = 2$. This is because the latter expression is defined at $x = 2$, whereas the former is not. This means f and g have different domains:

$$f(x) = \frac{x^2 - 5x + 6}{x^2 - 4} \qquad \text{domain is all values except } x = \pm 2$$

$$g(x) = \frac{x - 3}{x + 2} \qquad \text{domain is all values except } x = -2.$$

Thus, these functions are not exactly the same, even though their graphs are almost identical. See Figures 13.2 and 13.3. Notice that graph of f in Figure 13.2 does not have a vertical asymptote at $x = 2$, even though the denominator is zero there. Rather it has a hole, where f is undefined.

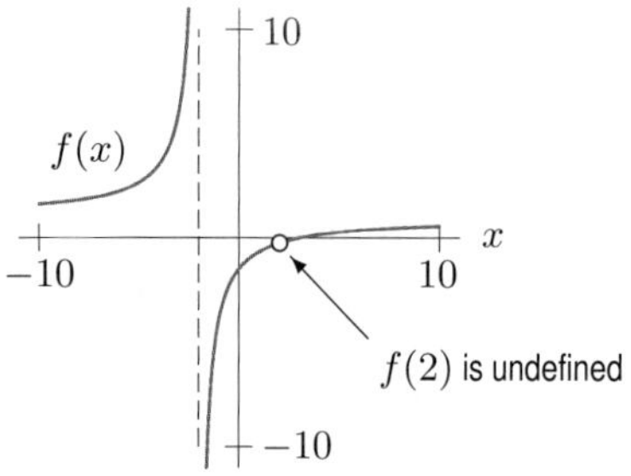

Figure 13.2: Graph of
$f(x) = (x^2 - 5x + 6)/(x^2 - 4)$

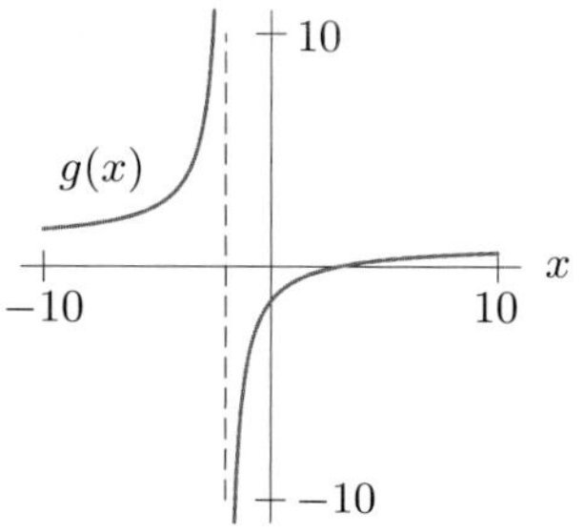

$f(2)$ is undefined

Figure 13.3: Graph of
$g(x) = (x - 3)/(x + 2)$

Example 7   Cancel common factors in each of the following rational expressions. Is the resulting expression equivalent to the original one?

(a) $\dfrac{x^2 - 2x + 1}{x - 1}$

(b) $\dfrac{2x^2 + 4x}{2x + 6}$

(c) $\dfrac{x^4 - 1}{(x^2 + 1)(x - 3)}$

Solution   (a) We have
$$\frac{x^2 - 2x + 1}{x - 1} = \frac{(x - 1)(x - 1)}{x - 1} = x - 1 \quad \text{provided } x \neq 1.$$

The resulting expression is not equivalent to $x - 1$, because the original expression is undefined at $x = 1$.

(b) We have
$$\frac{2x^2 + 4x}{2x + 6} = \frac{2(x^2 + 2x)}{2(x + 3)} = \frac{x^2 + 2x}{x + 3}.$$

Canceling the 2 is valid no matter what the value of x, so the two expressions are equivalent.

(c) We have
$$\frac{x^4 - 1}{(x^2 + 1)(x - 3)} = \frac{(x^2 - 1)(x^2 + 1)}{(x^2 + 1)(x - 3)} = \frac{x^2 - 1}{x - 3}.$$

Since $x^2 + 1$ is positive for all values of x, cancelation does not involve possible division by 0, so the two expressions are equivalent.

Not All Rational Functions Have Vertical Asymptotes

It is tempting to assume that the graph of any rational function has a vertical asymptote. However, this is not always the case. For instance, the graph of

$$f(x) = \frac{x^3 - 1}{x^2 + 1}$$

in Figure 13.4 has no vertical asymptote. Because the denominator is everywhere positive, the function's value never "blows up" as it would if we were dividing by zero.

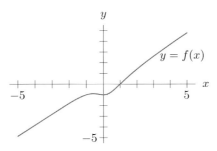

Figure 13.4: Graph of $f(x) = \dfrac{x^3 - 1}{x^2 + 1}$

Exercises and Problems for Section 13.1

EXERCISES

Put each expression in Exercises 1–4 into the form $a(x)/b(x)$ for polynomials $a(x)$ and $b(x)$.

1. $\dfrac{3}{2x + 6} + 5$

2. $\dfrac{1 + \frac{1}{x}}{2 - \frac{3}{x}}$

3. $\dfrac{1}{1 + \frac{1}{1 + \frac{1}{x}}}$

4. $\dfrac{2}{x^5 - 3x^2 + 7}$

In Exercises 5–8, find the zeros.

5. $f(x) = \dfrac{5x + 3}{2x + 2}$

6. $f(x) = \dfrac{10x + 4}{7x - 10}$

7. $f(x) = \dfrac{5x^2 - 4x - 1}{10 + 3x}$

8. $f(x) = \dfrac{-5x^2 + 1}{3x^2 + 4}$

In Exercises 9–15, find the domain.

9. $f(x) = \dfrac{9 - x}{2x + 5}$

10. $f(x) = \dfrac{x^3 - 8x + 1}{x^4 + 1}$

11. $f(x) = \dfrac{1}{x - \sqrt{2}}$

12. $f(x) = \dfrac{11x}{x^3 + 27}$

13. $f(x) = \dfrac{x^2 + x}{15x}$

14. $f(x) = \dfrac{4 - x}{2}$

15. $g(x) = \dfrac{x^2 - 4x + 3}{x^2 - 4}.$

In Exercises 16–18, find the vertical asymptotes.

16. $f(x) = \dfrac{2x - 2}{x + 1}$

17. $g(r) = \dfrac{r - 6}{r^2 - 3r - 4}$

18. $h(x) = \dfrac{x^2 + x - 6}{x^2 - 7x + 10}$

Divide common factors from the numerator and denominator in the rational expressions in Exercises 19–22. Are the resulting expressions equivalent to the original expressions?

19. $\dfrac{x^2 - 6x + 5}{x - 5}$

20. $\dfrac{x^3 + 6x^2 + 9x}{x^2 + 10x + 16}$

21. $\dfrac{2x^2 + 4}{(3x^2 + 5)(x^2 + 2)}$

22. $\dfrac{x^4}{x^3 - x^2}$

In Exercises 23–25, the graphs of the two given functions are identical, except for a hole. What are the coordinates of the hole?

23. $f(x) = x + 4$ and $g(x) = \dfrac{x^2 + x - 12}{x - 3}$

24. $h(x) = x^2 - 2x$ and $k(x) = \dfrac{x^3 - x^2 - 2x}{x + 1}$

25. $r(t) = 5t^3$ and $s(t) = \dfrac{10t^3 + 5t^4}{t + 2}$

PROBLEMS

26. A rental car is driven d miles per day. The cost c is

$$c = \frac{2500 + 8d}{d} \text{ cents per mile.}$$

   (a) Find the cost per mile for a 100 mile per day trip.

   (b) How many miles must be driven per day to bring the cost down to 13 cents or less per mile?

   (c) Rewrite the formula for c to make it clear that c decreases as d increases. What is the limiting value of c as d gets very large?

27. The cost of printing n posters is $10 + 0.02n$ dollars.

   (a) Write a formula for the average cost, a, per poster for printing n posters.

   (b) How many posters must be printed to make the average cost 4¢ per poster?

   (c) Rewrite the formula for the average cost to show why it is impossible to print posters for less than 2¢ per poster.

28. When doctors test a patient for a virus, such as HIV, the results are not always accurate. For a particular test, the proportion, P, of the people who test positive and who really have the disease is given by

$$P = \frac{0.9x}{0.1 + 0.8x},$$

where x is the fraction of the whole population that has the disease.

   (a) If 1% of the population has the disease, what is P?

   (b) What value of x leads to $P = 0.9$?

   (c) If doctors do not want to use the test unless at least 99% of the people who test positive really have the disease, for what values of x should it be used?

29. Let $R(z) = \dfrac{22 - z}{(z - 11)(1 - z)}$. Which of the following forms is equal to $R(z)$, and which is equal to $-R(z)$?

   (a) $\dfrac{z - 22}{(z - 11)(z - 1)}$ **(b)** $\dfrac{z - 22}{(z - 11)(1 - z)}$

   (c) $\dfrac{22 - z}{(11 - z)(z - 1)}$ **(d)** $\dfrac{22 - z}{(11 - z)(1 - z)}$

■ In Problems **30–33**, solve for x.

30. $5 = \dfrac{3x - 5}{2x + 3}$ **31.** $12 = \dfrac{6x - 3}{5x + 2}$

32. $13 = \dfrac{2x^2 + 5x - 1}{3 + x}$ **33.** $7 = \dfrac{5x^2 - 2x + 23}{2x^2 + 2}$

■ In Problems **34–37**, imagine solving the equation by multiplying by the denominator to convert it to a polynomial equation. What is the degree of the polynomial equation?

34. $\dfrac{2x - 5}{x - 7} = 7$ **35.** $\dfrac{2x + x^2}{2x^2 - 5} = 7$

36. $\dfrac{x^2 + 5x}{2x^2 + 3} = \dfrac{1}{2}$ **37.** $\dfrac{x^2 - 5}{x^3 + 2} = \dfrac{1}{x + 2}$

■ In Problems **38–40**, write an expression for the function whose graph is identical to the graph of $y = x^2$ except for a hole at the given x-value. What are the coordinates of the hole?

38. $x = 3$ **39.** $x = -3$ **40.** $x = 0$

13.2 LONG-RUN BEHAVIOR OF RATIONAL FUNCTIONS

When we write a rational function as the ratio of two polynomials in factored form, we can see its zeros and its vertical asymptotes or holes. What form is useful for seeing the long-run behavior?

The Quotient Form of a Rational Expression

A rational number can be thought of as the numerator divided by the denominator. For example, since 4 goes into 13 three times with a remainder of 1, we have

$$\frac{13}{4} = \frac{3 \cdot 4 + 1}{4} = \frac{3 \cdot 4}{4} + \frac{1}{4} = 3 + \frac{1}{4} = 3\frac{1}{4}.$$

The form on the right is useful for estimating the size of the number. Similarly, the rational function

$$\frac{x^2 - 5x + 7}{x - 2}$$

can be thought of as dividing $x - 2$ into $x^2 + 5x + 7$. Since

$$x^2 - 5x + 7 = x^2 - 5x + 6 + 1 = (x - 3)(x - 2) + 1,$$

we see that the number of times $x - 2$ goes into $x^2 - 5x + 7$ is $x - 3$ with a remainder of 1, so

$$\frac{x^2 - 5x + 7}{x - 2} = \frac{(x - 3)(x - 2) + 1}{x - 2} = (x - 3) + \frac{1}{x - 2}.$$

The form on the right is the analog for rational functions of the mixed-fraction form for rational numbers. Just as the remainder has to be less than the divisor when we divide integers, the remainder in polynomial division has to either have degree less than the degree of the divisor or be the zero polynomial.

Quotient Form of a Rational Expression

The quotient form of a rational expression is

$$f(x) = \frac{a(x)}{b(x)} = q(x) + \frac{r(x)}{b(x)}, \quad \text{where}$$

- The polynomial q is the *quotient*.
- The polynomial r is called the *remainder*, and must have smaller degree than q or be the zero polynomial. The term $r(x)/b(x)$ in the quotient form is called the *remainder term*. We have $a(x) = q(x)b(x) + r(x)$, so we can think of this as saying that the number of times $b(x)$ goes into $a(x)$ is $q(x)$, with remainder $r(x)$.

For example, we saw above that

$$\underbrace{x - 3}_{q(x)} + \underbrace{\frac{\overbrace{1}^{r(x)}}{x - 2}}_{b(x)} \quad \text{is the quotient form of} \quad \frac{\overbrace{x^2 - 5x + 7}^{a(x)}}{\underbrace{x - 2}_{b(x)}},$$

with $q(x) = x - 3$ and $r(x) = 1$. Here the degree of $r(x) = 1$ is 0, and the degree of $b(x) = x - 2$ is 1, so the remainder has degree less than the divisor, as required.

Quotient Form and Long-Run Behavior

First we look at what happens when the quotient $q(x)$ is a constant.

Example 1   For the rational function in quotient form

$$f(x) = 5 + \frac{3}{x+4},$$

(a) Graph f and describe its long-run behavior.
(b) Explain the long-run behavior using the quotient form.

Solution   (a) See Figure 13.5. We see that, in the long run, the rational function behaves like the constant function $y = 5$. We see that $y = 5$ is a horizontal asymptote.
(b) The function is in quotient form with $q(x) = 5$. As x gets larger and larger, $f(x)$ gets closer and closer to $q(x)$ because the remainder term gets smaller and smaller. For example, in evaluating $f(100)$, we get

$$f(100) = 5 + \frac{3}{100+4} = 5 + \frac{3}{104} = 5.029.$$

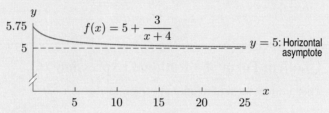

Figure 13.5: For large positive values of x, the graph looks like the horizontal line $y = 5$

In general, the quotient form of a rational function tells us its long-run behavior. This is because, for large values of x, the value of $q(x)$ is much larger than the value of $r(x)/b(x)$. For example, in evaluating

$$\frac{x^2 - 5x + 7}{x - 2} = x - 3 + \frac{1}{x - 2}$$

at $x = 1000$, we get

$$\frac{x^2 - 5x + 7}{x - 2} = (1000 - 3) + \frac{1}{1000 - 2} = \underbrace{997}_{\substack{\text{largest} \\ \text{contribution}}} + \underbrace{\frac{1}{998}}_{\substack{\text{small} \\ \text{correction}}} = 997.001.$$

Notice how close the function's value at $x = 1000$ is to the value of the quotient: 997.001 versus 997. In general:

For a rational function expressed in quotient form

$$f(x) = q(x) + \frac{r(x)}{b(x)},$$

the long-run behavior of f is the same as the long-run behavior of the polynomial function q. If $q(x) = k$, a constant, then the graph has a horizontal asymptote at $y = k$.

Putting a Function into Quotient Form When the Quotient is Constant: Horizontal Asymptotes

To see if a rational function has constant quotient, we try to write the numerator as a constant times the denominator plus a remainder. For example, to write the rational function

$$y = \frac{5x + 23}{x + 4}$$

in quotient form, we want to write the numerator as

$$\overbrace{5x + 23}^{\text{numerator}} = \text{constant} \overbrace{(x + 4)}^{\text{denominator}} + \text{remainder}.$$

Comparing coefficients of x on both sides tells us that the constant must be 5, so

$$\overbrace{5x + 23}^{\text{numerator}} = 5 \overbrace{(x + 4)}^{\text{denominator}} + \text{remainder},$$

and then comparing both sides tells us that remainder $= 3$. Now we can write:

$$\frac{5x + 23}{x + 4} = \frac{5(x + 4) + 3}{x + 4}$$

$$= \frac{5(x + 4)}{x + 4} + \frac{3}{x + 4} \quad \text{split numerator}$$

$$= 5 + \frac{3}{x + 4} \quad \text{quotient form.}$$

In general, if the numerator and denominator have the same degree, the ratio of the leading terms gives us the constant quotient when the rational function is in quotient form.

Example 2   Find the long-run behavior and horizontal asymptote of $f(x) = \dfrac{30x^3 - 9x^2 + 1}{10x^3 - 3x^2 - 1}$ by putting it in quotient form.

Solution   Notice that the leading term of the numerator is 3 times the leading term of the denominator. We write

$$\frac{30x^3 - 9x^2 + 1}{10x^3 - 3x^2 - 1} = \frac{3 \overbrace{(10x^3 - 3x^2 - 1)}^{\text{denominator}} + 4}{10x^3 - 3x^2 - 1} \qquad \text{write numerator as 3 times denominator plus remainder}$$

$$= \frac{3(10x^3 - 3x^2 - 1)}{10x^3 - 3x^2 - 1} + \frac{4}{10x^3 - 3x^2 - 1} \qquad \text{split numerator}$$

$$= 3 + \frac{4}{10x^3 - 3x^2 - 1} \qquad \text{quotient form.}$$

Thus, the quotient is $q(x) = 3$ and the remainder is $r(x) = 4$. In the long run $f(x)$ behaves like the constant function $q(x) = 3$. Figure 13.6 shows the graph of f approaching the graph of $y = 3$.

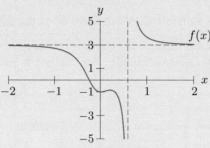

Figure 13.6: Graph of
$$f(x) = \frac{30x^3 - 9x^2 + 1}{10x^3 - 3x^2 - 1}$$

A special case of horizontal asymptotes occurs when the numerator has degree less than the denominator.

Example 3   Find the long-run behavior and horizontal asymptote of the rational function $y = \dfrac{3x + 1}{x^2 + x - 2}$.

Solution   In this case the function is already in quotient form with quotient zero:

$$y = \frac{3x + 1}{x^2 + x - 2} = \underbrace{0}_{q(x)} + \underbrace{\frac{3x + 1}{x^2 + x - 2}}_{r(x)/b(x)}.$$

The long-run behavior is given by $q(x) = 0$, so the horizontal asymptote is $y = 0$, or the x-axis.

Slant Asymptotes and Other Types of Long-Run Behavior

What if the quotient polynomial $q(x)$ is not a constant? One possibility is that $q(x)$ is a linear function. In this case, we say the graph has a *slant asymptote*, as illustrated by Example 4.

Example 4   Graph $f(x) = x - \dfrac{x + 1}{x^2 + 1}$ and describe its long-run behavior.

Solution   The function is in quotient form with quotient $q(x) = x$, so for large values of x it is close to the graph of the linear function $y = x$. See Figure 13.7.

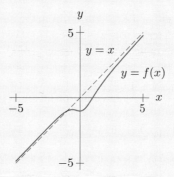

Figure 13.7: Graph of $f(x) = x - \dfrac{x+1}{x^2+1}$

Example 5   The rational function $f(x) = \dfrac{2x^3 + 3x^2}{x^2 + 1}$ has quotient form

$$\frac{2x^3 + 3x^2}{x^2 + 1} = 2x + 3 - \frac{2x + 3}{x^2 + 1}.$$

Graph f and interpret its long-run behavior in terms of the quotient form.

Solution   The quotient is $q(x) = 2x + 3$, so the long-run behavior of f is the same as the line $y = 2x + 3$. The graph of $y = 2x + 3$ is a a slant asymptote for f. See Figure 13.8.

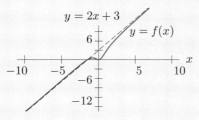

Figure 13.8: Graph of $f(x) = \dfrac{2x^3 + 3x^2}{x^2 + 1}$.

In Section 13.3 we describe a general method for putting functions in quotient form.

Long-Run Behavior and the Ratio of the Leading Terms

In Example 2, the quotient of

$$f(x) = \frac{30x^3 - 9x^2 + 1}{10x^3 - 3x^2 - 1}$$

is $q(x) = 3$, the ratio of the leading terms in the numerator and denominator, $30x^3$ and $10x^3$. Taking the ratio of leading terms works in general to give the long-run behavior of a rational function, since the ratio of leading terms is the leading term of the quotient.

Example 6   Graph $y = \dfrac{6x^4 + x^3 + 1}{-5x + 2x^2}$, and describe its long-run behavior.

Solution   Considering the ratio of leading terms, we see that the long-run behavior of this function is

$$y = \frac{6x^4 + x^3 + 1}{-5x + 2x^2} \approx \frac{6x^4}{2x^2} = 3x^2.$$

This means that on large scales, its graph resembles the parabola $y = 3x^2$. See Figure 13.9.

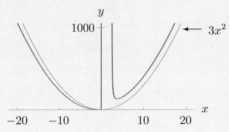

Figure 13.9: In the long run, the graph of $y = \dfrac{6x^4 + x^3 + 1}{-5x + 2x^2}$ looks like the graph of the parabola $y = 3x^2$

Graphing Rational Functions

Putting together everything we have learned in the last two sections, we can get a comprehensive picture of the graph of a rational function.

Example 7   Graph the following rational functions, showing all the important features.

(a) $y = \dfrac{x + 3}{x + 2}$

(b) $y = \dfrac{25}{(x + 2)(x - 3)^2}$

Solution   (a) See Figure 13.10.
- The numerator is zero at $x = -3$, so the graph crosses the x-axis here.
- The denominator is zero at $x = -2$, so the graph has a vertical asymptote here.
- At $x = 0$, $y = (0 + 3)(0 + 2) = 1.5$, so the y-intercept is $y = 1.5$.
- The long-run behavior is given by the ratio of leading terms, $y = x/x = 1$, so the graph

has a horizontal asymptote at $y = 1$.

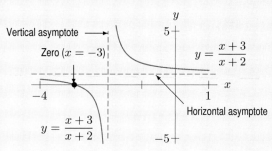

Figure 13.10: A graph of the rational function $y = \dfrac{x+3}{x+2}$
showing its asymptotes and zero

(b) See Figure 13.11.

- Since the numerator of this function is never zero, this rational function has no zeros, which means that the graph never crosses the x-axis.
- The graph has vertical asymptotes at $x = -2$ and $x = 3$ because this is where the denominator is zero.
- At $x = 0$, $y = 25/(0+2)(0-3)^2 = 25/18$, so the y-intercept is $y = 25/18$.
- The long-run behavior of this rational function is given by the ratio of the leading term in the numerator to the leading term in the denominator. The numerator is 25, and if we multiply out the denominator, we see that its leading term is x^3. Thus, the long-run behavior is given by $y = 25/x^3$, which has a horizontal asymptote at $y = 0$.

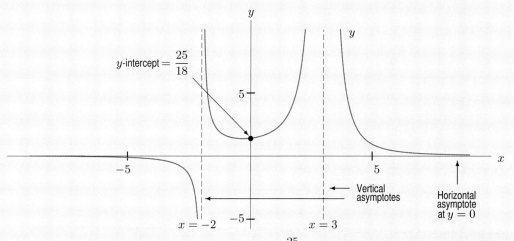

Figure 13.11: A graph of the rational function $y = \dfrac{25}{(x+2)(x-3)^2}$, showing intercept and asymptotes

Example 8   Find a possible formula for the rational function shown in Figure 13.12.

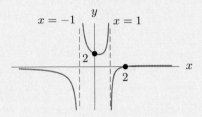

Figure 13.12: A rational function

Solution   The graph has vertical asymptotes at $x = -1$ and $x = 1$. At $x = 0$, we have $y = 2$, and at $y = 0$, we have $x = 2$. The graph of $y = \dfrac{(x-2)}{(x+1)(x-1)}$ satisfies each of these requirements. In the long run, as the x-values get larger and larger, the y-values get closer and closer to 0. Looking at the equation of the function, we see that the ratio of the leading terms, $y = \dfrac{x}{x^2} = \dfrac{1}{x}$ does get closer and closer to the horizontal line $y = 0$ as x gets very large.

Example 9   Find a possible formula for the rational function shown in Figure 13.13.

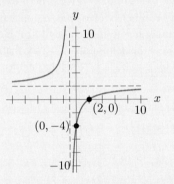

Figure 13.13: A rational function

Solution   The graph has a vertical asymptote at $x = -1$, and a zero at $x = 2$, so we try

$$y = \frac{x-2}{x+1}.$$

This has the right zero and vertical asymptote, but when $x = 0$ it has the value $y = -2$, whereas the graph has a vertical intercept at $y = -4$. We multiply by 2 to get

$$y = \frac{2(x-2)}{x+1}.$$

This has horizontal asymptote at $y = 2$ and y-intercept at $y = -4$, as required.

We now summarize what we have learned about the graphs of rational functions:

For a rational function given by $y = \dfrac{a(x)}{b(x)}$, where a and b are polynomials with different zeros, then:

- The **long-run behavior** is given by the ratio of the leading terms of $a(x)$ and $b(x)$.
- The **zeros** are the same as the zeros of the numerator, $a(x)$.
- A **vertical asymptote** or **hole** occurs at each of the zeros of the denominator, $b(x)$.
- A **hole** occurs at an x-value where both $a(x)$ and $b(x)$ are zero.

Can a Graph Cross an Asymptote?

The graph of a rational function never crosses a vertical asymptote. However, the graphs of some rational functions cross their horizontal asymptotes. The difference is that a vertical asymptote occurs where the function is undefined, so there can be no y value there, whereas a horizontal asymptote explains what happens to the y-values as the x-values get larger and larger (in either a positive or negative direction). There is no reason why the function cannot take on this y-value at a finite x-value. For example, the graph of $y = \dfrac{x^2 + 2x - 3}{x^2}$ crosses the line $y = 1$, its horizontal asymptote; the graph does not cross the vertical asymptote, the y-axis. See Figure 13.14.

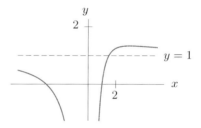

Figure 13.14: A rational function can cross its horizontal asymptote

Exercises and Problems for Section 13.2

EXERCISES

In Exercises **1–3**, find the horizontal asymptotes of the functions given.

1. $g = \dfrac{x - 4}{x^2 + 16}$

2. $s = \dfrac{4z^3 - z + 9}{1000 + 3z^3}$

3. $r = \dfrac{3t^4 - 3t^3 + t - 500}{3t^2 + 500t - 12}$

In Exercises **4–12**, use the method of Example 2 on page 415 to find the quotient $q(x)$ and the remainder $r(x)$ so that $a(x) = q(x)b(x) + r(x)$.

4. $\dfrac{x + 3}{x + 2}$

5. $\dfrac{2x + 3}{x + 1}$

6. $\dfrac{x^3 + x + 5}{x^2 + 1}$

7. $\dfrac{x^3 + 5x^2 + 6x + 7}{x^2 + 3x}$

8. $\dfrac{10x^3 - 3x^2 + 1}{10x^3 - 3x^2 - 1}$

9. $\dfrac{8x^5 - 3x^2 + 2x + 3}{8x^5 - 3x^2 + 2x + 2}$

10. $\dfrac{8x^5 - 3x^2 + 3x + 3}{8x^5 - 3x^2 + 2x + 2}$

11. $\dfrac{16x^5 - 6x^2 + 4x + 7}{8x^5 - 3x^2 + 2x + 2}$

12. $\dfrac{8x^5 + 2x^4 - 3x^2 + 2x + 2}{8x^5 - 3x^2 + 2x + 2}$

Match the following rational functions to the statements in Exercises **13–16**. A statement may match none, one, or several of the given functions. You are not required to draw any graphs to answer these questions.

(a) $y = \dfrac{1}{x^2 + 1}$

(b) $y = \dfrac{x - 1}{x + 1}$

(c) $y = \dfrac{x - 2}{(x - 3)(x + 1)}$

(d) $y = \dfrac{(x - 2)(x - 3)}{x^2 - 1}$

13. This function has no zeros (x-intercepts).

14. The graph of this function has a vertical asymptote at $x = 1$.

15. This function has long-run behavior that y approaches 0 as x gets larger and larger (either positive or negative). Its graph has a horizontal asymptote at $y = 0$.

16. The graph of this function has no vertical asymptotes.

Match the following rational functions to the statements in Exercises **17–20**. A statement may match none, one, or several of the given functions. You are not required to draw any graphs to answer these questions.

(a) $y = \dfrac{x^2 + 1}{x^2 - 1}$

(b) $y = \dfrac{x - 1}{x^2 - x - 2}$

(c) $y = \dfrac{x}{x + 1}$

(d) $y = \dfrac{1 + x^2}{2 + x^2}$

17. This function has no zeros (x-intercepts).

18. This function has the long-run behavior that y approaches 1 as x gets larger and larger (positive or negative). Its graph has a horizontal asymptote at $y = 1$.

19. The graph of this function has a vertical asymptote at $x = -1$.

20. The graph of this function contains the point $(0, 1/2)$.

PROBLEMS

21. The rational function $r(x)$ can be written in two forms:

I. $r(x) = \dfrac{(x - 1)(x - 2)}{x^2}$

II. $r(x) = 1 - \dfrac{3}{x} + \dfrac{2}{x^2}$

Which form most readily shows

(a) The zeros of $r(x)$? What are they?

(b) The value of $r(x)$ as r gets large, either positive or negative? What is it?

22. The rational function $q(x)$ can be written in two forms:

I. $q(x) = 1 + \dfrac{2x - 7}{x^2 - 1}$

II. $q(x) = \dfrac{(x - 2)(x + 4)}{(x - 1)(x + 1)}$

(a) Show that the two forms are equivalent.

(b) Which form most readily shows

   (i) The zeros of $q(x)$? What are they?

   (ii) The vertical asymptotes? What are they?

   (iii) The horizontal asymptote? What is it?

23. Find the horizontal asymptotes of

$$y = \frac{6x + 1}{2x - 3}$$

and

$$y = \frac{4x + 4}{2x - 3} + 1.$$

What do you observe? Use algebra to explain your observation. [Hint: Note that $6x + 1 = (4x + 4) + (2x - 3)$.]

24. Let $p(x) = \dfrac{x + 3}{2x - 5}$ and let $q(x) = \dfrac{3x + 1}{4x + 4}$.

(a) Find the horizontal and vertical asymptotes of $p(x)$ and $q(x)$.

(b) Let $f(x) = p(x) + q(x)$. Write $f(x)$ as a single rational expression.

(c) Find the horizontal and vertical asymptotes of $f(x)$. Describe the relationship between the asymptotes of $p(x)$ and $q(x)$ and the asymptotes of $f(x)$.

25. Explain why there are no vertical asymptotes for the graph of $y = \dfrac{1}{x^2 + 3}$.

Find possible formulas for the functions in Problems **26–31**.

26.

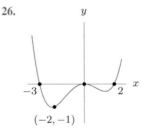

27.

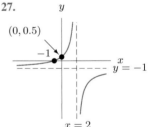

28.

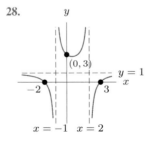

29.

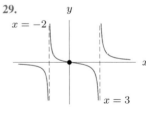

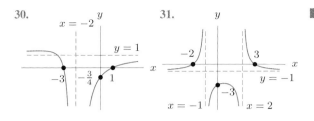

30.

31.

For the rational functions in Problems **32–35**, find all zeros and vertical asymptotes and describe the long-run behavior, then graph the function.

32. $y = \dfrac{x+3}{x+5}$

33. $y = \dfrac{x+3}{(x+5)^2}$

34. $y = \dfrac{x-4}{x^2-9}$

35. $y = \dfrac{x^2-4}{x-9}$

13.3 PUTTING A RATIONAL FUNCTION IN QUOTIENT FORM

In Example 2 on page 415, we saw that we could rewrite the rational function

$$f(x) = \frac{30x^3 - 9x^2 + 1}{10x^3 - 3x^2 - 1} \quad \text{as} \quad f(x) = 3 + \frac{4}{10x^3 - 3x^2 - 1}$$

by rewriting the numerator as a constant times the denominator plus a remainder. What if this is not possible? In this section we develop an algorithm for dividing one polynomial into another, similar to long division in arithmetic.

The Division Algorithm

In Example 9 on page 391, we saw that

$$3x^3 + 7x^2 + 2x - 3720 = (x - 10)(\text{Another polynomial}),$$

because the left-hand side is zero at $x = 10$. To find the other polynomial, we must divide $x - 10$ into $3x^3 + 7x^2 + 2x - 3720$. We start by looking at the leading term $3x^3$. Since $3x^3 = 3x^2 \cdot x$, we can write

$$3x^3 + 7x^2 + 2x - 3720 = (x - 10)\left(3x^2 \qquad\right).$$

Notice we have left space on the right for the lower-degree terms in the second factor on the right. We set this up in more traditional format as

$$
\begin{array}{r}
3x^2 \qquad\qquad \\[-2pt]
\hline
x - 10 \,\big)\; 3x^3 + 7x^2 + 2x - 3720
\end{array}.
$$

To find the next term after $3x^2$, we see what is left to divide after we subtract $3x^2(x - 10) = 3x^3 - 30x^2$:

$$
\begin{array}{r}
3x^2 \qquad\qquad \\[-2pt]
\hline
x - 10 \,\big)\; 3x^3 + 7x^2 + 2x - 3720 \\
-3x^3 + 30x^2 \qquad\qquad \\[-2pt]
\hline
37x^2 + 2x
\end{array}.
$$

Now we can find the next term in the quotient by comparing the divisor, $x - 10$, with the new dividend $37x^2 + 2x - 3720$. The new leading term is $37x^2 = 37x \cdot x$, so the next term is $37x$, giving

$$
\begin{array}{r}
3x^2 + 37x \qquad\qquad \\[-2pt]
\hline
x - 10 \,\big)\; 3x^3 + 7x^2 + 2x - 3720 \\
-3x^3 + 30x^2 \qquad\qquad \\[-2pt]
\hline
37x^2 + 2x
\end{array}.
$$

Continuing in this way, we get the completed calculation

$$
\begin{array}{r}
3x^2 \;+37x\;+372. \\
x-10\overline{)3x^3\;+7x^2\quad\;+2x-3720} \\
-3x^3+30x^2 \\
\hline
37x^2\quad\;+2x \\
-37x^2+370x \\
\hline
372x-3720 \\
-372x+3720 \\
\hline
0
\end{array}
$$

Notice that the remainder is zero, as we expected.

Polynomial Division with a Remainder

In general, we have to allow for the possibility that the remainder might not be zero.

| Example 1 | Divide $x^2 + x + 1$ into $3x^4 - x^3 + x^2 + 2x + 1$. |

| Solution | We write |

$$3x^4 \;-x^3\;+x^2+2x+1 = \left(x^2+x+1\right)\left(3x^2 \qquad\right)\qquad,$$

leaving space for the remainder as well as the quotient. Or, in traditional format

$$
\begin{array}{r}
3x^2\qquad\qquad\;. \\
x^2+x+1\overline{)3x^4\;-x^3\;+x^2+2x+1}
\end{array}
$$

We subtract $3x^2(x^2+x+1) = 3x^4 + 3x^3 + 3x^2$:

$$
\begin{array}{r}
3x^2\qquad\qquad \\
x^2+x+1\overline{)3x^4\;-x^3\;+x^2+2x+1} \\
-3x^4-3x^3-3x^2 \\
\hline
-4x^3-2x^2+2x
\end{array}
$$

Again, we find the next term in the quotient by comparing the divisor, $x^2 + x + 1$, with the new dividend $-4x^3 - 2x^2 + 2x$. The ratio of the leading terms is now $-4x^3/x^2 = -4x$, so the next term is $-4x$, giving

$$
\begin{array}{r}
3x^2-4x\qquad \\
x^2+x+1\overline{)3x^4\;-x^3\;+x^2+2x+1} \\
-3x^4-3x^3-3x^2 \\
\hline
-4x^3-2x^2+2x
\end{array}
$$

Continuing in this way, we get the completed calculation

$$
\begin{array}{r}
3x^2-4x+2 \\
x^2+x+1\overline{)3x^4\;-x^3\;+x^2+2x+1} \\
-3x^4-3x^3-3x^2 \\
\hline
-4x^3-2x^2+2x \\
4x^3+4x^2+4x \\
\hline
2x^2+6x+1 \\
-2x^2-2x-2 \\
\hline
4x-1
\end{array}
$$

This means that $4x - 1$ is the remainder, so

$$3x^4 - x^3 + x^2 + 2x + 1 = (x^2 + x + 1)(3x^2 - 4x + 2) + 4x - 1.$$

The Division Algorithm and Quotient Form

We can use the division algorithm to write rational functions in quotient form.

Example 2 Write in quotient form and state the quotient:

(a) $\dfrac{x^3 - 2x + 1}{x - 2}$ (b) $\dfrac{2x^3 + 3x^2}{x^2 + 1}$ (c) $y = \dfrac{x^3 - 1}{x^2 + 1}$

Solution (a) Dividing the denominator into the numerator gives

$$
\begin{array}{r}
x^2 + 2x + 2 \\
x - 2 \overline{)x^3 - 2x + 1} \\
\underline{-x^3 + 2x^2} \\
2x^2 - 2x \\
\underline{-2x^2 + 4x} \\
2x + 1 \\
\underline{-2x + 4} \\
5
\end{array}
$$

Thus, we can write

$$(x^3 - 2x + 1) = (x - 2)(x^2 + 2x + 2) + 5.$$
$$\frac{x^3 - 2x + 1}{x - 2} = x^2 + 2x + 2 + \frac{5}{x - 2}, \qquad \text{divide by } b(x) = x - 2$$

so $q(x) = x^2 + 2x + 2$.

(b) Dividing the denominator into the numerator gives

$$
\begin{array}{r}
2x + 3 \\
x^2 + 1 \overline{)2x^3 + 3x^2} \\
\underline{-2x^3 - 2x} \\
3x^2 - 2x \\
\underline{-3x^2 - 3} \\
-2x - 3
\end{array}
$$

Thus, we can write

$$2x^3 + 3x^2 = (2x + 3)(x^2 + 1) + (-2x - 3)$$
$$\frac{2x^3 + 3x^2}{x^2 + 1} = 2x + 3 + \frac{-2x - 3}{x^2 + 1}, \qquad \text{divide by } b(x) = x^2 + 1$$

so $q(x) = 2x + 3$.

(c) Dividing the denominator into the numerator gives

$$x^2+1) \overline{\begin{array}{r} x \\ x^3 \phantom{{}-x} -1 \\ \underline{-x^3-x} \\ -x \end{array}}$$

Thus, we can write

$$x^3-1 = (x)(x^2+1)+(-x-1)$$
$$\frac{x^3-1}{x^2+1} = x + \frac{-x-1}{x^2+1}, \qquad\qquad \text{divide by } b(x)=x^2+1$$

so $q(x)=x$.

Example 3   Determine the equation of the slant asymptote of the rational function $f(x)=\dfrac{2x^3+6x^2-26x-30}{x^2+2x-8}$.

Solution   Dividing x^2+2x-8 into $2x^3+6x^2-26x-30$ gives

$$x^2+2x-8) \overline{\begin{array}{r} 2x +2 \\ 2x^3+6x^2-26x-30 \\ \underline{-2x^3-4x^2+16x} \\ 2x^2-10x-30 \\ \underline{-2x^2 -4x+16} \\ -14x-14 \end{array}}$$

We see that

$$f(x) = \frac{2x^3+6x^2-26x-30}{x^2+2x-8} = 2x+2+\frac{-14x-14}{x^2+2x-8}.$$

Thus, the quotient is $q(x)=2x+2$, and for large values of x the graph is close to the slant asymptote $y=2x+2$.

The Remainder Theorem

A particularly simple division is when we divide the polynomial $x-a$, for a constant a, into some other polynomial $p(x)$. In that case the remainder is a constant, since it must have degree less than the degree of $x-a$. Thus,

$$p(x) = (x-a)q(x) + \text{ constant}.$$

What is the constant? If we put $x=a$ in the equation above, we get

$$p(a) = 0 \cdot q(a) + \text{ constant } = \text{ constant}.$$

This tells us that the constant is $p(a)$. We have shown:

The Remainder Theorem

For any polynomial $p(x)$ and any constant a, we have

$$p(x) = (x - a)q(x) + p(a),$$

where $q(x)$ is the quotient of $p(x)/(x-a)$. In particular, if $p(a) = 0$ then $p(x) = (x-a)q(x)$, so $x - a$ is a factor of $p(x)$.

On page 387 we saw that if $x - a$ is a factor of $p(x)$, then $p(a) = 0$. Now we have proved the converse, that if $p(a) = 0$ then $x - a$ is a factor of $p(x)$.

Example 4   (a) Is $x - 4$ a factor of the polynomial $p(x) = 5x^3 - 13x^2 - 30x + 8$?
(b) Is $x + 3$ a factor of the polynomial $p(x) = 2x^3 + 10x^2 + 11x - 6$?

Solution   We could answer both these questions by using the division algorithm to see if the remainder is zero. However, the Remainder Theorem gives us a short cut.

(a) We have $p(4) = 5(4)^3 - 13(4)^2 - 30(4) + 8 = 5(64) - 13(16) - 120 + 8 = 0$, so by the Remainder Theorem $x - 4$ is a factor of $p(x)$.

(b) We have $p(3) = 2(-3)^3 + 10(-3)^2 + 11(-3) - 6 = 2(-27) + 10(9) - 33 - 6 = -3$, so $x - 3$ cannot be a factor of $p(x) = 2x^3 + 10x^2 + 11x - 6$.

Exercises and Problems for Section 13.3

EXERCISES

In Exercises **1–10**, use the division algorithm to find the quotient $q(x)$ and the remainder $r(x)$ so that $a(x) = q(x)b(x) + r(x)$.

1. $\dfrac{5x^2 - 23x + 16}{x - 4}$

2. $\dfrac{3x^3 - 2x^2 - 3x + 5}{x + 1}$

3. $\dfrac{2x^2 - 3x + 11}{x + 7}$

4. $\dfrac{2x^2 - 3x + 11}{x^2 + 7}$

5. $\dfrac{x^4 - 2}{x - 1}$

6. $\dfrac{x^4 - 2}{x^2 - 1}$

7. $\dfrac{6x^2 + 5x + 3}{3x - 2}$

8. $\dfrac{20x^2 - 33x + 6}{4x - 5}$

9. $\dfrac{6x^3 - 11x^2 + 18x - 7}{3x - 1}$

10. $\dfrac{15x^3 + 4x^2 - 19x + 9}{5x - 2}$

Put the rational expressions in Exercises **11–15** into quotient form, identify any horizontal or slant asymptotes, and sketch the graph.

11. $\dfrac{x^2 - 5x + 7}{x - 2}$

12. $\dfrac{6x^3 + 3x^2 + 17x + 8}{2x^2 + x + 5}$

13. $\dfrac{6x^3 + 3x^2 + 17x + 8}{x + 2}$

14. $\dfrac{4x^3 + 2x + 9}{2x^3 + x + 3}$

15. $\dfrac{2x^3 + 5x^2 + 4x + 4}{2x^2 + 5x + 4}.$

In Exercises **16–19**, use the Remainder Theorem to decide whether the first polynomial is a factor of the second. If so, what is the other factor?

16. $x - 4,\, 3x^3 - 11x^2 - 10x + 24$

17. $x + 2,\, 5x^3 + 9x^2 - 8x - 10$

18. $x - 3,\, 3x^3 - 13x^2 + 10x + 9$

19. $x + 3,\, 2x^3 + x^2 - 13x + 6$

PROBLEMS

20. Given that
$$\frac{p(x)}{x^2 - 5x + 6} = x + 3,$$
find
$$\frac{p(x)}{x - 3}.$$

21. Given that
$$\frac{p(x)}{x - 4} = x^2 + 7x + 10,$$
find
$$\frac{p(x)}{x^2 + x - 20}.$$

22. Given that
$$\frac{x^4 + x + 1}{x + 1} = x^3 - x^2 + x \quad \text{with remainder } r = 1,$$
find
$$\frac{x^4 + x + 2}{x + 1}.$$

■ Refer to the Remainder Theorem (page 427) to answer Problems **23–26**.

23. Given that
$$\frac{x^3 + 2x + 3}{x - 2} = x^2 + 2x + 6 \quad \text{with remainder } r,$$
find r.

24. Given that
$$\frac{p(x)}{x - 3} = x^2 + 5 \quad \text{with remainder } 2,$$
find $p(x)$.

25. Given that
$$\frac{p(x)}{x - 4}$$
has remainder $r = 7$, find $p(4)$.

26. Given that
$$\frac{x^2 + 3x + 5}{x - a}$$
has remainder $r = 15$, find a.

■ Use the Remainder Theorem (page 427) to find the value of the constant r making the equations in Problems **27–30** identities.

27. $x^2 + 1 = (x - 1)(x + 1) + r$

28. $x^3 + 2x + 3 = (x - 4)(x^2 + 4x + 18) + r$

29.
$$p(x) = (x + 1)(x^4 - 2x^3 + 4x^2 - 5x + 4) + r$$
where $p(x) = x^5 - x^4 + 2x^3 - x^2 - x - 2$.

30.
$$\frac{x^7 - 2x^3 + 1}{x - 2} = q(x) + \frac{r}{x - 2}$$
where $q(x) = x^6 + 2x^5 + 4x^4 + 8x^3 + 14x^2 + 28x + 56$.

REVIEW EXERCISES AND PROBLEMS FOR CHAPTER 13

EXERCISES

■ In Exercises **1–3**, simplify the rational expressions. Assume a is a nonzero constant.

1. $\dfrac{x^2 - 2x - 15}{6 - x - x^2}$

2. $\dfrac{ax^2 + 4x}{2ax}$

3. $\dfrac{3x^2 - 2ax - a^2}{x - a}$

■ In Exercises **4–5**, find the zeros.

4. $f(x) = \dfrac{x^3 + 4x^2 + 3x}{2x - 1}$

5. $f(x) = \dfrac{x^3 + x^2 + 12x}{5x^2 - 2}$

■ In Exercises **6–10**, find the domain.

6. $f(x) = \dfrac{3x + 4}{3x^2 + 7x + 4}$

7. $f(x) = \dfrac{x + 7}{2\pi}$

8. $f(x) = \dfrac{2x + 1}{2x^2 - 5x - 3}$

9. $f(x) = \dfrac{32}{32 - x^5}$

10. $f(x) = \dfrac{2 - x^2}{x^4 - x^2 - 6}$

11. Give the zeros, holes, vertical asymptotes, long-run behavior, and the y-intercept of

$$y = \frac{x^2 - 4x + 3}{x^2 - x - 6}.$$

You are not required to graph this function.

▮Cancel any common factors in the rational expressions in Exercises **12–16**. Are the resulting expressions equivalent to the original expressions?

12. $\dfrac{x^4 - 1}{x^3 + x}$

13. $\dfrac{x^2 + 9x + 20}{x^2 + x - 6}$

14. $\dfrac{11 + x^2}{11x + x^3}$

15. $\dfrac{1 - x}{x^2 + x - 2}$

16. $\dfrac{4x^2 - a^2}{2ax - a^2}$, a a nonzero constant

PROBLEMS

17. At a fund raiser, Greg buys a raffle ticket. If he is the only one to buy a raffle ticket, he is sure to win the prize. If only one other person buys a ticket, his chance of winning is $1/2$. In general, if x tickets are sold, Greg's chance of winning is given by $w = 1/x$.

   (a) Find the horizontal asymptote of $w = 1/x$.
   (b) What is the practical interpretation of the horizontal asymptote?

18. Figure 13.15 gives the graph of the rational function

$$y = \frac{k(x - p)(x - q)}{(x - r)(x - s)}.$$

Based on the graph, find values of k, p, q, r, s given that $p < q$ and $r < s$.

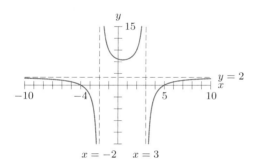

$x = -2 \quad x = 3$

Figure 13.15

19. The rational function $r(x)$ can be written in two forms:

   I. $r(x) = \dfrac{x^2 - 9}{x^2 - 4}$

   II. $r(x) = \dfrac{(x - 3)(x + 3)}{(x - 2)(x + 2)}$

   Which form most readily shows

   (a) The zeros of $r(x)$? What are they?
   (b) The vertical intercept? What is it?
   (c) The vertical asymptotes? What are they?

20. **(a)** Show that the rational function

$$r(x) = \frac{3x + 10}{4x + 8}$$

   can also be written in the form

$$r(x) = \frac{3}{4} + \frac{1}{x + 2}.$$

   (b) Explain how you can determine the horizontal asymptote from the second form, and find the asymptote.

▮Match the following rational functions to the statements about their graphs in Problems **21–23**. A statement may match none, one, or several of the given functions. You are not required to draw any graphs to answer these questions.

(a) $y = \dfrac{x^2 - 4}{x^2 - 9}$

(b) $y = \dfrac{x^2 + 4}{x^2 - 9}$

(c) $y = \dfrac{x^2 - 4}{x^2 + 9}$

(d) $y = \dfrac{x^2 + 4}{x^2 + 9}$

21. The graph has at least one x-intercept.

22. The graph has at least one vertical asymptote.

23. The graph lies entirely above the x-axis.

▮Put the rational expressions in Problems **24–26** into quotient form, identify any horizontal or slant asymptotes, and sketch the graph.

24. $\dfrac{3x^2 + 5x + 7}{3x^2 + 5x + 6}$

25. $\dfrac{x^3 - 2x + 7}{x - 2}.$

26. $\dfrac{10x^4 + 14x^3 - 8x^2 + 10x + 7}{5x^3 + 7x^2 - 4x + 5}.$

27. Without a calculator, match the functions (a)–(f) with their graphs in (i)–(vi) by finding the zeros, asymptotes, and end behavior for each function.

(a) $y = \dfrac{-1}{(x-5)^2} - 1$ **(b)** $y = \dfrac{x-2}{(x+1)(x-3)}$

(c) $y = \dfrac{2x+4}{x-1}$ **(d)** $y = \dfrac{1}{x+1} + \dfrac{1}{x-3}$

(e) $y = \dfrac{1-x^2}{x-2}$ **(f)** $y = \dfrac{1-4x}{2x+2}$

(i)
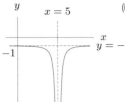

(ii)
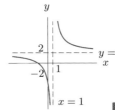

(iii)
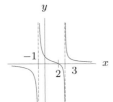

(iv)
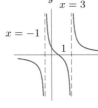

(v)
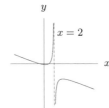

(vi)

■ Graph the functions in Problems **28–29** without a calculator.

28. $y = 2 + \dfrac{1}{x}$ **29.** $y = \dfrac{2x^2 - 10x + 12}{x^2 - 16}$

30. Express as a ratio of two polynomials:

$$\frac{\dfrac{2x^2+1}{x-3} - \dfrac{x-1}{x-4}}{\dfrac{x+5}{x-3} \cdot \dfrac{x^2+2}{x-4}}.$$

31. Find a rational function whose domain is all $x, x \neq 2, x \neq -3$.

■ In Problems **32–34**, write an expression for the function whose graph is identical to the graph of $y = 1/x$ except for a hole at the given x-value. What are the coordinates of the hole?

32. $x = 2$ **33.** $x = 0.02$ **34.** $x = 0.002$

Chapter 14

Summation Notation

CONTENTS

14.1 USING SUBSCRIPTS AND SIGMA NOTATION

Subscript Notation

A survey of 9 local gas stations finds the following list of gas prices in dollars per gallon:

$$1.82, 1.84, 1.78, 1.85, 1.86, 1.87, 1.90, 1.88, 1.86.$$

In working with a list like this it is often useful to use subscript notation, where we let g_1 stand for the first price in the list, g_2 for the second price, and so on up until g_9 for the last price. Thus, for example, we say $g_4 = 1.85$ to indicate that the fourth price is \$1.85, or $g_5 = g_9$ to indicate that the prices at the fifth and ninth gas stations surveyed are the same (\$1.86).

Example 1   Write expressions using subscript notation for

(a) The cost of 5 gallons of gas at the second gas station.
(b) The average price of gas at all stations.

Solution   (a) The cost of gas at the second station is g_2 dollars per gallon, so

$$\text{Cost of 5 gallons at second station} = 5g_2.$$

(b) The average price is obtained by adding up all the prices and dividing by the number of stations:

$$\text{Average cost} = \frac{g_1 + g_2 + g_3 + g_4 + g_5 + g_6 + g_7 + g_8 + g_9}{9}.$$

We often shorten an expression like this using three dots to indicate the missing terms:

$$\text{Average cost} = \frac{g_1 + g_2 + \cdots + g_9}{9}.$$

In the previous example there were 9 items in the list. Sometimes we use a letter to stand for the number of items. For example, we might ask a friend in another city to survey prices at their local gas stations. Assuming there are n stations surveyed, and using dots again to indicate missing values, we can list the prices as

$$g_1, g_2, \ldots, g_n,$$

where

$$g_i = \text{Price of gas at the } i^{\text{th}} \text{ gas station, for } i = 1, 2, \ldots, n.$$

Notice that in this case we have used a subscripted letter i to vary over the numbers from 1 to n.

Example 2   There are n snowstorms in a town one winter. The first storm delivers d_1 inches of snow, the second delivers d_2 inches, and so on until the last storm, which delivers d_n inches.

(a) Write an expression for the total amount of snow that fell all winter.
(b) Write an expression for the total amount of snow that fell in the final three storms.

Solution   (a) We have:

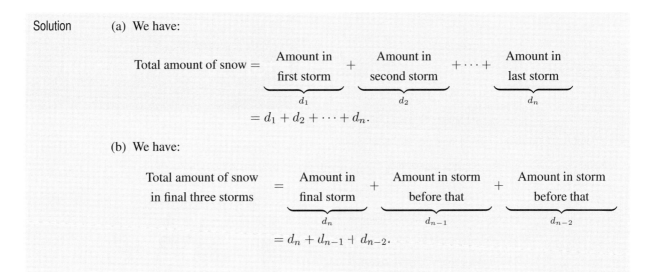

$$= d_1 + d_2 + \cdots + d_n.$$

(b) We have:

$$= d_n + d_{n-1} + d_{n-2}.$$

The word *index* is often used interchangeably with the term subscript.

Sigma Notation

To represent sums, we often use a special symbol, the capital Greek letter Σ (pronounced "sigma"). Using this notation, we write:

$$\sum_{i=1}^{n} a_i = a_1 + a_2 + a_3 + \cdots + a_n.$$

The Σ tells us we are adding up some numbers. The a_i tells us that the numbers we are adding are called a_1, a_2, and so on. The sum begins with a_1 and ends with a_n because the subscript i starts at $i = 1$ (at the bottom of the Σ sign) and ends at $i = n$ (at the top of the Σ sign). Sometimes the a_i are given by a list, as in Example 1, and sometimes they are given by a formula in i, as in the next example.

Example 3   Write out the sum $\displaystyle\sum_{i=1}^{5} i^2$.

Solution   We have

Last value of i is 5
↓

$$\sum_{i=1}^{5} i^2 = 1^2 + 2^2 + 3^2 + 4^2 + 5^2.$$

↑
First value of i is 1

We find the terms in the sum by sequentially giving i the values 1 through 5 in the expression i^2.

Using summation notation, we can write the solution to Example 1(b) as

$$\text{Average price} = \frac{g_1 + g_2 + \cdots + g_9}{9} = \frac{1}{9}(g_1 + g_2 + \cdots + g_9) = \frac{1}{9}\sum_{i=1}^{9} g_i.$$

Example 4   There are n airlines. Airline i owns q_i planes and carries p_i passengers per plane, for $i = 1, 2, \ldots, n$. Using sigma notation, write an expression for the average number of passengers per plane at all n airlines.

Solution   To find the average, we divide the total number of passengers by the total number of planes. Since airline i has q_i planes,

$$\text{Total number of planes from all airlines} = \sum_{i=1}^{n} q_i.$$

Since airline i carries p_i passengers on each of its q_i planes, it carries $p_i q_i$ passengers in total. Thus

$$\text{Total number of passengers from all airlines} = \sum_{i=1}^{n} p_i q_i.$$

So

$$\text{Average number of passengers per plane} = \frac{\sum_{i=1}^{n} p_i q_i}{\sum_{i=1}^{n} q_i}.$$

Example 5   Evaluate the following expressions based on Table 14.1, which gives the first 9 prime numbers.

Table 14.1

p_1	p_2	p_3	p_4	p_5	p_6	p_7	p_8	p_9
2	3	5	7	11	13	17	19	23

(a) $\displaystyle\sum_{i=1}^{5} p_i$ (b) $\displaystyle\sum_{i=4}^{8} p_i$ (c) $\displaystyle 3 + \sum_{i=1}^{5} p_i$

(d) $\displaystyle\sum_{i=1}^{5} (3 + p_i)$ (e) $\displaystyle\sum_{i=2}^{4} 2p_i$ (f) $\displaystyle\sum_{i=1}^{4} (-1)^i p_i$

Solution   (a) Here, we find the terms by assigning successive integers starting at $i = 1$ and going up to $i = 5$. Then we add the terms:

$$\sum_{i=1}^{5} p_i = p_1 + p_2 + p_3 + p_4 + p_5$$

$$= 2 + 3 + 5 + 7 + 11 = 28.$$

(b) Here, we are adding p_i terms starting at $i = 4$ and going up to $i = 8$:

$$\sum_{i=4}^{8} p_i = p_4 + p_5 + p_6 + p_7 + p_8$$

$$= 7 + 11 + 13 + 17 + 19 = 67.$$

(c) Here, we are adding 3 to the sum of p_i terms starting at $i = 1$ and going up to $i = 5$:

$$3 + \sum_{i=1}^{5} p_i = 3 + \overbrace{p_1 + p_2 + p_3 + p_4 + p_5}^{\text{from part (a), equals 28}}$$

$$= 3 + 28 = 31.$$

(d) Here, we are adding terms that look like $3 + p_i$, starting at $i = 1$ and going up to $i = 5$:

$$\sum_{i=1}^{5} (3 + p_i) = (3 + p_1) + (3 + p_2) + \cdots + (3 + p_5)$$

$$= (3 + 2) + (3 + 3) + (3 + 5) + (3 + 7) + (3 + 11)$$

$$= 5 + 6 + 8 + 10 + 14 = 43.$$

Another approach would be to write

$$\sum_{i=1}^{5} (3 + p_i) = (3 + p_1) + (3 + p_2) + \cdots + (3 + p_5)$$

$$= (3 + 3 + 3 + 3 + 3) + \overbrace{p_1 + p_2 + p_3 + p_4 + p_5}^{\text{from part (a), equals 28}} \quad \text{regroup}$$

$$= 15 + 28 = 43.$$

Notice that this is different from the answer we got in (c), so

$$3 + \sum_{i=1}^{5} p_i \quad \text{is not equivalent to} \quad \sum_{i=1}^{5} (3 + p_i).$$

(e) Here, we are adding terms that look like $2p_i$, starting at $i = 2$ and going up to $i = 4$:

$$\sum_{i=2}^{4} 2p_i = 2p_2 + 2p_3 + 2p_4$$

$$= 2 \cdot 3 + 2 \cdot 5 + 2 \cdot 7 = 30.$$

(f) We are adding terms which alternate between negative and positive values:

$$\sum_{i=1}^{4} (-1)^i p_i = -p_1 + p_2 - p_3 + p_4$$

$$= -2 + 3 - 5 + 7 = 3.$$

Example 6   Using sigma notation, write expressions standing for:

(a) The square of the sum of the first 5 prime numbers.

(b) The sum of the squares of the first 5 prime numbers.

Do these expressions have the same value?

Solution   (a) We have:

$$\begin{array}{l}\text{The square of the sum} \\ \text{of the first 5 prime numbers}\end{array} = \left(\text{The sum of the first 5 prime numbers}\right)^2$$

$$= \left(\sum_{k=1}^{5} p_i\right)^2 = (2 + 3 + 5 + 7 + 11)^2 = 28^2 = 784.$$

(b) We have:

$$\begin{array}{l}\text{The sum of the squares} \\ \text{of the first 5 prime numbers}\end{array} = p_1^2 + p_2^2 + p_3^2 + p_4^2 + p_5^2$$

$$= \sum_{k=1}^{5} p_i^2 = 2^2 + 3^2 + 5^2 + 7^2 + 11^2$$

$$= 4 + 9 + 25 + 49 + 121 = 208.$$

The values are not the same. In general, $\left(\displaystyle\sum_{k=1}^{n} p_i\right)^2 \neq \displaystyle\sum_{k=1}^{n} p_i^2$.

Exercises and Problems for Section 14.1

EXERCISES

In Exercises **1–4**, write an expression in sigma notation for each situation.

1. The total expenditure for a week if the daily expenditures are $e_1, e_2, \ldots, e_6, e_7$.

2. The sum of the cubes of the first 5 odd integers.

3. The total rainfall for the month of September if the rainfall on the ith day is r_i.

4. The total weekly caloric intake for a diet that starts with 2000 calories on the first day and decreases by 10% each day, so that on the ith day it is $2000(0.9)^{i-1}$.

Write out the sums in Exercises **5–10**. (You do not need to evaluate them.)

5. $\displaystyle\sum_{i=0}^{5} i^2$

6. $\displaystyle\sum_{i=-2}^{2} (i+1)^2$

7. $\displaystyle\sum_{k=0}^{5} (3k+1)$

8. $\displaystyle\sum_{j=1}^{6} 5(j-3)$

9. $\displaystyle\sum_{j=1}^{4} (-1)^j$

10. $\displaystyle\sum_{n=1}^{5} (-1)^{n-1} 2^{-n}$

Evaluate the sums in Exercises **11–15**.

11. $\displaystyle\sum_{i=1}^{4} 2i$

12. $\displaystyle\sum_{i=1}^{4} 2$

13. $\displaystyle\sum_{i=1}^{4} 2^i$

14. $\displaystyle\sum_{i=1}^{4} (2+i)$

15. $\displaystyle\sum_{i=1}^{4} \frac{i}{2}$

PROBLEMS

16. In a survey of n gas stations, the price of gas at the ith gas station is p_i dollars/gallon. Write an expression for

   (a) The cost of buying 12 gallons of gas at the ith gas station.

   (b) The price of gas at the second to last gas station surveyed.

   (c) The average price at all n gas stations.

■ Are the expressions in Problems **17–25** equivalent?

17. $\displaystyle\sum_{i=1}^{5} a_i$ and $a_5 + a_4 + a_3 + a_2 + a_1$

18. $\displaystyle\sum_{i=0}^{5} a_i$ and $a_1 + a_2 + a_3 + a_4 + a_5$

19. $\displaystyle\sum_{i=0}^{5} 5$ and 30

20. $\displaystyle\sum_{i=0}^{5}(3i - 2)$ and $\displaystyle\sum_{i=1}^{6}(3i - 5)$

21. $2 + \displaystyle\sum_{i=0}^{5}(3i - 2)$ and $\displaystyle\sum_{i=0}^{5} 3i$

22. $\displaystyle\sum_{i=1}^{1} 10$ and 10

23. $\displaystyle\sum_{i=1}^{8}(3i + 2)$ and $\displaystyle\sum_{k=1}^{8}(2 + 3k)$

24. $\displaystyle\sum_{j=1}^{9} 3a_j$ and $3a_1 + 3a_2 + 3a_3 + \cdots + 3a_8 + 3a_9$

25. $\displaystyle\sum_{i=1}^{20}(-1)^i b_n$ and $-b_1 + b_2 + b_3 + \cdots + b_{19} + b_{20}$

Chapter 15

Sequences and Series

CONTENTS

15.1 SEQUENCES

Any ordered list of numbers, such as the years of the Winter Olympics,

$$1924, 1928, 1932, 1936, 1948, 1952, 1956, \ldots, 1988, 1992, 1994, 1998, 2002, 2006$$

is called a *sequence*, and the individual numbers are the *terms* of the sequence. This sequence tells us that the Winter Olympics started in 1924 and continues at four year intervals, except for 1940 and 1944 during World War II and a shift in 1994.[1] A sequence can be a finite list, such as the sequence of past Winter Olympic years, or it can be an infinite list, such as the sequence of positive even integers

$$2, 4, 6, 8, \ldots.$$

Example 1
(a) $0, 1, 4, 9, 16, 25, \ldots$ is the sequence of squares of non-negative integers.
(b) $2, 4, 8, 16, 32, \ldots$ is the sequence of positive integer powers of 2.
(c) $2, 7, 1, 8, 2, 8, \ldots$ is the sequence of digits in the decimal expansion of e, the base of the natural logarithm.
(d) $0.06, 0.07, 0.11, 0.11, 0.14, 0.15$ is the sequence of US wind energy production amounts, in trillions of Btu, for the first 6 years of the 21st century.[2]

Notation for Sequences

We denote the terms of a sequence by

$$a_1, a_2, a_3, \ldots, a_n, \ldots$$

so that a_1 is the first term, a_2 is the second term, and so on. We use a_n to denote the nth or *general term* of the sequence. If there is a pattern in the sequence, we may be able to find a formula for a_n.

Example 2
For the sequence $5, 10, 15, 20, 25, \ldots$

(a) What is a_3? a_6?
(b) If $a_i = 20$ what is i?
(c) If $a_j = 50$ what is j?
(d) What is a possible general formula for the nth term?

Solution
(a) The third term is 15, so $a_3 = 15$. The sixth term requires that we continue the pattern of adding 5 to the previous term, and we see that the sixth term is 30, so $a_6 = 30$.
(b) The value 20 is the fourth term, so $a_4 = 20$, which means $i = 4$.
(c) The value 50 is not shown, but by continuing the pattern to the tenth term, we reach 50, so $a_{10} = 50$, which means $j = 10$.
(d) A possible general formula for the nth term is $a_n = 5n$. The pattern is to add 5 to reach the next term.

$$a_1 = 5 = 1 \cdot 5 = 5$$
$$a_2 = 5 + 5 = 2 \cdot 5 = 10$$

---

[1] www.databaseolympics.com, accessed June 1, 2008.
[2] http://www.census.gov/compendia/statab/tables/08s0897.pdf, accessed June 1, 2008.

$$a_3 = 5 + 5 + 5 = 3 \cdot 5 = 15$$

$$\vdots$$

$$a_n = \underbrace{5 + 5 + \cdots + 5}_{n \text{ times}} = n \cdot 5 = 5n.$$

Example 3   Find the first three terms and the 20th term of each sequence.

(a) $a_n = 2n - 5$ (b) $b_n = \dfrac{n+1}{n}$ (c) $c_n = (-1)^n n^2$

Solution   (a) We see that the terms are

$$a_1 = 2 \cdot 1 - 5 = -3, \quad a_2 = 2 \cdot 2 - 5 = -1, \quad a_3 = 2 \cdot 3 - 5 = 1, \quad a_{20} = 2 \cdot 20 - 5 = 35.$$

(b) Here, the terms are

$$b_1 = \frac{1+1}{1} = 2, \quad b_2 = \frac{2+1}{2} = \frac{3}{2}, \quad b_3 = \frac{3+1}{3} = \frac{4}{3}, \quad b_{20} = \frac{20+1}{20} = \frac{21}{20}.$$

(c) We see that

$$c_1 = (-1)^1 1^2 = -1, \quad c_2 = (-1)^2 2^2 = 4, \quad c_3 = (-1)^3 3^2 = -9, \quad c_{20} = (-1)^{20} 20^2 = 400.$$

This sequence is called *alternating* because the terms alternate in sign.

A sequence can be thought of as a function whose domain is the set of positive integers. Each term of the sequence is an output value for the function, so $a_n = f(n)$.

Example 4   List the first 5 terms of the sequence $a_n = f(n)$, where $f(x) = 3x^2 - 1$.

Solution   In order to find a_1 and a_2 we evaluate $f(1)$ and $f(2)$:

$$a_1 = f(1) = 3(1)^2 - 1 = 2 \quad \text{and} \quad a_2 = f(2) = 3(2)^2 - 1 = 11.$$

Similarly, $a_3 = f(3) = 26$, $a_4 = f(4) = 47$, and $a_5 = f(5) = 74$.

Arithmetic Sequences

The lease of a new car normally has a restriction on the number of miles you can drive each year. A restriction of 15,000 annual miles has end-of-year odometer restrictions that form a sequence, a_n, whose terms are

$$15{,}000, \quad 30{,}000, \quad 45{,}000, \quad 60{,}000, \quad 75{,}000, \quad \ldots.$$

Each term of the sequence is obtained from the previous term by adding 15,000; that is, the difference between successive terms is 15,000. A sequence in which the difference between pairs of successive terms is a fixed quantity is called an *arithmetic sequence*.

Example 5   Which of the following sequences are arithmetic?

(a) $9, 5, 1, -3, -7$

(b) $2, 4, 8, 16, 32$

(c) $2 + p, 5 + p, 8 + p, 11 + p$

(d) $10, 5, 0, 5, 10$

Solution   (a) Each term is obtained from the previous term by subtracting 4. This sequence is arithmetic.

(b) This sequence is not arithmetic because each term is twice the previous term. The differences are 2, 4, 8, 16.

(c) This sequence is arithmetic because 3 is added to each term to obtain the next term.

(d) This is not arithmetic. The difference between the second and first terms is -5, but the difference between the fifth and fourth terms is 5.

We can write a formula for the general term of an arithmetic sequence. Look at the sequence $2, 6, 10, 14, 18, \ldots$ in which the terms increase by 4, and observe that

$$a_1 = 2$$
$$a_2 = 6 = 2 + 1 \cdot 4$$
$$a_3 = 10 = 2 + 2 \cdot 4$$
$$a_4 = 14 = 2 + 3 \cdot 4.$$

When we get to the nth term, we have added $(n - 1)$ copies of 4, so that $a_n = 2 + (n - 1)4$. In general:

For $n \geq 1$, the nth term of an arithmetic sequence is

$$a_n = a_1 + (n - 1)d,$$

where a_1 is the first term, and d is the difference between consecutive terms.

Example 6   For the arithmetic sequence $5, 9, 13, \ldots$, find the 10th term.

Solution   Using the formula, $a_n = a_1 + (n - 1)d$, where $a_1 = 5$ $d = 4$ and $n = 10$, we have

$$a_{10} = 5 + (10 - 1)(4) = 5 + 36 = 41.$$

Example 7   A car dealer leases a slightly-used demo car with 5000 miles on the odometer and includes an annual mileage restriction of 15,000 miles per year.

(a) Write a formula for odometer restriction at the end of the nth year.

(b) If the lease expires after six years what is the odometer restriction when the car is returned?

Solution   (a) At the end of the first year the odometer restriction is the original 5000 miles plus the annual 15,000 allowed miles, so $a_1 = 5000 + 15,000 = 20,000$. Each year the odometer reading increases by 15,000, so $d = 15,000$. Thus, $a_n = 20,000 + (n - 1)15,000$.

(b) At the end of the sixth year the odometer restriction is the first year's 20,000 mile restriction plus five more annual 15,000 restricted miles.

$$a_6 = 20,000 + (6 - 1) \cdot 15,000 = 95,000 \text{ miles.}$$

Arithmetic Sequences and Linear Functions

You may have noticed that the arithmetic sequence for the demo car's odometer reading looks like a linear function. The formula for the nth term, $a_n = 20{,}000 + (n-1)15{,}000$, can be simplified to $a_n = 5000 + 15{,}000n$. This is a linear function with slope $m = 15{,}000$ and initial value $b = 5000$. However, for a sequence we consider only positive integer inputs, whereas a linear function is defined for all values of n. We can think of an arithmetic sequence as a linear function whose domain has been restricted to the positive integers. Therefore the graph of an arithmetic sequence is a set of points that lie on a straight line.

Example 8 Plot the values of the sequence $a_n = 5000 + 15{,}000n$ in Example 7, and the graph of the function of the function $f(x) = 5000 + 15{,}000x$ on the same axes.

Solution See Figure 15.1.

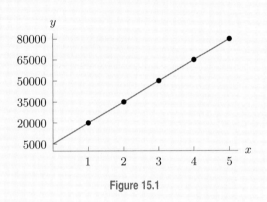

Figure 15.1

Recursively Defined Sequences

A recursively defined sequence is one where each term is defined by an expression involving previous terms.

Example 9 Find the first four terms of the sequence $a_n = a_{n-1} + 3$, and the nth term, when $a_1 = 5$. What sort of sequence is it?

Solution We know the first term $a_1 = 5$, and we use that to find the second term

$$a_2 = a_1 + 3 = 5 + 3 = 8.$$

Now use a_2 to find a_3

$$a_3 = a_2 + 3 = 8 + 3 = 11.$$

Finally, use a_3 to find a_4

$$a_4 = a_3 + 3 = 11 + 3 = 14.$$

Since we add 3 each time to get the next term, the difference between terms is always 3, so this is an arithmetic sequence. We have $a_2 = 5 + 3 \cdot 1$, $a_3 = 5 + 3 \cdot 2$, $a_4 = 5 + 3 \cdot 3$, and, in general,

$$a_n = 5 + 3(n-1).$$

Example 10   Find the first four terms of the sequence $a_n = 3a_{n-1}$, and the nth term, when $a_1 = 4$.

Solution   We use the first term, $a_1 = 4$, to find the second term

$$a_2 = 3a_1 = 3 \cdot 4 = 12.$$

Now use a_2 to find a_3

$$a_3 = 3a_2 = 3 \cdot 12 = 36.$$

Finally, use a_3 to find a_4

$$a_4 = 3a_3 = 3 \cdot 36 = 108.$$

Note that $a_2 = 3 \cdot 4$, $a_3 = 3^2 \cdot 4$, and $a_4 = 3^3 \cdot 4$. In general, we have

$$a_n = 3^{n-1} \cdot 4 = 4 \cdot 3^{n-1}.$$

A well-known recursively defined sequence is the Fibonacci sequence F_n.[3] The sequence starts with $F_1 = F_2 = 1$ and each successive term is created by adding the previous two terms, so $F_n = F_{n-1} + F_{n-2}$. Thus $F_3 = F_2 + F_1 = 1 + 1 = 2$, $F_4 = F_3 + F_2 = 2 + 1 = 3$, $F_5 = F_4 + F_3 = 3 + 2 = 5$, and so on. The first 10 terms of the sequence are

$$1, 1, 2, 3, 5, 8, 13, 21, 34, 55.$$

The Fibonacci sequence satisfies many interesting algebraic identities.

Example 11   Confirm that the Fibonacci identity $F_{2n} = F_{n+1}^2 - F_{n-1}^2$ is true for $n = 3$ and $n = 5$.

Solution   For $n = 3$ we have

$$F_{2 \cdot 3} = F_{3+1}^2 - F_{3-1}^2$$
$$F_6 = F_4^2 - F_2^2$$
$$8 = 3^2 - 1^2$$
$$8 = 8.$$

For $n = 5$ we have

$$F_{2 \cdot 5} = F_{5+1}^2 - F_{5-1}^2$$
$$F_{10} = F_6^2 - F_4^2$$
$$55 = 8^2 - 3^2$$
$$55 = 55.$$

Exercises and Problems for Section 15.1

EXERCISES

■ Are the sequences in Exercises **1–4** arithmetic?

1. $3, 8, 13, 18, \ldots$

2. $4, -3, -9, -14, \ldots$

3. $3, 8, 12, 15, \ldots$

4. $4, -3, -10, -17, \ldots$

---

[3] http://en.wikipedia.org/wiki/Fibonacci, accessed July 20, 2008.

Are the sequences in Exercises **5–7** arithmetic? For those that are, give a formula for the nth term.

5. $2, -2, 4, -4, \ldots$

6. $8, 12, 16, 20 \ldots$

7. $-3, -3.1, -3.2, -3.3, \ldots$

In Exercises **8–11**, find the 5th, 10th, n^{th} term of the arithmetic sequences.

8. $4, 6, 8, \ldots$

9. $5, 7.2, \ldots$

10. $a_1 = 4.7, a_3 = 3.3$

11. $a_3 = 2.7, a_6 = 9$

In Exercises **12–15**, write the first 5 terms of the sequence $a_n = f(n)$.

12. $f(x) = x^2 + 1$

13. $f(t) = 3t - 1$

14. $f(n) = \dfrac{n(n-1)}{2}$

15. $f(a) = a^2 + a + 41$

PROBLEMS

16. For the sequence $2, 5, 8, 11, 14, \ldots$

    (a) What is a_4?
    (b) If $a_i = 5$, what is i?
    (c) If $a_n = a_{n-1} + c$, what is c?
    (d) What is a possible formula for a_n?

17. You buy a car with 35,000 miles on it and each year you drive 7000 miles.

    (a) Write a formula for the mileage at the end of the mth year.
    (b) If the car becomes unusable after 140,000 miles, how many years does it last?

Determine whether the graphs in Problems **18–21** represent an arithmetic sequence. If it is an arithmetic sequence, give the common difference, d, and find a_6.

18.

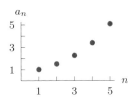

19.

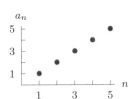

20.

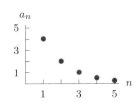

21.

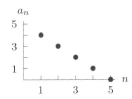

In Problems **22–25**, find the first four terms of the sequence and a formula for the general term.

22. $a_n = a_{n-1} + 2; a_1 = 1$

23. $a_n = a_{n-1} + 4; a_1 = 2$

24. $a_n = -a_{n-1}; a_1 = -1$

25. $a_n = 2a_{n-1}; a_1 = 1$

26. The Fibonacci sequence starts with $1, 1, 2, 3, 5, \ldots$, and each term is the sum of the previous two terms, $F_n = F_{n-1} + F_{n-2}$.

    (a) Find F_6, F_7, F_8 and F_{14}.
    (b) Check the identity: $F_{2n} = F_n(F_{n+1} + F_{n-1})$ for $n = 7$.

27. For a positive integer n, let a_n be the fraction of the US population with income less than or equal to $\$n$ thousand dollars.

    (a) Which is larger, a_{25} or a_{50}? Why?
    (b) What does the quantity $a_{50} - a_{25}$ represent in terms of US population?
    (c) Is there any value of n with $a_n = 0$? Explain.
    (d) What happens to the value of a_n as n increases?

15.2 ARITHMETIC SERIES

United States Municipal Waste Accumulation

Table 15.1 shows the amount of municipal waste generated each year in the US from 1990 to 2001, where a_n is the amount in millions of tons for year n, and $n = 1$ corresponds to 1990.[4]

Table 15.1   *Municipal Waste 1990 to 2001*

Year	1990	1991	1992	1993	1994	1995
Year, starting 1990	1	2	3	4	5	6
Waste (million tons)	269.0	293.6	280.7	291.7	306.9	322.9
Year	1996	1997	1998	1999	2000	2001
Year, starting 1990	7	8	9	10	11	12
Waste (million tons)	326.7	327.5	340.5	374.6	382.6	409.0

For each year, we can calculate the total amount of waste since the start of 1990. For example, at the end of 1992:

Total amount of waste (1990–1992) $= a_1 + a_2 + a_3 = 269.0 + 293.6 + 280.7 = 843.3$ million tons.

In general, we write S_n to denote the sum of the first n terms of any sequence. In this example, 1992 corresponds to $n = 3$, so we have $S_3 = 843.3$ million tons.

Example 1   Find and interpret S_8.

Solution   Since S_8 is the sum of the first 8 terms of the sequence, we have

$$S_8 = a_1 + a_2 + a_3 + a_4 + a_5 + a_6 + a_7 + a_8$$
$$= 269.0 + 293.6 + 280.7 + 291.7 + 306.9 + 322.9 + 326.7 + 327.5$$
$$= 2419.$$

Here, S_8 is the total waste in millions of tons from 1990 to 1997.

Arithmetic Series

The sum of the terms of a sequence is called a *series*. We write S_n for the sum of the first n terms of the sequence, called the nth *partial sum*. We see that S_n is a function of n, the number of terms in the partial sum. In this section we see how to evaluate functions defined by sums.

Example 2   To force a company to comply to pollution standards, judges sometimes impose daily fines that increase. Suppose that a judge imposes a 3 million dollar fine on the first day and increases the fine each day by 2 million dollars until compliance is reached. Using sequence notation, and expressing the daily fine in millions, we write $a_1 = 3$, $a_2 = 5$, $a_3 = 7$, and so on. Find S_4, the total fine in 4 days, and S_n, the sum of the fines for n days.

---

[4]http://www.ZeroWasteAmerica.org/Statistics.htm, Accessed June 5, 2008.

Solution   The total fines in four days is

$$\text{Total fines} = S_4 = a_1 + a_2 + a_3 + a_4 = 3 + 5 + 7 + 9 = 24 \text{ million dollars.}$$

For a longer period of 5 days, the total fine is given by

$$S_5 = a_1 + a_2 + a_3 + a_4 + a_5 = 3 + 5 + 7 + 9 + 11 = 35 \text{ million dollars.}$$

For a fine of n days, S_n, the total fine, is

$$S_n = f(n) = 3 + 5 + 7 + \cdots + (2n + 1).$$

Notice that each daily fine contains two more million dollars than the previous one. Thus, the daily fines,

$$3, 5, 7, 9, 11, 13, \ldots,$$

form an arithmetic sequence. This means that S_n is the sum of the terms of an arithmetic sequence. Such a sum is called an *arithmetic series*. We can recursively define the sequence of fines as follows:

$$\text{Fine on a day} = \text{Fine on previous day} + 2,$$

or

$$a_n = a_{n-1} + 2, \quad \text{where } a_1 = 3.$$

The Municipal Waste series in Example 1 was not arithmetic and must be added term by term. The daily fine for pollution in Example 2 is an arithmetic series and while we could add it term by term, there is a formula to calculate the sum S_n.

The Sum of an Arithmetic Series

We now find a formula for the sum of an arithmetic series. A famous story concerning a series[5] is told about the great mathematician Carl Friedrich Gauss (1777–1855), who as a young boy was asked by his teacher to add the numbers from 1 to 100. He did so almost immediately:

$$S_{100} = 1 + 2 + 3 + \cdots + 100 = 5050.$$

Of course, no one really knows how Gauss accomplished this, but he probably did not perform the calculation directly, by adding 100 terms. He might have noticed that the terms in the sum can be regrouped into pairs, as follows:

$$S_{100} = 1 + 2 + \cdots + 99 + 100 = \underbrace{(1 + 100) + (2 + 99) + \cdots + (50 + 51)}_{\text{50 pairs}}$$
$$= \underbrace{101 + 101 + \cdots + 101}_{\text{50 terms}} \quad \text{each pair adds to 101}$$
$$= 50 \cdot 101 = 5050.$$

[5] As told by E.T. Bell in *The Men of Mathematics*, p. 221 (New York: Simon and Schuster, 1937), the series involved was arithmetic, but more complicated than this one.

The approach of pairing numbers works for the sum from 1 to n, no matter how large n is. Provided n is an even number, we can write

$$S_n = 1 + 2 + \cdots + (n-1) + n = \underbrace{(1+n) + (2+(n-1)) + (3+(n-2)) + \cdots}_{\frac{1}{2}n \text{ pairs}}$$

$$= \underbrace{(1+n) + (1+n) + \cdots}_{\frac{1}{2}n \text{ pairs}}, \qquad \text{each pair adds to } 1 + n$$

so we have the formula:

$$S_n = 1 + 2 + \cdots + n = \frac{1}{2}n(n+1).$$

A similar derivation shows that this formula for S_n also holds for odd values of n.

Example 3 Check this formula for S_n with $n = 100$.

Solution Using the formula, we get the same answer, 5050, as before:

$$S_{100} = 1 + 2 + \cdots + 100 = \frac{1}{2} \cdot 100(100 + 1) = 50 \cdot 101 = 5050.$$

To find a formula for the sum of a general arithmetic series, we first assume that n is even and write

$$S_n = a_1 + a_2 + \cdots + a_n$$
$$= \underbrace{(a_1 + a_n) + (a_2 + a_{n-1}) + (a_3 + a_{n-2}) + \cdots}_{\frac{1}{2}n \text{ pairs}}.$$

In general:

The **sum**, S_n, of the first n terms of the **arithmetic series** with $a_n = a_1 + (n-1)d$ is

$$S_n = \frac{1}{2}n(a_1 + a_n) = \frac{1}{2}n(2a_1 + (n-1)d).$$

The first of the two formulas for $S_n = \frac{1}{2}n(a_1 + a_n)$ is used when we know the first and last terms of the sequence being added. For example, to find S_4, a four day pollution fine, we use $a_1 = 3$ and $a_4 = 9$ for a_1 and a_4 with $n = 4$ to get the same answer as before:

$$S_4 = \frac{1}{2}n(a_1 + a_n) = \frac{1}{2} \cdot 4(a_1 + u_4) = 2(3 + 9) = 24 \text{ million dollars.}$$

The second arithmetic series sum formula, $S_n = \frac{1}{2}n(2a_1 + (n-1)d)$, is useful when the last term is not immediately known. This formula requires that we know a_1, n and d.

Example 4   Using the formula for S_n, calculate the total fine for polluting 7 days .

Solution   We know the first term is $a_1 = 3$, and we have $n = 7$ and $d = 2$. We get

$$S_7 = \frac{1}{2}n\left(2a_1 + (n-1)d\right) = \frac{1}{2} \cdot 7 \left(2 \cdot 3 + (7-1) \cdot 2\right) = \frac{7}{2} \cdot (6 + 12) = 63 \text{ million dollars.}$$

Writing Arithmetic Series Using Sigma Notation

In Chapter 14 we saw how to to represent sums using the Greek letter Σ:

$$\sum_{i=1}^{n} a_i = a_1 + a_2 + a_3 + \cdots + a_n.$$

We can use sigma notation to represent arithmetic series.

Example 5   Use sigma notation to write the sum of the first 17 positive even numbers, and evaluate the sum.

Solution   The positive even integers form an arithmetic sequence: 2, 4, 6, 8, ... with $a_1 = 2$ and $d = 2$. The ith even number is $a_i = 2 + (i-1)2 = 2i$.

$$\text{Sum of the first 17 even numbers} = \sum_{i=1}^{17} a_i = \sum_{i=1}^{17} 2i.$$

Using the formula for the sum of an arithmetic series, with $n = 17$, $a_1 = 2$, and $d = 2$, we get

$$\text{Sum} = \frac{1}{2}n\left(2a_1 + (n-1)d\right) = \frac{1}{2} \cdot 17 \left(2 \cdot 2 + (17-1) \cdot 2\right) = 306.$$

Example 6   Use sigma notation to write the accumulated pollution fine from Example 2 over a period of n days, and evaluate the sum over a period of 5 days.

Solution   The fines form an arithmetic sequence: 3, 5, 7, 9, ... with $a_1 = 3$ and $d = 2$. The ith fine is $a_i = a_1 + (i-1)d = 3 + (i-1)2 = 2i + 1$.

$$\text{Sum of the first } n \text{ fines} = \sum_{i=1}^{n} a_i = \sum_{i=1}^{n} (2i + 1).$$

We evaluate the sum using a formula where $n = 5$, $a_1 = 3$, and $d = 2$:

$$\text{Sum} = \frac{1}{2}n\left(2a_1 + (n-1)d\right) = \frac{1}{2} \cdot 5 \left(2 \cdot 3 + (5-1) \cdot 2\right) = 35 \text{ million dollars.}$$

As illustrated in Example 6, it is sometimes necessary to enclose terms in parentheses to avoid ambiguity. Compare

$$\sum_{i=1}^{5} (2i + 1) = 3 + 5 + 7 + 9 + 11 = 35 \text{ million dollars}$$

with

$$\sum_{i=1}^{5} 2i + 1 = \left(\sum_{i=1}^{5} 2i\right) + 1 = (2 + 4 + 6 + 8 + 10) + 1 = 31 \text{ million dollars.}$$

Exercises and Problems for Section 15.2

EXERCISES

1. Find the sum of the first 500 integers: $1 + 2 + 3 + \cdots + 500$.

■ In Exercises 2–4, complete the tables with the terms of the arithmetic series $a_1, a_2, \ldots, a_n$, and the sequence of partial sums, $S_1, S_2, \ldots, S_n$. State the values of a_1 and d where $a_n = a_1 + (n-1)d$.

2. The table for the sequence $3, 8, 13, 18, \ldots$ is

n	1	2	3	4	5	6	7	8
a_n	3	8	13	18				
S_n	3	11	24	42				

3. The table for the terms of the series $2, 6, 10, \ldots$, is

n	1	2	3	4	5	6	7	8
a_n	2	6	10					
S_n	2	8						

4. The table for the terms of the series is

n	1	2	3	4	5	6	7	8
a_n								
S_n				86	135	195		

■ Evaluate the sums in Problems 5–12 using the formula for the sum of an arithmetic series.

5. $\sum_{i=1}^{15} 3i$

6. $\sum_{i=1}^{30} (2i + 10)$

7. $2.3 + 2.6 + 2.9 + \cdots + 5$

8. $-102 - 92 - 82 - 72 - 62 - \cdots - 12 - 2$

9. $22.5 + 20.5 + 18.5 + \cdots + 2.5 + 0.5$

10. $-4.01 - 4.02 - 4.03 - \cdots - 4.35$

11. $\sum_{n=0}^{15} \left(3 + \frac{1}{2}n\right)$

12. $\sum_{n=1}^{10} (5 - 4n)$

■ In Exercises 13–16, write the sum using sigma notation.

13. $4 + 8 + 12 + 16 + 20 + 24$

14. $7 + 10 + 13 + 16 + 19 + 22$

15. $\frac{3}{2} + 3 + \frac{9}{2} + 6 + \frac{15}{2}$

16. $40 + 35 + 30 + 25 + 20 + 15 + 10 + 5$

17. (a) Use sigma notation to write the sum $2 + 4 + 6 + \cdots + 18$.
    (b) Use the arithmetic series sum formula to find the sum in part (a).

PROBLEMS

18. Table 15.2 shows data of AIDS deaths[6] that occurred in the US where a_n is the number of deaths in year n, and $n = 1$ corresponds to 2001.

    (a) Find the partial sums S_3, S_4, S_5, S_6.
    (b) Find $S_5 - S_4, S_6 - S_5$.
    (c) Use your answer to part (b) to explain the value of $S_{n+1} - S_n$ for any positive integer n.

Table 15.2

Year	2001	2002	2003	2004	2005	2006
n	1	2	3	4	5	6
Deaths	17,402	16,948	16,690	16,395	16,268	14,016

[6]http://www.cdc.gov/hiv/resources/factsheets/print/us.htm, accessed August 10, 2008.

19. Table 15.3 shows US Census figures, in millions. Interpret these figures as partial sums, S_n, of a sequence, a_n, where n is the number of decades since 1940, so $S_1 = 150.7$, $S_2 = 179.3$, $S_3 = 203.3$, and so on.

Table 15.3

Year	1950	1960	1970	1980	1990	2000
n	1	2	3	4	5	6
Population	150.7	179.3	203.3	226.6	248.7	281.4

Find and interpret the following in terms of populations:

(a) S_5, S_6 **(b)** a_5, a_6 **(c)** $a_5/10$

20. Find the 30th positive multiple of 6 and the sum of the first 30 positive multiples of 6.

◾ In Problems **21–24**, for each of the four series (a)-(d), identify which have the same number of terms and which have the same value.

21. (a) $\displaystyle\sum_{i=1}^{5} i^2$ **(b)** $\displaystyle\sum_{j=0}^{4}(j+1)^2$
(c) $5 + 4 + \ldots + 0$ **(d)** $5^2 + 4^2 + \ldots + 1^2$

22. (a) $\displaystyle\sum_{i=4}^{20} i$ **(b)** $\displaystyle\sum_{j=1}^{16}(-j)$
(c) $0 - 1 - 2 - \ldots - 16$ **(d)** $48 + 50 + 52 + 54$

23. (a) $\displaystyle\sum_{i=1}^{20} 3$ **(b)** $\displaystyle\sum_{j=1}^{20}(3j)$
(c) $0 + 3 + 6 \ldots + 60$ **(d)** $60 + 57 + 54 + \cdots + 6 + 3$

24. (a) $\displaystyle\sum_{i=1}^{10} 2i$ **(b)** $\displaystyle\sum_{j=0}^{10} 2j$
(c) $2 + 4 + 6 + \ldots + 20$ **(d)** $20 + 18 + \ldots + 2$

25. The Fibonacci sequence starts with $1, 1, 2, 3, 5, \ldots$, and each term is the sum of the previous two terms, $F_n = F_{n-1} + F_{n-2}$. Write the formula for F_n in sigma notation.

26. Two brothers are dividing M&Ms between themselves. The older one gives one to his younger brother and takes one for himself. He gives another to his younger brother and takes two for himself. He gives a third one to younger brother and takes three for himself, and so on.

(a) On the nth round, how many M&Ms does the older boy give his younger brother? How many does he take himself?

(b) After n rounds, how many M&Ms does his brother have? How many does the older boy have?

27. An auditorium has 20 seats in the first row, 24 seats in the second row, 28 seats in the third row, and so on. If there are fifteen rows in the auditorium, how many seats are there in the last row? How many seats are there in the auditorium?

28. The Temple of Heaven in Beijing is a circular structure with three concentric tiers. At the center of the top tier is a round flagstone. A series of concentric circles made of flagstone surrounds this center stone. The first circle has nine stones, and each circle after that has nine more stones than the last one. If there are nine concentric circles on each of the temple's three tiers, then how many stones are there in total (including the center stone)?

15.3 GEOMETRIC SEQUENCES AND SERIES

Geometric Sequences

Suppose you are offered a job at a salary of \$40,000 for the first year with a 4% pay raise every year. Under this plan, your annual salaries form a sequence with terms

$$a_1 = 40{,}000$$
$$a_2 = 40{,}000(1.04) = 41{,}600$$
$$a_3 = 41{,}600(1.04) = 40{,}000(1.04)^2 = 43{,}264$$
$$a_4 = 43{,}264(1.04) = 40{,}000(1.04)^3 = 44{,}994.56,$$

and so on, where each term is obtained from the previous one by multiplying by 1.04. A sequence in which each term is a constant multiple of the preceding term is called a *geometric sequence*. In a geometric sequence, the ratio of successive terms is constant.

Example 1   Which of the following sequences are geometric?

(a) $10, 100, 1000, 10{,}000, \ldots$

(b) $10, 110, 210, 310, 410, \ldots$

(c) $100, 50, 25, 12.5, \ldots$

(d) $5, 25, 125, 625, \ldots$

Solution

(a) This sequence is geometric. Each term is 10 times the previous term. Note that the ratio of any term to its predecessor is 10. For example, $1000/100 = 10$.

(b) This sequence is not geometric. The difference between consecutive terms is 100. It is arithmetic. Notice the ratio between consecutive terms is not constant. For example, $110/10 = 11$ and $210/110 = 1.909 \neq 11$.

(c) This sequence is geometric. Each term is $1/2$ the previous one, so the ratio of successive terms is always $1/2$:

$$\frac{a_2}{a_1} = \frac{50}{100} = \frac{1}{2}, \quad \text{and} \quad \frac{a_3}{a_2} = \frac{25}{50} = \frac{1}{2}, \ldots.$$

(d) This sequence is geometric. Each term is 5 times the previous term. Note that the ratio of successive terms is 5.

As with arithmetic sequences, there is a formula for the general term of a geometric sequence. Consider the sequence $30, 300, 3000, 30{,}000, \ldots$ in which each term is 10 times the previous term. We have

$$a_1 = 30$$
$$a_2 = 300 = 30(10)$$
$$a_3 = 3000 = 30(10)^2$$
$$a_4 = 30{,}000 = 30(10)^3.$$

When we get to the nth term, we have multiplied 30 by $(n - 1)$ factors of 10, so that $a_n = 30(10)^{n-1}$. In general,

---

For $n \geq 1$, the nth term of a geometric sequence is

$$a_n = a_1 r^{n-1},$$

where a_1 is the first term, and r is the ratio of successive terms.

Example 2   For the geometric sequence 6, 18, 54, 162, 486, ..., identify a_1 and r, then write the nth term of the sequence.

Solution   The first term is 6, so $a_1 = 6$, the ratio of terms is 3, so $r = 3$ and we write

$$a_n = 6(3)^{n-1}.$$

Example 3

(a) Write a formula for the general term of the salary sequence which starts at \$40,000 and increases by 4% each year.

(b) What is your salary after 10 years on the job?

Solution   (a) Your starting salary is $40,000, so $a_1 = 40{,}000$. Each year your salary increases by 4%, so $r = 1.04$. Thus, $a_n = 40{,}000(1.04)^{n-1}$.

(b) After 10 years on the job, you are at the start of your 11th year, so $n = 11$. Your salary is

$$a_{11} = 40{,}000(1.04)^{11-1} \approx 59{,}210 \text{ dollars.}$$

Geometric Sequences and Exponential Functions

The formula for the salary sequence, $a_n = 40{,}000(1.04)^{n-1}$, looks like a formula for an exponential function, $f(t) = ab^t$. A geometric sequence is an exponential function whose domain is restricted to the positive integers. For many applications, this restricted domain is more realistic than an interval of real numbers. For example, salaries are usually increased once a year, rather than continuously.

Geometric Series

In the previous section, we studied arithmetic series, obtained by summing the terms in an arithmetic sequence. In this section, we study *geometric series*, obtained by adding the terms of a geometric sequence.

Bank Balance

A person saving for retirement deposits $5000 every year in an IRA (Individual Retirement Account) that pays 6% interest per year, compounded annually. After the first deposit (but before any interest has been earned), the balance in the account in dollars is

$$B_1 = 5000.$$

After 1 year has passed, the first deposit has earned interest, so the balance becomes $5000(1.06)$ dollars. Then the second deposit is made and the balance becomes

$$B_2 = \underbrace{2^{\text{nd}} \text{ deposit}}_{5000} + \underbrace{1^{\text{st}} \text{ deposit with interest}}_{5000(1.06)}$$
$$= 5000 + 5000(1.06) \text{ dollars.}$$

After 2 years have passed, the third deposit is made, and the balance is

$$B_3 = \underbrace{3^{\text{rd}} \text{ deposit}}_{5000} + \underbrace{2^{\text{nd}} \text{ dep. with 1 year interest}}_{5000(1.06)} + \underbrace{1^{\text{st}} \text{ deposit with 2 years interest}}_{5000(1.06)^2}$$
$$= 5000 + 5000(1.06) + 5000(1.06)^2.$$

Let B_n be the balance in dollars after n deposits. Then we see that

After 4 deposits $B_4 = 5000 + 5000(1.06) + 5000(1.06)^2 + 5000(1.06)^3$

After 5 deposits $B_5 = 5000 + 5000(1.06) + 5000(1.06)^2 + 5000(1.06)^3 + 5000(1.06)^4$

$$\vdots$$

After n deposits $B_n = 5000 + 5000(1.06) + 5000(1.06)^2 + \cdots + 5000(1.06)^{n-1}.$

Example 4   How much money is in this IRA at the start of year 6, right after a deposit is made? How much money is in this IRA at the start of year 26, right after a deposit is made?

Solution   At the beginning of year 6, we have made 6 deposits. We have

$$B_6 = 5000 + 5000(1.06) + 5000(1.06)^2 + \cdots + 5000(1.06)^5$$
$$= \$34,876.59.$$

At the beginning of year 26, we have made 26 deposits. Even using a calculator, it would be tedious to evaluate B_{26} by adding 26 terms. Fortunately, there is an algebraic shortcut. Start with the formula for B_{26}:

$$B_{26} = 5000 + 5000(1.06) + 5000(1.06)^2 + \cdots + 5000(1.06)^{25}.$$

Multiply both sides of this equation by 1.06 and then add 5000, giving

$$1.06B_{26} + 5000 = 1.06\left(5000 + 5000(1.06) + 5000(1.06)^2 + \cdots + 5000(1.06)^{25}\right) + 5000.$$

We simplify the right-hand side to get

$$1.06B_{26} + 5000 = 5000(1.06) + 5000(1.06)^2 + \cdots + 5000(1.06)^{25} + 5000(1.06)^{26} + 5000.$$

Notice that the right-hand side of this equation and the formula for B_{26} have almost every term in common. We can rewrite this equation as

$$1.06B_{26} + 5000 = \underbrace{5000 + 5000(1.06) + 5000(1.06)^2 + \cdots + 5000(1.06)^{25}}_{B_{26}} + 5000(1.06)^{26}$$
$$= B_{26} + 5000(1.06)^{26}.$$

Solving for B_{26} gives

$$1.06B_{26} - B_{26} = 5000(1.06)^{26} - 5000$$
$$0.06B_{26} = 5000(1.06)^{26} - 5000$$
$$B_{26} = \frac{5000(1.06)^{26} - 5000}{0.06}.$$

Using a calculator to evaluate this expression for B_{26}, we find that $B_{26} = 295{,}781.91$ dollars.

Formula for a Geometric Series

We see that the formula for the IRA balance after 25 years, or 26 deposits, is an example of a geometric series with $a_1 = 5000$ and $r = 1.06$,

$$B_{26} = \sum_{i-0}^{25} 5000(1.06)^i = 5000 + 5000(1.06) + 5000(1.06)^2 + \cdots + 5000(1.06)^{25},$$

In general, a geometric series is the sum of the terms of a geometric sequence—that is, in which each term is a constant multiple of the preceding term. Each term contains a_1, and for convenience, we drop the subscript notation and use a in place of a_1.

A **geometric series** is a sum of the form

$$S_n = a + ar + ar^2 + \cdots + ar^{n-1} = \sum_{i=0}^{n-1} ar^i.$$

Notice that S_n is defined to contain exactly n terms. Since the first term is $a = ar^0$, we stop at ar^{n-1}. For instance, the series

$$\sum_{i=0}^{25} 5000(1.06)^i = 5000 + 5000(1.06) + 5000(1.06)^2 + \cdots + 5000(1.06)^{25}$$

contains 26 terms, so $n = 26$. For this series, $r = 1.06$ and $a = 5000$.

Example 5 Given the geometric series,

$$2000 + 2000(1.03) + 2000(1.03)^2 + \cdots + 2000(1.03)^{29}$$

find a, r, and n and $\displaystyle\sum_{i=0}^{n-1} ar^i$.

Solution The first term is 2000 so $a_1 = 2000$, the ratio is $r = 1.03$, and $n = 30$. We can write

$$2000 + 2000(1.03) + 2000(1.03)^2 + \cdots + 2000(1.03)^{29} = \sum_{i=0}^{29} 2000(1.03)^i.$$

The Sum of a Geometric Series

Suppose we have a geometric series such as Example 4: can we find a shortcut formula to algebraically find the sum? We use the algebraic technique from Example 4. Let S_n be the sum of a geometric series of n terms, so that

$$S_n = a + ar + ar^2 + \cdots + ar^{n-1}.$$

Multiply both sides of this equation by r and add a, giving

$$rS_n + a = r\left(a + ar + ar^2 + \cdots + ar^{n-1}\right) + a$$
$$= \left(ar + ar^2 + ar^3 + \cdots + ar^{n-1} + ar^n\right) + a.$$

The right-hand side can be rewritten as

$$rS_n + a = \underbrace{a + ar + ar^2 + \cdots + ar^{n-1}}_{S_n} + ar^n$$
$$= S_n + ar^n.$$

Solving the equation $rS_n + a = S_n + ar^n$ for S_n gives

$$rS_n - S_n = ar^n - a$$

$$S_n(r - 1) = ar^n - a \qquad \text{factoring out } S_n$$

$$S_n = \frac{ar^n - a}{r - 1}$$

$$= \frac{a(r^n - 1)}{r - 1}.$$

By multiplying the numerator and denominator by -1, this formula can be rewritten as follows:

The sum of a **geometric series of n terms** is given by

$$S_n = a + ar + ar^2 + \cdots + ar^{n-1} = \frac{a(1 - r^n)}{1 - r}, \qquad \text{for } r \neq 1.$$

Example 6   In Example 4 we found $B_6 = 34{,}876.59$ and $B_{26} = 295{,}781.91$. Use the general formula for the sum of a geometric series to solve for B_6 and B_{26}.

Solution   We need to find B_6 and B_{26} where

$$B_n = 5000 + 5000(1.06) + 5000(1.06)^2 + \cdots + 5000(1.06)^{n-1}.$$

Using the formula for S_n with $a = 5000$ and $r = 1.06$, we get the same answers as before:

$$B_6 = \frac{5000(1 - (1.06)^6)}{1 - 1.06} = 34{,}876.59,$$

$$B_{26} = \frac{5000(1 - (1.06)^{26})}{1 - 1.06} = 295{,}781.91.$$

Exercises and Problems for Section 15.3

EXERCISES

■Are the sequences in Exercises **1–6** geometric? For those that are, give a formula for the nth term.

1. $8, 4, 2, 1, \frac{1}{2}, \frac{1}{4}, \ldots$

2. $4, 20, 100, 500, \ldots$

3. $1, \frac{1}{4}, \frac{1}{8}, \frac{1}{32}, \ldots$

4. $-2, 4, -8, 16, \ldots$

5. $2, 0.2, 0.02, 0.002, \ldots$

6. $1, \dfrac{1}{1.5}, \dfrac{1}{(1.5)^2}, \dfrac{1}{(1.5)^3}, \ldots$

■In Exercises **7–10**, find the 6th and n^{th} terms of the geometric sequence.

7. $1, 3, 9, \ldots$

8. $9, 6.75, \ldots$

9. $a_1 = 3, a_3 = 27$

10. $a_2 = 6, a_4 = 96$

■In Exercises **11–14**, is the series geometric? If so, give the number of terms and the ratio between successive terms. If not, explain why not.

11. $2 + 1 + \dfrac{1}{2} + \dfrac{1}{4} + \dfrac{1}{8} + \cdots + \dfrac{1}{128}$

12. $1 - \dfrac{1}{2} + \dfrac{1}{4} - \dfrac{1}{8} + \dfrac{1}{16} - \cdots + \dfrac{1}{256}$

13. $1 + \dfrac{1}{2} + \dfrac{1}{3} + \dfrac{1}{4} + \dfrac{1}{5} + \cdots + \dfrac{1}{50}$

14. $5 - 10 + 20 - 40 + 80 - \cdots - 2560$

■ Find the sum of the series in Exercises **15–19**.

15. $5 + \dfrac{5}{2} + \dfrac{5}{2^2} + \dfrac{5}{2^3} + \cdots + \dfrac{5}{2^{10}}$

16. $3 + 15 + 75 + 375 + \cdots + 3(5^8)$

17. $1/81 + 1/27 + 1/9 + \cdots + 243$

18. $\displaystyle\sum_{n=1}^{10} 5(2^n)$

19. $\displaystyle\sum_{k=0}^{6} 2\left(\dfrac{3}{4}\right)^k$

■ How many terms are there in the series in Exercises **20–21**? Find the sum.

20. $\displaystyle\sum_{j=6}^{18} 3 \cdot 2^j$

21. $\displaystyle\sum_{k=2}^{20} (-1)^k 4(0.8)^k$

PROBLEMS

22. In 2007, US natural gas consumption was 652.9 billion cubic meters. Asian consumption was 447.8 billion cubic meters.[7] During the previous decade, US consumption increased by only 0.14% a year, while Asian consumption grew by 5.93% a year. Assume these rates continue into the future.

   (a) Give the first four terms of the sequence, a_n, giving US consumption of natural gas n years after 2006.

   (b) Give the first four terms of a similar sequence b_n showing Asian gas consumption.

   (c) According to this model, when will Asian yearly gas consumption exceed US consumption?

23. Some people believe they can make money from a chain letter (they are usually disappointed). A chain letter works roughly like this: A letter arrives with a list of five names attached and instructions to mail a copy to five friends and to send $1 to the top name on the list. When you mail the five letters, you remove the top name (to whom the money was sent) and add your own name to the bottom of the list.

   (a) If no one breaks the chain, how much money do you receive?

   (b) Assuming no one broke the chain, how many total letters were written with your name on them?

24. Before email made it easy to contact many people quickly, groups used telephone trees to pass news to their members. In one group, each person is in charge of calling 3 people. One person starts the tree by calling 3 people. At the second stage, each of these 3 people calls 3 new people. In the third stage, each of the people in stage two calls 3 new people, and so on.

   (a) How many people have the news by the end of the 5th stage?

   (b) Write a formula for the total number of members in a tree of 10 stages.

   (c) How many stages are required to cover a group with 1000 members?

25. A ball is dropped from a height of 20 feet and bounces. Each bounce is $3/4$ of the height of the bounce before. Thus after the ball hits the floor for the first time, the ball rises to a height of $20(3/4) = 15$ feet, and after it hits the floor for the second time, it rises to a height of $15(3/4) = 20(3/4)^2 = 11.25$ feet.

   (a) Find an expression for the height to which the ball rises after it hits the floor for the nth time.

   (b) Find an expression for the total vertical distance the ball has traveled when it hits the floor for the first, second, third, and fourth times.

   (c) Find an expression for the total vertical distance the ball has traveled when it hits the floor for the nth time. Express your answer in closed form.

15.4 APPLICATIONS OF SERIES

There are many applications of series, in particular when a sequence of data values needs to be added to get a total or partial total.

Falling Objects

Example 1 If air resistance is neglected, a falling object travels 16 ft during the first second after its release, 48 ft during the next, 80 ft during the next, and so on. These distances form the arithmetic sequence $16, 48, 80, \ldots$. In this sequence, $a_1 = 16$ and $d = 32$.

[7]www.bp.com, Statistical Review of World Energy 2005, accessed July 17, 2008.

(a) Find a formula for the nth term in the sequence of distances. Calculate the fourth and fifth terms.

(b) Calculate S_1, S_2, and S_3, the total distance an object falls in 1, 2, and 3 seconds, respectively.

(c) Give a formula for S_n, the distance fallen in n seconds.

Solution

(a) The nth term is $a_n = a_1 + (n-1)d = 16 + 32(n-1) = 32n - 16$. Thus, the fourth term is $a_4 = 32 \cdot 4 - 16 = 112$. The value of $a_5 = 32 \cdot 5 - 16 = 144$.

(b) Since $a_1 = 16$ and $d = 32$,

$$S_1 = a_1 = 16 \text{ feet}$$
$$S_2 = a_1 + a_2 = a_1 + (a_1 + d) = 16 + 48 = 64 \text{ feet}$$
$$S_3 = a_1 + a_2 + a_3 = 16 + 48 + 80 = 144 \text{ feet}.$$

(c) The formula for S_n is $S_n = \frac{1}{2}n\left(2a_1 + (n-1)d\right)$. Since $a_1 = 16$ and $d = 32$,

$$S_n = \frac{1}{2}n(2 \cdot 16 + (n-1)32) = \frac{1}{2}n(32 + 32n - 32)$$
$$= \frac{1}{2}n(32n) = 16n^2.$$

We can check our answer to part (b) using this formula:

$$S_1 = 16 \cdot 1^2 = 16, \quad S_2 = 16 \cdot 2^2 = 64, \quad S_3 = 16 \cdot 3^2 = 144.$$

Compound Interest with Payments

Example 2

A bank account in which interest is earned at 3% per year, compounded annually, starts with a balance of $50,000. Payments of $1000 are made out of the account once a year for ten years, starting today. Interest is earned right before each payment is made. What is the balance in the account right after the tenth payment is made?

Solution

Let B_n be the balance in the account right after the nth payment is made. Then

$$B_1 = 50{,}000 - 1000 = 49{,}000$$
$$B_2 = B_1(1.03) - 1000 = 49{,}000(1.03) - 1000$$
$$B_3 = B_2(1.03) - 1000 = 49{,}000(1.03)^2 - 1000(1.03) - 1000$$
$$B_4 = B_3(1.03) - 1000 = 49{,}000(1.03)^3 - 1000((1.03)^2 + 1.03 + 1)$$
$$B_5 = 49{,}000(1.03)^4 - 1000((1.03)^3 + (1.03)^2 + 1.03 + 1)$$
$$\vdots$$
$$B_{10} = 49{,}000(1.03)^9 - 1000((1.03)^8 + (1.03)^7 + \cdots + 1).$$

Excluding the first term, we can use the formula for the sum of a finite geometric series, where

$a = -1000$, $r = 1.03$ and $n = 9$, so we have

$$B_{10} = 49{,}000(1.03)^9 - \frac{1000(1 - (1.03)^9)}{1 - 1.03} = 53{,}774.78 \text{ dollars.}$$

Drug Levels in the Body

Geometric series arise naturally in many different contexts. The following example illustrates a geometric series with decreasing terms.

Example 3 A patient is given a 250 mg injection of a therapeutic drug. Each day, the patient's body metabolizes 20% of the drug present, so that after 1 day 80%, or 4/5, of the original amount remains, after 2 days only 16/25 remains, and so on. The patient is given a 250 mg injection of the drug every day at the same time. Write a geometric series that gives the drug level in this patient's body right after the nth injection.

Solution Immediately after the 1st injection, the drug level in the body is given by

$$Q_1 = 250.$$

One day later, the original 250 mg has fallen to $250 \cdot \frac{4}{5} = 200$ mg and the second 250 mg injection is given. Right after the second injection, the drug level is given by

$$Q_2 = \underbrace{2^{\text{nd}} \text{ injection}}_{250} + \underbrace{\text{Residue of } 1^{\text{st}} \text{ injection}}_{\frac{4}{5} \cdot 250}$$

$$= 250 + 250 \left(\frac{4}{5} \right) = 450.$$

Two days later, the original 250 mg has fallen to $(250 \cdot \frac{4}{5}) \cdot \frac{4}{5} = 250(\frac{4}{5})^2 = 160$ mg, the second 250 mg injection has fallen to $250 \cdot \frac{4}{5} = 200$ mg, and the third 250 mg injection is given. Right after the third injection, the drug level is given by

$$Q_3 = \underbrace{3^{\text{rd}} \text{ injection}}_{250} + \underbrace{\text{Residue of } 2^{\text{nd}} \text{ injection}}_{250 \cdot \frac{4}{5}} + \underbrace{\text{Residue of } 1^{\text{st}} \text{ injection}}_{250 \cdot \frac{4}{5} \cdot \frac{4}{5}}$$

$$= 250 + 250 \left(\frac{4}{5} \right) + 250 \left(\frac{4}{5} \right)^2 = 610.$$

Continuing, we see that

$$\text{After } 4^{\text{th}} \text{ injection} \quad Q_4 = 250 + 250 \left(\frac{4}{5} \right) + 250 \left(\frac{4}{5} \right)^2 + 250 \left(\frac{4}{5} \right)^3$$

$$\text{After } 5^{\text{th}} \text{ injection} \quad Q_5 = 250 + 250 \left(\frac{4}{5} \right) + 250 \left(\frac{4}{5} \right)^2 + \cdots + 250 \left(\frac{4}{5} \right)^4$$

$$\vdots$$

$$\text{After } n^{\text{th}} \text{ injection} \quad Q_n = 250 + 250 \left(\frac{4}{5} \right) + 250 \left(\frac{4}{5} \right)^2 + \cdots + 250 \left(\frac{4}{5} \right)^{n-1}.$$

This is another example of a geometric series. Here, $a = 250$ and $r = 4/5$ in the geometric series formula

$$Q_n = a + ar + ar^2 + \cdots + ar^{n-1}.$$

To calculate the drug level for a specific value of n, we use the formula for the sum of a geometric series.

Example 4   What quantity of the drug remains in the patient's body of Example 3 after the 10th injection?

Solution   After the 10th injection, the drug level in the patient's body is given by

$$Q_{10} = 250 + 250\left(\frac{4}{5}\right) + 250\left(\frac{4}{5}\right)^2 + \cdots + 250\left(\frac{4}{5}\right)^9.$$

Using the formula for S_n with $n = 10$, $a = 250$, and $r = 4/5$, we get

$$Q_{10} = \frac{250(1 - (\frac{4}{5})^{10})}{1 - \frac{4}{5}} = 1115.782 \text{ mg}.$$

By continuing for another 10 days the level of remaining drug does not increase dramatically and in the long run reaches a plateau near 1250 mg.

Suppose the patient from Example 4 receives injections over a long period of time. Can we find this long-run drug level in the patient's body? One way to think about the patient's drug level over time is to consider an *infinite geometric series*. After n injections, the drug level is given by the sum of the finite geometric series

$$Q_n = 250 + 250\left(\frac{4}{5}\right) + 250\left(\frac{4}{5}\right)^2 + \cdots + 250\left(\frac{4}{5}\right)^{n-1} = \frac{250(1 - (\frac{4}{5})^n)}{1 - \frac{4}{5}}.$$

What happens to the value of this sum as the number of terms approaches infinity? It does not seem possible to add up an infinite number of terms. However, we can look at the *partial sums*, Q_n, to see what happens for large values of n. For large values of n, we see that $(4/5)^n$ is very small, so that

$$Q_n = \frac{250\,(1 - \text{Small number})}{1 - \frac{4}{5}} = \frac{250\,(1 - \text{Small number})}{1/5}.$$

We write $\to$ to mean "approaches." Thus, as $n \to \infty$, we know that $(4/5)^n \to 0$, so

$$Q_n \to \frac{250(1 - 0)}{1/5} = \frac{250}{1/5} = 1250 \text{ mg}.$$

The Sum of an Infinite Geometric Series

Consider the geometric series $S_n = a + ar + ar^2 + \cdots + ar^{n-1}$. In general, if $|r| < 1$, then $r^n \to 0$ as $n \to \infty$, so

$$S_n = \frac{a(1 - r^n)}{1 - r} \to \frac{a(1 - 0)}{1 - r} = \frac{a}{1 - r} \text{ as } n \to \infty.$$

Thus, if $|r| < 1$, the partial sums S_n approach a finite value, S, as $n \to \infty$. In this case, we say that the series *converges* to S.

For $|r| < 1$, the **sum of the infinite geometric series** is given by

$$S = a + ar + ar^2 + \cdots + ar^n + \cdots = \sum_{i=0}^{\infty} ar^i = \frac{a}{1-r}.$$

On the other hand, if $|r| \geq 1$, the above formula does not apply and we say that the series does not converge. For example, with $|r| > 1$, the terms get larger and larger as $n \to \infty$, so adding infinitely many of them does not give a finite sum. If $r = 1$ then $S_n = a \cdot n$, or if $r = -1$ the partial sums alternate between a and 0, thus neither case converges.

Example 5   Suppose you decide to cut back on your chocolate chip cookie consumption and every day you vow to eat half as much as the day before. If you eat 2 cookies on the first day, how many whole cookies do you need so you can follow this diet for the rest of your life?

Solution   Assume that the sum is infinite, since we do not now the exact number of days. We have the series:

$$S = 2 + 1 + \frac{1}{2} + \frac{1}{4} + \frac{1}{8} + \cdots.$$

Using the formula for S with $a = 2$, and $r = 1/2$, we get

$$S = \frac{2}{1 - \frac{1}{2}} = 4 \text{ cookies.}$$

Exercises and Problems for Section 15.4

EXERCISES

In Exercises 1–5, write each of the repeating decimals as a fraction using the following technique. To express $0.232323\ldots$ as a fraction, write it as a geometric series $0.232323\ldots = 0.23 + 0.23(0.01) + 0.23(0.01)^2 + \cdots$, with $a = 0.23$ and $r = 0.01$. Use the formula for the sum of an infinite geometric series to find

$$S = \frac{0.23}{1 - 0.01} = \frac{0.23}{0.99} = \frac{23}{99}.$$

1. $0.235235235\ldots$

2. $6.19191919\ldots$

3. $0.12222222\ldots$

4. $0.4788888\ldots$

5. $0.7638383838\ldots$

PROBLEMS

Problems 6–9 refer to the falling object of Example 1 on page 457, where we found that the total distance, in feet, that an object falls in n seconds is given by $S_n = 16n^2$.

6. **(a)** Find the total distance that an object falls in 4, 5, 6 seconds.
   (b) If the object falls from 1000 feet at time $t = 0$. Calculate its height at $t = 4$, $t = 5$, $t = 6$ seconds.

7. **(a)** Find the total distance that an object falls in 7, 8, 9 seconds.
   (b) If the object falls from 1000 feet at time $t = 0$. Calculate its height at $t = 7$, $t = 8$, $t = 9$ seconds.
   (c) Does part (b) make sense physically?

8. The formula for S_n is defined for positive integers but also can be written as a function $f(n)$, where $n \geq 0$. Find and interpret $f(3.5)$ and $f(7.9)$.

9. Using the function $f(n)$ from Problem 8, determine the following

   (a) If the object falls from 1000 feet, how long does it take to hit the ground?
   (b) If the object falls from twice the height, 2000 feet, how long does it take to hit the ground?
   (c) Is the falling time twice as long?

10. You have an ear infection and are told to take a 250 mg tablet of ampicillin (a common antibiotic) four times a day (every six hours). It is known that at the end of six hours, about 1.6% of the drug is still in the body.[8] Let Q_n be the quantity, in milligrams, of ampicillin in the body right after the nth tablet. Find Q_3 and Q_{40}.

11. In Problem 10 we found the quantity Q_n, the amount (in mg) of ampicillin left in the body right after the nth tablet is taken.

   (a) Make a similar calculation for P_n, the quantity of ampicillin (in mg) in the body right *before* the nth tablet is taken.
   (b) Find a simplified formula for P_n.
   (c) What happens to P_n in the long run? Is this the same as what happens to Q_n? Explain in practical terms why your answer makes sense.

12. The graph in Figure 15.2 shows quantity over time for 250 mg of ampicillin taken every 6 hours, starting at time $t = 0$.

   (a) Label the graph values of $Q_0, Q_1, Q_2, \ldots$ from Problem 10.
   (b) Label the values of $P_1, P_2, P_3, \ldots$ calculated in Problem 11.
   (c) Label the t-axis values at medication times.

q, quantity (mg)

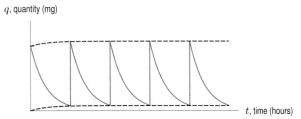

t, time (hours)

Figure 15.2

13. Figure 15.3 shows the quantity of the drug atenolol in the body as a function of time, with the first dose at time $t = 0$. Atenolol is taken in 50 mg doses once a day to lower blood pressure.

   (a) If the half-life of atenolol in the body is 6 hours, what percentage of the atenolol[9] present at the start of a 24-hour period is still there at the end?
   (b) Find expressions for the quantities $Q_0, Q_1, Q_2, Q_3, \ldots$, and Q_n shown in Figure 15.3. Write the expression for Q_n.
   (c) Find expressions for the quantities $P_1, P_2, P_3, \ldots$, and P_n shown in Figure 15.3. Write the expression for P_n.

q (quantity, mg)

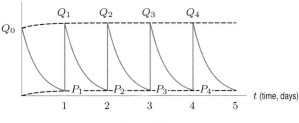

t (time, days)

Figure 15.3

14. A bank account with a $75,000 initial deposit is used to make annual payments of $1000, starting one year after the initial $75,000 deposit. Interest is earned at 3% a year, compounded annually, and paid into the account right before the payment is made.

   (a) What is the balance in the account right after the 24th payment?
   (b) Answer the same question for yearly payments of $3000.

15. A deposit of $1000 is made once a year, starting today, into a bank account earning 4% interest per year, compounded annually. If 20 deposits are made, what is the balance in the account on the day of the last deposit?

16. What effect does doubling each of the following quantities (leaving other quantities the same) have on the answer to Problem 15? Is the answer doubled, more than doubled, or less than doubled?

   (a) The deposit.
   (b) The interest rate.
   (c) The number of deposits made.

---

[8] http://en.wikipedia.org/wiki/ampicillin.
[9] http://www.answers.com/topic/atenolol, accessed August 20, 2008.

REVIEW EXERCISES AND PROBLEMS FOR CHAPTER 15

EXERCISES

■Are the sequences in Exercises **1–4** geometric?

1. $3, 6, 12, 18, \ldots$

2. $5, 15, 45, 135, \ldots$

3. $4, 0.4, 0.04, 0.004, \ldots$

4. $-2, 1, -\frac{1}{2}, \frac{1}{4}, -\frac{1}{8}, \ldots$

■Are the series in Exercises **5–8** arithmetic?

5. $1 + 3 + 4 + 6 + 7 + 8 + \cdots$

6. $\frac{1}{7} + \frac{8}{7} + \frac{15}{7} + \frac{22}{7} + \cdots$

7. $3 + 6 + 12 + 24 + \cdots$

8. $-10 - 8 - 6 - 4 - 2 - 0 - \ldots$

■In Exercises **9–10**, complete the tables with the terms of the arithmetic series $a_1, a_2, \ldots, a_n$, and the sequence of partial sums, $S_1, S_2, \ldots, S_n$. State the values of a_1 and d where $a_n = a_1 + (n-1)d$.

9.

n	1	2	3	4	5	6	7	8
a_n	5			14				
S_n								

10.

n	1	2	3	4	5	6	7	8
a_n	3							
S_n	3	15	36	66				

■In Exercises **11–15**, write the sum using sigma notation.

11. $1 + 3 + 9 + 27 + 81 + 243$

12. $4 - 16 + 64 - 256 + 1024$

13. $3 + 15 + 75 + 375 + 1875 + 9375 + 46875$

14. $128 - 64 + 32 - 16 + 8 - 4 + 2 - 1$

15. $1000 + 900 + 800 + 700 + \cdots + 0$

16. **(a)** Evaluate $\displaystyle\sum_{n=1}^{5} (3n - 4)$ by writing it out.

(b) Check the sum in part (a) using the formula for the sum of an arithmetic series.

17. Find the sum of the first ten terms of the arithmetic series: $8 + 11 + 14 + \cdots$. What is the tenth term?

18. Find the sum of the first nine terms of the arithmetic series: $6 + 11 + 16 + \cdots$.

■In Exercises **19–20**, give the first term and common ratio of the geometric series, write it in summation notation, and find the sum.

19. $1 + 3x + (3x)^2 + (3x)^3 + \cdots + (3x)^{11}$

20. $y^2 - y^4 + y^6 - \cdots + y^{14}$

PROBLEMS

21. A store clerk has 117 cans to stack. He can fit 25 cans on the bottom row and can stack the cans 9 rows high. Assuming the number of cans in each row form an arithmetic sequence, by how much should he decrease each row as he goes up?

22. A university with an enrollment of 8000 students in 2009 is projected to grow by 3% in each of the next three years and by 2% each of the following seven years. Find the sequence of the university's projected student enrollment for the next 10 years.

23. Each person in a group of 25 shakes hands with each other person exactly once. How many total handshakes take place?

24. Worldwide consumption of oil was about 85 billion

barrels in 2007.[10] Assume that consumption continues to increase at 1.2% per year, the rate for the previous decade.

(a) Write a sum representing the total oil consumption for 10 years, starting with 2007.

(b) Evaluate this sum.

25. You inherit $\$100,000$ and put the money in a bank account earning 4% per year, compounded annually. You withdraw $\$3000$ from the account each year, right after the interest is earned. Your first $\$3000$ is withdrawn before any interest is earned.

(a) Compare the balance in the account right after the first withdrawal and right after the second withdrawal. Which do you expect to be higher?

---

[10]www.bp.com/downloads, Statistical Review of World Energy 2007, accessed July 15, 2008.

(b) Calculate the balance in the account right after the 20th withdrawal is made.

(c) What is the largest yearly withdrawal you can take from this account without the balance decreasing over time?

26. After breaking her leg skiing, a patient retrains her muscles by going for walks. The first day, she manages to walk 300 yards. Each day after that she walks 100 yards farther than the day before.

(a) Write a sequence that represents the distances walked each day during the first week.

(b) How long until she is walking at least one mile?

27. Suppose $100 is deposited in a bank. The bank lends 80% to a customer who deposits it in another bank. Then 80% of $80, or $0.08(80) = $64, is loaned out again and eventually redeposited. Of the $64, the bank again loans out 80%, and so on.

(a) Each deposit is a term of a geometric sequence. Starting with $a_1 = 100$ write a general formula for each deposit.

(b) Find the last term that is more than a penny.

(c) Find the total amount of money deposited in a bank, rounded to the nearest dollar.

(d) The total amount of money deposited divided by the original deposit is called the *credit multiplier*. Calculate the credit multiplier for this example and explain what this number tells us.

28. Two graduates are about to enter the workforce and work for 40 years before retiring. The first one does nothing towards saving for retirement for the first 20 years of his career and then invests $1000 a year at 10% compounded annually for the last 20 years of his career. The second one starts saving immediately for retirement, and invests $1000 a year at 10% compounded annually for the first 20 years of her career, and then she

leaves the proceeds in the account for the next 20 years, earning interest, but no longer adding $1000 yearly. Both graduates make their contributions at the end of each year, so he makes his first contribution at the end of the twenty-first year, and she makes hers at the end of the first year. Ignoring the impact of inflation and taxes, answer the following questions.

(a) How much has each invested?

(b) How much does each have at retirement?

(c) If he wanted to have the same amount at retirement as she does, how much a year should he have invested? Under these circumstances, how much will he have invested?

29. You take a rectangular piece of chocolate with dimensions 1 inch by 2 inches. The first day you divide it into two equal pieces, and eat one of them. The next day you divide the uneaten half into two equal pieces and eat one of them. You continue in this way each day.

(a) Draw a picture that illustrates this process.

(b) Find a series that describes the eaten area amount after n divisions.

(c) What value does the series in (b) approach if it continued indefinitely?

30. **(a)** Show that
$$x^n - 1 = (x - 1)(1 + x + x^2 + \cdots + x^{n-1}).$$

(b) Use part (a) to show that
$$x^{2n} - 1 = (x^2 - 1)(1 + x^2 + x^4 + \cdots + x^{2n-2}).$$

(c) Use part (b) to show that the only real roots of $x^{2n} - 1 = 0$ are $x = \pm 1$.

(d) Show that $(x^{2n+1} - 1)(x^{2n+1} + 1) = (x^{4n+2} - 1)$.

(e) Use parts (c) and (d) to show that the only real root of $x^{2n+1} - 1 = 0$ is $x = 1$, and the only real root of $x^{2n+1} + 1 = 0$ is $x = -1$.

Chapter 16

Matrices and Vectors

CONTENTS

16.1 MATRICES

Two different plans are proposed for renovating a residential property. Plan A involves less carpentry but more plumbing and wiring work; plan B involves more carpentry but less plumbing and wiring. To keep track of the details, we can record them in a table-like array of numbers called a *matrix*:

$$\mathbf{W} = \text{Hours of work by type and plan} = \begin{array}{c} \text{carpentry} \quad \text{electrical} \quad \text{plumbing} \\ \left(\begin{array}{ccc} 200 & 120 & 80 \\ 275 & 80 & 60 \end{array} \right) \begin{array}{l} \text{Plan A} \\ \text{Plan B} \end{array} \end{array}$$

Notation for Matrices

It is customary to write matrices in parentheses, as we have done here, and to name them using capital boldface variables like $\mathbf{W}$. It is also conventional to refer to entries in a matrix using pairs of subscripts. For instance, the entry in row 2, column 3 is

$$w_{23} = 60.$$

This tells us that plan B requires 60 hours of plumbing work.

A matrix with m rows and n columns is an $m \times n$ *matrix*, and we say it has *dimensions $m \times n$* (pronounced "m by n"). For example, $\mathbf{W}$ is a 2×3 matrix.

Example 1	Evaluate the following expressions and say what they tell you about the two plans.

(a) w_{12} (b) w_{21} (c) $w_{13} - w_{23}$

Solution

(a) This is the entry in row 1, column 2, so $w_{12} = 120$, telling us that plan A requires 120 hours of electrical work.

(b) This is the entry in row 2, column 1, so $w_{21} = 275$, telling us that plan B requires 275 hours of carpentry work.

(c) This is the difference between the entry in row 1, column 3, and the entry in row 2, column 3:

$$w_{13} - w_{23} = 80 - 60 = 20.$$

This tells us that plan A requires 20 more hours of plumbing than plan B.

Algebraic Operations with Matrices.

Just as we use a variable like x to stand for a number, we use a matrix like $\mathbf{W}$ to stand for a collection of numbers. We can interpret algebraic expressions involving $\mathbf{W}$ in the same way that we interpret algebraic expressions in x.

$2x$:	Means the number you get by doubling x
$2\mathbf{W}$:	Means the matrix you get by doubling all the entries of $\mathbf{W}$
$1.1x$:	Means 10% more than x
$1.1\mathbf{W}$:	Means the matrix whose entries are 10% more than the entries of $\mathbf{W}$
$x + y$:	Means the sum of x and y
$\mathbf{W} + \mathbf{S}$:	Means the matrix whose entries are the sum of the corresponding entries of $\mathbf{W}$ and $\mathbf{S}$

Adding two matrices is called *matrix addition* and multiplying a matrix by a number is called *scalar multiplication*.

Example 2 Find (a) $2\mathbf{W}$ (b) $1.1\mathbf{W}$.

Solution (a) Here, each entry is twice the corresponding entry in $\mathbf{W}$:

$$2\mathbf{W} = 2\begin{pmatrix} 200 & 120 & 80 \\ 275 & 80 & 60 \end{pmatrix}$$

$$= \begin{pmatrix} 2 \cdot 200 & 2 \cdot 120 & 2 \cdot 80 \\ 2 \cdot 275 & 2 \cdot 80 & 2 \cdot 60 \end{pmatrix}$$

$$= \begin{pmatrix} 400 & 240 & 160 \\ 550 & 160 & 120 \end{pmatrix}.$$

(b) Here, each entry is 1.1 times, or 10% more than, the corresponding entry in $\mathbf{W}$:

$$1.1\mathbf{W} = 1.1\begin{pmatrix} 200 & 120 & 80 \\ 275 & 80 & 60 \end{pmatrix}$$

$$= \begin{pmatrix} 1.1 \cdot 200 & 1.1 \cdot 120 & 1.1 \cdot 80 \\ 1.1 \cdot 275 & 1.1 \cdot 80 & 1.1 \cdot 60 \end{pmatrix}$$

$$= \begin{pmatrix} 220 & 132 & 88 \\ 302.5 & 88 & 66 \end{pmatrix}.$$

Example 3 To budget for delays, we can allow for overruns in the work described by matrix $\mathbf{W}$. Let the matrix $\mathbf{S}$ give the number of hours of overruns we will budget for:

$$\mathbf{S} = \begin{pmatrix} 8 & 14 & 11 \\ 16 & 12 & 7 \end{pmatrix}.$$

Find $\mathbf{W} + \mathbf{S}$ and say what it tells you about the planned renovations.

Solution We have

$$\mathbf{W} + \mathbf{S} = \begin{pmatrix} 200 & 120 & 80 \\ 275 & 80 & 60 \end{pmatrix} + \begin{pmatrix} 8 & 14 & 11 \\ 16 & 12 & 7 \end{pmatrix}$$

$$= \begin{pmatrix} 200 + 8 & 120 + 14 & 80 + 11 \\ 275 + 16 & 80 + 12 & 60 + 7 \end{pmatrix}$$

$$= \begin{pmatrix} 208 & 134 & 91 \\ 291 & 92 & 67 \end{pmatrix}.$$

This tells us the number of hours budgeted under each plan, including overruns. For instance, the total number of hours budgeted for electrical work under plan A is $120 + 14 = 134$.

Properties of Scalar Multiplication and Matrix Addition

Matrix addition and multiplication of matrices by a scalar obey the same rules of arithmetic as addition and multiplication of numbers.

- *Commutativity of addition*: $\mathbf{A} + \mathbf{B} = \mathbf{B} + \mathbf{A}$.
- *Associativity of addition*: $(\mathbf{A} + \mathbf{B}) + \mathbf{C} = \mathbf{A} + (\mathbf{B} + \mathbf{C})$.
- *Associativity of scalar multiplication*: $k_1(k_2\mathbf{A}) = (k_1 k_2)\mathbf{A}$.
- *Distributivity of scalar multiplication*: $(k_1 + k_2)\mathbf{A} = k_1\mathbf{A} + k_2\mathbf{A}$ and
$$k(\mathbf{A} + \mathbf{B}) = k\mathbf{A} + k\mathbf{B}.$$

Exercises and Problems for Section 16.1

EXERCISES

Given the matrices below, evaluate the expressions in Exercises **1–18**, if possible. If it is not possible, explain why.

$$\mathbf{A} = \begin{pmatrix} 4 & 6 \\ 2 & 5 \end{pmatrix} \qquad \mathbf{B} = \begin{pmatrix} 3 & -2 \\ -1 & 6 \end{pmatrix},$$

$$\mathbf{C} = \begin{pmatrix} 2 & -1 & -3 \\ 6 & 0 & 2 \end{pmatrix} \qquad \mathbf{D} = \begin{pmatrix} 4 & -1 & 3 \\ -5 & -1 & -3 \end{pmatrix}.$$

1. The dimensions of $\mathbf{B}$
2. The dimensions of $\mathbf{D}$
3. The dimensions of $\mathbf{A} + \mathbf{B}$
4. The dimensions of $\mathbf{C} - \mathbf{D}$

5. a_{12} 6. b_{21} 7. c_{13}

8. d_{23} 9. $\mathbf{A} + \mathbf{B}$ 10. $\mathbf{A} - \mathbf{B}$

11. $\mathbf{A} + \mathbf{C}$ 12. $3\mathbf{D}$ 13. $-2\mathbf{B}$

14. $\mathbf{C} - \mathbf{D}$ 15. $\mathbf{D} - \mathbf{C}$ 16. $3\mathbf{A} + 2\mathbf{B}$

17. $\mathbf{D} - \mathbf{B}$ 18. $4\mathbf{B} - 5\mathbf{A}$

Check the statements in Exercises **19–23** using the matrices

$$\mathbf{U} = \begin{pmatrix} 2 & 3 \\ 1 & 2 \end{pmatrix}, \mathbf{V} = \begin{pmatrix} -1 & 4 \\ 0 & 2 \end{pmatrix}, \mathbf{W} = \begin{pmatrix} 5 & -5 \\ 4 & 7 \end{pmatrix}.$$

19. $\mathbf{U} + \mathbf{W} = \mathbf{W} + \mathbf{U}$ 20. $2\mathbf{W} + 3\mathbf{W} = 5\mathbf{W}$

21. $4(\mathbf{U}+\mathbf{V}) = 4\mathbf{U}+4\mathbf{V}$ 22. $2(3\mathbf{V}) = 6\mathbf{V}$

23. $(\mathbf{W} + \mathbf{U}) + \mathbf{V} = \mathbf{W} + (\mathbf{U} + \mathbf{V})$

PROBLEMS

In Problems **24–25**, refer to $\mathbf{R}$ and $\mathbf{M}$, matrices of mean SAT scores. The columns are mean SAT reasoning scores for the years 2001–2008. The first row is scores for males and the second row is scores for females. Matrix $\mathbf{R}$ is the Critical Reading scores, and matrix $\mathbf{M}$ is the Mathematics scores.[1]

$$\mathbf{M} = \begin{pmatrix} 533 & 534 & 537 & 537 & 538 & 536 & 533 & 533 \\ 498 & 500 & 503 & 501 & 504 & 502 & 499 & 500 \end{pmatrix}$$

$$\mathbf{R} = \begin{pmatrix} 509 & 507 & 512 & 512 & 513 & 505 & 504 & 504 \\ 502 & 502 & 503 & 504 & 505 & 502 & 502 & 500 \end{pmatrix}$$

24. Calculate $\mathbf{R} + \mathbf{M}$. What does this represent?

25. Calculate $\mathbf{M} - \mathbf{R}$. What does this represent?

[1]The College Board, New York, NY, *2008 College-Bound Seniors Total Group Profile Report*.

In Problems **26–28**, refer to the matrices of Olympic medal counts. The columns are the numbers of gold, silver, and bronze medals, and the rows are the number of medals for Australia, China, Germany, Russia and the United States.[2]

$$\mathbf{G} = 2004 \text{ Olympics} = \begin{pmatrix} 17 & 16 & 16 \\ 32 & 17 & 14 \\ 14 & 16 & 18 \\ 27 & 27 & 38 \\ 35 & 39 & 29 \end{pmatrix}$$

$$\mathbf{E} = 1996 \text{ Olympics} = \begin{pmatrix} 9 & 9 & 23 \\ 16 & 22 & 12 \\ 20 & 18 & 27 \\ 26 & 21 & 16 \\ 44 & 32 & 25 \end{pmatrix}$$

$$\mathbf{H} = 2008 \text{ Olympics} = \begin{pmatrix} 14 & 15 & 17 \\ 51 & 21 & 28 \\ 16 & 10 & 15 \\ 23 & 21 & 28 \\ 36 & 38 & 36 \end{pmatrix}$$

$$\mathbf{F} = 2000 \text{ Olympics} = \begin{pmatrix} 16 & 25 & 17 \\ 28 & 16 & 15 \\ 13 & 17 & 26 \\ 32 & 28 & 28 \\ 40 & 24 & 33 \end{pmatrix}$$

26. Calculate $\mathbf{H} - \mathbf{G}$. What does this represent?

27. Calculate $\mathbf{E} + \mathbf{F} + \mathbf{G} + \mathbf{H}$. What does this represent?

28. Calculate $\frac{1}{4}(\mathbf{E} + \mathbf{F} + \mathbf{G} + \mathbf{H})$. What does this represent?

16.2 MATRIX MULTIPLICATION

Three contractors are approached for bids for the renovation work described on page 466. Their hourly rates are recorded in the matrix below:

$$\mathbf{R} = \begin{array}{c} \text{Hourly rate by type} \\ \text{of work and contractor} \end{array} = \begin{array}{ccc} \text{first} & \text{second} & \text{third} \\ \text{contractor} & \text{contractor} & \text{contractor} \end{array} \begin{pmatrix} 55 & 70 & 60 \\ 80 & 90 & 100 \\ 85 & 110 & 100 \end{pmatrix} \begin{array}{c} \text{carpentry} \\ \text{electrical} \\ \text{plumbing} \end{array}$$

In this 3×3 matrix, each contractor has her own column, and each type of work has its own row. For instance, we see that the first contractor (in column 1) charges hourly rates of $55 for carpentry, $80 for electrical, and $85 for plumbing.

Finding the Total Cost

On page 466, we defined the matrix $\mathbf{W}$ as:

$$\mathbf{W} = \begin{array}{c} \text{Hours of work by} \\ \text{type and plan} \end{array} = \begin{array}{ccc} \text{carpentry} & \text{electrical} & \text{plumbing} \end{array} \begin{pmatrix} 200 & 120 & 80 \\ 275 & 80 & 60 \end{pmatrix} \begin{array}{c} \text{plan A} \\ \text{plan B} \end{array}$$

Here, each type of work (carpentry, electrical, plumbing) has its own row. Referring to the information in matrices $\mathbf{R}$ and $\mathbf{W}$, we see that the cost for the first contractor to complete the first plan is given by

[2]en.wikipedia.org/wiki/1996_Summer_Olympics, www.infoplease.com/ipsa/A0875902.html, news.bbc.co.uk/sport1/hi/olympics_2004/default.stm, results.beijing2008.cn/WRM/ENG/INF/GL/95A/GL0000000.shtml.

$$\begin{array}{l} \text{Cost of plan } A \\ \text{for first contractor} \end{array} = \text{Carpentry cost} + \text{Electrical cost} + \text{Plumbing cost}$$

$$= 200 \text{ hours} \times \$55/\text{hour} + 120 \text{ hours} \times \$80/\text{hour} + 80 \text{ hours} \times \$85/\text{hour}$$

$$= \$11{,}000 + \$9600 + \$6800 = \$27{,}400.$$

Notice that here, we are multiplying entries in the first row of $\mathbf{W}$, the *Plan-A row*, by the corresponding entries in the first column of $\mathbf{R}$, the *first-contractor column*, and then adding up the results:

Multiply entries in first row of $\mathbf{W}$… …by entries in the first column of $\mathbf{R}$… …then add up the results.

$$\begin{pmatrix} \boxed{\begin{matrix} 200 & 120 & 80 \end{matrix}} \\ 275 & 80 & 60 \end{pmatrix} \begin{pmatrix} \boxed{\begin{matrix} 55 \\ 80 \\ 85 \end{matrix}} & \begin{matrix} 70 \\ 90 \\ 110 \end{matrix} & \begin{matrix} 60 \\ 100 \\ 100 \end{matrix} \end{pmatrix} \rightarrow \text{Cost} = 200 \cdot 55 + 120 \cdot 80 + 80 \cdot 85$$

$$= 11{,}000 + 9600 + 6800$$
$$= 27{,}400.$$

Likewise, to find the find the cost for the second contractor to complete the first plan, we multiply the *Plan-A row* of $\mathbf{W}$ by the *second-contractor column* of $\mathbf{R}$:

Multiply entries in first row of $\mathbf{W}$… …by entries in the *second* column of $\mathbf{R}$… …then add up the results.

$$\begin{pmatrix} \boxed{\begin{matrix} 200 & 120 & 80 \end{matrix}} \\ 275 & 80 & 60 \end{pmatrix} \begin{pmatrix} 55 & \boxed{\begin{matrix} 70 \\ 90 \\ 110 \end{matrix}} & 60 \\ 80 & & 100 \\ 85 & & 100 \end{pmatrix} \rightarrow \text{Cost} = 200 \cdot 70 + 120 \cdot 90 + 80 \cdot 110$$

$$= 14{,}000 + 10{,}800 + 8800$$
$$= 33{,}600.$$

Extending this pattern, we see that:

If we multiply entries in row i of $\mathbf{W}$ by the corresponding entries in column j of $\mathbf{R}$, and then sum the results, we obtain the total cost to have plan i completed by contractor j.

Forming a New Matrix by Combining Rows and Columns

Since there are two plans and three contractors, we can calculate six totals in all. A natural way to record our results is in a new "cost" matrix, $\mathbf{C}$:

$$\mathbf{C} = \begin{array}{c} \text{Total cost for each contractor} \\ \text{to complete each plan} \end{array} = \begin{pmatrix} \overset{\substack{\text{first} \\ \text{contractor}}}{27{,}400} & \overset{\substack{\text{second} \\ \text{contractor}}}{33{,}600} & \overset{\substack{\text{third} \\ \text{contractor}}}{32{,}000} \\ 26{,}625 & 33{,}050 & 30{,}500 \end{pmatrix} \begin{array}{l} \text{plan A} \\ \text{plan B} \end{array}$$

For instance, we see that $c_{11} = 27{,}400$ and $c_{12} = 33{,}600$, because (as we calculated above), the cost for the plan A to be completed by contractors 1 and 2 is, respectively, \$27,400 and \$33,600.

Example 1   Show that $c_{22} = 33{,}050$.

Solution   We need to calculate the entry in row 2, column 2 of matrix $\mathbf{C}$, which corresponds to the cost for having the second contractor complete plan B. This means we need to multiply the entries of row 2 of $\mathbf{W}$ by the corresponding entries of column 2 of $\mathbf{R}$, then add up the results:

Multiply entries in second row of $\mathbf{W}$... ...by entries in the second column of $\mathbf{R}$... ...then add up the results to obtain $c_{22} = 275 \cdot 70 + 80 \cdot 90 + 60 \cdot 110$.

$$\begin{pmatrix} 200 & 120 & 80 \\ 275 & 80 & 60 \end{pmatrix} \begin{pmatrix} 55 & 70 & 60 \\ 80 & 90 & 100 \\ 85 & 110 & 100 \end{pmatrix} \rightarrow \begin{pmatrix} 27{,}400 & 33{,}600 & 32{,}000 \\ 26{,}625 & 33{,}050 & 30{,}500 \end{pmatrix}$$

The Significance of Matrix Multiplication

Matrix multiplication requires so much calculation that it can be easy to lose track of the broader significance. The key thing to notice is that we have rewritten a complicated relationship involving three contractors, two plans, and three types of work as a single equation:

$$\mathbf{C} = \mathbf{WR}.$$

Overview of Matrix Multiplication

In general, if matrix $\mathbf{A}$ has the same number of columns as matrix $\mathbf{B}$ has rows, we can multiply $\mathbf{A}$ and $\mathbf{B}$ as follows:

Matrix Multiplication

For an $m \times n$ matrix $\mathbf{A}$ and an $n \times p$ matrix $\mathbf{B}$, the product $\mathbf{C} = \mathbf{AB}$ is an $m \times p$ matrix. Notice that the number of columns of $\mathbf{A}$ and the number of rows of $\mathbf{B}$ are the same: both equal n. Entries of $\mathbf{C}$ are found by multiplying entries for a given row of $\mathbf{A}$ by the corresponding entries for a given column of $\mathbf{B}$, then summing the results. (This is why the number of columns of $\mathbf{A}$ must equal the number of rows of $\mathbf{B}$.) Using sigma notation, we can summarize this process by writing

$$c_{ij} = \sum_{r=1}^{n} a_{ir} b_{rj}.$$

Here, c_{ij} is the entry in row i, column j of the new matrix, $\mathbf{C}$.

It can be harder to *describe* matrix multiplication than to *perform* it. The key point is that to find an entry in row i, column j of the new matrix $\mathbf{C}$, we must combine the entries in row i of $\mathbf{A}$ with the entries in column j of $\mathbf{B}$.

Example 2   Find $\mathbf{Z} = \mathbf{XY}$ where $\mathbf{X} = \begin{pmatrix} 2 & 1 & 4 \\ 3 & -1 & 2 \\ 0 & 2 & -3 \end{pmatrix}$ and $\mathbf{Y} = \begin{pmatrix} 3 & 1 & 2 \\ -1 & 1 & 2 \\ 2 & -1 & 3 \end{pmatrix}$.

Solution   We first find z_{11}, the entry in row 1, column 1 of the new matrix $\mathbf{Z}$. To do this, we combine row 1 of $\mathbf{X}$ with column 1 of $\mathbf{Y}$:

Multiply entries in first row of $\mathbf{X}$. . . . . . by entries in the first column of $\mathbf{Y}$. . .

. . . then add up the results to obtain $z_{11} = 2 \cdot 3 + 1(-1) + 4 \cdot 2 = 13.$

$$\left(\begin{array}{ccc} \boxed{2 \quad 1 \quad 4} \\ 3 & -1 & 2 \\ 0 & 2 & -3 \end{array} \right) \quad \left(\begin{array}{ccc} \boxed{3} & 1 & 2 \\ \boxed{-1} & 1 & 2 \\ \boxed{2} & -1 & 3 \end{array} \right) \rightarrow \left(\begin{array}{ccc} \boxed{13} & ? & ? \\ ? & ? & ? \\ ? & ? & ? \end{array} \right)$$

One entry down, eight to go! As you can see, matrix multiplication can involve a *lot* of calculation. Next we find z_{12}, the entry in row 1, column 2 of the new matrix $\mathbf{Z}$. We have:

Multiply entries in first row of $\mathbf{X}$. . . . . . by entries in the second column of $\mathbf{Y}$. . .

. . . then add up the results to obtain $z_{12} = 2 \cdot 1 + 1 \cdot 1 + 4(-1) = -1.$

$$\left(\begin{array}{ccc} \boxed{2 \quad 1 \quad 4} \\ 3 & -1 & 2 \\ 0 & 2 & -3 \end{array} \right) \quad \left(\begin{array}{ccc} 3 & \boxed{1} & 2 \\ -1 & \boxed{1} & 2 \\ 2 & \boxed{-1} & 3 \end{array} \right) \rightarrow \left(\begin{array}{ccc} 13 & \boxed{-1} & ? \\ ? & ? & ? \\ ? & ? & ? \end{array} \right)$$

Skipping ahead to the final (bottom-right) entry of $\mathbf{Z}$, we have:

Multiply entries in third row of $\mathbf{X}$. . . . . . by entries in the third column of $\mathbf{Y}$. . .

. . . then add up the results to obtain $z_{33} = 0 \cdot 2 + 2 \cdot 2 - 3 \cdot 3 = -5.$

$$\left(\begin{array}{ccc} 2 & 1 & 4 \\ 3 & -1 & 2 \\ \boxed{0 \quad 2 \quad -3} \end{array} \right) \quad \left(\begin{array}{ccc} 3 & 1 & \boxed{2} \\ -1 & 1 & \boxed{2} \\ 2 & -1 & \boxed{3} \end{array} \right) \rightarrow \left(\begin{array}{ccc} 13 & -1 & ? \\ ? & ? & ? \\ ? & ? & \boxed{-5} \end{array} \right)$$

We can find the remaining six entries in the same fashion, or by using a calculator or computer:

$$\mathbf{Z} = \mathbf{XY} = \left(\begin{array}{ccc} 2 & 1 & 4 \\ 3 & -1 & 2 \\ 0 & 2 & -3 \end{array} \right) \left(\begin{array}{ccc} 3 & 1 & 2 \\ -1 & 1 & 2 \\ 2 & -1 & 3 \end{array} \right) = \left(\begin{array}{ccc} 13 & -1 & 18 \\ 14 & 0 & 10 \\ -8 & 5 & -5 \end{array} \right).$$

The next example shows another application of matrix multiplication.

Example 3   A large company consists of two main divisions, and each division consists of three departments. Let the number of employees in the company be described by the matrix

$$\mathbf{E} = \left(\begin{array}{ccc} 200 & 300 & 100 \\ 400 & 200 & 200 \end{array} \right),$$

where each row represents a division and each column represents a department. For instance, we see that there are $e_{21} = 400$ employees in the first department of the second division. The company is restructured during a takeover. The new structure is given by

$$\mathbf{G} = \mathbf{EF} = \begin{pmatrix} 200 & 300 & 100 \\ 400 & 200 & 200 \end{pmatrix} \begin{pmatrix} 0.6 & 0.4 \\ 0.3 & 0.7 \\ 0.1 & 0.9 \end{pmatrix}.$$

Find $\mathbf{G}$, and say what this tells you about the company.

Solution   The first entry of $\mathbf{G}$, g_{11}, is found by combining the first row of $\mathbf{E}$ with the first column of $\mathbf{F}$. We have:

$$g_{11} = 200 \cdot 0.6 + 300 \cdot 0.3 + 100 \cdot 0.1 = 120 + 90 + 10 = 220.$$

Continuing, we find the other three entries of $\mathbf{G}$ as follows:

$$g_{12} = 200 \cdot 0.4 + 300 \cdot 0.7 + 100 \cdot 0.9 = 380$$
$$g_{21} = 400 \cdot 0.6 + 200 \cdot 0.3 + 200 \cdot 0.1 = 320$$
$$g_{22} = 400 \cdot 0.4 + 200 \cdot 0.7 + 200 \cdot 0.9 = 480,$$

which gives

$$\mathbf{G} = \mathbf{EF} = \begin{pmatrix} 220 & 380 \\ 320 & 480 \end{pmatrix}.$$

This tells us that after the reorganization, there are two divisions each with two departments. The first division has 220 employees in the first department and 380 in the second. The second division has 320 employees in the first department and 480 in the second. Notice that the total number of employees in the two divisions has not changed.

Properties of Matrix Multiplication

Matrix multiplication shares certain properties with multiplication of ordinary numbers (also called *scalars*).

If $\mathbf{A}$, $\mathbf{B}$, and $\mathbf{C}$ are matrices and k is a constant:
- $\mathbf{A}(\mathbf{BC}) = (\mathbf{AB})\mathbf{C}$. This tells us we can regroup (though not necessarily reorder) the product of three or more matrices any way we like.
- $\mathbf{A}(\mathbf{B} + \mathbf{C}) = \mathbf{AB} + \mathbf{AC}$. This tells us that matrix multiplication distributes over matrix addition in the familiar way.
- $k(\mathbf{AB}) = (k\mathbf{A})\mathbf{B} = \mathbf{A}(k\mathbf{B})$. This tells us that scalar multiplication can also be regrouped as we see fit.

Matrix Multiplication is Not Commutative

Although matrix multiplication has certain things in common with ordinary (*scalar*) multiplication, the two operations differ in important ways. Perhaps most surprisingly, it is not always true that $\mathbf{AB} = \mathbf{BA}$ (see Problem 20). In other words, unlike ordinary multiplication, order matters when multiplying matrices, and we say that matrix multiplication does not *commute*. Recall that there are other familiar operations that do not commute, including subtraction (for instance, $3 - 2$ is not the same as $2 - 3$) and exponentiation (for instance, 3^2 is not the same as 2^3).

Exercises and Problems for Section 16.2

EXERCISES

■ Given the matrices below, evaluate the expressions in Exercises 1–12, if possible. If it is not possible, explain why.

$$A = \begin{pmatrix} 2 & 3 \\ 8 & 4 \end{pmatrix} \qquad B = \begin{pmatrix} 5 & -3 \\ -2 & 7 \end{pmatrix}$$

$$C = \begin{pmatrix} 4 & -2 & -5 \\ 0 & -4 & -3 \end{pmatrix} \qquad D = \begin{pmatrix} 2 & 4 & -4 \\ 3 & -10 & 2 \\ 2 & 4 & 5 \end{pmatrix}.$$

1. **AB**
2. **BA**
3. **CD**
4. **DC**
5. **AC**
6. **BC**
7. **CB**
8. **A(A + B)**
9. **AA + AB**
10. **C(C + D)**
11. **(AB)C**
12. **A(BC)**

PROBLEMS

13. What do you need to know about two matrices to know if their product exists?

14. What do you need to know about two matrices to know if their sum exists?

■ Given $G = \begin{pmatrix} 5 & 2 \\ 1 & 3 \end{pmatrix}$ and $H = \begin{pmatrix} -2 & 4 \\ 3 & -1 \end{pmatrix}$, verify the statements in Problems 15–16.

15. $3(\mathbf{GH}) = (3\mathbf{G})\mathbf{H}$
16. $\mathbf{GH} \neq \mathbf{HG}$

17. The number of meals, **N**, and the cost of the meals, **C**, for a weekend class reunion are given by the matrices

$$N = \begin{pmatrix} 20 & 35 & 70 \\ 30 & 35 & 50 \end{pmatrix}, \qquad C = \begin{pmatrix} 8 \\ 12 \\ 50 \end{pmatrix}.$$

The first column of **N** is the number of breakfasts, the second the number of lunches, and the third the number of dinners. The first row of **N** is the meals needed on Saturday, the second on Sunday. The first row of **C** is the cost of breakfast, the second row is the cost of lunch, and the last row is the cost of dinner.

(a) Calculate **NC**.
(b) What is the practical meaning of **NC**?

18. A manufacturer produces three different types of widgets and ships them to two different warehouses. The number of widgets shipped, **W**, and the price of the widgets, **P**, are given by

$$W = \begin{pmatrix} 900 & 3500 \\ 2250 & 1200 \\ 3310 & 1500 \end{pmatrix}, \qquad P = \begin{pmatrix} 25 & 15 & 10 \end{pmatrix}$$

The columns of **W** correspond to the two warehouses and the rows to the three types of widgets. The columns of **P** are the prices of the three types of widgets.

(a) Calculate **PW**.
(b) Explain the practical meaning of **PW**.
(c) How much more widget inventory (in dollars) does the second warehouse contain?

19. Given the matrices **R** and **W** from page 469, show that, unlike the product **WR**, the product **RW** is undefined.

20. Show that $\mathbf{V} = \mathbf{YX}$ does not equal $\mathbf{Z} = \mathbf{XY}$, where **X**, **Y**, and **Z** are the matrices from Example 2. [Hint: It is not necessary to evaluate the products completely.]

16.3   MATRICES AND VECTORS

In a certain town, the number of employed people is $e = 5000$, and the number of unemployed people is $u = 250$. Since both e and u are required for a complete description of the town's economic status, we can choose to treat this pair of values as a single algebraic object called a *vector*, writing

$$\vec{E} = (e, u) = (5000, 250).$$

Here, e and u are called the *components* of the vector $\vec{E}$. The arrow over the E signifies that $\vec{E}$ is a vector instead of a number (called a *scalar*).

Ordered Pairs

Notice that the vector $(250, 5000)$ describes a very different kind of town than does the vector $\vec{E} = (5000, 250)$. Such a town would have only 250 employed people and 5000 unemployed people. Since the order of the components of $\vec{E}$ matters, we refer to $(5000, 250)$ as an *ordered pair*.

Vectors in Physics and Geometry

The components (e, u) of $\vec{E}$ might remind you of the familiar (x, y)-coordinate notation, and in fact vectors like $\vec{E}$ are often described geometrically as points on the plane. Moreover, in physics and other disciplines, vectors are often represented graphically using arrows, where they describe directed quantities such as force, velocity, or electric and magnetic fields. In this book, though, we will focus less on physical and geometrical interpretations of vectors than on their use as algebraic objects representing the components of a total, such as the total number of workers (both employed and unemployed) in a town.

Vector Addition

Suppose 200 new people move to the town, and that only 150 of them have jobs. We can describe the status of the town after their arrival with the new vector

$$\vec{E_1} = \begin{matrix} \text{Employment status of} \\ \text{town after newcomers arrive} \end{matrix} = (\overbrace{5000 + 150}^{\text{employed}}, \overbrace{250 + 50}^{\text{unemployed}}) = (5150, 300).$$

Notice we can use vectors to describe the employment status status of both groups:

$$\vec{E} = \text{Original employment status} = (5000, 250)$$
$$\vec{D} = \text{Employment status of newcomers} = (150, 50).$$

Having described everything with vectors in this way, it seems natural to write

$$\vec{E_1} = \begin{matrix} \text{Employment status of} \\ \text{town after newcomers arrive} \end{matrix} = \overbrace{(5000, 250)}^{\vec{E}} + \overbrace{(150, 50)}^{\vec{D}}$$
$$= (5000 + 150, 250 + 50)$$
$$= (5150, 300).$$

When, as here, we add the corresponding components of two vectors to obtain a new vector, we say we are performing *vector addition*.

Scalar Multiplication

Now imagine that a nearby, larger town has twice as many employed people as the first town, and twice as many unemployed as the first town. Letting $\vec{F}$ describe the economic status of this town, we can write

$$\vec{F} = 2\vec{E} = 2(5000, 250)$$
$$= (2 \cdot 5000, 2 \cdot 250)$$
$$= (10{,}000, 500).$$

When we multiply the components of a vector like $\vec{E}$ by a scalar like 2, we say we are performing *scalar multiplication*.

Properties of Vector Operations

Sometimes it can be useful to think of vectors as matrices with only one row (or, in some cases, with only one column). In particular, we can think of vector addition as a special case of matrix addition, and scalar multiplication of a vector as a special case of scalar multiplication of a matrix. This means that the properties described on page 468 also apply to vectors.

Example 1   Evaluate the following expressions given that $\vec{w} = (2, 3)$ and $\vec{v} = (-1, 4)$.

(a) $\vec{w} + \vec{v}$ (b) $2\vec{w} - 3\vec{v}$ (c) $3(\vec{v} - \vec{w})$

Solution   (a) We have:

$$
\begin{aligned}
\vec{w} + \vec{v} &= (2, 3) + (-1, 4) \\
&= (2 - 1, 3 + 4) \\
&= (1, 7).
\end{aligned}
$$

(b) We have:

$$
\begin{aligned}
2\vec{w} - 3\vec{v} &= 2(2, 3) - 3(-1, 4) \\
&= (2 \cdot 2, 2 \cdot 3) - (3(-1), 3 \cdot 4) \\
&= (4, 6) - (-3, 12) \\
&= (4 - (-3), 6 - 12) \\
&= (7, -6).
\end{aligned}
$$

(c) We have:

$$
\begin{aligned}
3(\vec{v} - \vec{w}) &= 3((-1, 4) - (2, 3)) \\
&= 3(-1 - 2, 4 - 3) \\
&= 3(-3, 1) \\
&= (3(-3), 3 \cdot 1) \\
&= (-9, 3).
\end{aligned}
$$

Another approach is to write:

$$
\begin{aligned}
3(\vec{v} - \vec{w}) &= 3\vec{v} - 3\vec{w} \\
&= 3(-1, 4) - 3(2, 3) \\
&= (3(-1), 3 \cdot 4) - (3 \cdot 2, 3 \cdot 3) \\
&= (-3, 12) - (6, 9) \\
&= (-3 - 6, 12 - 9) \\
&= (-9, 3).
\end{aligned}
$$

Notice that we get the same answer as before.

Matrix Multiplication of a Vector

The economic status of the town on page 474 is described by the vector $\vec{E} = (5000, 250)$. Suppose that over the course of the year, 10% of the employed people become unemployed and 20% of the unemployed people become employed. To help keep everything straight, we write

$$\vec{E} = (e_{\text{old}}, u_{\text{old}})$$
$$\vec{E}_{\text{new}} = (e_{\text{new}}, u_{\text{new}}).$$

To find the new number of employed people, we write:

$$e_{\text{new}} = \begin{array}{c}\text{Number of previously employed} \\ \text{people who remain employed}\end{array} + \begin{array}{c}\text{Number of previously unemployed} \\ \text{people who become newly employed}\end{array}.$$

Since 10% of the 5000 previously employed people become unemployed, this means 90% remain employed, so

$$\begin{array}{c}\text{Number of previously employed} \\ \text{people who remain employed}\end{array} = 0.9(5000) = 4500.$$

Likewise, since 20% of the 250 previously unemployed people become employed, we have

$$\begin{array}{c}\text{Number of previously unemployed} \\ \text{people who become newly employed}\end{array} = 0.2(250) = 50.$$

This means

$$e_{\text{new}} = \underbrace{\begin{array}{c}\text{Number of previously employed} \\ \text{people who remain employed}\end{array}}_{0.9(5000)} + \underbrace{\begin{array}{c}\text{Number of previously unemployed} \\ \text{people who become newly employed}\end{array}}_{0.2(250)}$$
$$= 0.9(5000) + 0.2(250)$$
$$= 4500 + 50 = 4550.$$

Similarly, we see that the new number of unemployed people is given by

$$u_{\text{new}} = \underbrace{\begin{array}{c}\text{Number of previously employed} \\ \text{people who become newly unemployed}\end{array}}_{0.1(5000)} + \underbrace{\begin{array}{c}\text{Number of previously unemployed} \\ \text{people who remain unemployed}\end{array}}_{0.8(250)}$$
$$= 0.1(5000) + 0.8(250)$$
$$= 500 + 200 = 700.$$

Here, both components of the new employment vector $\vec{E}_{\text{new}}$ depend on both components of the original vector $\vec{E}$:

$$e_{\text{new}} = 0.9 \underbrace{(5000)}_{e_{\text{old}}} + 0.2 \underbrace{(250)}_{u_{\text{old}}} = 0.9e_{\text{old}} + 0.2u_{\text{old}}$$
$$u_{\text{new}} = 0.1 \underbrace{(5000)}_{e_{\text{old}}} + 0.8 \underbrace{(250)}_{u_{\text{old}}} = 0.1e_{\text{old}} + 0.8u_{\text{old}}.$$

This leads us to the crucial step. Notice that, thinking temporarily of $\vec{E}$ and $\vec{E}_{\text{new}}$ as one-column matrices, we can use matrix multiplication to write

$$\begin{pmatrix} 0.9 & 0.2 \\ 0.1 & 0.8 \end{pmatrix} \underbrace{\begin{pmatrix} 5000 \\ 250 \end{pmatrix}}_{\vec{E}} = \begin{pmatrix} 0.9(5000) + 0.2(250) \\ 0.1(5000) + 0.8(250) \end{pmatrix} = \underbrace{\begin{pmatrix} 4550 \\ 700 \end{pmatrix}}_{\vec{E}_{\text{new}}}.$$

In other words, matrix multiplication provides us with the exact tool we need to combine the components of $\vec{E}$ in order to obtain the components of $\vec{E}_{\text{new}}$:

$$\vec{E}_{\text{new}} = \begin{pmatrix} 0.9 & 0.2 \\ 0.1 & 0.8 \end{pmatrix} \vec{E}.$$

What Matrix Multiplication of a Vector Means

For convenience, we allowed ourselves to think of $\vec{E}$ as a one-column matrix in the above calculation. However, even though the calculations are similar, vectors are not in general the same as matrices. Instead, we often think of matrices as behaving more like functions, so that when we multiply a vector by a matrix, the matrix takes the vector as an "input" and yields a new vector as the "output."

Example 2   Suppose the employment in a town in year $t = 0$ is described by the vector $\vec{G}_0 = (8000, 400)$, and that each year,

- 2% of the employed people become unemployed
- 10% of the unemployed people become employed.

Find a matrix $\mathbf{M}$ such that $\vec{G}_1$, the employment vector in year $t = 1$, is given by $G_1 = \mathbf{M}\vec{G}_0$. Evaluate $\vec{G}_1$.

Solution   Each year, 98% of previously employed people remain employed, and 10% of unemployed people become newly employed. This means e_1, the number of employed people in year 1, is given by:

$$e_1 = \underbrace{\begin{array}{c} \text{Number of previously employed} \\ \text{people who remain employed} \end{array}}_{0.98e_0} + \underbrace{\begin{array}{c} \text{Number of previously unemployed} \\ \text{people who become newly employed} \end{array}}_{0.1u_0}.$$

Likewise, 2% of previously employed people become unemployed, and 90% of previously unemployed people remain unemployed. This means u_1, the number of unemployed people in year 1, is given by:

$$u_1 = \underbrace{\begin{array}{c} \text{Number of previously employed} \\ \text{people who become newly unemployed} \end{array}}_{0.02e_0} + \underbrace{\begin{array}{c} \text{Number of previously unemployed} \\ \text{people who remain unemployed} \end{array}}_{0.9u_0}.$$

We have:

$$e_1 = 0.98e_0 + 0.1u_0$$
$$u_1 = 0.02e_0 + 0.9u_0.$$

Using matrix multiplication of a vector, we can rewrite this as:

$$\underbrace{\begin{pmatrix} e_1 \\ u_1 \end{pmatrix}}_{\vec{G}_1} = \underbrace{\begin{pmatrix} 0.98 & 0.1 \\ 0.02 & 0.9 \end{pmatrix}}_{\mathbf{M}} \underbrace{\begin{pmatrix} e_0 \\ u_0 \end{pmatrix}}_{\vec{G}_0},$$

so we conclude that $\mathbf{M} = \begin{pmatrix} 0.98 & 0.1 \\ 0.02 & 0.9 \end{pmatrix}$. Since $\vec{G}_0 = (8000, 400)$, we find $\vec{G}_1$ as follows:

$$\vec{G}_1 = \begin{pmatrix} 0.98 & 0.1 \\ 0.02 & 0.9 \end{pmatrix} \begin{pmatrix} 8000 \\ 400 \end{pmatrix}$$

$$= \begin{pmatrix} 0.98(8000) + 0.1(400) \\ 0.02(8000) + 0.9(400) \end{pmatrix}$$

$$= \begin{pmatrix} 7840 + 40 \\ 160 + 360 \end{pmatrix}$$

$$= \begin{pmatrix} 7880 \\ 520 \end{pmatrix}.$$

Higher Dimensional Vectors

Vectors are not limited to just two components—we can define a vector with as many components as necessary for the problem at hand. We often refer to a vector with n components as an n-*dimensional vector* or simply an n-vector.

Example 3   A naturalist is studying three neighboring groups of animals. The sizes of the populations varies as individual animals move back and forth between the groups. Let the 3-vector $\vec{P}_0 = (100, 120, 200)$ give the sizes of three different animal populations in year $t = 0$. Suppose that after a year has passed,

$$\vec{P}_1 = \mathbf{T}\vec{P}_0 = \begin{pmatrix} 0.7 & 0.2 & 0.2 \\ 0.2 & 0.6 & 0.2 \\ 0.1 & 0.2 & 0.5 \end{pmatrix} \vec{P}_0.$$

(a) Find $\vec{P}_1$.
(b) Find $\vec{P}_2$ assuming $\vec{P}_2 = \mathbf{T}\vec{P}_1$.

Solution   (a) We have

$$\vec{P}_1 = \underbrace{\begin{pmatrix} 0.7 & 0.2 & 0.2 \\ 0.2 & 0.6 & 0.2 \\ 0.1 & 0.2 & 0.5 \end{pmatrix}}_{\mathbf{T}} \underbrace{\begin{pmatrix} 100 \\ 120 \\ 200 \end{pmatrix}}_{\vec{P}_0}$$

$$= \begin{pmatrix} 0.7 \cdot 100 + 0.2 \cdot 120 + 0.2 \cdot 200 \\ 0.2 \cdot 100 + 0.6 \cdot 120 + 0.2 \cdot 200 \\ 0.1 \cdot 100 + 0.2 \cdot 120 + 0.5 \cdot 200 \end{pmatrix} = \begin{pmatrix} 134 \\ 132 \\ 134 \end{pmatrix}.$$

This tells us that the first group of animals increases in size from 100 to 134, that the second increases from 120 to 132, and that the third decreases from 200 to 134.

(b) Here, we assume the same transition occurs between years 1 and 2 as between years 0 and 1:

$$\vec{P_2} = \underbrace{\begin{pmatrix} 0.7 & 0.2 & 0.2 \\ 0.2 & 0.6 & 0.2 \\ 0.1 & 0.2 & 0.5 \end{pmatrix}}_{\mathbf{T}} \underbrace{\begin{pmatrix} 134 \\ 132 \\ 134 \end{pmatrix}}_{\vec{P_1}}$$

$$= \begin{pmatrix} 0.7 \cdot 134 + 0.2 \cdot 132 + 0.2 \cdot 134 \\ 0.2 \cdot 134 + 0.6 \cdot 132 + 0.2 \cdot 134 \\ 0.1 \cdot 134 + 0.2 \cdot 132 + 0.5 \cdot 134 \end{pmatrix} = \begin{pmatrix} 147 \\ 132.8 \\ 106.8 \end{pmatrix}.$$

Rounding down, this tells us that the first group increases in size from 134 to 147, that the second does not change, and that the third decreases from from 134 to 106.

An Application of Matrices and Vectors to Economics

Economists divide a country's production into sectors. For example, a country might have three sectors: agricultural, industrial, and service. During the course of production, each sector consumes part of the production of all three sectors, in a manner described by a consumption matrix

$$\mathbf{C} = \begin{pmatrix} 0.10 & 0.25 & 0.12 \\ 0.33 & 0.11 & 0.16 \\ 0.21 & 0.12 & 0.42 \end{pmatrix}.$$

The first row gives the agricultural sector's share in the consumption of the other three sectors. For instance, the 0.25 in the middle of the first row says that the agricultural sector consumes 25% of the production of the industrial sector. The second row gives the industrial sector's consumption, and the third row gives the service sector's consumption.

Example 4   Suppose that a country's agricultural production is worth \$75 billion, its industrial production is worth \$45 billion, and its services are worth \$60 billion. We describe this using the production vector $\vec{P} = (75, 45, 60)$.

(a) Calculate $\mathbf{C}\vec{P}$ and interpret the result.

(b) Find the surplus vector, $\vec{S}$, which gives the amount of production remaining for each sector after accounting for what has been consumed in production. What does the surplus vector tell you about the country's economy?

Solution   (a) We calculate

$$\begin{pmatrix} 0.10 & 0.15 & 0.22 \\ 0.33 & 0.11 & 0.16 \\ 0.21 & 0.12 & 0.42 \end{pmatrix} \begin{pmatrix} 75 \\ 45 \\ 60 \end{pmatrix} = \begin{pmatrix} 27.45 \\ 39.30 \\ 46.35 \end{pmatrix}.$$

This tells us that \$25.95 billion in agricultural products, \$39.30 billion in industrial products, and \$46.35 billion in services are consumed during production.

(b) The surplus is what remains after what has been consumed in production, so

$$\vec{S} = \vec{P} - \mathbf{C}\vec{P} = \begin{pmatrix} 75 \\ 45 \\ 60 \end{pmatrix} - \begin{pmatrix} 27.45 \\ 39.30 \\ 46.35 \end{pmatrix} = \begin{pmatrix} 47.55 \\ 5.70 \\ 13.65 \end{pmatrix}.$$

This tells us that, after internal consumption is accounted for, \$47.55 billion in agricultural products, \$5.70 billion in industrial products and \$13.65 billion in services are available for consumption and export.

Exercises and Problems for Section 16.3

EXERCISES

In Exercises 1–8, find a single vector resulting from the operations.

1. $(5, 7, 13) + (2, 6, 8)$ **2.** $(100, 50) + (77, -3)$

3. $(\frac{1}{3}, 2, -\frac{5}{7}) + (1, 2, 3)$ **4.** $(x, y, z) + (2, 6, 8)$

5. $3(5, 7, 13)$ **6.** $\frac{1}{2}(10, 20, 30)$

7. $x(5, 7, 13)$ **8.** $2 \cdot 3(5, 7, 13)$

In Exercises 9–16, find a single vector resulting from the operations.

9. $5((2, 3, 4) + (3, 5, 7))$ **10.** $\frac{1}{2}((\frac{1}{7}, 3, 4) + (3, \frac{1}{2}, 9))$

11. $y((2, 3, 4) + (3, 5, 7))$ **12.** $\begin{pmatrix} 5 & 10 \\ 20 & 40 \end{pmatrix} \begin{pmatrix} 3 \\ 7 \end{pmatrix}$

13. $\begin{pmatrix} 9 & 15 \\ 2 & 6 \end{pmatrix} \begin{pmatrix} 4 \\ 5 \end{pmatrix}$

14. $ab((x, y, z) + (m, n, p))$

15. $\begin{pmatrix} 0.6 & 0.2 & 0 \\ 0.4 & 0.8 & 1 \end{pmatrix} \begin{pmatrix} 70 \\ 120 \\ 50 \end{pmatrix}$

16. $\begin{pmatrix} 0.6 & 0.2 & 0 \\ 0.3 & 0.4 & 0.5 \\ 0.1 & 0.4 & 0.5 \end{pmatrix} \begin{pmatrix} 100 \\ 200 \\ 300 \end{pmatrix}$

PROBLEMS

In Problems 17–22, the vectors

$$\vec{P} = (5, 22, 35, 18)$$
$$\vec{S} = (20, 33, 14, 40)$$
$$\vec{F} = (12, 28, 25, 20)$$
$$\vec{W} = (2, 19, 42, 12)$$

represent the average number of customers in the morning, early afternoon, late afternoon, and evening in a cafe during Spring ($\vec{P}$), Summer ($\vec{S}$), Fall ($\vec{F}$), and Winter ($\vec{W}$). For example, $\vec{P} = (5, 22, 35, 18)$ means that, in Spring, the cafe has an average of 5 customers in the morning, 22 in the early afternoon, 35 in the late afternoon, and 18 in the evening.

17. In what season does the cafe have the most customers on an average day?

18. In what season does the cafe have the most customers on the average late afternoon?

19. At what time of day does the cafe average the most customers in Summer?

20. Find $\vec{Q} = \vec{S} - \vec{P}$. What does $\vec{Q}$ represent?

21. Find $\vec{H} = 91\vec{W}$. What might $\vec{H}$ represent? [Hint: Assume there are 91 days in Winter.]

22. What might $\vec{A} = 91(\vec{P} + \vec{S} + \vec{F} + \vec{W})$ represent? [Hint: Assume each season is 91 days long.]

In a certain town, the number of Democrats, Republicans, and Independents is represented by a vector $\vec{V} = (d, r, i) = (450, 560, 110)$. Each group plans to use a voter drive in order to add voters, represented by the vector $\vec{E} = (100, 80, 0)$. Evaluate and interpret the expressions in Problems 23–26.

23. $\vec{V} + \vec{E}$

24. $1.05\vec{V}$

25. $\vec{V} + (60, 50, -110)$

26. $\vec{V} + 2\vec{E}$

In Problems 27–28, use the information in Example 4.

27. Which of the following production vectors gives a surplus closest to the goal of $50 billion in each sector?

(a) $(1100, 1250, 1500)$ (b) $(100, 100, 100)$

(c) $(110, 125, 150)$ (d) $(50, 100, 150)$

28. Suppose that production is reduced by 10%. Show algebraically that the surplus is also reduced by 10%.

16.4 MATRICES AND SYSTEMS OF LINEAR EQUATIONS

Consider the system of equations

$$\begin{cases} 3x + y = 14 \\ \phantom{3x + {}} y = 11. \end{cases}$$

Substituting the second equation, $y = 11$, into the first, we get $3x + 11 = 14$, so $x = 1$. Thus $(x, y) = (1, 11)$ is a solution to the system. We can think of this solution as a vector and use matrix multiplication to write:

$$\begin{pmatrix} 3 & 1 \\ 0 & 1 \end{pmatrix} \begin{pmatrix} x \\ y \end{pmatrix} = \begin{pmatrix} 14 \\ 11 \end{pmatrix} .$$

The coefficients   The variables   The required values
in our system   in our system   of our system

To see how this works, perform the matrix multiplication on the left-hand side in the usual way:

Coefficients Variables   Values

$$\begin{pmatrix} 3 & 1 \\ 0 & 1 \end{pmatrix} \begin{pmatrix} x \\ y \end{pmatrix} = \begin{pmatrix} 14 \\ 11 \end{pmatrix}$$

$$\begin{pmatrix} 3 \cdot x + 1 \cdot y \\ 0 \cdot x + 1 \cdot y \end{pmatrix} = \begin{pmatrix} 14 \\ 11 \end{pmatrix} \qquad \text{Multiply out left-hand side}$$

$$\begin{pmatrix} 3x + y \\ y \end{pmatrix} = \begin{pmatrix} 14 \\ 11 \end{pmatrix} . \qquad \text{This mirrors original system}$$

Left-hand side of   Values
original system

Notice in particular that:

- The first component of the vector on the left, $3x + y$, must equal the first component of the vector on the right, or 14.

- The second component of the vector on the left, y, must equal the second component of the vector on the right, or 11.

Thus, a single, matrix-based equation captures the essence of a system of two equations:

Single matrix based equation   System of two ordinary equations

$$\begin{pmatrix} 3 & 1 \\ 0 & 1 \end{pmatrix} \begin{pmatrix} x \\ y \end{pmatrix} = \begin{pmatrix} 14 \\ 11 \end{pmatrix} \qquad \longleftrightarrow \qquad \begin{cases} 3x + y = 14 \\ \phantom{3x + {}} y = 11. \end{cases}$$

In general:

> It is often useful to summarize an entire system of two or more equations with a single equation involving matrices and vectors.

Example 1   Given that

$$\mathbf{M} = \begin{pmatrix} 2 & 1 & 5 \\ 3 & 0 & 2 \\ 5 & -2 & 3 \end{pmatrix}, \vec{v} = \begin{pmatrix} x \\ y \\ z \end{pmatrix}, \quad \text{and} \quad \vec{w} = \begin{pmatrix} 20 \\ 12 \\ 17 \end{pmatrix},$$

state (but do not solve) the system of equations determined by $\mathbf{M}\vec{v} = \vec{w}$.

Solution   We have:

$$\overbrace{\begin{pmatrix} 2 & 1 & 5 \\ 3 & 0 & 2 \\ 5 & -2 & 3 \end{pmatrix}}^{\mathbf{M}} \overbrace{\begin{pmatrix} x \\ y \\ z \end{pmatrix}}^{\vec{v}} = \overbrace{\begin{pmatrix} 20 \\ 12 \\ 17 \end{pmatrix}}^{\vec{w}}$$

$$\begin{pmatrix} 2x + 1 \cdot y + 5z \\ 3x + 0 \cdot y + 2z \\ 5x + (-2)y + 3z \end{pmatrix} = \begin{pmatrix} 20 \\ 12 \\ 17 \end{pmatrix} \quad \text{Multiply out left-hand side}$$

$$\begin{pmatrix} 2x + y + 5z \\ 3x + 2z \\ 5x - 2y + 3z \end{pmatrix} = \begin{pmatrix} 20 \\ 12 \\ 17 \end{pmatrix}. \quad \text{Simplify}$$

Setting equal the first, second, and third components of both sides, we see that:

$$\overbrace{\begin{pmatrix} 2 & 1 & 5 \\ 3 & 0 & 2 \\ 5 & -2 & 3 \end{pmatrix}\begin{pmatrix} x \\ y \\ z \end{pmatrix} = \begin{pmatrix} 20 \\ 12 \\ 17 \end{pmatrix}}^{\text{Single matrix-based equation}} \longleftrightarrow \overbrace{\begin{cases} 2x + y + 5z = 20 \\ 3x + 2z = 12 \\ 5x - 2y + 3z = 17. \end{cases}}^{\text{System of three ordinary equations}}$$

Augmented Matrices

Another way to describe a system of equations is to use a special kind of matrix called an *augmented matrix*:

$$\underbrace{\begin{cases} \mathbf{3}x + \mathbf{1}y = \mathbf{14} \\ \mathbf{0}x + \mathbf{1}y = \mathbf{11}. \end{cases}}_{\substack{\text{A simultaneous} \\ \text{system of equations}}} \longleftrightarrow \underbrace{\left(\begin{array}{cc|c} 3 & 1 & 14 \\ 0 & 1 & 11 \end{array} \right)}_{\substack{\text{The corresponding} \\ \text{augmented matrix}}}$$

Here, we ignore the variables and focus only on the coefficients and values. A characteristic feature of augmented matrices is the vertical line used to separate the coefficients of the original system from the values on the right-hand side.

One advantage of describing a system of equations using an augmented matrix is that the matrix form is more compact and can be easier to work with.

Example 2   Describe the following systems using augmented matrices.

(a) $\begin{cases} 2v - 3w = 7 \\ 5v + 9w = 0 \end{cases}$ (b) $\begin{cases} 3p - 2q + 7r = 6 \\ 2p - r = 8 \\ 4q + \dfrac{r}{2} = 2 \end{cases}$ (c) $\begin{cases} 7x + 2y = 5 - 3z \\ 2y - 4z = 2x - 1 \\ 3(x - y) = 2(x - z - 3) \end{cases}$

Solution   (a) We can describe this system using the augmented matrix

$$\left(\begin{array}{cc|c} 2 & -3 & 7 \\ 5 & 9 & 0 \end{array} \right).$$

Notice that the variables v and w are not included in the description; as far as the matrix is concerned, they may as well be x and y or any other two variables.

(b) We first rewrite the system so that each equation is in the same form and the variables appear in the same order:

$$\begin{cases} 3p + (-2)q + 7r = 6 \\ 2p + 0 \cdot q + (-1)r = 8 \\ 0 \cdot p + 4q + 0.5r = 2. \end{cases}$$

We can now describe this using the augmented matrix

$$\left(\begin{array}{ccc|c} 3 & -2 & 7 & 6 \\ 2 & 0 & -1 & 8 \\ 0 & 4 & 0.5 & 2 \end{array} \right).$$

(c) We first rewrite the system so that each equation is in the same form and the variables appear in the same order:

$$\begin{cases} 7x + 2y + 3z = 5 \\ (-2)x + 2y + (-4)z = -1 \\ 1 \cdot x + (-3)y + 2z = -6. \end{cases}$$

We can now describe this using the augmented matrix

$$\left(\begin{array}{ccc|c} 7 & 2 & 3 & 5 \\ -2 & 2 & -4 & -1 \\ 1 & -3 & 2 & -6 \end{array} \right).$$

Echelon Form

The system on page 482 has a particularly simple form. The second equation, $y = 11$, has no x on the left, so it gives us the value of y directly. This makes the equation easy to solve, because we can substitute the value of y into the first equation and solve for x. The simple form of the system is reflected in the fact that the augmented matrix for that system,

$$\left(\begin{array}{cc|c} 3 & 1 & 14 \\ 0 & 1 & 11 \end{array} \right),$$

has a zero in the bottom left corner. For systems of equations in three variables, there is also a simple form of the augmented matrix that makes the equation easy to solve.

Example 3   Solve the following system and find the augmented matrix:

$$\begin{cases} x + y + z = 6 \\ \phantom{x + {}} y + z = 5 \\ \phantom{x + y + {}} z = 3. \end{cases}$$

Solution   The system can be solved by substitution. The third equation tells us $z = 3$. Substituting this into the second equation we get $y = 2$. Then substituting $y = 2$ and $z = 3$ into the first equation we get $x = 1$. Thus the solution is $(x, y, z) = (1, 2, 3)$. The augmented matrix for the system is

$$\left(\begin{array}{ccc|c} 1 & 1 & 1 & 6 \\ 0 & 1 & 1 & 5 \\ 0 & 0 & 1 & 3 \end{array} \right).$$

Example 3 was easy to solve because of the zeros below the diagonal in the augmented matrix, which made it possible to solve the system by successively substituting each equation into the one above. A similar form works for systems of 4 equations in 4 variables (see Problem 25). This leads to the following definition:

Echelon Form of an Augmented Matrix

An augmented matrix is said to be in *echelon form* if the first nonzero entry in each row occurs to the right of the first nonzero entry in the row above, and if all the rows which consist entirely of zeros to the left of the bar are at the bottom. A matrix in echelon form corresponds to a system that can be solved easily by substitution.

Solving Systems of Equations by Putting Matrices into Echelon Form

To solve complicated systems of equations, we use elimination to put them in echelon form.

Example 4   Solve the following system, keeping track of the augmented matrix:

$$\begin{cases} 4x + 3y = 18 \\ 2x - 5y = -4. \end{cases}$$

Solution   The following table shows steps in solving the system. Since the equations in the system correspond to rows of the augmented matrix, the steps correspond to operations on the rows, which we record in the right column of the table.

Step taken	Simultaneous system of equations	Corresponding matrix	Row operation used
(i) The original system	$\begin{cases} 4x + 3y = 18 \quad \text{(equation 1)} \\ 2x - 5y = -4 \quad \text{(equation 2)} \end{cases}$	$\begin{pmatrix} 4 & 3 & \vert & 18 \\ 2 & -5 & \vert & -4 \end{pmatrix}$	The original matrix
(ii) Multiply equation 2 by 2	$\begin{cases} 4x + 3y = 18 \\ 4x - 10y = -8 \quad \text{(equation 3)} \end{cases}$	$\begin{pmatrix} 4 & 3 & \vert & 18 \\ 4 & -10 & \vert & -8 \end{pmatrix}$	Multiply second row by 2
(iii) Subtract equation 1 from equation 3	$\begin{cases} 4x + 3y = 18 \\ -13y = -26 \quad \text{(equation 4)} \end{cases}$	$\begin{pmatrix} 4 & 3 & \vert & 18 \\ 0 & -13 & \vert & -26 \end{pmatrix}$	Subtract first row from second row

Solving equation 4 for y we get $y = 2$, and substituting this into equation 1 we get

$$4x + 3(2) = 18, \quad \text{so} \quad x = 3.$$

Thus the solution is $(x, y) = (3, 2)$.

Notice that the matrix in step (iii) of our solution to Example 4 is in echelon form, because the bottom left entry is 0. This corresponds to the fact that the x has been eliminated.

Using Row Operations

In Example 4 we keep track of the steps by recording **row operations** on the augmented matrix. Table 16.1 shows row operations corresponding to legal procedures when solving equations by elimination.

Table 16.1   *Row operations correspond to steps used in the process of elimination*

Step in the process of elimination	Corresponding row operation
Add or subtract any two equations	Add or subtract any two rows
Multiply an equation by a nonzero constant	Multiply a row by a nonzero constant
Switch the order the equations are written in	Switch any two rows

We can use this correspondence to solve equations by operating on the augmented matrix directly, not on the equations. Just as the goal of elimination is to isolate one or more of the variables, the goal of row operations is to put a matrix in echelon form.

Example 5   Use row operations to simplify the augmented matrix for the following system:

$$\begin{cases} 2x + y = 17 \\ 3x - 2y = 1. \end{cases}$$

Use the resulting matrix to solve the system.

Solution   The augmented matrix for this system is $\left(\begin{array}{cc|c} 2 & 1 & 17 \\ 3 & -2 & 1 \end{array} \right)$. Our goal is to put it into echelon form using row operations. In order to get a zero in the bottom left, we first multiply both rows by a constant so that they have the same left entry, then subtract the first row from the second.

$$\left(\begin{array}{cc|c} 2 & 1 & 17 \\ 3 & -2 & 1 \end{array} \right) \longrightarrow \left(\begin{array}{cc|c} 6 & 3 & 51 \\ 3 & -2 & 1 \end{array} \right) \qquad \text{multiply first row by 3}$$

$$\left(\begin{array}{cc|c} 6 & 3 & 51 \\ 3 & -2 & 1 \end{array} \right) \longrightarrow \left(\begin{array}{cc|c} 6 & 3 & 51 \\ 6 & -4 & 2 \end{array} \right) \qquad \text{multiply second row by 2}$$

$$\left(\begin{array}{cc|c} 6 & 3 & 51 \\ 6 & -4 & 2 \end{array} \right) \longrightarrow \left(\begin{array}{cc|c} 6 & 3 & 51 \\ 0 & -7 & -49 \end{array} \right) \qquad \text{subtract first row from second row.}$$

We have obtained a matrix in echelon form. This matrix corresponds to a system of equations that can be solved easily by substitution:

$$\left(\begin{array}{cc|c} 6 & 3 & 51 \\ 0 & -7 & -49 \end{array} \right) \longrightarrow \begin{cases} 6x + 3y = 51 \\ -7y = -49. \end{cases}$$

The second equation gives $y = 7$, and substituting into the first equation we get $6x + 21 = 51$, so $x = 5$. We verify that $(x, y) = (5, 7)$ is a solution by substituting values into the original system.

Using Echelon Form to Detect Systems with No Solutions

Putting a matrix into echelon form also helps us see when the equation has no solutions.

Example 6   Try to use row operations to solve the following system.

$$\begin{cases} 3x - 9y = 5 & \text{(equation 5)} \\ 4x - 12y = 6. & \text{(equation 6)} \end{cases}$$

What happens?

Solution   The augmented matrix for this system is $\left(\begin{array}{cc|c} 3 & -9 & 5 \\ 4 & -12 & 6 \end{array} \right)$. Following is one of many possible approaches we might use to simplify this matrix.

$$\begin{pmatrix} 3 & -9 & | & 5 \\ 4 & -12 & | & 6 \end{pmatrix} \longrightarrow \begin{pmatrix} 12 & -36 & | & 20 \\ 4 & -12 & | & 6 \end{pmatrix} \quad \text{multiply first row by 4}$$

$$\begin{pmatrix} 12 & -36 & | & 20 \\ 4 & -12 & | & 6 \end{pmatrix} \longrightarrow \begin{pmatrix} 12 & -36 & | & 20 \\ 12 & -36 & | & 18 \end{pmatrix} \quad \text{multiply second row by 3}$$

$$\begin{pmatrix} 12 & -36 & | & 20 \\ 12 & -36 & | & 18 \end{pmatrix} \longrightarrow \begin{pmatrix} 12 & -36 & | & 20 \\ 0 & 0 & | & -2 \end{pmatrix} \quad \text{subtract first row from second.}$$

Our new augmented matrix corresponds to the system

$$\begin{pmatrix} 12 & -36 & | & 20 \\ 0 & 0 & | & -2 \end{pmatrix} \longrightarrow \begin{cases} 12x - 36y = 20 \\ 0 \cdot x + 0 \cdot y = -2. \end{cases} \longrightarrow \begin{cases} 12x - 36y = 20 \\ 0 = -2. \quad \text{inconsistent!} \end{cases}$$

We see that the resulting system is *inconsistent*—it does not make sense. Since it is algebraically equivalent to the original system, we conclude that the original system has no solution.

The matrix at the end of Example 6 is in echelon form, because it has a zero at the bottom left. However, it also has a zero at the bottom right, leading to an equation with no solutions, $0 = -2$. Putting a matrix into echelon form helps detect systems with no solutions.

A Systematic Method for Putting Matrices into Echelon Form

In Example 5 we multiplied rows by a constant so that the x coefficients matched, then subtracted the first row from the second row. Although it is usually easy to see how to do this, for the purposes of describing a systematic method it is useful to introduce a new operation:

> Adding a multiple of one row to another row.

We illustrate it using the system in Example 5, which has augmented matrix

$$\begin{pmatrix} 2 & 1 & | & 17 \\ 3 & -2 & | & 1 \end{pmatrix}.$$

Since $3 = 1.5 \times 2$, we can get a zero at the bottom left by adding -1.5 times the first row to the second row:

$$\begin{pmatrix} 2 & 1 & | & 17 \\ 3 & -2 & | & 1 \end{pmatrix} \longrightarrow \begin{pmatrix} 2 & 1 & | & 17 \\ 0 & -3.5 & | & -24.5 \end{pmatrix} \quad \text{add } -1.5 \text{ times the first row to the second row.}$$

Although the matrix is different from the one we had before, it still has the same solution, since the equation $-3.5z = -24.5$ has the solution $z = 7$, as in Example 5.

Example 7   Solve

$$\begin{pmatrix} 0 & -1 & 2 \\ 1 & 2 & -1 \\ -4 & -1 & 1 \end{pmatrix} \begin{pmatrix} x \\ y \\ z \end{pmatrix} = \begin{pmatrix} 8 \\ -2 \\ 7 \end{pmatrix}.$$

Solution   The augmented matrix is

$$\left(\begin{array}{ccc|c} 0 & -1 & 2 & 8 \\ 1 & 2 & -1 & -2 \\ -4 & -1 & 1 & 7 \end{array} \right).$$

We want to start by getting zeros in the left column, but there is a problem with the matrix in its current form: adding multiples of the top row to the others will have no effect on the left column, since the top left entry is zero. So our our first step is to swap the top two rows (corresponding to swapping the order of equations in the system) so that the top left entry is 1.

$$\left(\begin{array}{ccc|c} 0 & -1 & 2 & 8 \\ 1 & 2 & -1 & -2 \\ -4 & -1 & 1 & 7 \end{array} \right) \longrightarrow \left(\begin{array}{ccc|c} 1 & 2 & -1 & -2 \\ 0 & -1 & 2 & 8 \\ -4 & -1 & 1 & 7 \end{array} \right) \quad \text{swap the first two rows}$$

Now we add a multiple of the top row to the third row, choosing the multiple so as to get a zero at the bottom left.

$$\left(\begin{array}{ccc|c} 1 & 2 & -1 & -2 \\ 0 & -1 & 2 & 8 \\ -4 & -1 & 1 & 7 \end{array} \right) \longrightarrow \left(\begin{array}{ccc|c} 1 & 2 & -1 & -2 \\ 0 & -1 & 2 & 8 \\ 0 & 7 & -3 & -1 \end{array} \right) \quad \text{add 4 times the first row to the third}$$

We now have all the zeros we want in the first column, and we have finished with subtracting the first row. Turning our attention to the second row, we want to add some multiple of it to the third row to get a zero in the middle column at the bottom. The correct multiple is 7.

$$\left(\begin{array}{ccc|c} 1 & 2 & -1 & -2 \\ 0 & -1 & 2 & 8 \\ 0 & 7 & -3 & -1 \end{array} \right) \longrightarrow \left(\begin{array}{ccc|c} 1 & 2 & -1 & -2 \\ 0 & -1 & 2 & 8 \\ 0 & 0 & 11 & 55 \end{array} \right) \quad \text{add 7 times the second row to the third row}$$

The matrix is now in echelon form, and we can solve the corresponding system of equations by substitution. The third row corresponds to the equation

$$11z = 55 \quad \longrightarrow z = 5.$$

Substituting this into the second equation we get

$$-y + 2 \cdot 5 = 8 \quad \longrightarrow \quad y = 2,$$

and substituting both these into the first equation we get

$$x + 2 \cdot 2 - 1 \cdot 5 = -2 \quad \longrightarrow \quad x = -1.$$

So the solution is $(x, y, z) = (-1, 2, 5)$.

The general method illustrated in the previous example is

---

1. Swap rows so there is a nonzero entry at the top left.

2. Add multiples of the top row to the others in order to get zeros below the top entry in the left column.

3. Repeat steps 1 and 2 with the rows below the top row to get zeros below the top two entries in the second column.

4. The method can be continued in the same way with systems that have more than 3 variables or more than 3 equations.

Exercises and Problems for Section 16.4

EXERCISES

■Exercises **1–6** give the augmented matrix in echelon form of a system of linear equations in x and y. Find x and y.

1. $\begin{pmatrix} 1 & 1 & | & 9 \\ 0 & 1 & | & 7 \end{pmatrix}$

2. $\begin{pmatrix} 1 & -1 & | & 9 \\ 0 & 2 & | & 3 \end{pmatrix}$

3. $\begin{pmatrix} 2 & -1 & | & 9 \\ 0 & -3 & | & 9 \end{pmatrix}$

4. $\begin{pmatrix} \frac{5}{2} & 1 & | & \frac{21}{2} \\ 0 & \frac{1}{2} & | & 4 \end{pmatrix}$

5. $\begin{pmatrix} \frac{1}{3} & -\frac{1}{2} & | & 10 \\ 0 & 2 & | & 10 \end{pmatrix}$

6. $\begin{pmatrix} 2 & -1 & | & 0.01 \\ 0 & 10 & | & 0.5 \end{pmatrix}$

■In Exercises **7–8**, the augmented matrices for a system of equations have been reduced to a form which makes them easy to solve. Find the values of x and y by inspection (that is, without writing anything down except the answer).

7. $\begin{pmatrix} 1 & 0 & | & 2 \\ 0 & 1 & | & 3 \end{pmatrix}$

8. $\begin{pmatrix} 2 & 0 & | & 4 \\ 0 & 3 & | & 6 \end{pmatrix}$

9. State the system of equations determined by $\mathbf{N}\vec{p} = \vec{q}$ for

$$\mathbf{N} = \begin{pmatrix} 5 & 7 \\ 2 & 4 \end{pmatrix}, \vec{p} = \begin{pmatrix} x \\ y \end{pmatrix}, \vec{q} = \begin{pmatrix} 9 \\ -5 \end{pmatrix}$$

10. State the system of equations determined by $\mathbf{M}\vec{v} = \vec{w}$ for

$$\mathbf{M} = \begin{pmatrix} 4 & 3 & 2 \\ 1 & -5 & -4 \\ -3 & -2 & -1 \end{pmatrix}, \vec{v} = \begin{pmatrix} x \\ y \\ z \end{pmatrix}, \vec{w} = \begin{pmatrix} 10 \\ 15 \\ 20 \end{pmatrix}$$

11. State the system of equations determined by $\mathbf{A}\vec{v} = \vec{r}$ for

$$\mathbf{A} = \begin{pmatrix} c & d & e \\ f & g & h \\ l & m & n \end{pmatrix}, \vec{v} = \begin{pmatrix} x \\ y \\ z \end{pmatrix}, \vec{r} = \begin{pmatrix} 1 \\ 2 \\ 3 \end{pmatrix}$$

■Write the system of equations described by the augmented matrices in Exercises **12–15**.

12. $\begin{pmatrix} 14 & 7 & | & 10 \\ 19 & 11 & | & 12 \end{pmatrix}$

13. $\begin{pmatrix} 1 & 2 & 3 & | & 4 \\ 5 & 6 & 7 & | & 8 \\ 9 & 10 & 11 & | & 12 \end{pmatrix}$

14. $\begin{pmatrix} a & b & c & | & d \\ e & f & g & | & h \\ i & j & k & | & l \end{pmatrix}$

15. $\begin{pmatrix} 7 & 0 & 3 & 5 & | & a \\ 6+m & 0 & 0 & 2 & | & b \\ 0 & 1 & 1 & 1 & | & c \\ 5 & 7 & 9 & 11 & | & d \end{pmatrix}$

■ If possible, use row operations to solve the systems in Exercises **16–23**.

16. $\begin{cases} x + y = 5 \\ 2x - y = 7 \end{cases}$

17. $\begin{cases} 5x + 3y = 4 \\ x + 7y = 20 \end{cases}$

18. $\begin{cases} 7x - 4y = 12 \\ -6x - 3y = -36 \end{cases}$

19. $\begin{cases} -13x + 26y = -13 \\ 3x - 8y = -1 \end{cases}$

20. $\begin{cases} 6x + 5y = 4 \\ x - 3y = 14 \end{cases}$

21. $\begin{cases} 4y - 2x = 7 \\ 2x + 4y = 9 \end{cases}$

22. $\begin{cases} 2x - 3y = 4 \\ 7 = -3y + 4x \end{cases}$

23. $\begin{cases} 2x + 5 = 4 + y \\ 7y = 4x - 3 \end{cases}$

PROBLEMS

24. For constants a, b, c, a system of linear equations has augmented matrix

$$\left(\begin{array}{cc|c} a & b & 4 \\ 0 & c & 3 \end{array} \right).$$

(a) If $a = 1$ and $b = 2$, are there any values of c for which it is impossible to find unique solutions for x and y? If so, what are these values of c?

(b) If $a = 1$ and $c = 3$, are there any values of b for which it is impossible to find unique solutions for x and y? If so, what are these values of b?

(c) If $b = 2$ and $c = 3$, are there any values of a for which it is impossible to find unique solutions for x and y? If so, what are these values of c?

25. (a) Solve

$$x + y + z + w = 10$$
$$y + z + w = 9$$
$$z + w = 7$$
$$w = 4.$$

(b) Write the augmented matrix for this system.

(c) What special form of the augmented matrix makes the system easy to solve?

■ In Problems **26–29**, write the system of equations

(a) In the form $\mathbf{M}\vec{v} = \vec{w}$.

(b) As an augmented matrix.

26. $\begin{cases} x + y + 6z = 15 \\ 5x + 2y + 3z = 19 \\ -2x - 3y + 4z = 7. \end{cases}$

27. $\begin{cases} 3x + 6z = 7 \\ -5x + 2y = 8 \\ 4y - z = 9. \end{cases}$

28. $\begin{cases} 2x + 3y + 6z = 6z + 15 \\ 3z = x - y + 19 \\ -3y + 17 = 4x - 7. \end{cases}$

29. $\begin{cases} x = ay - z + 4 \\ 3z - ax = x - y + 19 \\ bz - 3y + 15 = 4x - cz. \end{cases}$

30. State the system of equations determined by $\mathbf{M}\vec{v} = \vec{w}$ for

$$\mathbf{M} = \begin{pmatrix} 1 & 5 & 7 & 2 \\ 1 & 0 & 0 & 4 \\ 0 & 2 & 1 & 3 \\ 5 & 6 & 0 & 0 \end{pmatrix}, \vec{v} = \begin{pmatrix} x \\ y \\ z \\ t \end{pmatrix}, \vec{w} = \begin{pmatrix} 10 \\ 5 \\ 7 \\ 9 \end{pmatrix}$$

31. A new music player is ready for market and has the following economic equations, where Q is in thousands of units, and p is price in dollars.

$$\text{Demand } Q = 500 - 8p$$
$$\text{Supply } Q = 10 + 2p.$$

(a) Find the augmented matrix for this system.

(b) Use row operations to find the equilibrium point, where supply equals demand.

32. A rental company has 350 cars. Among these cars D of them are at its downtown location, P of them at the port and A of them at the airport. Each year, 10% of the cars rented downtown are returned at the port, 20% of the cars rented at the port are returned at the airport, and 5% of the cars rented at the airport are returned downtown. All other cars are returned to where they were rented. The number of cars at each location remains the same each year.

(a) Write four equations from the given information.

(b) Write an augmented matrix for this system of equations.

(c) Use row operations to find the number of cars at each location.

33. A Double-Double fast food diet has approximately double fat to protein and double carbohydrates to fat. One day, you eat B burgers, S shakes, and F servings of fries, for a total of 125 gm of protein, 250 gm of fat, and 505 gm of carbohydrates. Table 16.2 shows the nutritional information for three fast foods.

Table 16.2

	Burger	Shake	Fries
Fat (gm)	40	30	25
Carbohydrates (gm)	50	115	60
Protein (gm)	30	15	5

(a) Write a system of equations describing this situation.

(b) Write an augmented matrix for the system.

(c) Use row operations to find how many of each item you eat.

REVIEW EXERCISES AND PROBLEMS FOR CHAPTER 16

EXERCISES

■ Evaluate the expressions in Exercises **1–6** using

$$\mathbf{A} = \begin{pmatrix} 0.1 & 0.9 \\ -1.1 & 1 \end{pmatrix}, \mathbf{B} = \begin{pmatrix} 3 & 0 \\ 1 & -4 \end{pmatrix}$$

1. $10\mathbf{A} + \mathbf{B}$

2. $10(\mathbf{A} + 0.1\mathbf{B})$

3. $\mathbf{A} - 0.5\mathbf{B}$

4. $\mathbf{A}^2$

5. $\mathbf{AB} + \mathbf{BA}$

6. $\mathbf{ABA}$

■ In Exercises **7–10**, does the expression $3\mathbf{A} + 2\mathbf{B}$ make sense for the given $\mathbf{A}$ and $\mathbf{B}$? You do not need to evaluate the expression.

7. $\mathbf{A} = \begin{pmatrix} 0 & 0 \\ 0 & 0 \end{pmatrix}, \mathbf{B} = \begin{pmatrix} 1 & 1 \\ 1 & 1 \end{pmatrix}$

8. $\mathbf{A} = \begin{pmatrix} 1 & 0 & 1 \\ 2 & -1 & 2 \end{pmatrix}, \mathbf{B} = \begin{pmatrix} 1 & 2 \\ 0 & -1 \\ 1 & 2 \end{pmatrix}$

9. $\mathbf{A} = \begin{pmatrix} 2 \\ 0 \end{pmatrix}, \mathbf{B} = \frac{1}{2}\begin{pmatrix} 1 \\ 0 \end{pmatrix}$

10. $\mathbf{A} = \begin{pmatrix} 3 & 4 \\ 4 & 3 \\ 0 & 1 \end{pmatrix}, \mathbf{B} = 0.5\mathbf{A}$

■ In Exercises **11–14**, does the expression $\mathbf{A}(\mathbf{BC})$ make sense for the given $\mathbf{A}, \mathbf{B}$, and $\mathbf{C}$?

11. $\mathbf{A} = \begin{pmatrix} 10 & 1 \\ -1 & 23 \end{pmatrix}, \mathbf{B} = \begin{pmatrix} 3 & 5 \\ 0 & -2 \\ 6 & 8 \end{pmatrix},$

$\mathbf{C} = \begin{pmatrix} 5 & 1 & 4 \\ 6 & 2 & -9 \end{pmatrix}$

12. $\mathbf{A} = \begin{pmatrix} 10 & 1 \\ -1 & 23 \end{pmatrix}, \mathbf{B} = \begin{pmatrix} 5 & 1 & 4 \\ 6 & 2 & -9 \end{pmatrix},$

$\mathbf{C} = \begin{pmatrix} 3 & 5 \\ 0 & -2 \\ 6 & 8 \end{pmatrix}$

13. $\mathbf{A} = \begin{pmatrix} 10 & 1 \\ -1 & 23 \end{pmatrix}, \mathbf{B} = \begin{pmatrix} 3 & 5 \\ 0 & -2 \\ 6 & 8 \end{pmatrix},$

$\mathbf{C} = \begin{pmatrix} 5 & 6 \\ 1 & 2 \\ 4 & -9 \end{pmatrix}$

14. $\mathbf{A} = \begin{pmatrix} 10 & 1 \\ -1 & 23 \end{pmatrix}, \mathbf{B} = \begin{pmatrix} 3 & 0 & 6 \\ 5 & -2 & 8 \end{pmatrix},$

$\mathbf{C} = \begin{pmatrix} 5 & 1 & 4 \\ 6 & 2 & -9 \end{pmatrix}$

Exercises **15–16** give the augmented matrix in echelon form of a system of linear equations in x, y, and z. Find x, y, and z.

15. $\begin{pmatrix} 1 & 2 & 3 & | & 5 \\ 0 & 1 & 1 & | & 2 \\ 0 & 0 & 1 & | & -6 \end{pmatrix}$

16. $\begin{pmatrix} 2 & -1 & 3 & | & 2 \\ 0 & 3 & -2 & | & 0.5 \\ 0 & 0 & 5 & | & 2.5 \end{pmatrix}$

In Exercises **17–18**, the augmented matrices for a system of equations have been reduced to a form which makes them easy to solve. Find the values of x, y and z by inspection (that is, without writing anything down except the answer).

17. $\begin{pmatrix} 1 & 0 & 0 & | & 2 \\ 0 & 1 & 0 & | & -3 \\ 0 & 0 & 1 & | & 4 \end{pmatrix}$

18. $\begin{pmatrix} 2 & 0 & 0 & | & 10 \\ 0 & -1 & 0 & | & 7 \\ 0 & 0 & 1/2 & | & 3 \end{pmatrix}$

If possible, use row operations to solve the systems in Exercises **19–22**.

19. $\begin{cases} x + y + 2z = 9 \\ 2x - y + z = 3 \\ -x + y - z = -2 \end{cases}$

20. $\begin{cases} 3x + y + 4z = 13 \\ -x - 2y + 3z = 8 \\ x + 3y - z = 3 \end{cases}$

21. $\begin{cases} 2y + 5z = 9 \\ x - 3y = 12 \\ 2x - z = 67 \end{cases}$

22. $\begin{cases} y + z = 9 - y + x \\ 4x - y = y + 2 \\ z = 3x + 4y - 8 \end{cases}$

PROBLEMS

Problems **23–25** refer to Table 16.3, which shows the grams of fat, carbohydrates, and protein in three foods.[3] A gram of fat has 9 calories, while each gram of carbohydrate or protein has 4 calories. Let $\mathbf{M}$ be the matrix whose rows represent different types of foods and whose columns represent fat, carbohydrates, or protein.

Table 16.3

	Fat (gm)	Carbohydrates (gm)	Protein (gm)
Cookie	10	30	2
Sandwich	4.5	50	20
Salad	0	11	3

23. Let $\vec{V}$ be the vector whose components are the calories per gram for fat, carbohydrates, and protein.

  (a) Evaluate the product $\vec{C} = \mathbf{M}\vec{V}$. What is the practical meaning of $\vec{C}$?
  (b) Which has more calories, a salad and a chocolate chip cookie, or a sandwich?

24. Suppose you add a new food, pizza.

  (a) What are the dimensions of the matrix $\mathbf{M}$?
  (b) What are the dimensions of the calorie vector $\vec{C}$?

25. Suppose you add a new ingredient, salt.

  (a) What are the dimensions of the matrix $\mathbf{M}$?

  [3] http://www.subway.com, Accessed May 3, 2009

  (b) What extra information is necessary to express $\vec{C}$ as a product $\mathbf{M}\vec{V}$?

26. Show that the matrices

$$\begin{pmatrix} 1 & 0 \\ 0 & 1 \end{pmatrix}, \begin{pmatrix} 0 & 0 \\ 0 & 0 \end{pmatrix}, \quad \text{and} \quad \begin{pmatrix} 1 & 0 \\ 0 & 0 \end{pmatrix}$$

all satisfy $\mathbf{A}^2 = \mathbf{A}$. Give a fourth matrix that satisfies the equation.

27. Using the information in Example 2 on page 478, find $\vec{G}_2$, the employment vector in year $t = 2$.

28. In an election for town dog-catcher, 320 voters are committed to candidate A and 679 are committed to candidate B. After a brutal campaign, 5% of candidate A's supporters and 20% of candidate B's supporters switch sides.

  (a) Write a vector $\vec{V}$ representing the number of supporters of each candidate before the campaign.
  (b) Write a matrix $\mathbf{M}$ so that $\mathbf{M}\vec{V}$ represents the number of supporters of each candidate after the campaigning.
  (c) Evaluate $\mathbf{M}\vec{V}$. Who wins the election?

29. A coffee roaster produces a blend where the fraction of coffee from Africa, at \$8 per lb, is A, the fraction from Brazil, at \$6 per lb, is B, and the fraction from Columbia, at \$12 per lb, is C. The mix contains half as much coffee from Brazil as from Columbia. The price of the blend is \$9 per lb.

  (a) Write three equations from the given information.
  (b) Write an augmented matrix for this system of equations.
  (c) Use row operations to find the fraction of each source that is in the blend.

30. The Pauli matrices[4] are a set of 2×2 matrices used by physicists, defined as

$$\sigma_x = \begin{pmatrix} 0 & 1 \\ 1 & 0 \end{pmatrix}, \sigma_y = \begin{pmatrix} 0 & -i \\ i & 0 \end{pmatrix}, \sigma_z = \begin{pmatrix} 1 & 0 \\ 0 & -1 \end{pmatrix}.$$

  (a) Find σ_x^2, σ_y^2, and σ_z^2.
  (b) Write $\sigma_x \cdot \sigma_y$ in terms of σ_z.
  (c) Write $\sigma_y \cdot \sigma_x$ in terms of σ_z.
  (d) What is the effect of changing the multiplication order for any pair?

---

[4]Wikipedia: Pauli matrices.

Chapter 17

Probability and Statistics

CONTENTS

17.1   THE MEAN

According to the American Automobile Association, on October 8, 2008, the average price of a gallon of regular gas in North Carolina (NC) was $3.765, compared to $3.047 in Oklahoma (OK).[1] We can conclude that, at least on this particular day, gas was more expensive "on average" in North Carolina than in Oklahoma. But what does this really mean? After all, certain high-priced Oklahoma stations might charge more than their low-priced North Carolina counterparts. So how can we say gas in Oklahoma is cheaper?

Comparing Totals

Let us first compare the *total* amount spent on gas by two different people in these two states. Table 17.1 gives the price per gallon of regular gas at five different stations in each state. Imagine that a person in each state buys 1 gallon of gas at each of five stations. The last column of Table 17.1 gives the total that each person spends, and it appears that the person in North Carolina spends more than the person in Oklahoma.

Table 17.1   *Spending on gasoline at different stations*

Station	1	2	3	4	5	Total spent
OK price ($/gal)	3.099	3.019	2.979	3.169	2.969	$15.235
NC price ($/gal)	3.959	3.679	3.789	3.879	3.519	$18.825

A Drawback to Comparing Totals

A drawback to comparing totals is that a larger total does not necessarily mean higher prices—instead, it may mean a person buys more than one gallon of gas. Suppose the person in Oklahoma drives more than the person in North Carolina and buys two gallons of gas at each station, instead of one. In this case, the person in Oklahoma spends more, even though gas is cheaper there:

$$\text{Total amount spent in OK} = 2(3.099) + 2(3.019) + 2(2.979) + 2(3.169) + 2(2.969) = \$30.47.$$

Comparing Averages

We know that the total amount spent depends both on the price and on the amount:

$$\text{Total spent on gas} = \text{Price of one gallon} \times \text{Number of gallons}.$$

We can rewrite this equation as

$$\text{Price per gallon} = \frac{\text{Total spent on gas}}{\text{Number of gallons}}.$$

Here, we imagine we are buying gas at a fixed price. But we can use basically the same formula *even if the person buys gas at different prices*:

---

[1] http://www.fuelgaugereport.com/sbsavg.asp, page accessed October 8, 2008. Note that gas prices normally include 9/10 of a cent, so the average is reported with three decimals.

$$\text{Average price per gallon in OK} = \frac{\text{Total spent on gas in OK}}{\text{Number of gallons}}$$
$$= \frac{2(3.099) + 2(3.019) + 2(2.979) + 2(3.169) + 2(2.969)}{2 + 2 + 2 + 2 + 2}$$
$$= \frac{\$30.47}{10} = \$3.047.$$

Here, we say "average price per gallon" instead of "price per gallon," because the person is paying various amounts for gas. Likewise:

$$\text{Average price per gallon in NC} = \frac{\text{Total spent on gas in NC}}{\text{Number of gallons}}$$
$$= \frac{3.959 + 3.679 + 3.789 + 3.879 + 3.519}{1 + 1 + 1 + 1 + 1}$$
$$= \frac{\$18.825}{5} = \$3.765.$$

Notice that the average price paid in NC is more than the average price paid in OK, even though the person in OK spent more in total.

A Formula for the Mean

The term "average" is imprecise. In everyday speech, people say things like "This restaurant is better than average" or "He's only average looking." Restaurants or good looks are not being summed and divided like gas prices. Rather, "average" implies a norm or an accepted level of what is being measured—in this case, restaurant quality or physical attractiveness.

For this reason, mathematicians and statisticians prefer the term "mean" for what we have been calling "average." Even so, many people—including mathematicians—sometimes use the term "average" when they intend "mean."[2]

The **mean value** of a data set is given by the formula

$$\text{Mean value} = \frac{\text{Sum of values}}{\text{Number of values}}.$$

Example 1   Referring to Table 17.1 on page 496, find the new mean gas price if:

(a) The price at all five Oklahoma stations goes up by $0.20.
(b) The price at the fifth station goes up by $1.00, but stays the same elsewhere.
(c) The price doubles at all five stations.

---

[2]The term "average" is also common when using computers or calculators. For instance, Microsoft Excel has a built-in AVERAGE() function, but no built-in MEAN() function, but many calculators use the term "mean" rather than "average."

Solution

(a) If the price at all five stations goes up by $0.20, we have

$$\text{Mean price} = \frac{\overbrace{(3.099 + 0.20)}^{\$0.20 \text{ higher}} + \overbrace{(3.019 + 0.20)}^{\$0.20 \text{ higher}} + \cdots + \overbrace{(2.969 + 0.20)}^{\$0.20 \text{ higher}}}{5}$$

$$= \frac{3.299 + 3.219 + 3.179 + 3.369 + 3.169}{5}$$

$$= 3.247 \text{ dollars.}$$

Notice that this is $0.20 higher than the original mean price. We see that if *each* price goes up by $0.20, so does the mean.

(b) If the price at the fifth station goes up by $1.00, we have

$$\text{Mean price} = \frac{3.099 + 3.019 + 2.979 + 3.169 + \overbrace{(2.969 + 1.00)}^{\$1 \text{ higher}}}{5}$$

$$= \frac{3.099 + 3.019 + 2.979 + 3.169 + 3.969}{5}$$

$$= 3.247 \text{ dollars.}$$

Notice this is the same answer as in part (a). This is because adding $0.20 to five values gives the same total as adding $1.00 to one of the values.

(c) If the price at all five stations doubles, we have

$$\text{Mean price} = \frac{\overbrace{2(3.099)}^{\text{Price doubles}} + \overbrace{2(3.019)}^{\text{Price doubles}} + \cdots + \overbrace{2(2.969)}^{\text{Price doubles}}}{5}$$

$$= \frac{6.198 + 6.038 + 5.958 + 6.338 + 5.938}{5}$$

$$= 6.094 \text{ dollars.}$$

Notice that this is twice the original mean price. This is because when we double each price, we double the total, so the mean itself is doubled.

Statistical Notation and the Mean

We often use subscripts to indicate different values of a variable. For instance, from the previous section, if p is the price of gas in $/gallon, we write p_1 for the price at the first station, p_2 for the price at the second station, and so on.

It is customary to write the mean value as a variable with a bar over it, like this: $\bar{x}$, pronounced "x-bar." For instance, we can write $\bar{p}$ for the mean price of gas in Oklahoma. We now define the mean using subscripts:

Given n values labeled $x_1, x_2, \ldots, x_n$, we have:

$$\text{Mean value} = \bar{x} = \frac{\text{Sum of values}}{\text{Number of values}} = \frac{x_1 + x_2 + \cdots + x_n}{n}.$$

Example 2   Let p_1, p_2, p_3, p_4, p_5 be the gas prices at five different stations. Then

$$\text{Mean gas price} = \bar{p} = \frac{p_1 + p_2 + p_3 + p_4 + p_5}{5}.$$

Find the new mean price if:

(a) The price at all five stations goes up by $0.20.
(b) The price at the fifth station goes up by $1.00.
(c) The price doubles at all five stations.

Solution   (a) If the price at all five stations goes up by $0.20, we have:

$$\text{Mean price} = \frac{(p_1 + 0.20) + (p_2 + 0.20) + \cdots + (p_5 + 0.20)}{5}$$

$$= \frac{(p_1 + p_2 + p_3 + p_4 + p_5) + (0.2 + 0.2 + 0.2 + 0.2 + 0.2)}{5} \qquad \text{regroup}$$

$$= \frac{(p_1 + p_2 + p_3 + p_4 + p_5) + 1.00}{5}$$

$$= \underbrace{\frac{p_1 + p_2 + p_3 + p_4 + p_5}{5}}_{\bar{p}} + \frac{1.00}{5} \qquad \text{split numerator}$$

$$= \bar{p} + 0.20.$$

Notice that this is $0.20 higher than the original mean price, $\bar{p}$. Thus, as we saw in Example 1 of Section 17.1, if each price goes up by $0.20, then so does the mean.

(b) If the price at the fifth station goes up by $1.00, we have:

$$\text{Mean price} = \frac{p_1 + p_2 + p_3 + p_4 + (p_5 + 1.00)}{5}$$

$$= \underbrace{\frac{p_1 + p_2 + p_3 + p_4 + p_5}{5}}_{\bar{p}} + \frac{1.00}{5} \qquad \text{split numerator}$$

$$= \bar{p} + 0.20.$$

Notice this is the same answer as in part (a). We see that adding $0.20 to five values gives the same mean as adding $1.00 to one of the values.

(c) If the price doubles at each station, we have:

$$\text{Mean price} = \frac{2p_1 + 2p_2 + 2p_3 + 2p_4 + 2p_5}{5}$$

$$= \frac{2(p_1 + p_2 + p_3 + p_4 + p_5)}{5} \qquad \text{factor}$$

$$= 2 \cdot \underbrace{\frac{p_1 + p_2 + p_3 + p_4 + p_5}{5}}_{\bar{p}}$$

$$= 2\bar{p}.$$

We see that when we double each price, we double the total, so the mean itself is doubled.

Sigma Notation and the Mean

Given n values labeled $x_1, x_2, \ldots, x_n$, we can write our definition of the mean using sigma notation:

$$\text{Mean value} = \bar{x} = \frac{\text{Sum of values}}{\text{Number of values}}$$

$$= \frac{1}{n} \cdot (\text{Sum of values})$$

$$= \frac{1}{n} \cdot \sum_{i=1}^{n} x_i.$$

Sometimes people use an abbreviated form of sigma notation, writing

$$\sum^{n} x_i \quad \text{or even} \quad \sum x_i \quad \text{instead of} \quad \sum_{i=1}^{n} x_i.$$

Using this abbreviated form, we define:

Given n values $x_1, x_2, \ldots, x_n$, we have

$$\text{The mean} = \bar{x} = \frac{1}{n} \cdot \sum x_i.$$

Notice that we are still summing from x_1 to x_n.

The Difference between Population Parameters and Sample Statistics

Suppose we want to know the mean age of all 5000 people in a certain town. The straightforward, though laborious, approach would be:

- Determine the first person's age a_1, the second person's age a_2, all the way up to the last person's age, a_{5000}.

- Sum these ages and divide by 5000 to obtain

$$\bar{a} = \frac{a_1 + a_2 + \cdots + a_{5000}}{5000}.$$

Provided we correctly determine the age of all 5000 people, we can find the exact mean age of the town. Such a measurement is called a *population parameter*, since it tells us something about the overall population. Customarily, we write population parameters using Greek letters. The mean of a population is often denoted by the Greek letter μ, pronounced "mu".

Another less exact, though also less laborious, approach is to take a *random sample* in which each resident of the town has the same chance of being selected as any other resident. Here, we might randomly select 100 people from the town and determine their ages: a_1, a_2, up to a_{100}. We can then use the mean age of these 100 people to estimate the mean age of the entire town. We refer to such a measurement as a *sample statistic* because it applies to a small sample instead of to the overall population. Customarily, we write sample statistics using bar notation, like $\bar{a}$.

Sometimes it is not possible to determine the true value of a population parameter. For instance, calculating the average height of every tree in a forest would likely involve an impractical amount of work. By making a *statistical inference*, we can estimate a population parameter (like mean tree height) based on a sample statistic (like the mean height of a sample of 100 trees). However, statistical inference is a topic more suited to a statistics book than an algebra book.

Exercises and Problems for Section 17.1

EXERCISES

In Exercises 1–6, find the mean of the data set.

1. $2, 4, 6, 8$

2. $2, 4, 6, 8, 100$

3. $102, 104, 106, 108$

4. $-5, -2, 0, 5, 2$

5. $5, 2, 19, 6, 5, 2$

6. $5, 5, 5, 0, 0, 0, 0, 0, 5, 5$

In Exercises 7–11, find $\bar{a}$.

7. $a_i = i^2, i = 1, \ldots, 6$

8. $a_i = 2i, i = 1, \ldots, 10$

9. $a_i = 2, i = 1, \ldots, 10$

10. $a_i = 2^i, i = 1, \ldots, 5$

11. $a_i = i/2$ and $i = 1, \ldots, 5$

PROBLEMS

12. Table 17.2 shows the number of passengers taking a particular daily flight from Boston to Washington over the course of a week. Find the mean number of passengers for the week.

Table 17.2

Day	Mon	Tue	Wed	Thur	Fri	Sat	Sun
Passengers	228	110	215	178	140	72	44

13. Figure 17.1 shows the temperature (in °F) every two hours over the course of an autumn day. What is the approximate mean temperature on this day?

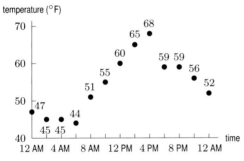

Figure 17.1: The temperature over the course of an autumn day

14. Catherine has the following phone bills over a twelve-month period: $32, $27, $20, $40, $33, $20, $32, $30, $36, $31, $37, $22.

   (a) What is the average phone bill?
   (b) Suppose Catherine spends $5 more on phone bills each month. What happens to her average phone bill? What if she spends $10 more each month?

   (c) Suppose she spends $60 more on the highest phone bill, but the same amount on the other 11 bills. What happens to her average phone bill? What if she spends $120 more on the highest bill?

15. Table 17.3 gives the mean discharge of a river over a two-week period in ft^3/sec.

   (a) What was the mean discharge for the
      (i) first week?   (ii) second week?
   (b) During which week is there a larger total amount of water?
   (c) Comparing the means for the two weeks, can we tell which week had a larger total flow of water?

Table 17.3

Day	Mon	Tue	Wed	Thu	Fri	Sat	Sun
Wk 1	230	256	240	247	245	224	238
Wk 2	233	240	251	234	238	477	427

16. On the back cover of the classic jazz album *Kind of Blue* by Miles Davis, the lengths of the five songs are shown in parentheses: (9:02), (9:33), (5:26), (11:31), (9:25). What is the mean length of a song on this album?

17. Find the mean of $1, 2, 3, \ldots, n$, if n is

   (a) 3   (b) 4   (c) 5
   (d) 6   (e) 2k   (f) 2k+1

18. A sample of n frogs has a total weight of W grams.

   (a) What is the mean weight of a frog in the sample?
   (b) The largest frog eats the smallest one. What is the mean weight now?

19. A sample of 20 frogs has a total weight of W grams.

   (a) What is the mean weight (in grams per frog) of the sample?
   (b) One of the frogs has been mis-weighed. Instead of x grams, its weight is y grams. What is the corrected mean weight of the sample?

20. A sample of n_1 frogs has a total weight of W_1 grams. A second sample of n_2 frogs has a total weight of W_2 grams.

   (a) What is the mean weight (in grams per frog) of each sample?
   (b) What is the mean weight of all the frogs in both samples? Does it equal the average of the mean weights of the samples taken separately?

21. Consider the following list of sale prices (in \$1000s) for eight houses on a certain road: \$820, \$930, \$780, \$950, \$3540, \$680, \$920, \$900. One of the houses is worth much more than the other seven because it is much larger, it is set well back from the road, and it is adjacent to the shore of a lake to which it has private access.

   (a) What is the mean price of these eight houses?
   (b) Is the mean a good description of the value of the houses on this block? Explain your reasoning.

22. In 10 packages the number of M&M's was

$$56, 53, 54, 54, 52, 55, 52, 53, 55, 55.$$

   (a) What is the mean number of M&M's per package?
   (b) Is the mean a good description of the count for a package of M&M's? Explain your reasoning.

23. Suppose you record the hours of daylight each day for a year in Tucson, Arizona, and find the mean.

   (a) What do you expect for an approximate mean?
   (b) How would your data compare with a student doing the same project in Anchorage, Alaska?
   (c) How would your mean compare with a student doing the same project in Anchorage, Alaska?

24. Table 17.4 shows how much money different income groups made in the US during 2007, according to the US Census. The poorest 20% of households made below an annual income of \$20,291 and the fourth quintile made between \$62,000 and \$100,000. The Census gives no limit on the wealthiest quintile, but it reports a lower limit of \$177,000 on the top 5%.

   (a) Using this table, can you find the overall mean household income of the US in 2007?
   (b) Use the data in some way to find an estimate of the mean. Hint: Find an estimate within each quintile.

Table 17.4

Quintiles (fifths)	Lowest	Second	Middle	Fourth	Fifth
Upper limit	20,291	39,100	62,000	100,000	Not given

25. Find the mean of each data set:

   (a) Five readings equaling (not totaling) 120, three readings equaling 130, two readings equaling 140, four readings equaling 150, and one reading equaling 160.
   (b) Three readings equaling x_1, six readings equaling x_2, seven readings equaling x_3, five readings equaling x_4, and four readings equaling x_5.
   (c) n_1 readings equaling x_1, n_2 readings equaling x_2, and so on, up to n_5 readings equaling x_5.

26. Suppose two samples of 5 values are taken from a population

$$a : 8, 9, 4, 7, 5 \text{ and } b : 6, 9, 10, 5, 7.$$

   (a) Find $\bar{a}$ and $\bar{b}$.
   (b) Find the mean of the sample you get by combining the two samples.
   (c) Is the mean of the combined sample equal to the mean of the two values $\bar{a}$ and $\bar{b}$?
   (d) Explain your answer in (c) algebraically.

27. Suppose two samples of values are taken from a population

$$a : 8, 2, 4, 7, 5 \text{ and } b : 6, 9, 10, 5, 7, 8, 2, 5.$$

   (a) Find $\bar{a}$ and $\bar{b}$.
   (b) Find the mean of the sample you get by combining the two samples.
   (c) Is the mean of the combined sample equal to the mean of the two values $\bar{a}$ and $\bar{b}$?
   (d) Explain why the means in (c) are the same or different.

28. A researcher has a colony of ants with a mean weight of μ, and takes a sample of 10 ants with weights $w_1, \ldots, w_{10}$. Decide whether each of the following is the mean of the sample. Explain.

   (a) $\sum w_i$
   (b) $\bar{w}$
   (c) $\dfrac{1}{10} \sum_{i=1}^{10} w_i$
   (d) μ
   (e) $(w_1 + w_2 + \cdots + w_{10})/10$
   (f) $\dfrac{\mu}{10}$

17.2 THE STANDARD DEVIATION

We call the mean a *summary statistic* because it summarizes a data set. For instance, although gas prices might vary widely across the state of Oklahoma, knowing that the mean price is $3.047 per gallon tells us something about gas prices in the entire state. But the summary provided by the mean is often incomplete because two data sets can have the same mean and yet be very different.

Example 1 A naturalist measures the *clutch size* (the number of eggs in the nest) for two different subspecies of ptarmigan, a game bird, recording her results in Table 17.5. Do the mean clutch sizes of these two data sets give a reasonable description of the differences between them?

Table 17.5 *Clutch sizes in various nests*

Subspecies A	7	8	10	8	7	7	9	8
Subspecies B	9	5	11	6	6	7	8	12

Solution Computing the mean clutch size for each subspecies, we have

$$\text{Mean clutch size for subspecies } A = \frac{7+8+\cdots+8}{8} = \frac{64 \text{ eggs}}{8 \text{ nests}} = 8 \text{ eggs/nest}$$

$$\text{Mean clutch size for subspecies } B = \frac{9+5+\cdots+12}{8} = \frac{64 \text{ eggs}}{8 \text{ nests}} = 8 \text{ eggs/nest.}$$

Based on this data, these two subspecies have the same mean clutch size. However, this summary conceals an important difference between the two subspecies: Clutch sizes for A fall between 7 and 10 eggs, whereas for B they are more spread out, ranging from 5 to 12 eggs.

Using a Range to Measure the Spread of a Data Set

To provide a more thorough summary of the data, we can summarize the spread in a variety of ways. One way is simply to report a range of values, as we did in Example 1.

A difficulty with this approach is that an "outlier" (or atypical data value) can dramatically affect the range. For this reason, statisticians often rely on a more sophisticated method to measure spread, called the *standard deviation*.

Calculating the Standard Deviation

The standard deviation involves the notion of *deviation from the mean*, or simply *deviation*:

Deviation of a data value = Difference between that value and the mean value
= Data value − Mean value.

The deviation tells us how far our data value lies from the mean or "average" value.

Example 2   Find the deviations of the data sets in Table 17.5.

Solution   We know the mean of both data sets is 8. For subspecies A, the deviations are:

$$\text{Deviation for first data value} = 7 - 8 \ = -1$$
$$\text{Deviation for second data value} = 8 - 8 \ = 0$$
$$\text{Deviation for third data value} = 10 - 8 = +2$$
$$\vdots$$

Likewise, for subspecies B, the deviations are:

$$\text{Deviation for first data value} = 9 - 8 \ = +1$$
$$\text{Deviation for second data value} = 5 - 8 \ = -3$$
$$\text{Deviation for third data value} = 11 - 8 = +3$$
$$\vdots$$

These results are in Table 17.6. Above-average clutch sizes have positive deviations, while below-average clutch sizes have negative deviations. A nest holding the mean number of eggs has zero deviation.

Table 17.6   *Deviations in clutch sizes of two subspecies*

Deviations for subspecies A	−1	0	2	0	−1	−1	1	0
Deviations for subspecies B	1	−3	3	−2	−2	−1	0	4

Table 17.6 shows that the deviations for subspecies B tend to be bigger than for subspecies A. This indicates the data for B is more "spread out." But how much more spread out is set B than set A? We might try to answer this by calculating the "average spread," or mean deviation, for both sets:

$$\text{Mean deviation for subspecies } A = \frac{\text{Sum of deviations}}{\text{Number of values}} = \frac{-1 + 0 + \cdots + 0}{8} = \frac{0}{8} = 0$$

$$\text{Mean deviation for subspecies } B = \frac{\text{Sum of deviations}}{\text{Number of values}} = \frac{1 + (-3) + \cdots + 4}{8} = \frac{0}{8} = 0.$$

In both cases, the deviations cancel out to zero! This is no coincidence. It is a property of the mean, and the same thing happens for *any* data set.

The Variance: The Mean Squared Deviation

In order to prevent the individual deviations from canceling out, we square the individual deviations before finding the mean.

$$\text{Mean squared deviation for subspecies } A = \frac{\overbrace{(-1)^2 + 0^2 + 2^2 + 0^2 + (-1)^2 + (-1)^2 + 1^2 + 0^2}^{\text{sum of squared deviations}}}{8} = \frac{8}{8} = 1$$

$$\text{Mean squared deviation for subspecies } B = \frac{\overbrace{1^2 + (-3)^2 + 3^2 + (-2)^2 + (-2)^2 + (-1)^2 + 0^2 + 4^2}^{\text{Sum of squared deviations}}}{8} = \frac{44}{8} = 5.5.$$

After squaring each deviation, the terms no longer cancel. We see that the mean squared deviation for subspecies B, at 5.5, is quite a bit larger than it is for subspecies A, at 1. The mean squared deviation provides a reliable measure of how spread out, or varied, a data set is, and is called the *variance*.

The **variance** of a data set is the mean of the squared deviations:

$$\text{Variance} = \frac{\text{Sum of squared deviations}}{\text{Number of values}}.$$

The Standard Deviation: The Square Root of the Variance

Thinking about units, we know that a clutch size is a number of eggs. The deviation of a data value is also a number of eggs. For instance, a nest of 6 eggs has a deviation of $6 - 8 = -2$ eggs. Being negative, this tells us this nest has 2 fewer eggs than the mean nest.

When we find the variance, we square all the deviations and then divide by the number of nests. Thus, a nest of 6 eggs has a squared deviation of $(6 - 8)^2 = 4$ eggs^2. In order for our measure of spread to have the same units as the original data—in this case, eggs—we take a final step in our calculations by finding the square root of the variance. The square root of the variance is known as the *standard deviation*. It is denoted by the lowercase Greek letter σ, pronounced "sigma".

$$\sigma_A = \text{Standard deviation for subspecies } A = \sqrt{\text{variance}} = \sqrt{1} = 1 \text{ egg}$$
$$\sigma_B = \text{Standard deviation for subspecies } B = \sqrt{\text{variance}} = \sqrt{5.5} = 2.345 \text{ eggs}.$$

Here, we see that σ for species A is quite a bit smaller than for species B, which is consistent with the fact that the clutch sizes for A are quite a bit less spread out.

We define the **standard deviation** of a data set with n values as

$$\sigma = \text{Standard deviation} = \sqrt{\text{Variance}} = \sqrt{\frac{1}{n} \cdot (\text{Sum of squared deviations})}.$$

Example 3 Using the data in Table 17.1 on page 496, calculate the standard deviations to determine which state has a higher spread in gas prices, Oklahoma or North Carolina.

Solution   The mean gas price in Oklahoma is \$3.047, so:

$$\text{Variance in OK} = \frac{1}{5} \cdot \text{Sum of squared deviations}$$

$$= \frac{(3.099 - 3.047)^2 + (3.019 - 3.047)^2 + \cdots + (2.969 - 3.047)^2}{5} = 0.005816,$$

and σ in OK $= \sqrt{\text{variance}} = \sqrt{0.005816} = \$0.0763$.

Similarly, the mean in North Carolina is \$3.765, so

$$\text{Variance in NC} = \frac{(3.959 - 3.765)^2 + (3.679 - 3.765)^2 + \cdots + (3.519 - 3.765)^2}{5} = 0.023824,$$

and σ in NC $= \sqrt{\text{variance}} = \sqrt{0.023824} = \$0.1544$.

Since $\sigma = \$0.1544$ for North Carolina and $\sigma = \$0.0763$ for Oklahoma, the prices in North Carolina are more spread out.

Using Sigma Notation in the Definition of Standard Deviation

Let $x_1, x_2, \ldots, x_n$ stand for our n data values. If $\bar{x}$ is the mean value, then:

$$\text{Deviation of data value } x_1 = x_1 - \bar{x}$$
$$\text{Deviation of data value } x_2 = x_2 - \bar{x},$$

and, for data value x_i,

$$\text{Deviation of data value } x_i = x_i - \bar{x}.$$

The squared deviations are given by:

$$\text{Squared deviation of data value } x_1 = (x_1 - \bar{x})^2$$
$$\text{Squared deviation of data value } x_2 = (x_2 - \bar{x})^2$$
$$\text{and, for data value } x_i, \quad \text{Squared deviation of data value } x_i = (x_i - \bar{x})^2.$$

The variance is the mean of the squared deviations from x_1 up to x_n:

$$\text{Variance} = \frac{1}{n} \cdot \overbrace{\left((x_1 - \bar{x})^2 + (x_2 - \bar{x})^2 + \cdots + (x_n - \bar{x})^2 \right)}^{\text{Sum of squared deviations}}$$

$$= \frac{1}{n} \sum_{i=1}^{n} (x_i - \bar{x})^2.$$

Thus, since the standard deviation σ is the square root of the variance, we have:

The **standard deviation** of a data set $x_1, x_2, \ldots, x_n$ with mean $\bar{x}$ is given by

$$\sigma = \sqrt{\frac{1}{n} \sum_{i=1}^{n} (x_i - \bar{x})^2}.$$

Calculating the Standard Deviation Using n Versus $n - 1$

Many calculators and computers provide two different ways of calculating the standard deviation.[3] The first way is to divide by n when finding the variance, as we have done in this chapter. The second approach is to divide instead by $n - 1$. The reason has to do with statistical inference. When we only intend to measure how spread out our data set is, we can use n. But if we intend to use our data to predict (or *infer*) the standard deviation of the overall population, we should use $n - 1$. This is because, in contrast to the mean of a data set, which gives an unbiased estimate of the mean of the population, the standard deviation of a data set tends slightly to underestimate the standard deviation of the population. To correct for this bias, we divide by $n - 1$ instead of n when making statistical inferences.

Exercises and Problems for Section 17.2

EXERCISES

In Exercises **1–5**, find the mean and standard deviation of the data set.

1. $13, 14, 19, 28, 30, 31, 50$

2. $25, 30, 32, 32, 41, 45, 57, 62$

3. $81, 57, 14, 98, 20, 20, 6$

4. $16, 66, 30, 99, 74, 50, 35, 7$

5. $12, -8, 13, -15, 60, -72, 23, -13$

6. Find the mean and standard deviation for each of the following data sets.

   (a) $1, 2, 3, 4, 5, 6, 7$
   (b) $4, 4, 4, 4, 4, 4, 4$
   (c) $2, 2, 4, 4, 4, 6, 6$

PROBLEMS

7. The clutch size of a bird is the number of eggs laid by the bird. Table 17.7 shows the clutch size of six different birds labeled (i)–(vi). What is the

   (a) Mean clutch size?
   (b) Standard deviation of these clutch sizes?

   Table 17.7

Bird	(i)	(ii)	(iii)	(iv)	(v)	(vi)
Clutch size	6	7	2	3	7	5

8. The sale prices (in $1000s) for eight houses on a certain road are: $820, $930, $780, $950, $3540, $680, $920, $900. Find the mean and standard deviation of the

   (a) Eight houses.

   (b) Seven similar houses (leave out the top-priced house).

A naturalist collects samples of a species of lizard and measures their lengths. For the samples in **9–11**, give the

   (a) sample size   **(b)** mean   **(c)** range
   (d) standard deviation.

9.

   Table 17.8

Lizard no.	1	2	3	4	5
Length (cm)	5.8	6.8	6.9	6.9	7.0
Lizard no.	6	7	8	9	10
Length (cm)	7.1	7.1	7.1	7.2	8.1

---

[3] For instance, Microsoft Excel provides STDEV(), which uses $n - 1$, and STDEVP(), which uses n.

10.

Table 17.9

Lizard no.	1	2	3	4	5
Length (cm)	5.8	5.9	5.9	6.0	6.5
Lizard no.	6	7	8	9	10
Length (cm)	7.9	7.9	8.0	8.0	8.1

11.

Table 17.10

Length (cm)	4	5	6	7	8	9	10
No. lizards	1	6	26	36	23	6	2

12. Table 17.11 shows five different data sets numbered (i) through (v). Each data set has the same mean value of 20. Rank the data sets in order of their standard deviation from lowest to highest. You are not required to make any calculations to answer this question, but you should explain your reasoning.

Table 17.11

(i)	(ii)	(iii)	(iv)	(v)
10	10	10	0	18
15	20	10	15	19
20	20	20	20	20
25	20	30	25	21
30	30	30	40	22

13. Suppose you record the hours of daylight in Tucson, Arizona, each day for a year and find the mean amount.

   (a) What do you expect for an approximate mean?

   (b) How would your data compare with a student doing the same project in Anchorage, Alaska?

   (c) How would your standard deviation compare with a student doing the same project in Anchorage, Alaska?

■In Problems **14–18**, find the standard deviation of the data set.

14. 20, 30, 40, 80, 130.

15. $x_1, x_2, \ldots, x_5$, mean is m.

16. Five readings each equaling 120, three readings each equaling 130, two readings each equaling 140, four readings each equaling 150, and 1 reading equaling 160.

17. Three readings each equaling x_1, six readings each equaling x_2, seven readings each equaling x_3, five readings each equaling x_4, and four readings each equaling x_5. The mean of this data set is m.

18. n_1 readings each equaling x_1, n_2 readings readings each equaling x_2, and so on, up to n_5 readings each equaling x_5. The mean of this data set is m.

17.3 PROBABILITY

The *probability* of an outcome is a measure of how likely the outcome is. Winning the lottery, getting struck by lightning, or finding a pearl in an oyster—these events all seem extremely unlikely, so we assign to each of them a low probability, that is, a probability near zero. Some events are downright impossible, such as drawing five aces from a (fair) poker deck. Such events are said to have no likelihood of occurring, or a probability of 0%. At the other end of the scale, some events are extremely likely. It seems virtually certain that the sun will rise tomorrow, so we assign this event a probability near 100%.

Probabilities in Terms of Percentages

We can often think of probabilities in terms of percentages or *relative frequencies*. For example, if the weather forecast says there is a 50% chance of rain, this means, according to historical weather data such as temperature, humidity, prevailing winds, and cloud cover, that for all of the days with conditions similar to this day, 50% of them actually saw rain:

$$\text{Probability of rain} = \frac{\text{Number of days with similar conditions that saw rain}}{\text{Total number of days with similar conditions}} = 50\%.$$

Here, a meteorologist is measuring the relative frequency of rain for all days having similar conditions. The assumption is that if in the past certain conditions favored rain, then similar conditions should favor rain today.

Example 1   Shannon buys 10 tickets in a lottery with 500 entries. What is the probability of her winning?

Solution   She will win the lottery if any one of her 10 tickets is drawn. Therefore

$$\text{Probability of drawing one of Shannon's tickets} = \frac{\text{Number of tickets Shannon buys}}{\text{Total number of tickets}} = \frac{10}{500} = 2\%.$$

In general, probability values must range between 0 and 1, that is between 0% and 100%. For example, if Shannon does not buy a ticket, the numerator in the probability fraction is zero, and thus

$$\text{Probability of selecting Shannon's ticket} = \frac{0}{500} = 0\%.$$

On the other hand, if Shannon buys all 500 tickets:

$$\text{Probability of selecting Shannon's ticket} = \frac{500}{500} = 1 = 100\%.$$

Example 2   There are 540 members of the United States Congress, composed of 100 members in the US Senate (2 from every state), and 440 members of the US House of Representatives. The state of California (CA) has 53 representatives in the House.[4] What is the probability that:

(a) A randomly picked member of the House represents California?
(b) A randomly picked senator represents California?
(c) A randomly picked member of Congress represents California?

Solution   There are 440 different ways to pick a representative from the House, 100 ways to pick a senator from the Senate, and 540 ways to pick a congressperson from the congress.

(a) There are 53 different ways to pick a California representative from the House:

$$\text{Probability of picking a CA representative from the House} = \frac{\text{Number of ways to pick a CA representative}}{\text{Number of ways to pick a representative from the House}}$$
$$= \frac{53}{440} = 12.045\%.$$

(b) There are only 2 ways to pick a California senator from the Senate:

$$\text{Probability of picking a CA senator from the Senate} = \frac{\text{Number of ways to pick a CA senator}}{\text{Number of ways to pick a senator from the Senate}}$$
$$= \frac{2}{100} = 2\%.$$

(c) There are $53 + 2 = 55$ ways to pick a California congressperson from the US Congress:

$$\text{Probability of picking a CA congressperson from Congress} = \frac{\text{Number of ways to pick a CA congressperson}}{\text{Number of ways to pick a congressperson from Congress}}$$
$$= \frac{55}{540} = 10.185\%.$$

---

[4] http://en.wikipedia.org/wiki/Current_members_of_the_United_States_Congress, page accessed November 5, 2008. To be precise, there are 435 voting members of the house and 5 non-voting delegates.

Two-Way Tables

Clinical research typically involves allotting subjects into groups and measuring any differences between them. *Two-way tables*, also known as *contingency tables*, are a useful way to organize this sort of data.[5]

Example 3	A researcher investigates a medication for preventing the common cold. Starting with a group of 120 people, none of whom has a cold, she gives the medication to 80 of them and withholds it from the rest, recording her results in Table 17.12. Discuss the following claims.	

Claim 1 A competing researcher states: *The medication has no effect: of the people who got colds, 50% took it and 50% did not.*

Claim 2 The investigating researcher counterclaims: *The medication seems effective: taking it appears to lower the rate of getting a cold from 20% to 10%.*

Table 17.12 *Results of a medical experiment*

	Gets cold	Does not get cold	Total
Takes medication	8	72	80
Does not take medication	8	32	40
Total	16	104	120

Solution **Claim 1** While it is true that of the 16 people who got colds, 8 (50%) took the medication and 8 (50%) did not, these calculations do not support the competing researcher's claim. Focusing only on people in the first column of Table 17.12—that is, people who develop colds—leaves out an important group of people: those in the second column, the ones who did *not* get colds.

Claim 2 In order for the investigating researcher to support her counterclaim, she must compare the probability of getting a cold for people taking the medication to the probability of getting a cold for people *not* taking it. Referring again to Table 17.12, we see that:

- 8 out of 80 or 10% of people who took the medication got colds.
- 8 out of 40 or 20% of people who did not take it also got colds.

We see that these calculations do lend support to her counterclaim: Taking the medication appears to lower the rate of getting a cold from 20% to 10%.

Joint and Conditional Probability

To denote the probability of an event E, we write $P(E)$. If event E_1 is rolling a six on a fair six-sided die, and event E_2 is flipping a fair coin and getting heads, then

$$P(E_1) = \text{Probability of rolling a 6 on a fair six-sided die} = \frac{1}{6}.$$

$$P(E_2) = \text{Probability of flipping a coin and getting heads} = \frac{1}{2}.$$

This looks like function notation, and in a way it *is* function notation: there are inputs and outputs, and for every input we assign a unique output. The difference is that, while the outputs are numbers, the inputs are not—they're *events*: getting a six on a die or heads on a coin.

---

[5]In this book, we omit discussion of a number of important issues, including experimental design, control groups, random assignment, the placebo effect, and statistical significance.

Joint Probability

The *joint probability* of two events is the probability of their occurring together. Consider the following events:

Event A A person picked at random from Table 17.12 took the medication.

Event B A person picked at random from Table 17.12 got a cold.

Here, the joint probability of events A and B is the probability that a person picked at random took the medication *and* got a cold. This probability is written $P(A \cap B)$:

$$P(A \cap B) = \frac{8 \text{ people took medication and got colds}}{120 \text{ people altogether}} = 6.67\%.$$

Note that $P(B \cap A)$ means the same thing as $P(A \cap B)$: both expressions describe the probability that events A and B happen together.

Conditional Probability

Suppose we know that event B has happened and want to know how likely it is now for event A to happen. This is known as the *conditional probability of event A given event B*, and is written $P(A|B)$. Although the conditional probability of A given B sounds similar to the joint probability of events A and B, they are very different. For instance, out of the 16 people who got colds, 8 took the medication, so:

$$P(A|B) = \frac{8 \text{ people who got colds also took medication}}{16 \text{ people got colds}} = 50\%.$$

We see that $P(A|B) = 50\%$ whereas $P(A \cap B) = 6.67\%$. It is also worth noticing that $P(B|A)$ is not at all the same as $P(A|B)$. Of the 80 people who take the medication, 8 get colds, so:

$$P(B|A) = \frac{8 \text{ people who got colds also took medication}}{80 \text{ people took medication}} = 10\%.$$

Probabilities and Two-Way Tables

To understand the difference between joint and conditional probabilities, it can be helpful to think in terms of two-way tables.

- The **joint probability** of events A and B, written $P(A \cap B)$, is the probability of their occurring together. To find a joint probability, we divide a table entry by the grand total.
- The **conditional probability** of event A given event B, written $P(A|B)$, is the probability that event A occurs given that event B occurs. To find a conditional probability, we divide a table entry by a row or column total, *not* by the grand total.

Example 4 A geneticist seeks to establish a link between a certain gene and a certain type of cancer. She studies 250 mice, recording her results in Table 17.13. A mouse is selected at random from this group:

- Event A stands for picking a mouse having the gene.

- Event B stands for picking a mouse not having the gene.
- Event C stands for picking a mouse developing the cancer.
- Event D stands for picking a mouse not developing the cancer.

Evaluate the following expressions.

(a) $P(A \cap C)$ (b) $P(C \cap A)$ (c) $P(C \cap B)$

(d) $P(A|C)$ (e) $P(C|A)$ (f) $P(C|B)$

Table 17.13   *Link between a gene and a cancer in mice*

	Has gene	Lacks gene	Total
Develops cancer	10	15	25
Does not develop cancer	43	182	225
Total	53	197	250

Solution

(a) This is a joint probability, so we divide a table entry by the grand total:

$$P(A \cap C) = \frac{10 \text{ mice have the gene and develop cancer}}{250 \text{ mice total}} = 4\%.$$

(b) This is the joint probability of picking a mouse developing cancer and having the gene, so it is the same as $P(A \cap C)$. Thus, from part (a), we see that $P(C \cap A) = 4\%$.

(c) Again we divide a table entry by the grand total:

$$P(C \cap B) = \frac{15 \text{ mice develop cancer and lack the gene}}{250 \text{ mice total}} = 6\%.$$

(d) This is a conditional probability, so we divide a table entry by a row or column total (and not by the grand total):

$$P(A|C) = \frac{10 \text{ mice have the gene and develop cancer}}{25 \text{ mice develop cancer}} = 40\%.$$

(e) In part (d), we divided by a row total. This time, we divide by a column total:

$$P(C|A) = \frac{10 \text{ mice develop cancer and have the gene}}{53 \text{ mice have the gene}} = 18.9\%.$$

Although we saw that $P(C \cap A)$ is the same as $P(A \cap C)$, notice that $P(C|A)$ is *not* the same as $P(A|C)$.

(f) We have

$$P(C|B) = \frac{15 \text{ mice develop cancer and lack the gene}}{197 \text{ mice lack the gene}} = 7.61\%.$$

Independent Events

Under ordinary circumstances, flipping a coin does not affect the card we draw from a deck of cards. These two events are independent: The outcome of one (heads or tails) does not affect the outcome of the other (ace of spades, jack of hearts, etc.). In contrast, in Example 4, we see that

$$P(C|A) = \frac{\text{Conditional probability for a mouse to develop cancer given that it has a certain gene}}{} = 18.9\%$$

$$P(C|B) = \frac{\text{Conditional probability for a mouse to develop cancer given that it lacks the gene}}{} = 7.61\%.$$

This tells us that the probability for a mouse to develop cancer is *not* independent from its having a particular gene. These two events are intertwined: Having the gene appears to make it more than twice as likely (18.9% versus 7.61%) that the mouse will develop cancer. (Whether the gene actually *causes* the cancer is another question.) In general:

> Two events are **independent** if the occurrence of one of them has no effect on the probability of the occurrence of the other.

Example 5   The researcher in Example 4 investigates a second gene. See Table 17.14. Is the probability of a mouse developing cancer independent of its having this second gene?

Table 17.14   *Link between another gene and a cancer in mice*

	Has gene	Lacks gene	Total
Develops cancer	18	7	25
Does not develop cancer	162	63	225
Total	180	70	250

Solution   To check for independence, we compute three probabilities:

$$\text{Cancer probability for any mouse} = \frac{25 \text{ mice develop cancer}}{250 \text{ mice total}} = 10\%$$

$$\text{Cancer probability for a mouse with the gene} = \frac{18 \text{ mice with gene develop cancer}}{180 \text{ mice have the gene}} = 10\%$$

$$\text{Cancer probability for a mouse without the gene} = \frac{7 \text{ mice lacking gene develop cancer}}{70 \text{ mice lack the gene}} = 10\%.$$

Thus, the probability of a mouse developing cancer is 10% whether or not it has the gene, which tells us that developing cancer and having the gene are statistically independent.

Independence and Joint Probability

Suppose that, for mice, having a certain gene is independent of developing cancer. This means the probability, c, of developing cancer is the same for mice with the gene as it is for mice without the gene. If g is the probability that a mouse has the gene, then out of N mice we expect:

$$\text{Number of mice with gene} = \underbrace{\text{Probability of having gene}}_{g} \times \underbrace{\text{Number of mice}}_{N} = gN.$$

Of these mice, we expect:

$$\begin{matrix}\text{Number of mice with} \\ \text{gene getting cancer}\end{matrix} = \underbrace{\text{Probability of getting cancer}}_{c} \times \underbrace{\text{Number of mice with gene}}_{gN} = cgN.$$

Therefore,

$$\begin{matrix}\text{Joint probability of having gene} \\ \text{and developing cancer}\end{matrix} = \frac{\overbrace{\text{Number of mice with gene getting cancer}}^{cgN}}{\underbrace{\text{Number of mice}}_{N}}$$

$$= \frac{cgN}{N} = cg.$$

Therefore:

$$\begin{matrix}\text{Joint probability of having gene} \\ \text{and developing cancer}\end{matrix} = \begin{matrix}\text{Probability of} \\ \text{having gene}\end{matrix} \times \begin{matrix}\text{Probability of} \\ \text{developing cancer}\end{matrix}.$$

We can rewrite this statement using our notation for joint probability: If event A stands for developing cancer and B stands for having the gene, then

$$P(A \cap B) = P(A)P(B).$$

This relation is always true for statistically independent events, not just for genes and cancer in mice:

If two events A and B are **independent**, then

$$P(A \cap B) = P(A)P(B).$$

Example 6   Show that for Table 17.13, $P(A \cap B)$ is *not* the same as $P(A)P(B)$.

Solution   From the table, we see that

$$P(A) = \begin{matrix}\text{Probability for a mouse} \\ \text{to develop cancer}\end{matrix} = \frac{25 \text{ mice develop cancer}}{250 \text{ mice total}} = 10\%$$

$$P(B) = \begin{matrix}\text{Probability for a mouse} \\ \text{to have gene}\end{matrix} = \frac{53 \text{ mice have gene}}{250 \text{ mice total}} = 21.2\%$$

$$P(A \cap B) = \begin{array}{c} \text{Probability for a mouse to} \\ \text{have gene and develop cancer} \end{array} = \dfrac{\left(\begin{array}{c} 10 \text{ mice have gene} \\ \text{and develop cancer} \end{array} \right)}{250 \text{ mice total}} = 4\%.$$

We see that

$$P(A)P(B) = 10\%(21.2\%) = 2.12\% \neq P(A \cap B).$$

This confirms what we already knew: Events A and B are not statistically independent.

Conditional Probability and Joint Probability

For independent events, there is a connection between the joint probability $P(A \cap B)$ and the individual probabilities $P(A)$ and $P(B)$, namely that $P(A \cap B) = P(A)P(B)$. But what if events A and B are not independent? Returning to our earlier argument, we again let g be the probability that a mouse has a certain gene. This time, though, we let c be the probability of the mouse developing cancer *given that it has the gene*. Once again, we see that

$$\text{Number of mice with gene } = \underbrace{\text{Probability of having gene}}_{g} \times \underbrace{\text{Number of mice}}_{N} = gN.$$

Since c is the probability these gN mice get cancer,

$$\text{Number of mice with gene getting cancer } = c \cdot gN.$$

Thus,

$$\begin{array}{c} \text{Joint probability of having gene} \\ \text{and developing cancer} \end{array} = \dfrac{\overbrace{\text{Number of mice with gene getting cancer}}^{cgN}}{\underbrace{\text{Number of mice}}_{N}}$$

$$= \dfrac{cgN}{N} = cg.$$

This is the same result we got before, except that this time, c is a conditional probability: It is the probability of developing cancer given the presence of the gene. Thus, we can write $c = P(A|B)$. Since $g = P(B)$, we have

$$\underbrace{\begin{array}{c} \text{Joint probability of having gene} \\ \text{and developing cancer} \end{array}}_{P(A \cap B)} = \underbrace{\begin{array}{c} \text{Conditional probability} \\ \text{of getting cancer} \end{array}}_{c = P(A|B)} \cdot \underbrace{\begin{array}{c} \text{Probability of} \\ \text{having gene} \end{array}}_{g = P(B)}$$

$$P(A \cap B) = P(A|B)P(B).$$

Dividing both sides by $P(B)$, which we assume to be nonzero, we have:[6]

---

[6]This formula is commonly used as a definition for $P(A|B)$.

Provided $P(B) > 0$, the conditional probability of event A given event B is related to the joint probability of these events by the formula

$$P(A|B) = \frac{P(A \cap B)}{P(B)}.$$

Example 7   Suppose $P(A) = 0.5$, $P(B) = 0.8$, and $P(A \cap B) = 0.2$.

(a) Are events A and B independent?
(b) Compute $P(A|B)$.
(c) Compute $P(B|A)$.

Solution   (a) We know that $P(A \cap B) = 0.2$. However, $P(A)P(B) = (0.5)(0.8) = 0.4$. Since $P(A \cap B) \neq P(A)P(B)$, events A and B are not independent.

(b) Since $P(A|B) = \dfrac{P(A \cap B)}{P(B)}$, we have

$$P(A|B) = \frac{0.2}{0.8} = 0.25.$$

(c) In a similar fashion,

$$P(B|A) = \frac{P(A \cap B)}{P(A)} = \frac{0.2}{0.5} = 0.4.$$

Medical Screening Tests

According to *The New York Times*,[7] a new screening test for ovarian cancer, called OvaSure, detected "95 percent of the cancers, and its false positive rate—detecting a cancer that was not there—was 0.6%." The article goes on to say "only 1 of 3000 women has ovarian cancer."

Suppose event A corresponds to having ovarian cancer, and event B corresponds to getting a positive result on this test. Based on this information, we know that $P(B|A) = 95\%$: That is, given that a woman has ovarian cancer, there is a 95% chance she will test positive. Thus, one might conclude that getting a positive result almost certainly means one has ovarian cancer. This turns out not to be the case. The probability we are given is $P(B|A)$, but the probability we would like to know is $P(A|B)$: How likely is it that a woman has ovarian cancer given that she has a positive result?

Let us make some calculations and record them in a two-way table. See Table 17.15. Suppose the test is given to 3,000,000 women.

- Given that 1 in 3000 women has ovarian cancer, we expect:

$$\text{Number of women with cancer} = \frac{1}{3000} \times \text{Number of women}$$

$$= \frac{1}{3000} \cdot 3{,}000{,}000 = 1000.$$

---

[7]http://www.nytimes.com/2008/08/26/health/26ovar.html, page accessed August 26, 2008.

This also means we should expect $3,000,000 - 1000 = 2,999,000$ women *not* to have ovarian cancer. We have recorded these as the row totals in the last column of Table 17.15.

- Of the 1000 women with cancer, the test should detect 95%, or 950 women. These results are called "true positives" because they are positive and they are correct. Notice that the test gives the remaining $1000 - 950 = 50$ women negative results even though they actually have cancer. Such results are called "false negatives." These calculations go into the first row of the table.

- We expect the test to give "false positives" to 0.6% of the 2,999,000 women not having ovarian cancer, or

$$0.006(2,999,000) = 17,994 \text{ women}.$$

The remaining $2,999,000 - 17,994 = 2,981,006$ women are given "true negative" test results. These calculations go into the second row of our table.

- Finally, we sum the positive test results as well as the negative test results to complete the column totals in our table.

Table 17.15   *Test results for ovarian cancer*

	Test positive	Test negative	Total
Have cancer	950	50	1000
Do not have cancer	17,994	2,981,006	2,999,000
Total	18,944	2,981,056	3,000,000

We now find the conditional probability that a woman has cancer *given that she gets a positive result*, that is, $P(A|B)$. We have:

$$P(A|B) = \frac{\text{Number of true positive results}}{\text{Total number of positive results}}$$

$$= \frac{950}{18,944} = 0.0501 = 5.01\%.$$

This means, given a positive test result, a woman has only a 5.01% probability of actually having ovarian cancer. This number may seem surprisingly low considering the fact that the test correctly identifies 95% of all ovarian cancers. The reason the probability is low is because so many women in our table—2,999,000 of them—*don't* have cancer. Thus, even though the test misidentifies these women only 0.6% of the time, the resulting 17,994 false positives swamp the relatively few 950 true positives.

Example 8   In the general population, some people are left-handed. Table 17.16 shows an investigator's result after polling both men and women. Let event M stand for selecting a male, event F for selecting a female, and event L for selecting a left-handed person. Find

(a) $P(M \cap L)$

(b) $P(F \cap L)$

(c) $P(L|M)$

(d) $P(L|F)$

(e) Use this information to decide if left-handedness is independent of gender.

Table 17.16 *Left-handedness and gender*

	Left	Not left	Total
Male	6	54	60
Female	4	36	40
Total	10	90	100

Solution

(a) The expression $P(M \cap L)$ is the joint probability of selecting a male who is left-handed. Since there are 6 left-handed males and a total of 100 people, we have $P(M \cap L) = 6/100 = 6\%$.

(b) The expression $P(F \cap L)$ is the joint probability of selecting a female who is left-handed. Since there are 4 left-handed females and a total of 100 people, we have $P(F \cap L) = 4/100 = 4\%$.

(c) The expression $P(L|M)$ is the probability that a person is left-handed given that the person is male. There are 60 males, 6 of whom are left-handed, so $P(L|M) = 6/60 = 10\%$.

(d) The expression $P(L|F)$ is the probability that a person is left-handed given that the person is female. There are 40 females, 4 of whom are left-handed, so $P(L|F) = 4/40 = 10\%$.

(e) It appears that left-handedness is independent of gender.

Exercises and Problems for Section 17.3

EXERCISES

In Exercises **1–6**, give the probability, as a percentage, of picking the indicated card from a deck.

1. Queen

2. King of Hearts

3. Red card

4. Face card

5. 2 or 3

6. Spade

7. Maine has the same number of representatives as senators. What is the probability, given as a percentage, that:

  (a) A randomly picked member of the House represents Maine?

  (b) A randomly picked senator represents Maine?

  (c) A randomly picked member of Congress represents Maine?

8. Table 17.17 shows the number of passengers taking a particular daily flight from Boston to Washington over the course of a week. Picking at random, what is the probability, given as a percentage, that

  (a) A flight carried at least 150 passengers?

  (b) A passenger flew on Friday?

Table 17.17

Day	Mon	Tue	Wed	Thur	Fri	Sat	Sun
Passengers	228	110	215	178	140	72	44

9. If your music player has four playlists, Rock (233 songs), Hip-Hop (157 songs), Jazz (107 songs) and Latin (258 songs), and you select the shuffle mode, what is the probability, given as a percentage, of starting with a song from

  (a) Rock

  (b) Hip-Hop

  (c) Jazz

  (d) Latin.

In Exercises **10–17**, the probability expressions refer to drawing a card from a standard deck of cards. State in words the meaning of the expression and give the probability as a fraction.

10. $P(\text{Red})$

11. $P(\text{Red} \cap \text{King})$

12. $P(\text{King} \cap \text{Red})$

13. $P(\text{King} \mid \text{Red})$

14. $P(\text{Red} \mid \text{King})$

15. $P(\text{Heart} \mid \text{Red})$

16. $P(\text{Red} \mid \text{Heart})$

17. $P(\text{Spade} \mid \text{Red})$

PROBLEMS

18. Table 17.18 shows the number of flights carrying a given number of passengers over a ten-week time period.

Table 17.18

Passengers	0-50	51-100	101-150	151-200
Flights	3	10	16	20
Passengers	201-250	251-300	301-350	
Flights	15	4	2	

(a) What is the probability, given as a percentage, that a flight picked at random from this group of flights carried more than 200 passengers?

(b) What is the probability that a passenger picked at random from among this group of passengers was on a plane carrying more than 200 passengers?

19. Table 17.19 gives the vehicle occupancy for people driving to work in 1990 as determined by the US Census. For instance, 84,215,000 people drove alone and 12,078,000 people drove in 2-person car pools. Picking at random, what is the probability, given as a percentage, that:

(a) A commuter drives to work alone?

(b) A vehicle carries 4 or more people?

Table 17.19

Occupancy	1	2	3	4	5	6	7+
People, 1000s	84,215	12,078	2,001	702	209	97	290

20. There are 54 M&Ms in a packet: 14 blue, 4 brown, 6 green, 14 orange, 7 red, and 9 yellow.

(a) For each color, find the probability, as a percentage, of randomly picking that color from the packet.

(b) Find the probability, as a percentage, of randomly picking a blue if someone has eaten all the reds.

21. The classic album *Kind of Blue* by Miles Davis lists five songs with their lengths in parentheses. If your music player is currently playing this album in shuffle mode, what is the probability, given as a percentage, that when you plug in your earphones you hear:

(a) So What (9:02)

(b) Freddie Freeloader (9:33)

(c) Blue in Green (5:26)

(d) All Blue (11:31)

(e) Flamenco Sketches (9:25).

22. A city is divided into 4 voting precincts, A, B, C, and D. Table 17.20 shows the results of mayoral election held for two candidates, a Republican and a Democrat.

Table 17.20

Precinct	Number voters	Republican	Democrat
A	10,000	4,200	5,800
B	15,000	7,100	7,900
C	17,000	8,200	8,800
D	18,000	12,400	5,600

Assuming random selection, what is the probability, given as a percentage, that a voter:

(a) Lives in precinct B?

(b) Is a Republican?

(c) Both (a) and (b)

(d) Is Republican given that he or she lives in precinct B?

(e) Lives in precinct B given that he or she is Republican?

23. For his term project in biology, Robert believed he could increase the weight of mice by feeding them a hormone. Do his results, in Table 17.21, support the claim that the hormone increases weight?

Table 17.21

	Weight increase	No weight increase	Total
Fed hormone	120	30	150
Not fed hormone	25	25	50
Total	145	55	200

Face recognition systems pick faces out of crowds at airports to see if any matches occur with law enforcement databases. Performance of the systems can be affected by lighting, gender and age of the target, and age of the database. In Problems 24–27, the tables give identification rates for faces under various conditions. Decide whether the rate of recognition is independent of the given factor.[8]

[8]From a news release issued on March 13, 2003, by the National Institute of Standards and Technology, http://www.nist.gov/public_affairs/releases/n03-04.htm. The complete report is available at http://www.itl.nist.gov/iad/894.03/face/face.html#FRVT2002.

24. Table 17.22 compares face recognition under different lighting conditions.

Table 17.22

Lighting	Did recognize	Did not recognize
Indoors	900	100
Outdoors	300	300

25. Table 17.23 compares face recognition for men and women.

Table 17.23

Gender	Did recognize	Did not recognize
Men	78	22
Women	117	33

26. Table 17.24 compares face recognition for different ages.

Table 17.24

Age	Did recognize	Did not recognize
18–22	248	152
38–42	222	78

27. Table 17.25 compares face recognition when using a fresh, same-day database image to face recognition when using an older database image.

Table 17.25

Image	Did recognize	Did not recognize
Fresh	47	3
Older	68	12

28. A high-tech company makes silicon wafers for computer chips, and tests them for defects. The test identifies 90% of all defective wafers, and misses the remaining 10%. In addition, it misidentifies 20% of all non-defective wafers as being defective.

(a) Suppose 5000 wafers are made. Of the 5% of these wafers that contain defects, how many are correctly identified by the test as being defective?

(b) How many of the non-defective wafers are incorrectly identified by the test as being defective?

(c) What is the probability, given as a percentage, that a wafer identified as defective is actually defective?

REVIEW EXERCISES AND PROBLEMS FOR CHAPTER 17

EXERCISES

In Exercises **1–5**, find the mean of the data set.

1. $1, 3, 5, 7$
2. $2, 6, -3, 7, 7, -3, 6, 2$
3. $4, 4, 4, 0, 0, 0, 0, 0, 0, 4, 4, 4$
4. $-100, -78, -34, 0, 34, 78, 100$
5. $5.6, 5.2, 4.6, 4.9, 5.7, 6.4$

In Exercises **6–8**, find $\overline{a}$.

6. $a_i = i^3 - 1,\ i = 1, \dots, 5.$
7. $a_i = i^2/2,\ i = 1, \dots, 5.$
8. $a_i = 4,\ i = 1, \dots, 12.$

In Exercises **9–13**, find the mean and standard deviation of the data set.

9. $10, 12, 17, 24, 30, 32, 50$
10. $-10, -8, -6, -4, -4, 0, 0, 0, 0, 4, 6$
11. $5, 5, 5, 5, 5, 5, 5, 5$
12. $3, 3, 5, 5, 5, 7, 7$
13. $1, 2, 3, 4, 5, 6, 7, 8, 9, 10$

In Exercises **14–19**, give the probability, as a percentage, of picking the indicated card from a standard deck of cards.

14. Jack

15. Queen of Spades

16. Black card

17. Face card

18. 5 or 6

19. Diamond

In Exercises 20–27, the probability expressions refer to drawing a card from a standard deck of cards. State in words the meaning of the expression and give the probability as a fraction.

20. $P(\text{Black})$

21. $P(\text{Black} \cap \text{Queen})$

22. $P(\text{Jack} \cap \text{Red})$

23. $P(\text{King} \mid \text{Black})$

24. $P(\text{Black} \mid \text{King})$

25. $P(\text{Spade} \mid \text{Black})$

26. $P(\text{Red} \mid \text{Diamond})$

27. $P(\text{Spade} \mid \text{Club})$

PROBLEMS

In Problems 28–29, refer to Table 17.26, which shows the amount spent on groceries by customers at two stores, A and B, on a given afternoon.

Table 17.26

Data set for Store A and B		
Amount	No. A	No. B
$0 - 10$	83	20
$10 - 20$	109	44
$20 - 30$	142	65
$30 - 40$	151	74
$40 - 50$	117	75
$50 - 60$	77	66
$60 - 70$	42	55
$70 - 80$	30	42
$80 - 90$	21	24
$90 - 100$	14	15
$100 - 110$	10	11
$110 - 120$	6	8
$120 - 130$	3	6
$130 - 140$	4	5
$140 - 150$	1	3

28. (a) What is the mean amount spent by customers at store A? Store B? Which is higher?
   (b) Which store has more revenue over the course of the afternoon?
   (c) One of these stores is in a busy urban area with heavy foot traffic. The other is in a suburban area and is most easily reached by car. Which store is which, and why?

29. Which store do you expect to have the higher standard deviation? Explain your reasoning.

30. A person's metabolic rate is the rate at which the body consumes energy. Consider the following list of metabolic rates (in calories per day) of eight men who took part in a study of dieting: 1432, 1668, 1838, 1428, 1560, 1634, 1380, 1420.

   (a) What is the mean metabolic rate of these eight men?

   (b) The man with the highest metabolic rate decided to quit the study. What is the mean metabolic rate of the remaining 7?

31. Kristopher has the following gas bills over a twelve-month period: $54, $56, $49, $47, $43, $34, $32, $32, $34, $35, $37, $39.

   (a) What is the average gas bill?
   (b) Suppose Kristopher spends $5 more on gas bills each month. What happens to his average gas bill? What if he spends $10 more each month?
   (c) Suppose he spends $12 more on the highest gas bill, but the same amount on the other 11 bills. What happens to his average gas bill? What if he spends $36 more on the highest bill?

32. Table 17.27 shows the top ten movies at the box office in 1996 (figures are in millions of dollars).

   (a) Estimate the standard deviation of this data set.
   (b) Of these ten movies, three made over $150 million. Not counting these three movies, what is the standard deviation of the remaining seven? Is it higher or lower than for all ten movies?

Table 17.27

Title	Gross	Title	Gross
Independence Day	306	101 Dalmatians	131
Twister	242	The Nutty Professor	129
Mission: Impossible	181	The Birdcage	124
The Rock	134	Jerry Maguire	110
Ransom	132	A Time to Kill	109

33. Table 17.28 shows the number of passengers taking a particular daily flight from Boston to Washington over the course of a week. Picking at random, what is the probability, as a percentage, that

   (a) A flight carried at most 130 passengers?
   (b) A passenger flew on Tuesday?

Table 17.28

Day	Mon	Tue	Wed	Thur	Fri	Sat	Sun
Passengers	228	110	215	178	140	72	44

34. A company has 1,000 employees, 600 of whom are women and 80 of whom are managers. Among managers there are 45 women.

   (a) Choosing at random, what is the probability, as a percentage, that an employee is a manager? An employee is a woman? An employee is a female manager? A manager is a woman? A woman is a manager?
   (b) Who is more likely to be a manager, a woman or a man?

35. Suppose a packet of M&M's contains a total of 52 candies with the following colors and counts: Blue: 10, Brown: 4, Green: 8, Orange: 12, Red: 7, Yellow: 11.

   (a) For each color, find its probability, as a percentage, of being randomly picked as the first candy.
   (b) Suppose someone has eaten just the orange ones, find the probability, as a percentage, of randomly picking a yellow.

36. A student takes a four-question quiz. The first two questions are True/False (T/F) and the other two are multiple choice with four possible answers.

   (a) For a T/F problem, what is the probability of guessing the correct answer?
   (b) For a multiple choice problem, what is the probability of guessing the correct answer?
   (c) What is the probability of guessing a multiple choice answer, if you know one choice is not the answer?
   (d) Is guessing an answer for a question independent from guessing on a different question?
   (e) If you know the correct answer to the first T/F, guess on the second and third, and know that one of the choices on the last question is incorrect, what is the probability that you get them all correct?

37. A drug test for professional athletes correctly identifies 95% of all drug users and 95% of all non-drug users. In a population of 1,000 athletes, 4% are drug users. The question is: If an athlete gets a positive result, how likely is it that he or she is actually a drug user? The following steps should help you answer this question.

   (a) How many of the 1,000 athletes are drug users? How many do not use drugs?
   (b) How many of the drug users test positive? Negative?
   (c) How many of the athletes who do not use drugs test positive? Negative?
   (d) How many athletes altogether test positive? What fraction of them actually use drugs?
   (e) If an athlete tests positive, how likely is it that he or she is actually a drug user?

38. Federal sentencing laws passed in 1986 and 1988 established a sharp distinction between two forms of cocaine, crack cocaine and powder cocaine. A person convicted of possession of 5 grams of powder cocaine would likely receive probation, whereas a person convicted of possession of 5 grams of crack cocaine would receive an automatic five-year prison term. According to The Sentencing Project,[9] "a dealer charged with trafficking 400 grams of powder, worth approximately $40,000, could receive a shorter sentence than a user he supplied with crack valued at $500." Critics have questioned the fairness of these laws, both in terms of the relative danger of the two forms of cocaine, and also in terms of the impact of these laws on different demographic groups. Table 17.29, taken from a report to the Congress by the United States Sentencing Commission, gives race and ethnicity data for federal cocaine offenders in 1992 and 2000.[10]

   (a) In 1992, what percent of federal cocaine offenders were involved with powder cocaine? Crack cocaine? What about 2000?
   (b) In 1992, what percent of federal cocaine offenders were white? Black? What about 2000?
   (c) In 2000, is a crack cocaine offender more likely to be black, white, or Hispanic? What about a powder cocaine offender?
   (d) Based on these calculations, discuss the possible effects of more severe penalties for crack cocaine offenses in terms of demographics. Do trends suggest that these penalties are reducing crack cocaine offenses?

[9]http://www.sentencingproject.org/brief/pub1003.htm.
[10]Report to the Congress: Cocaine and Federal Sentencing Policy May 2002, http://www.ussc.gov/r_congress/02crack/2002crackrpt.htm.

Table 17.29

Race/ethnicity	1992		2000	
	Powder	Crack	Powder	Crack
White	2113	74	932	269
Black	1778	2096	1596	4069
Hispanic	2601	121	2662	434
Other	44	3	49	33

ANSWERS TO ODD-NUMBERED PROBLEMS

Section 1.1

1 29

3 1/2

5 4

7 0

13 Multiply by 2, add 1

17 $(P + Q)/2$

19 $Q + 50t$ is larger

21 $2(x - 1) + 3$

23 $2(x + 3) - 1$

25 $20,000r$

27 $0.9rp$

29 $0.99a + 1.25p$ dollars

31 (a) $50s + 10t$
    (b) \$220

33 (a) 5 children
    (b) \$40
    (c) \$250
    (d) $c/3$

35 Production same as wells 1 and 2 combined

37 Production half that of well 2

39 Production average of wells 1 and 2

43 The debt remaining

45 $5n$; $4n$ remain

47 $k^2 + 3$

49 $12/(z^2 + 3)$

51 (a) $4p + 3e + 9c$ dollars
    (b) $4p/(4p + 3e + 9c)$
    (c) $Pp + Ee + Cc$; $Pp/(Pp + Ee + Cc)$

53 Yes; $a = b^2$, $x = \theta^2$

55 Yes; $a = 5$, $x = y^2 + 3$

57 $0.01p + 0.05n + 0.10d + 0.25q$

59 $5 + 870/v$ hour, 1.8 hour

61 $9.45A$

63 cA is total cost for A acres;
    pb is revenue for b bushels

Section 1.2

1 $2p = 18$

3 $p + 0.20p = 10.80$

5 A doubled number is 16

7 Ten more than a number is twice the number

9 Four less than number is 3 times number

11 Solution

13 Not a solution

15 Not a solution

17 Yes

19 No

21 (a), (b), (d)

23 (a)

a	0	1	2	3	4
$3 - a^2$	3	2	-1	-6	-13

(b) $a = 3$

25 $x = 5$

27 $x = 3$

29 $w = 4$

31 $x = 7$

33 $t = 15$

35 $y = 25$

37 (a) 3500
    (b) $3500 - 700t = 0$

39 $100 + 75 + x = 300$; \$125

41 $600,000 = 40,000x + 20,000x$, 10 firemen

43 $15 = 6 + 40t - 16t^2$

45 $90 = -394 + 5v$

47 $-1, 1, 2$

49 1

51 Yes

53 No

55 No

61 6

63 Positive

Section 1.3

1 Many possible answers

3 Many possible answers

5 Equivalent

7 Equivalent

9 Not equivalent

11 Not equivalent

13 First two

15 (a) $(1/5)b$; $b/5$
    (b) Yes

17 (a) $0.8b$; $b/(8/10)$
    (b) No

19 Both equal 1; not equivalent

21 Identity

23 Not an identity

25 Equal

27 $I - R - F + 100$ is larger

29 (a) $0.5n - 0.1(0.5n)$, $0.5(n - 0.1n)$
    (b) Yes, $0.45n$

31 No, both $0.5p + 0.5q$

33 $2x + 1 \neq 2(x + 1)$

35 Identity

37 $2x + 3x$ and $5x$ equivalent

-11	-7	0	7	11
-22	-14	0	14	22
-33	-21	0	21	33
-55	-35	0	35	55
-55	-35	0	35	55

39 $2m^2 + 2m^2$ and $4m^2$ are equivalent

m	-1	0	1
m^4	1	0	1
$2m^2$	2	0	2
$2m^2 + 2m^2$	4	0	4
$4m^4$	4	0	4
$4m^2$	4	0	4

41 $-(x + 3)$ and $-x - 3$ are equivalent

x	-2	-1	0	1	2
$x + 3$	1	2	3	4	5
$-(x + 3)$	-1	-2	-3	-4	-5
$-x$	2	1	0	-1	-2
$-x + 3$	5	4	3	2	1
$-x - 3$	-1	-2	-3	-4	-5

Section 1.4

1 Subtract 0.1, -0.2

3 Add t, 8

5 Multiply by -1, 4

7 Mult. by -5, -20

9 Subtract 7; $x = 3$

11 Multiply by 9; $x = 153$

13 Subtract 3 then divide by 2; $x = 5$

15 Subtract 5 then multiply by 3; $x = 45$

17 Valid, add $2x$ to both sides

19 Valid, subtract 1 from both sides and rearrange

21 Invalid

23 Valid, divide both sides by 5

25 Not equivalent

27 Equivalent; Add $3x$ to both sides

29 Equivalent; Add 6 to both sides

31 Not equivalent

33 Same solution: I, V, VI
    Same solution: II IV

35 $x = 15.6$

37 $x = 9$

39 $x = 6$

41 $z = 22/3 = 7.333$

43 $y = 28$

45 $x = 4$

47 (a)

49 (d)

51 (c)

53 (c)

55 Increases

57 Unchanged

59 (a) Yes
   (b) No

61 (a) No
   (b) Yes
   (c) Yes
   (d) No

63 Divide by -2; $z = -6.5$

65 Multiply by $7/3$; $M = 28/9$

67 Cube root; $y = -2$

69 $(b(x - r))^2 + 4 = 13$; $x = 5$

Chapter 1 Review

1 174.80

3 $704.5w/h^2$

5 150π ft^2

7 108π ft^2

9 $2(x + 2) - 4$

11 $(x - 1)^2 + 1$

13 -8

15 -12

17 Many possible answers

19 Many possible answers

21 Not equivalent

23 Equivalent

25 Not equivalent

27 $t = 0, t = 4$

29 $t = 0$

31 $t = 0, t = 1$

33 $t = -8, t = -6, t = 6$

35 $2x = 24$

37 $x + 6 = -x$

39 $2x + x = 99$

41 0 and 1

43 6.5

45 $8, -3$

47 -2

49 Cube of number is same as its square

51 One-quarter a number is 100

53 $r = 3$

55 $J = 32$

57 $d = -4$

59 Add 11 to both sides

61 Multiply both sides by $3/2$

63 Expression, $2n + (n + 3)$

65 Expression, $J + S$

67 Equation, $225 = w + 10$, or $225 - 10 = w$

69 (a) Total number of widgets company receives
   (b) Proportion of widgets supplied by A
   (c) Nothing
   (d) Number of defective widgets supplied by A
   (e) Total number of defective widgets company receives
   (f) Proportion of defective widgets company receives

(g) Proportion of non-defective widgets company receives

71 1 gram of fiber

73 Total amount spent on tickets

75 $1.5wq$

77 Increases

79 Decreases

81 Increases

83 $A = r, B = s$

85 $A = n + m, B = z^2$

87 $A = x, B = x$

89 $A = 1, B = -2$ (many other solutions)

91 $a = r$ and $x = n + 1$

93 $a = 3$ and $x = 2d$

97 $2w^3 + 10w^2$

99 (a) $2p + 3q$
   (b) $4q$
   (c) $5p$
   (d) $Ap + Cq$

101 $6q$, no

103 $2p + q + 4r + s$

105 (D)

107 (a)

h	0	1	2	3
$40 - 8h$	40	32	24	16
h	4	5	6	
$40 - 8h$	8	0	-8	

(b) 40
(c) 2 pm

109 (a) No
   (b) Yes

111 (c)

113 $100,000,000 = 75,000,000 + 300y$

115 $90,000,000 = 2,015,000w$

117 False

Section 2.1

1 No

3 Yes

5 Yes

7 Incorrect

9 Correct

11 $7x + 9$

13 $3x + 6$

15 $8x^2 + 5x$

17 $24xy + 4x + 5y$

19 22

21 3000

23 $-a - b - 3c$

25 68.75 square meters

Section 2.2

1 $2x + 6y$

3 $-10x + 15$

5 $2x^3 - 6x^2 + 8x$

7 $10x - 6y + 5$

9 $x(2a - 3b)$

11 $\frac{1}{5}(x + y)$

13 $-mn(m + 3n)$

15 $-2ab(2a + 3b + 1)$

17 $(b + 3)(b - 6)$

19 $2x(x + 4)(2a - 1)$

21 No

23 No

25 No

27 No

29 33

31 $3np + 2n$

33 $2(x + 25)$; $k = 2, A = 25$

35 $-5(x + (-3))$; $k = -5, A = -3$

37 $0.2(x + (-300))$; $k = 0.2, A = -300$

39 Yes

41 $-(1/2)(a + 1) + 1$ and $-(1/2)a + (1/2)$ are equivalent

43 $a = x^2, b = x + r, c = 3$

Section 2.3

1 $x^2 + 7x + 10$

3 $z^2 - 11z + 30$

5 $3b^2 + 7bc + 2c^2$

7 $3x^2 - 16x$

9 $x^2 - 16x + 64$

11 $x^2 - 26x + 169$

13 $x^2 + 2xy + y^2$

15 $25p^4 - 10p^2q + q^2$

17 $8a^3 - 36a^2b + 54ab^2 - 27b^3$

19 $x^2 - 64$

21 $3x^2 - 31x - 22$

23 $15x^3 + 11x^2 - 56x$

25 $2s^2 - 15$

27 $(y - 6)(y + 1)$

29 $(g - 10)(g - 2)$

31 $(q + 10)(q + 5)$

33 $(x + 3y)(x + 8y)$

35 $(z + 4)(4z + 3)$

37 $3(w + 6)(w - 2)$

39 Cannot be factored

41 $(s - 6t)^2$

43 $(x + 6)(x - 6)$

45 $(x + 5)(x + 5)$

47 $(x + 7)(x + 8)$

49 $(5x + 8)(x - 9)$

51 $(2x - y)(4x - 3)$

53 $x(x - 8)(x - 8)$

55 $r(r - 7s^3)^2$

57 $2x(3x^3 + 4z^2)^2$

59 $2x^2(3 - xz^3)(3 + xz^3)$

61 $-3(3t + 1)(3t + 5)$

63 $q^4(1 + q^2)(1 + q)(1 - q)$

65 $2w(w - 4)^2$

67 $st(4s - 3t)^2$

69 $x^2 + 2xh + h^2 - 1$

71 $2y^7 - y/2$

Section 2.4

1 $5m/6$

3 $-1/((x - 2)(x - 3))$

5 $1/x$

7 $e/8$

9 $2a/((a - b)(a + b))$

11 $55/12$

13 $p/(2q)$

15 $(2a^2b^2)/3$

17 $r/(2s)$

19 $(a + b)/ab$

21 $r/8$

23 $3(w + 1)/(w - 1)$

25 $z/2 + 1/2$

27 $\dfrac{2}{c + 2} + \dfrac{2}{c + 2}$

29 $p - (1/2)$

31 $-4 + (5x)/2$

33 $2x + h$

35 $\dfrac{c + 1}{a + b} + \dfrac{-1}{a + b}$

37 $(3/t)(1/(r + s))$

39 $(1/4) + (a/2) + (3b/4)$

41 $(x + 1)^2/(xy) - (1/x)$

43 $\dfrac{x + 1}{x - 1} - \dfrac{1}{x - 1}$

45 (c)

47 (b)

49 None

51 $2a^3b^2/3$

53 $36(p + q)/(3p + 2q)$

55 $2/x$

57 $-(x + 5)/(5x)$

59 $(d + c)/d$

61 $m/(m - 1)$

63 $(w + 6)/(w + 3)$

Chapter 2 Review

1 Yes

3 No

5 -30

7 $2p^2 - 2q^2 + 6pq$

9 $(5/12)A$

11 $-3t^2 + 8t + 4$

13 $5x - 2$

15 $27 - 19x$

17 $9x^2 - 7x + 18$

19 $2a^2 - 3ab - 5b^2$

21 $12m^2n + 5mn^2$

23 $(z - 3)(z - 2)$

25 $(v - 8)(v + 4)$

27 Cannot be factored

29 $(2x + 3)(3x - 2)$

31 $(q - 3z)(q - z)$

33 $2(3w - 4)(2w + 1)$

35 Cannot be factored

37 $x(x^2 + 4)$

39 $(x + 12)(x - 12)$

41 $(x - 11)(x - 11)$

43 $(3x + 7)(x + 5)$

45 $(a + b)(x - y)$

47 $x(x + 12)(x + 11)$

49 $y(y + 9)(y - 2)$

51 $-3t^3(t + 2v)^2(t - 2v)^2$

53 $(a - 2)/4$

55 $(1 - 3w^2)/2w$

57 $3t/4$

59 $(x - y)/(x + y)$

61 $pq/(p - q)$

63 $-3z/(z + 3)$

65 $(r^2 + 1)/rp$

67 $(r - s)/rs$

69 1

71 270 square meters

73 $2(3x^2 + 6); r = 3x^2 + 6$

75 $2 \cdot 3x^2 + (2 + 1)4; r = 2, v = 3x^2, w = 4$
or
$3 \cdot 2x^2 + (3 + 1) \cdot 3; r = 3, v = 2x^2, w = 3$
other answers possible

77 $3 - (7x)/(x - (-5)); k = 3, m = 7, n = -5$

79 (a) $A + B$
(b) $1.05(A + B)$
(c) $1.05A + B$
(d) $A + 1.05B$
(e) $1.05A + 1.05B$
(f) (b) and (e) are equivalent

81 $a = 3x^2, b = x, c = 2$

Section 3.1

1 $x = 28$

3 $a = 2$

5 $p = -1.54$

7 $r = -14/5 = -2.8$

9 $m = 15/8 = 1.875$

11 $n = -23/4 = -5.75$

13 $B = -3$

15 $z = 5$

17 $x = -1$

19 $x = 1$

21 $x = 5/6 = 0.833$

23 $a = -5$

25 $y = 18$

27 $t = y/(3\pi)$

29 $a = 2(s - v_0t)/t^2$

31 $t = (-5x - 1)/(3x - 2)$

33 $y = z - x$

35 $r = 3t$

37 $g = -3h$

39 $x = 3$

Section 3.2

1 $1 \le s \le 240$, s an integer

3 $30 \le V \le 200$

5 $n, 16 \le n < 24$, n an integer

7 Divide by 12; no change; $x \ge 5$

9 Add 4.1; no change; $c \le 6.4$

11 Multiply by $-7/3$; changes direction; $P > -1$

13 Subtract 7, divide by 5; $y \le 3$

15 Subtract 25, divide by -3, direction change; $a \le 4$

17 Collect like terms, multiply by $2/3$; $r < 4$

19 (a) $z = 2, z = 3, z = 4$
(b) $z = 0$
(c) $z = 1$

21 $173.60 \le c \le 224$

23 (a) (i) Total number of trips
(ii) Amount of sand if each truck makes one trip
(iii) Total amount of sand delivered
(iv) Average amount of sand delivered per trip
(v) First truck delivers more total sand than second truck
(vi) Per delivery, second truck transports more sand than first truck
(b) $S > T$

25 $x < -6$

27 $-2 < x < 1$

Section 3.3

1 Yes

3 No

5 Yes

7 No

9 (a) False
(b) False
(c) False
(d) True
(e) False
(f) True

11 $x = 7$ and $x = -5$

13 $t = 2$ and $t = -1$

15 $p = 7$ and $p = -9$

17 $x \le -27$ or $x \ge -15$

19 $z < 5/3$ or $z > 13/3$

21 $|x| \le 3$

23 $|x - 1| = 2$

25 $|x - 2| \ge 1$

Chapter 3 Review

1 $x = 28$

3 $k = -9/2 = -4.5$

5 $x = (c + d)/(b - a)$

7 $m = (p - n)/2$

9 $j = -29/5 = -5.8$

11 $r = 1/8 = 0.125$

13 Multiply by -1; changes direction; $q \le -14$

15 Multiply by $-5/2$; changes direction; $r < -10/7$

17 Add $5n$, subtract 7; $n \le -2$

19 Add $7k$, divide by 9; $k > 4/3$

21 Add 13, divide by -12 with direction change; $s > 1/3$

23 Divide by -4 with direction change, add 5; $x < 7$

25 Collect like terms, multiply by $3/4$; $r > 6$

27 No

29 No

31 $t = -1/3, 3$

33 $w = -5/4, 11/4$

35 $x = 0, 4$

37 $j \le -9$ or $j \ge 33$

39 $c \ge 4$ or $c \le -4/3$

41 $p = k/(\sqrt{q} + n)$

43 $s = 1 + ((zr - 1)/(r - z))^2$

45 $x > 7$

47 $x < -3$ or $x > 12$

49 $31/64 \le p \le 33/64$; probably not

Section 4.1

1 $w = f(c)$

3 Dep: N, Ind: C

5 $f(-7) = -9/2$

7 (a) 99
  (b) -15
  (c) 145
  (d) 177

9 $1/3$

11 $1/\sqrt{2} = \sqrt{2}/2$

13 4

15 $10 - 3r$

17 $19 - 3k$

19 $10 - 15t^2$

21 $-x^2 + 3x - 1$

23 (a) 40: $^\circ$C
      3: liters
  (b) 3 liters

25 (a) -2
  (b) -2 and 3
  (c) 0
  (d) 0 and 2

27 (a) 2
  (b) -2 and 1

29 (a) 4
  (b) 2 and -2

31 (a) At price $15 revenue is $112,500
  (b) At price a revenue is $0
  (c) At price $1 revenue is b
  (d) c is revenue in dollars at price p

33 Daily downloads at $0.99/song

35 Average daily downloads after price drops 10 cents

37 Number of downloads in one year at current price

39 Average hourly downloads

41 5050

45 Mileage if tires are 10% underinflated

Section 4.2

1 (a) $f(1) = 8, f(3) = 6, f(1)$ greater

3 (a) $C(100) = -20, C(200) = -40, C(100)$ greater

5 Equivalent

7 Not equivalent

9 Same

11 Same

13 Different.

15 Different.

17 Variable: r; Constant: π

19 Variable: t, Constant: r, A

21 (a) $G(50) > G(100)$

23 (a) $325 - 65t, 65(5 - t)$
  (b) Yes

25 $Q = (1/4)t; k = 1/4$

27 $Q = (b + r)t; k = b + r$

29 $Q = (\alpha - \beta)t/\gamma; k = (\alpha - \beta)/\gamma$

31 (a) Variable: d Constant: A
  (b) Variable: A Constant: d
  (c) Variable: A and d

Section 4.3

1 $x = 18$

3 $x = 9/4$

5 $x = 25$

7 $t = 5$

9 $t = 0$

11 $t = -1$

13 20

15 (a) 75
  (b) 5

17 (a) $t = 6$
  (b) $t = 1, t = 2$

19 (a) $15 - d/20 = 10$
  (b) $d = 100$

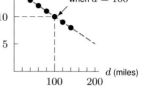
gallons of gas
10 gallons when $d = 100$
d (miles)

21 $b = 40/61$

23 $t = -1$

25 (a) $x = -2, x = -1$ and $x = 3$
  (b) $x = 2$ and $x = 4$
  (c) $x = 0$
  (d) No values

27 (a) $s = 0.5, s = 0.75, s = 1$
  (b) $s = 0$
  (c) $s = 0.25$

29 (a) $218
  (b) $39
  (c) $a = 60$
  (d) $a = 55$
  (e) 35, 40, and 45
  (f) 50, 55, 60, 65, and 70

31 (a) $t = 7$ years

Section 4.4

1 9

3 -3

5 14

7 170

9 154.98 dollars/year

11 0.00591 meters/year

13 Debt increased at average rate of $531.3 billion/year

15 US debt not going down

17 Value increased by $30,000

19 Value decreased at average rate of $1000/year

21 $a \ne 3$

23 Area decreases 25,000 acres between $t = 0$ and $t = 25$

25 Greater decrease from $t = 20$ to $t = 30$ than $t = 10$ to $t = 20$

27 $h > 0$

29 Average increase 11 ppb/yr from 1985 to 1990

31 No, increasing, but less fast

Section 4.5

1 Yes; $k = 55$

3 Yes; $k = 0.7$

5 Yes, 5

7 No

9 Yes, $1/9$

11 No

13 Yes, 42

15 $q = 8$

17 $t = 10$

19 $k = 3/2; y = (3x)/2; x = 5.33$

21 No

23 Yes, $k = 2/3$

25 Yes, $k = -4/3$

27 No

29 Yes; $k = 1/50$

31 $k = 4$

33 (a) 3.33, \$/gallon
     (b) \$49.95

35 (a) 1.219
     (b) Yes; 0.820

37 (a) $C = kx$
     (b) $k = 9.50$ dollars per yard
     (c) \$52.25

39 Blood mass $= 0.05$(Body mass);
     3.5 kilograms

Chapter 4 Review

1 (a) 180 lbs 4000 miles from center
    (b) 36 lbs a thousand miles from center
    (c) b lbs 36,000 miles from center
    (d) w lbs r thousand miles from center

3 1

5 0.75

7 4/5

9 (a) 1.25
    (b) 0.5
    (c) $(a^2 + 1)/(5 + a)$.

11 (a) -20
     (b) 30
     (c) -5
     (d) 80

13 (a) 205.0 mn people in 1970
     (b) $t = 50$; in 2000 there were 281.4 mn people

15 (a) $1/4 = 0.25$
     (b) 16
     (c) 0
     (d) 5

17 Not equivalent

19 Equivalent

21 $x = 8/5 = 1.6$

23 $x = 5/6 = 0.833$

25 (a) $8 + 4t = 24$
     (b) $t = 4$

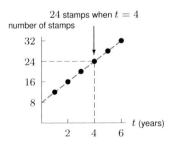
24 stamps when $t = 4$

27 (a) $100,000 - 10,000t = 70,000$

(b) $t = 3$

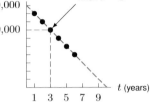

29 10

31 (a) 150 glasses

33 Proportional, 1

35 Not proportional

37 Proportional, $a - 2$

41 (a) Initial depth water $= 36$ inches
     (b) Cauldron empty after 75 hours

43 (a) $f(0) = 0$ and $g(0) = 15$
     (b) $x = 0$
     (c) $x = 7.5$
     (d) $x = -5$ and $x = 3$
     (e) Between $x = -5$ and $x = 3$

45 $g(3) = 14$

47 $g(-2) = 6$

49 $8t + 2$

51 x^2

53 $-7z + 6r - s$

55 $v = 26/3$

57 53.482; \$53.48 worth 50 CAD

59 Change in gas spending from preceding week

61 $(p(2)q(2) + p(3)q(3) + p(4)q(4) + p(5)q(5))/4$

63 $w(18) = 1762$

65 Investment's bond value in year 5 twice its stock value in year 4

67 $w(t) = v(t) + 3000$

69 $g(14) = 37$

71 $\dfrac{g(15) - g(12)}{15 - 12}$ is larger

73 (a) 150 miles
     (b) $D = 30t$, $k = 30$,

75 Yes

77 96 grams; carbs/cracker

79 (a) $M = 167H$
     (b) $M = 20b$
     (c) $B = 8.35H$, yes

81 $C = 81.667b$,
     326.667 calories

Section 5.1

1 $D = 1000 - 50d$ miles

3 $T = 30 - 0.04d$

5 $b = -5300, m = 250$

7 $b = 30, m = 0$

9 $b = 0.007, m = -0.003$

11 (a) Initial rental cost
     (b) \$100 per hour

13 600, 5

15 $b = 25, m = 0.06$

17 $b = 50, m = 1.2$

19 $b = 100, m = -3$

21 $b = 20,000, m = 1500$

23 $b = 120, m = -5$

25 $4, -1$

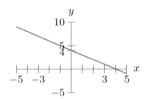

27 $-2, 3$

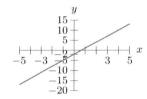

29 $-0.2, -0.5$

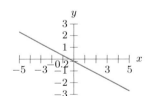

31 (a) \$0.14/kWh over 250
     (b) \$43.60

33 (a) (iii)
     (b) (i)
     (c) (ii)
     (d) (iv)
     (e) (v)

35 Yes, slope a, vertical intercept $5a$

37 Yes

39 $m = 1/3; b = -11$

41 $m = -2/3; b = 20/3$

43 $m = \pi; b = 0$

45 Ounces per bird per hour; weight of seed one bird consumes in one hour

530

Section 5.2

1 Linear

3 Linear

5 Not linear

7 Linear

9 4, 3

11 $w + 1, w$

13 $mn + m + 7, m + 5$

15 $y = 9 + 3x$

17 $g(n) = 22 - 2/3n$

19 $(1, 5), 3$

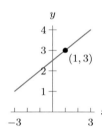

21 $(1, 3), 1/2$

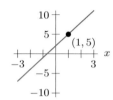

23 (a) 1/2
   (b) -4

27 $y = 20 - 2x$

29 $y = 5.4 - 0.5x$

31 Not linear

33 Not linear

35 (a) Cost of gas; expenses; car rental
   (b) Dollars; miles
   (c) Yes

37 Not linear

39 Linear

41 Linear

43 Linear

45 Not linear

47 $R = 0.1S$, rain equivalent to 1 inch of snow, no rain = no snow

49 $T = 40 - 3h$, rate of decrease of temperature, temperature at midnight

51 (iii)

55 (a) after 20 years collection has 50 butterflies, 2 butterflies per year
   (b) $B(x) = 10 + 2x$

57 $-1000 + 12q$ dollars

59 $520 + 40x$ dollars

61 $x^2 + 5x - 2$, not linear

63 $g(x) = 20 + 0.1x$

Section 5.3

1 (a) \$1900
   (b) 7 credits

3 (a) 5
   (b) 7/2

5 1000

7 1000

9 750

11 $t = 47/13 = 3.615$

13 $r = 11$

15 $t = 11$

17 Positive

19 Negative

21 Positive

23 Negative

25 Positive

27 Negative

29 One

31 Infinite

33 One

35 Infinite

37 \$100

39 (a) \$62
   (b) \$211
   (c) 800 miles

41 18 ft

43 (a) $A = 0$
   (b) $A > 0$
   (c) None

45 (a) $A = 5$
   (b) $A > 5$
   (c) None

47 (a) Any value of A
   (b) No value of A
   (c) None

49 (a) None
   (b) $A > 0$
   (c) $A = 0$

51 One

53 Infinite

55 One

57 Infinite

Section 5.4

1 $y = 160 - 3x, m = -3, b = 160$

3 $y = 300 - 3x, m = -3, b = 300$

5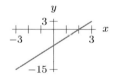

7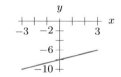

9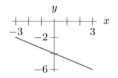

11 (a) (V)
   (b) (IV)
   (c) (I)
   (d) (VI)
   (e) (II)
   (f) (III)
   (g) (VII)

13 $y = 7 + 6(x + 1)$

15 $y = -9 - (2/3)(x - 2)$

17 $y = 5 - 4(x - 6)$ or $y = 1 - 4(x - 7)$

19 $y = -7 - (1/2)(x - 6)$ or $y = -1 - (1/2)(x + 6)$

21 $y = -6 + (5/4)(x - 3)$

23 $x - 3y = -2$

25 $5x + 2y = 7$

27 $x - 3y = 7$

29 $9x + 9y = 5$

31 $-5x + y = 2a$

33 (a) $y = 10 + 5x$
   (b) $y = 12 - 2x$
   (c) $y = 1 + 7x$
   (d) $y = 60 - 9x$

35 Not parallel

37 Not parallel

39 Parallel

41 Not parallel

43 I, V, and VI

45 III and V

47 $y = 3 + 3x$

49 $y = 5x - 7.1$

51 Slope-intercept

53 Point-slope

55 Point-slope

57 $b = \beta, m = -1/\alpha$

59 $b = b_1 + b_2, m = m_1 + m_2$

61 (a) $y = 3 + 6x$
   (b) $y = 5 - 3x$

63 (a) $2x = 4y + 3$
   (b) $2x = 4y + 3$

65 (a) $y + 2 = 3(x - 1)$
   (b) $y = 6 - 50x$

67 (a)

Section 5.5

1 Yes

3 No

5 Yes

7 Linear, $Q = 5.13 + 0.22t$

9 (a) $1.29c + 3.49s = 100$
 (b) $(10, 25), (25, 19.4)$
 (c) s

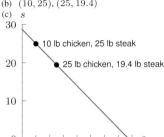

11 (a) $4.5t + 7.87i = 5000$
 (b) $(600, 292), (350, 435)$
 (c) i

13 (a) Yes, $I = 46 + 453.6w$
 (b) Number of grams in a pound (instrument weighs in grams), bad calibration of instrument

15 (a) $9f + 4c = 2000$
 (b) $f = 66.7$

17 (a) 50
 (b) 48
 (c) $h = 100 - (5/4)a$

19 $750a + 1200r = 1,000,000$

Section 5.6

1 $x = 6, \; y = -1$

3 $x = 1, \; y = -1$

5 $a = 5, b = -2$

7 $p = -1, r = 4$

9 $(x, y) = (2, 3)$

11 $(m, n) = (0.1, 0.5)$

13 $(x, y) = (10, -20)$

15 $(x, y) = (-1, 3)$

17 $(x, y) = (3, -7)$

19 $(e, f) = (6, -7)$

21 $x = 1, y = 1$

23 $x = 4, y = 8$

25 $x = 3, y = 11$

27 Approximately $x = 0.47, y = 1.21$

29 One

31 One

33 Infinite

35 $(x, y) = (3, -1)$

37 $(\alpha, \beta) = (6, 5)$

39 $(x, y) = (-2, -5)$

41 $(\kappa, \psi) = (2, -1)$

43 Same solutions: I, III, IV

45 $29/2$ and $5/2$

47 (a) $4.05
 (d) Yes

49 $x = 1, y = -1, z = 2$

Chapter 5 Review

1 4.29, 3.99

3 250, 1/36

9 III

11 $50 + 45h$ dollars

13 $2400 - 500y$ dollars

15 $80 + 2p$

17 $350 + 30m$ dollars

19 $d = 77 - 3.2t$

21 $b = 200, m = 14$

23 $b = 0, m = 1/3$

25 $b = 7/3, m = 2/3$

27 -7

29 $4h$

31 Linear

33 Linear

35 Not linear

37 Linear

39 Linear

41 Linear

43 Positive

45 Positive

47 Negative

49 Positive

51 Negative

53 No solution

55 $r = I/(Pt)$

57 $y = -a(b+1)/(ad - c)$

59 (a) $y = 5 - 2x$
 (b) $y = 8 - 4x$

61

$$y$$

63 $y = -14 + \frac{1}{8}x; b = -14, m = 1/8$

65 $y = -1 + (10/3)x; b = -1, m = 10/3$

67 $y = 90 + 0 \cdot x; b = 90, m = 0$

69 $-3x + y = -2$

71 $y = 5 + 2x$

73 $y = 8 - (3/4)x$

75 $y = -16 - 3x$

77 $y = 13 - (2/3)x$

79 $y = 6 + (1/3)x$

81 $y = 9 - (3/5)x$

83 $y = -13 - (4/5)x$

85 $y = -60 - 3x$

87 $s(t) = 8200 + 150t$

89 $w(x) = -3x + 32$

91 $w(x) = 0.115 - 0.15x$

93 $y = 24 - 4x$

95 $y = -160/9 + (5/9)x$

97 $u = (1/12)n$

99 $y = 459.7 + 1x$

101 Yes

103 No

105 Yes

107 $a = 5, b = -4$

109 $(x, y) = (0.5, -2)$

111 Number of gallons initially in tank; number of gallons per minute flowing into tank; 200 gallons

113 $30 + 15.84i$

115 Length of vacation in days

117 $f(x) = 6x + 4$

119 No

121 No

123 $15 + (15/100)t$ dollars, yes

125 $23 + 10N$ dollars, linear

127 $20 - 0.25d$ ft^3; $0 \le d \le 80$

129 (a) $F = 32 + (9/5)C$
 (b) (i) $50°$F

  (ii) $30°$C

131 (a) $C = 12.5 + 1.5(n - 3)$
 (b) $C = 8 + 1.5n$

133 $500; $2500

135 (v)

137 (ii)

139 (iv)

141 (ii)

143 (iii)

145 (v)

147 (i)

149 (iii)

151 (i)

153 (vi)

155 (a) $F = 2C$
 (b) $320°$F $= 160°$C

157 $f(t) = 4500 - 0.25t$

159 2.5

163 (d)

165 (c)

167 $x = 4, y = 9$

169 $x = 6.5, y = -4$

171 If $0 < x < 5$ use A, if $5 < x \le 12$ use B, if $x = 5$ use either company

173 $u = 6/5$, $d = 1/20$; UK cup is $6/5$ US cup, UK dessertspoon is $1/20$ UK cup

Chap. 5 Solving

1 $x = 2$

3 $t = 67/9$

5 $x = -9/2 = -4.5$

7 $w = 130/29 = 4.483$

9 $t = 78.4$

11 $x = -11/2 = -5.5$

13 $p = -17/2 = -8.5$

15 $t = 6.110$

17 $t = (q - s)/(r - p)$

19 $r = -(0.1sw + 1.8s)/(sw - 0.2w - 5)$

21 $s = (25t + ABr + Brt - Cr^2)/(Ar - At)$

23 $y' = (5x - x^2y^2 - 10)/(2xy + x^2 + 2)$

25 $[T] = (3A_0V + A_0V_0 - 10H^2 - 10V_0)/(25V_0S^2)$

Section 6.1

1 24

3 -8

5 100

7 -500

9 72

11 1

13 -30

15 -9

17 $-8/125$

19 -32

21 $729/1000$

23 x^8

25 x^{10}

27 x^2

29 x^{15}

31 x^5

33 2^{n+2}

35 $(a/b)^x$

37 2^{2n-m}

39 B^{2a+1}

41 $(x + y)^{20}$

43 $(g + h)^3$

45 $(a + b)^3$

47 Negative

49 Negative

51 Negative

53 Positive

55 Negative

57 Negative

59 Positive

61 c^{12}/d^4

63 $64r^6/125s^{12}$

65 $36g^{10}/49h^{14}$

67 $32p^5$

69 $16^t b^{4t}$

71 $3 \cdot 16^x e^{4x}$

73 (a), (c), (d) equivalent; (b), (e) equivalent

75 (a), (c), (e) equivalent; (b), (d) equivalent

77 (a), (b) equivalent; (c), (d), (e) equivalent

79 $a^{4+1} = a^4 \cdot a$

81 $10^4/10^z$

83 $4^p \cdot 4^3$

85 $(-n)^a(-n)^b$

87 $p/(p^a p^b)$

89 $(p + q)^a/(p + q)^b$

91 $(x + 1)^{ab}(x + 1)^c$

93 $(2^3)^x = 8^x$

95 $\left(3^4\right)^a = 81^a$

97 $\left(\sqrt{3}e^t\right)^2$

99 w^3

101 $c/64$

Section 6.2

1 2

3 $3/2$

5 100^x

7 $\dfrac{x^2}{y}$

9 $4ab^3\sqrt{3ab}$

11 $x^{-1/2}$; $n = -1/2$

13 $x^{3/2}$; $n = 3/2$

15 x^{-2}, $n = -2$

17 $10\sqrt{11}$

19 $5\sqrt{3}/2$

21 $2\left(3\sqrt[3]{3} - \sqrt{3}\right)$

23 $15a^2\sqrt{3}$

25 $\dfrac{3\sqrt{5}}{5}$

27 10

29 -1

31 $ab^2 - a^2b$

33 $\sqrt{3} - 1$

35 $2(\sqrt{6} + 1)$

37 $2 + \sqrt{3}$

Chapter 6 Review

1 72

3 64

5 $1/25$

7 $(a/3)^b$

9 $(4xy)^3$

11 $(a + b)^7$

13 $((x + y + z)(u + v + w))^{21}$

15 Negative

17 Negative

19 Negative

21 $\dfrac{1}{49}$

23 $\dfrac{2}{7}$

25 k^8/g^8

27 r^5/s^{20}

29 $-5a^6b^{12}$

31 $6x^5$

33 $9/d^2$

35 s^n/r^n

37 $x^{a+b+1}/2$

39 $2c^2/3$

41 $-5y^6/x^{12}$

43 $2b^3c^5$

45 x^4/a

47 $\dfrac{x^2}{9c^5}$

49 $1/5$

51 7

53 $-h$

55 $(a - b\sqrt{a})/(a - b^2)$

57 $1 + \sqrt{x}$

59 Positive

61 Negative

63 Positive

65 Negative

67 $(a + b)^2(a + b)^s/(a + b)^r$

69 $p^2p^t(p + q)^2/(p + q)^t$

Section 7.1

1 2; 1

3 $1/3$; 1

5 2; 4π

7 (a) Coefficient: $\pi/5$; exponent: 3
   (b) $8\pi/5$ cm^3
   (c) $64\pi/5$ cm^3

9 $p > 1$

11 $p = 1$

13 $p > 1$

15 (a) Proportional to $x^{1/2}$
    (b)

x	1	10	100	1000
y	3	9.49	30	94.87

(c) Increases

17 (a) Inversely proportional to x^2
    (b)

x	1	10
y	5	0.05
x	100	1000
y	0.0005	0.000005

(c) Decreases

19 (a) Odd

(b) Positive

21 a^4

23 b^{-4}

25 (a) Above when $x > 1$ or $-1 < x < 0$;
    below when $0 < x < 1$ or $x < -1$

(b) $x^5 > x^3$ when $x > 1$ or $-1 < x < 0$;
    $x^5 < x^3$ when $0 < x < 1$ or $x < -1$

27 $x^4 > x^2$ when $x > 1$; $x^4 < x^2$ when
    $0 < x < 1$

29 (a) 600 hours, 240 hours

(b) Decreases

(c) $w = 2400/h$, inversely proportional

31 (a) $-3/2$; 0.15

(b) 2.3438; 1.2

33 (a)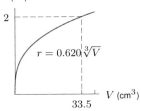

(b) 33.5 cm^3

35 $-x^4 \neq (-x)^4$

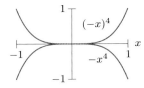

Section 7.2

1 Power function

3 Not a power function

5 Power function

7 Coefficient: 3; exponent: $1/2$

9 Coefficient: 4; exponent: $-1/4$

11 Coefficient: $1/4$; exponent: 3

13 Coefficient: 2; exponent: -2

15 Coefficient: $3^{1/4}/2$; exponent: $1/2$

17 $k = 2/3, p = -1/2$

19 $k = 49, p = 6$

21 $k = 27, p = 6$

23 $k = 1/5, p = -1$

25 Not possible

27 $k = 1/8, p = -3/2$

29 $k = 64, p = 6$

31 $f(a)$

33 $f(b)$

35 $f(b)$

37 1; (III)

39 3; (I)

41 -1; (II)

43 -2; (II)

45 3; (I)

47 0.3; (IV)

49 Negative

51 Positive

53 (a) Not possible

(b) $P = 8w; k = 8; p = 1$

55 $k = 2/3^{1/4}, p = 1/2$

57 Increases

59 Increases

61 Decreases

63 Remains unchanged

65 Decreases

67 A

69 (a) 22,289

(b) 58,045

(c) 5,377

Section 7.3

1 $x = 50^{1/3} = 3.684$

3 $x = 1/16$

5 No solutions

7 $a = 81$

9 $y = 123$

11 $x = 83/3$

13 $c = 2.438$

15 $p = -2$

17 $L = \pm 1/4$

19 $x = -1/27$

21 $z = 720\pi$

23 $r = \pm\sqrt{2A/\pi}$

25 $x = \pm 3y$

27 (i)

29 (iv)

31 (vi)

33 (vi)

35 (i)

37 (i)

39 (i)

41 (ii)

43 Squaring both sides; $x = 0$ is extraneous

45 Dividing by r; lost solution $r = 0$

47 Multiplying by p; $p = 0$ is extraneous

49 Both solutions are positive

51 3 cm

53 (a) About 84 kg

(c) $M = 0.0000276 S^{3/2}$

55 (a) $(4/3)\pi r^3 = (1/3)\pi r^2 h$

(b) $h = 4r$

57 7.177%

59 (a) $A = 0$

(b) $A > 0$

(c) $A < 0$

61 (a) $A = 0$

(b) $A < 0$

(c) $A > 0$

63 (a) $A = 0$

(b) Any $A \neq 0$

(c) No A

65 (a) $A = 0$

(b) No A

(c) No A

67 (a) $A = 1$

(b) $A > 1$

(c) $A < 1$

69 (a) Any A

(b) $A = 0$

(c) $A = 0$

71 (a) $A = 4$

(b) $A > 4$

(c) $A < 0$

73 (a) $A = -4$

(b) $A < -4$

(c) $A > 0$

75 (a)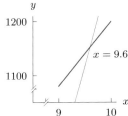

(b) $x = 9.608$

Section 7.4

1 $y = 23.25 x^5$

3 $y = 40 x^2$

5 $s = 14.142\sqrt{t}$

7 $k = 2; c = 2d^2; c = 98$

9 $S = kh^2$

11 $r = kA^{1/2}$

13 Multiplied by a factor of 8

15 Halved

17 (a) $T = kR^2 D^4$

(b) Yes

(c) Yes

19 (a) $T = kB^{1/4}$

(b) $k = 17.4$

(c) 50.3 seconds

21 $N = k/L^2$; small

23 $y = 4x^2$

25 $y = -(1/5)x^3$

27 (a) 100

(b) 100

29 (a) 8

(b) 27

(c) 1/8

(d) 0.001

31 Increases by 21%

33 (a) Life span increases with body size. Bird.

(c) (ii) 9.78×10^{37} kg. Unrealistic.

35 Yes

37 No

Chapter 7 Review

1 $y = 3x^{-2}; k = 3, p = -2$

3 $y = (3/8)x^{-1}; k = 3/8, p = -1$

5 $y = (5/2)x^{-1/2}; k = 5/2, p = -1/2$

7 $y = 0.2x^2; k = 0.2, p = 2$

9 $y = 125x^3; k = 125, p = 3$

11 $y = (1/5)x; k = 1/5, p = 1$

13 Base: w; exponent: -1; coefficient: $-1/7$

15 Base: v; exponent: 2; coefficient: 5

17 Base: x; exponent: 1; coefficient: 12

19 Base: x; exponent: -2; coefficient: 3

21 Base: r; exponent: 2; coefficient: 48

23 Base: a; exponent: 2; coefficient: π^2

25 $x = 2.614$

27 $x = 16$

29 $x = 7$ and $x = -3$

31 (a) $L = (R^2 C^4)/(4\pi^2)$

(b) $C = \pm\sqrt{2\pi\sqrt{L}/R}$

33 Two

35 Zero

37 One

39 (a) 3.172 cm

(b) $r = 0.671 R^{1/4}$

(c) Yes, 1/4

41 Inversely proportional to cube root of P

43 (i)

45 (iii)

47 (i)

49 (i)

51 (iii)

53 (i)

55 (iii)

61 (c) and (d)

63 (a) $P = 365(d/93,000,000)^{3/2}$

(b) 689 earth days

Chap. 7 Solving

1 $x = 10^{1/3} = 2.154$

3 $x = 1.75$

5 $p = 10$

7 $t = 0.1024$

9 $p = \pm 50^{1/4} = \pm 2.659$

11 $x = \pm 2.217$

13 $q = (2.5)^{1/3} = 1.357$

15 $t = (0.196)^{1/2.2} = 0.477$

17 $t = -1.558$

19 $r = \pm\sqrt{5.8} = \pm 2.408$

21 $x = \sqrt[3]{b/a}$

23 $p = \pm\sqrt{0.4q - 3.4}$ if the solution exists

25 $y = -10x^2/(24x - 20)$

27 $t = (-6r^2 s^3 - 12r^2 - 15)/(30r^3 - 8r)$

29 $x = ((D - A)^3 - C)/B$

Section 8.1

1 (a) All real numbers
(b) All real numbers

3 (a) All real numbers
(b) $y = 7$

5 (a) All real numbers
(b) All real numbers

7 (a) $x > 4$
(b) $y > 0$

9 (a) $x \geq -1$
(b) $y \geq 0$

11 (a) $x \neq -1$
(b) $y \neq 3$

13 Domain: $z \geq 25$;
Range: $h(z) \geq 5$

15 $5 \leq v \leq 75$ (others possible)

17 (a) $-8 \leq x \leq 1$
(b) $-5 \leq y \leq 20$

19 (a) $-3 \leq x \leq 1$ and $3 \leq x \leq 7$
(b) $0 \leq y \leq 7$

21 (a) $0 \leq x \leq 2000$
(b) $0 \leq C \leq 10,000$

23 $k = -1$; range is $x = -1$

25 $y \geq -9$

27 $0 \leq y \leq 3$

29 Domain: $0 \leq q \leq 14$; Range: $0 \leq d \leq 336$

31 Domain: $28 \leq t \leq 54$; Range: $203 \leq P \leq 391.5$

33 (b) $x \neq 3$

35 (b) All x

37 (b) $x \geq 3$

39 All real numbers

41 $y \geq 0$

43 $b = 14$

45 Domain: all x; Range: $0 < y \leq 100$

47 Domain: $1 \leq x \leq 9$; Range: $0 \leq y \leq 4$

Section 8.2

1 $y = (x + 1)^4$

3 $w = t^6 + 5$

5 $q = 3 + 10s^3$

7 $y = x^4 + x^2 + 1$

9 $u = x^2 + 1$ and $y = \sqrt{u}$;
Other answers are possible

11 $u = x^3$ and $y = 3u - 2$;
Other answers are possible

13 $h(x) = 1 - x^3, k = 17, p = -4$

15 $h(x) = 1 + 1/x, k = 2, p = -1/2$

17 $h(x) = 1 + x + x^2, k = 1, p = 3$

19 $h(x) = x + 3, k = 7.5, p = 2$

21 Outside function: take the square root

23 1.25

25 $f(x) = 2x^2 + 1$

27 (a) $f(g(x)) = (5 + 2x)^3$
(b) $g(f(x)) = 5 + 2x^3$

29 (a) $f(g(t)) = 3(2t + 1)^2$
(b) $g(f(t)) = 6t^2 + 1$
(c) $f(f(t)) = 27t^4$
(d) $g(g(t)) = 4t + 3$

31 Possible answer: $y = (3x + 1)^3$

33 $y = 5\sqrt{x} + 2$

35 $R = 25 - 0.08\sqrt{t}$

Section 8.3

1 $g(x) = x^3 + 3$

3 $g(x) = (x + 1)^3$

5 $g(x) = (x + 1)^3 - 3$

7 $g(x) = 2 - x$

9 $g(x) = 7 - x$

11 $y = x^2 + 2$

13 $y = (x - 1)^2$

15 $y = (x - 3)^2 + 1$

17

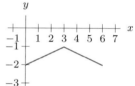

19

21

23 $n = 16m$

25 $n = 0.001m$

27 $n = 0.01m$

29 (a) $H = f(t - 5)$ for $t \geq 5$
(b) $H = f(t) - 10$

31 (a) $y = x + 5$
(b) $y = x + 5$
(c) They are the same

33 $g(t) = 12f(t)$

35 $g(n) = 1,000,000 f(n)$

37 $g(t) = (1/3.262)f(t)$

39 $w(r) = f((1/60)r)$

41 $w(r) = f(1024r)$

43 $w(s) = f((1/2.471)s)$

45 $g(t) = 2.54f(t)$

47 $g(s) = (1/3.785)f(s)$

49 $g(m) = 0.454f((1/0.454)m)$

Section 8.4

1 Add 8

3 Raise to the $1/7^{\text{th}}$ power

5 Add 2 then divide by 5

7 Divide by 2 then raise to $1/5^{\text{th}}$ power

13 (a) $y = 8x^5 + 4$
   (b) Subtract 4, divide by 8, then take the 5th root.

15 $n = g(T)$ is MB processed in T seconds

17 $r = g(Y)$ is number of acres yielding Y bushels

19 $s = g(P)$ gives altitude at pressure P

27 $x = 25$

29 $x = 3$

31 $g(y) = \sqrt[5]{(y-7)/9}$

33 $g(x) = ((3x-5)/(2x+1))^2$

Chapter 8 Review

1 Domain: $0 \le C \le 10$
  Range: $0 \le T \le 2300$

3 Domain: $x \ne 0$,
  Range: $y \ne 0$

5 Domain: $t \ne 0$,
  Range: $A \ne 0$

7 Domain: All real numbers,
  Range: All real numbers

9 Domain: All real numbers
  Range: All real numbers

11 Domain: $x \ne 3$; range: $y \ne 4$

13 $y > 0$

15 $y \ne 0$

17 $y \ge 8$

19 All real numbers

21 (a) $-1 \le x \le 9$
   (b) $0 \le y \le 4$

23 $u = x - 1$ and $y = 1 + 2u + 5u^2$;
   Other answers are possible

25 (a) $f(g(x)) = 45x^2$
   (b) $g(f(x)) = 15x^2$

27 $g(x) = 2(x+1)^2 - 1$

29 $g(x) = 2(x-3)^2 + 1$

31 $g(s) = f(s/3600)$

33 $h(m) = 1.609f(m/60)$

35 $g(y) = (y-5)/7$

37 $g(y) = \sqrt[5]{(3y-2)/(7-7y)}$

39 Divide by 77

41 Raise to the 7th power

43 Domain: integers $0 \le n \le 200$
   Range: $0, 4, 8, \dots, 800$

45 (a) all real
   (b) $s \ge 2$

47 $0 < y \le 3$

49 $0 < y \le 4$

51 $y \ge 0$

53 $0 \le G \le 0.75$

55 Outside function: Square, then multiply by 5

57 $g(h(x)) = (x+3)^5$, $h(g(x)) = x^5 + 3$

59 (a) $f(g(t)) = -3 - 5t^2$
   (b) $g(f(t)) = (2 - 5t)^2 + 1$
   (c) $f(f(t)) = 25t - 8$
   (d) $g(g(t)) = (t^2 + 1)^2 + 1$

61 There are many possible answers

63 $y = (2x - 1)^5$

65 (a) $y = b + mx + k$; the y-intercept is $b + k$
   (b) $y = b + m(x - k)$; the y-intercept is $b - mk$

69 $g(y) = (3y + 2)/(1 - 2y)$

71 $g(y) = (11y - 3)^2/(1 + y)^2$

73 (a) $y = \sqrt[9]{x/7 + 4}$
   (b) Raise to the ninth power, subtract 4, multiply by 7

Section 9.1

1 At $t = 8$ secs

3 (a) $0, 30$
   (b) $0 \le p \le 30$

5 Positive: $x < -7$ or $x > 8$; negative: $-7 < x < 8$

7 Positive: $1 < x < 2$; negative: $x < 1$ or $x > 2$

9 $h = 0$; $k < 0$

11 $h > 0$; $k < 0$

13 (b)

15 (f)

17 (g)

19 (d)

21 (b) Tree
   (c) Same height; falls faster

23 (a) $a > 0, c > 0$
   (b) $h < 0, k < 0$
   (c) Yes; $r < s < 0$

25 (a) $a < 0, c > 0$
   (b) $h = 0, k > 0$
   (c) Yes; $r < 0, s > 0$

27 (a) $u > 0, c = 0$
   (b) $h > 0, k < 0$
   (c) Yes; $r = 0, s > 0$

29 (a) $R(x) = x(1500 - 3x)$
   (b) $C(x) = 5(1500 - 3x)$
   (c) $P(x) = (1500 - 3x)(x - 5)$

31 a changes from positive to negative, and k changes from negative to positive

33 $y \ge 5$

Section 9.2

1 $y = -(x - 6)^2 + 15$

3 $y = (x - 3)^2 - 5$

5 $y = (-7/4)(x - 2)^2 + 3$.

7 $y = -(x - 3)^2 + 9$

9 $l(l - 6)$ square feet

11 (a) $x^2 + 6x + 8$, $(x + 2)(x + 4)$
   (b) $8, 8, 8; 35, 35, 35$

13 $f(x) = x^2 - 3x$;
   $a = 1, b = -3, c = 0$

15 $f(n) = n^2 + 3n - 28$;
   $a = 1, b = 3, c = -28$

17 $m(t) = 2t^2 - 4t + 14$;
   $a = 2, b = -4, c = 14$

19 $h(x) = x^2 - (r + s)x + rs$;
   $a = 1, b = -(r + s), c = rs$

21 $q(p) = 4p^2 - 5p + 6$;
   $a = 4, b = -5, c = 6$

23 $(x + 3)(x + 5)$

25 $(x - 2)(x - 6)$

27 $(2z - 7)(2z + 7)$

29 $(x - 3)(x + 4)$

31 $a = -2, h = 5, k = 5$

33 $a = 4, h = 2, k = 6$

35 $a = 1, h = -6, k = -16$

37 $y = (x + 1)^2 + 5; a = 1, h = -1, k = 5$

39 $y = (x + 3)^2 - 5; a = 1, h = -3, k = -5$

41 $y = (x - 3/2)^2 - 17/4; a = 1, h = 3/2, k = -17/4$

43 y: 12, x: $-4, -1$

45 y: 2, x: $\pm\sqrt{17/4} - 5/2$

47 $(1/2, -1/4)$

49 $(0, 2)$

51 10

53 -4

55 (b) Factored; $p = 3$ or $p = 9$
   (c) Standard; $54,000$ loss
   (d) Vertex; profit is $18,000$ when $p = 6$

57 $f(x) = (x - (a + 1))(x - 3a)$;
   $f(x) = x^2 - (4a + 1)x + 3a(a + 1)$

59 (a) $y = x^2 - 10x + 21, a = 1, b = -10, c = 21$
   (b) $y = (x - 3)(x - 7), a = 1, r = 3, s = 7$
   (c) $y = (x - 5)^2 + (-4), a = 1, h = 5, k = -4$

61 $F(x) = -(x - 3)^2 + 100$; $F(x) = -x^2 + 6x + 91$

63 (a) 20 ft by 40 ft
   (b) 2560

65

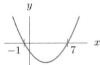

67 $y = 8(x - 1/8)^2 + (-121/8)$;
   $a = 8, h = 1/8, k = -121/8$

69 -3

71 $5x^2 - 15x + 10$

Section 9.3

1   31.707 meters

3   $a = -2, b = 3, c = 7$

5   $a = \pi/4, b = 0, c = -A$

7   $a = 7, b = -18, c = 10$

9   $a = t - t^3 - 4, b = t^2 - 3, c = -t^3 - 5$

11   $x = \pm 3$

13   $x = \pm\sqrt{14} = \pm 3.742$

15   $x = 7, x = -1$

17   $x = 5 \pm \sqrt{6}$

19   No solutions

21   $x = 3 \pm \sqrt{3}$

23   $x = -1, -5$

25   $x = (-3 \pm \sqrt{17})/4$

27   $x = (-5 \pm \sqrt{53})/2$

29 (a) $x = -6, -2$
   (b) $x = -6, -2$

31 (a) $x = -4 \pm \sqrt{28}$
   (b) $x = -4 \pm \sqrt{28}$

33 (a) $x = 9/2 \pm \sqrt{73}/2$
   (b) $x = 9/2 \pm \sqrt{73}/2$

35 (a) $x = 11 \pm \sqrt{111}$
   (b) $x = 11 \pm \sqrt{111}$

37 (a) $x = -3 \pm \sqrt{25/3}$
   (b) $x = -3 \pm \sqrt{25/3}$

39 (a) $x = -11/12 \pm \sqrt{361}/12$
   (b) $x = -11/12 \pm \sqrt{361}/12$

41 (a) $x = 8/7, x = -1$
   (b) $x = 8/7, x = -1$

43 (a) $x = 1/3$
   (b) $x = 1/3$

45 (a)   No solutions
   (b)   No solutions

47   No solutions

49   Two solutions

51   One solution

53   No solutions

55   $t = \sqrt{k}/4$ seconds

57   None

59   $A > 10$

61 (a) $A = lw + \pi w^2/8$
   (b) $w = 2.9$ ft, $l = 5.8$ ft, $r = 1.5$ ft

63   10 in by 12 in

67   $c < 1/3$

71   $b \le -20$ or $b \ge 20$

73   $x^2 - 2x - 1 = 0$

Section 9.4

1   $x = 2, 3$

3   $x = -2, -3$

5   $x = 5, -1$

7   No solution

9   $x = 3, -2, -7$

11   $x = 3, x = -5$

13   $x = 0, x = -2$

15   $x = 4, x = -1$

17   $x = 2, x = 6$

19   $x = (-6 \pm \sqrt{52})/2 = -6.606$ and 0.606

21   $x = 1, x = 7$

23   $x = (1 \pm \sqrt{34})/3$

25   $x = 3/2, 1/3$

27   $x = 7/2, 5/3$

29   $(x - 2)(x + 3) = 0$

31   $(x - 2)^2 = 3$

33   $(x - p)^2 = q$

35   $x^2 - 2ax + a^2 = 0$

37   $0, 2, -1$

39   1

41   1

43   $t = \pm\sqrt{5}$

45   $t = 3 + \sqrt[3]{2}$ or $t = 3 + \sqrt[3]{3}$

47 (a) $x = -3$ and $x = -4$
   (b) $x = -3$ and $x = -4$

49   Two solutions

51   No solutions

53   One solution

55   Two solutions

57 (a)   1.5 sec
   (b)   12.25 ft, 0.625 sec

59   1 ft

61 (a) $p = -q \pm \sqrt{q^2 - 5q}$
   (b) $q = -p^2/(2p + 5)$

63 (a) $p = (1 \pm \sqrt{1 - 8q^2})/(2q^2)$
   (b) $q = \pm(\sqrt{p - 2})/p$

65 (b) $x = 0$ and $x = -b/a$
   (c) $x = 0$ and $x = -b/a$

67   No

69   No; they have different values at $x = 0$

Section 9.5

1   $25 - 20i$

3   $\frac{4}{5} - \frac{3}{5}i$

5   $7i$

7   $3i$

9   $-5 + 16i$

11   $21 + 20i$

13   $11 - 6i$

15   $18i\sqrt{5}$

17   $11 - 13i$

19   $a = 8, b = 4$

21   $a = 2, b = 0$

23   $a = 4, b = -4$

25   $4 \pm i$

27   $3 \pm i\sqrt{5}$

29   $\frac{5}{2} \pm \frac{1}{2}i$

31   $\frac{1}{4} \pm \frac{\sqrt{3}}{4}i$

Chapter 9 Review

1
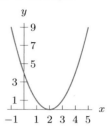

3
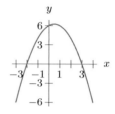

5
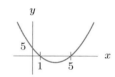

7
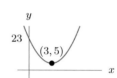

9   $-5x^2 - 2x + 3$

11   $(1/5)z^2 + (4/5)z + 7/5$

13   $2s^2 - 6s - 2$

15   Min: -8

17   Max: q

19   $2(x - 3)^2 + 4, a = 2, h = 3, k = 4$

21   $4(x + 2)^2 - 7, a = 4, h = -2, k = -7$

23   $-(x - b)^2, a = -1, h = b, k = 0$

25   $(1/4)(t - 6)^2 - 3/4, a = 1/4, h = 6, k = -3/4$

27   $(x + 2)(x + 1) = 0; -1, -2$

29   $(3z + 1)(2z + 1) = 0; -1/3, -1/2$

31 (a) $y = 6x^2 - 29x + 35, a = 6, b = -29, c = 35$
   (b) $6(x - 7/3)(x - 5/2), a = 6, r = 7/3, s = 5/2$
   (c) $y = 6(x - 29/12)^2 - 1/24, a = 6, h = 29/12, k = -1/24$

33   $x = \pm 2$

35   $x = 6$

37   $x = 1, x = 2$

39 $x = -59, x = -47$

41 $x = -14, x = -7$

43 $x = -8$

45 $x = -6 \pm \sqrt{29}$

47 $x = -5 \pm \sqrt{22}$

49 $x = 0$

51 $x = -1/2, x = -2$

53 $x = 4, x = -3/2$

55 $y = (-1 \pm \sqrt{109})/6$

57 $t = (5 \pm \sqrt{19})/2$

59 $y = -5 \pm \sqrt{27}$

61 $r = (1 \pm \sqrt{21})/2$

63 $v = 1, v = -4$

65 If x is positive, left-hand side is positive

67 $9t^2 - 3t - 2 = 0$

69 $x^2 + (\sqrt{3} - \sqrt{5})x - \sqrt{15} = 0$

71 $\frac{11}{10} - \frac{7}{10}i$

73 $38i$

75 $-10i\sqrt{3}$

77 $18 + 4i$

79 $7 - 4i$

81 $\frac{31}{2} + \frac{7}{2}i$

83 $5i$

85 (a) Positive, $c = 21$
   (b) $r = 3, s = 7$
   (c) $h = 5, k = -4$

87 $a = 2, b = 4, c = -30$

89 $a = 2, h = -1, k = -32$

91 10

93 $(x - 3)^2$

95 (a) 1483 feet
   (b) 9.627 secs

97 (a) $A = -x^2 + bx$
   (b) $x = b/2$
   (c) Square

99 (a) $A = (200 - 3y)y/2$
   (b) 1666.667 sq ft

101 Two x-intercepts,

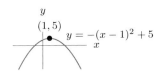

$y = -3(x + 3)(x - 5)$

103 Two x-intercepts,

$(1, 5)$

$y = -(x - 1)^2 + 5$

105 $y = 8$

107 $y = x^2$

109 $A < 0$

111 $A > 0$

113 $c = 0$

115 $ah^2 + k = 0$

117 $a = x$

119 $x = 1$

121 $a \neq -6$

123 All a

125 (c) k is positive

127 $(x + a)(x - 1) = 0; x = -a, x = 1$

Chap. 9 Solving

1 $x = 3, x = 1$

3 $t = 1.6$

5 $r = (3 \pm \sqrt{13})/2$

7 $w = -1.5, w = -4$

9 $(1 \pm i\sqrt{11})/2$

11 $x = 0$ or $x = 1.817$

13 $q = -0.673, q = 3.033$

15 $x = 2, x = -2$

17 $p = -1.275, p = 6.275$

19 $s = -2.273$

21 $p = (-5q \pm \sqrt{25q^2 - 32qr^3})/2$

23 $r = ((-p^2 - 5pq)/(8q))^{1/3}$

25 $a = (-s - b \pm \sqrt{(s + b)^2 + 4bs^2 + 8bs})/2$

27 $p = (q \pm \sqrt{q^2 + 48q})/2$

29 $H = ((2[T]V_0 + 15VV_0)/(5V))^{1/5}$

Section 10.1

1 No

3 Yes; Init: 0.75; growth: 0.2

5 $Q = 300 \cdot 3^t; a = 300, b = 3$

7 $200 \cdot 9^t; a = 200, b = 9$

9 Not exponential

11 Exponential

13 $P = MZ^t$

15 $800 \cdot d^k$

17 $V_0(2/3)^n$

19 Nz^t

21 (III)

23 (a) (III)
   (b) (IV)
   (c) (II)
   (d) (I)

25

27 (a) (i) 10,194,900

     (ii) 10,189,803
     (iii) 10,184,708
     (iv) 10,174,525
   (b) $f(t) = 10,200,000(0.9995)^t$

29 220,000; 1.016

31 Increases by factor of 1.3728

33 $P = 400(0.8)^t$

35 $Q = 50(1/25)^t; a = 50, b = 1/25$

37 $Q = 40 \cdot 4^t; a = 40, b = 4$

39 $Q = 0.2 \cdot 10^t, a = 0.2, b = 10$

41 $y = 5 + 3(x - (-1)); m = 3, x_0 = -1$

43 $a = 5, b = 4, c = -2, d = 3$

Section 10.2

1 Growth

3 Decay

5 Decay

7 1.085

9 0.54

11 70%

13 -73%

15 $a = 200, b = 1.031, r = 3.1\%$

17 $a = \sqrt{3}, b = \sqrt{2}, r = b - 1 = 41.42\%$

19 $a = 5, b = 2, r = 100\%$

21 Initial value $2200; growth rate 21.1%/yr

23 Initial value $8800; decreases by 4.6%/yr

25 $5(1.3)^t$ million

27 $2(1 + r)^{10}$ million dollars; r is growth rate

29 Radioactive substance decaying at 3%

31 Machine depreciating at 2%

35 (c), (d)

37 $a = 1/40 = 0.025, b = 10, r = 900\%$

39 $a = 50, b = 0.9727, r = -2.73\%$

41 Decreases by 25%

43 (a) 50 mg; 77%/day
   (b) 23.979 mg

45 1890.6 mg, 1787.2 mg

47 Bismark's shrink factor per year

49 Yes, $y = 1 \cdot 7^x$

51 Yes, $y = 5(0.9)^{(x-4)/5} = 5.440(0.979)^x$

Section 10.3

1 (a)

Ja	Fe	Ma	Ap	Ma	Ju
1	1	2	4	8	16

   (b) 2^{n-2} for $2 \leq n \leq 6$
   (c) 32 inches

3 $a = 500, b = 2^{1/7} = 1.1041, r = 10.41\%$

5 Flower 1 grows faster

7 $a = -5, b = 1/3$

9 $a = 32, b = 27$

11 Initial value \$3500; doubles every 7 years

13 $a = 400, b = 3, T = 4$ if t in years

15 $a = 2000, b = 2/3, T = 2$ if t in months

17 $a = 120{,}000, b = 4/3, T = 2$ if t in years

19 $a = N, b = 10/9, T = 5$ if t in days

21 $a = 5, b = 2, t_0 = 3$

23 $a = 200, b = 0.825, t_0 = 8$

25 (a) 1 week
    (b) 5 days
    (c) 2 weeks

27 45 min

29 Population q

31 (a) $a = 400, b = 2, T = 4$; doubles every 4 years
    (b) 18.92%

33 (a) $a = 80, b = 3, T = 5$; triples every 5 years
    (b) 24.57%

35 (a) $a = 50, b = 1/2, T = 6$; halves every 6 years
    (b) −10.91%

37 $a = 2000, b = 1.0058, r = 0.58\%$

39 $a = 5000, b = 0.9873, r = -1.27\%$.

41 14.03%

43 22.95%

45 $\$6(1.05)^7$

47 $(1.05)^{25} - 1$

49 $\$20(1.10)^7$

51 $((1.05)^{10} - 1) \cdot 100\%$

53 $\$250(0.99)(1.05)^{10}$ million

55 (a) 2.1% interest every quarter
    (b) 8.67%

57 (a) Earns 0.7% monthly interest
    (b) 8.73%

59 (a) Grows by 20% every 2 years
    (b) 9.54%

Section 10.4

1 $t = 2$

3 $t = 0$

5 $t = -2$

7 $t = 0$

9 $t = 2.6$

11 $2.6 < t < 2.8$

13 Positive

15 Negative

17 (a) Domain: all real; Range: $Q > 0$
    (b) Domain: $t \geq 0$; Range: $0 < Q \leq 200$

19 (a) (IV)
    (b) (V)
    (c) (III)

21 (b)   (i) $x = 0, 1$
          (ii) $x = 2, 3, 4$
         (iii) $x = 3$

23 (b) $x = -1.5, 0.5, 2.5$

25 (a) 57.2; 52.1; 51.1; 50.2; 49.3; 48.4;

    47.5; 39.4
    (b) About 3.7 hours

27 (b) $t = 30$
    (c) $t = 5, 10$

29 (a) Years 0, 1, 2
    (b) $t = 1$

31 (a)

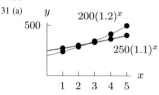

    (b)   (i) $x \leq 2$
          (ii) $x \geq 3$
    (c) Use more points

33 No solution

35 Positive

37 Positive

39 Negative

41 Negative

43 Positive

45 (a) (III)
    (b) (II)
    (c) (IV)

47 (a) $A > 0$
    (b) $A = 1$
    (c) $A > 1$

49 (a) $A > 0$
    (b) $A = 1$
    (c) $0 < A < 1$

51 (a) $A > 0$
    (b) $A = 1/2.1$
    (c) $A > 1/2.1$

53 (a) $A < 0$
    (b) $A = -1$
    (c) $A < -1$

55 $x = 3u$

57 $x = (1/2)(3u - 1)$

59 $x = -2u$

61 $x = 1.176$

63 $x = 1.505$

Section 10.5

1 $Q_0(1.08)^t$

3 $B_0(1.056)^t$

5 0.486%

7 10.409%

9 12.94%

11 8.16%

13 −33.13%

15 −3.37%

17 $g(t) = 80(0.9202)^t$

19 $p(x) = 16.6433(0.8851)^x$

21 $w(x) = 31.752(0.9117)^x$

23 $g(t) = 8000(0.9215)^t$

25 $V = 3450(1.0475)^t$

27 $P = 800(1/2)^{t/19}$

29 $f(x) = 503.153(0.9551205)^x$

31 $V = 3000 \cdot 1.0539^t$

33 $g(x) = 83.687(0.9823)^x$

35 (a) $0.5Q_0$ after 62 days; $0.25Q_0$ after 124 days
    (b) After 186 days
    (c) $0.9889Q_0$

37 (a) 0.352%
    (b) 3.004 million

39 (a) Decay rate: 29.856%
    (b) 117.949 mg
    (c) 14.025 mg

Section 10.6

1 $a = 1200, k = 4.1\%$,
   $b = 1.0419, r = 4.19\%$

3 $a = 7500, k = -5.9\%$,
   $b = 0.9427, r = -5.73\%$

5 $a = 20{,}000, k = -44\%$,
   $b = 0.6440, r = -0.3560$

7 $Q = 20e^{-0.2t}, a = 20, k = -0.2$

9 $Q = 0.025e^{-0.4t}, a = 0.025$,
   $k = -0.4$

11 $Q = 23.1407e^{-5t}$,
    $a = e^{\pi} = 23.1407, k = -5$

13

t (yr)	Inv A, \$	Inv B, \$
0	1,000.00	1,000.00
10	2,593.74	2,718.28
20	6,727.50	7,389.06
30	17,449.40	20,085.54
40	45,259.26	54,598.15
50	117,390.85	148,413.16

15 \$223,169.85

17 Decaying, −1.67%

19 P_0

21 $L/(P_0 + A)$

23 $36.945e^{3t}; a = 5e^2 = 36.945, k = 3$

25 $e^{6x} + 4e^{3x} + 4; r = 6, a = 4, s = 3, b = 4$

27 $(\cosh x)^2 - (\sinh x)^2 = 1$

Chapter 10 Review

1 Linear

3 Exponential

5 Exponential

7 500, 23.2%

9 900, −1.1%

11 $w = 10(0.99)^t$

13 Yes; $a = 1, b = 0.1$

15 Yes; $a = 80, b = 4$

17 Yes; $a = 3/25, b = 1/5$

19 $a = 1250, b = 0.923, r = -7.7\%$

21 $a = 80, b = 0.113, r = -88.7\%$

23 Exponential, constant 1, base $2/3$

25 1.6

27 2

29 9.5%

31 116%

33 9.05% per year

35 −20.63% per day

37 2.73%

39 $V(1.04)^n$

41 $V_0 \cdot k^{20/h}$

43 $P_0 r^4 s^7$

45 $y = \sqrt{2}^x = (1.414)^x$

47 $Q = (8/5)(5/2)^{t/3}$

49 $Q = (6/3^{1/3})3^{t/9}$

51 $Q = 2(7/2)^{t/3} = 2(1.518)^t$

53 $Q = 7.537(0.937)^t$

55 Quadratic

57 Exponential

59 Linear

61 $90(0.83)^t$ mg

63 $\$30(1.022)^t$

65 $1500(1.03)^t - 1500(1.02)^t$

67 Decreasing; 4%

69 Decreasing; 47.5%

71 $f(t) = 8.5859(1.11806)^t$

73 $v(x) = 10.4613(1.06936)^x$

75 $f(t) = 2000(1.201)^t$

77 $P = 7 \cdot 2^{t/5}$

79 $a = 2^{1/T}$

81 $P - 282(1.01)^{20} = 0$

83 Second investment

85 (a) (I), (IV), (V), (VI)
   (b) (II), (III)
   (c) (V), (VI)

87 (a) (III)
   (b) (II)

89 (b)

91 (e)

93 (d)

95 (c)

97 (c)

99 Less than

101 Less than

103 Less than

105 Greater than

107 Greater than

109 Linear: a, b, c if $n = 1$, none if $n \neq 1$
     Exponential: none

111 Linear: A and b if $t = 1/2$
     Exponential: t, base b^2

Section 11.1

1 $10^{-2} = 0.01$

3 $10^{1.301} = 20$

5 $10^{3.699} = 5000$

7 $10^{3x^2 + 2y^2} = \alpha\beta$

9 $\log 100{,}000 = 5$

11 $\log 200 = 2.301$

13 $\log 39{,}994 = 4.602$

15 $\log 97 = a^2 b$

17 4

19 −0.145

21 1.248

23 1.756

25 −3

27 3/2

29 Undefined

31 1

33 7

35 12

37 (a) 0.301, 1.301, 2.301, 3.301
   (b) 4.301, −0.699

39 $3 < \log 8991 < 4$

41 $4 < \log(0.99 \cdot 10^5) < 5$

43 $-2 < \log 0.012 < -1$

45 Between 1.42 and 1.44

47 2.7; 2.6990

49 −1.5; −1.5229

51 3; 3.0244

53 Yes

55 Yes

57 $\log 2 + \log A - \log B$

59 $-\log A - \log B$

61 Not possible

63 $\log A$

65 $u - w$

67 $1.5u/(v - 2w)$

69 Not possible

71 $y = b^x, 0 < b < 1$

73 $y = 2\log x$

75 $y = -b^x, b > 1$

Section 11.2

1 $x = 2$

3 $x = 10$

5 $x = 9$

7 $\log 3.25/\log 1.071 = 17.183$

9 $\log(5/7)/\log(2/3) = 0.830$

11 $\log 0.54/\log 0.088 = 0.254$

13 $\log 90/\log 2 = 6.492$

15 $\log 55/\log 0.988 = -331.937$

17 0.845

19 No solution

21 $\log(5.2)/\log 1.041 = 41.030$

23 $\log 15/\log 1.033 = 83.409$

25 $\log 2.5/\log(2/3) = -2.260$

27 $\log(8/70)/\log 0.882 = 17.275$

29 $17\log 0.125/\log(2/3) = 87.185$

31 $\log(220/130)/\log(1.031/1.022) =$
    60.003

33 $\log(114/82)/\log(1.031/1.029) =$
    169.682

35 $\log(170/132)/\log(1.045/1.067) =$
    −12.143

37 $\log 1.5/\log 1.032 = 12.872$

39 $\log(800/2215)/\log 0.944 = 17.6716$

41 0.001

43 14.859

45 0.0316

47 13

49 No

51 Yes

53 No

55 Yes

57 6.136 years

59 $t = 6.213$ months

61 20 months

63 $t = -13.853$ years

Section 11.3

1 $t = 3.636$

3 $t = 0.431$

5 628 days

7 1.996 years

9 2.64 years

11 6.642 days; 13.284 days

13 $t = 16.3411$ years

15 (a) 11.786 years
   (b) 20.149 years

17 (a) 3.094 years
   (b) 5.290 years

19 (a) 20 million
   (b) 180 million
   (c) 3.561 hours
   (d) 38 minutes

21 (a) $1/2$ cm^3
   (b) 9 min
   (c) 7.644 min

23 (a) 35.710 years
   (b) 91.289 years

25 (a) 2.726 mg
   (b) 18.574 days

27 (a) False
   (b) 3.322 inches

29 $(\log 2/\log b)$ hours

Section 11.4

1 $t = -2\ln 0.375$

3 $t = (\ln 4)/-0.117$

5 $t = 90\ln(1700/23)$

7 (a) $t = \log 3/(0.231\log e)$
   (b) $t = \ln 3/0.231$

9 (a) $t = (\log 0.2)/(\log 0.926)$
   (b) $t = (\ln 0.2)/(\ln 0.926)$

11 $a = 210, b = 0.8098, r = -0.1902, k = -0.211$

13 $a = 350, b = 1.318, r = 0.318, k = 0.2761$

15 $a = 27.2, b = 1.399, r = 0.399, k = 0.3358$

17 19.903 years

19 6

21 -2

23 0.5

25 2

27 -1

29 2

31 $\log_2 12$

33 $\log_5 0.75$

35 $t = 16.447$; 10 words unchanged after 16,447 yrs

37 $f(200) = 134.296$; about 134,300 articles after 200 days

39 Growing 542 articles/day

41 $\phi = -6.252$

43 ϕ_2 16 times as large as ϕ_1

45 $\log \sqrt{17}, x = \sqrt{17}$

47 $\log_{25} \sqrt{90}, x = \sqrt{90}$

49 $\log_5 81, x = 81$

Chapter 11 Review

1 2

3 0

5 Undefined

7 1/3

9 3.68

11 $2n + 1$

13 Undefined

15 5.9

17 $10^1 = 10$

19 $10^{1.733} = 54.1$

21 $10^r = w$

23 $\log 1{,}000{,}000 = 6$

25 $\log 0.1 = -1$

27 $\log 0.558 = -0.253$

29 (a) $\log 100 = 2$
   (b) $\sqrt{100} = 10$

31 $5x - 3$

33 $6x^3$

35 $-x/3$

37 $3x/2$

39 $\log 100/\log \pi = 4.023$

41 5

43 1.850

45 2

47 5.060

49 -0.511

51 3.594

53 2, 5

55 4

57 $\log 0.4/(2 \log 0.6)$

59 9

61 $x = -2/3$

63 $x = \log(9/8)$

65 $\log 1.5$

67 $t = \log 0.5/\log 1.117$

69 $t = \log(1/6)/(\log 0.5)$

71 $t = \log(z - r)/\log b$

73 $(\log 800 - \log 250)/(\log 0.8 - \log 1.1)$

75 $\log(0.0422/0.0315)/\log(0.988/0.976)$

77 $x = 3$

79 1000

81 $x = 10^{1/3}$

83 $x = 10^{2.5} = 316.228$

85 $x = 10^{4/3} = 21.544$

87 7.273 years

89 $e^0 = 1$

91 $e^y = 10$

93 $e^4 = 7r$

95 $\ln 1 = 0$

97 $\ln 0.018 = -4$

99 $\ln 1.221 = 0.2$

101 0

103 -3

105 $-1/2$

107 Undefined

109 0

111 1/2

113 Cannot be simplified

115 4

117 -2

119 (a) $\log 100/\log 2 = 6.644$
   (b) $\ln 100/\ln 2 = 6.644$

121 (a) $\log 5/\log 3 = 1.465$
   (b) $\ln 5/\ln 3 = 1.465$

123 3.170

125 1.585

131 1.176

133 (III)

135 (I)

137 (IV)

139 $u + v$

141 $\log 2 + u + 3w + 0.5v$

143 $u + \log(\log 3)$

145 $2x + 1$

147 $(x + 1)^2$

149 $t = (4 \log 3.5)/\log 3$

151 $x = 200$

153 0.485

155 -0.2

157 1000

159 500,000.5

161 $x = 10{,}000{,}000{,}000$

163 The year 2206

165 23.449 years

167 7.17 weeks

Section 12.1

1 Monomial

3 Monomial

5 Monomial

7 Polynomial

9 Polynomial

11 Not polynomial

13 Yes

15 No

17 Yes

19 0

21 -3

23 $4t^3 + 12t^2 + 11t + 3$

25 $0, 1/2, -1/2$

27 $800 + 3200r + 4800r^2 + 3200r^3 + 800r^4$

29 Yes,

x	0.1	0.2	0.3
$f(x)$	1.054	1.118	1.195
$g(x)$	1.054	1.118	1.192

31 $g(1/2) = 177/128$

Section 12.2

1 $5x^7 + 4x^5 - 2x^2 + 3x$

3 $-x^3 + 4x^2 - 2x$

5 $a_0 = 17$

7 $a_0 = 15$

9 $a_0 = -1$

11 6

13 3

15 $-4x^{11}$

17 x^{13}

19 x^8

21 $-1/2$

23 $\sqrt{7}$

25 $a_4 = 3, a_3 = 6, a_2 = -3, a_1 = 8, a_0 = 1$

27 $a_{13} = 1/3$

29 $2x^3 + x - 2; n = 3;$
   $u_3 = 2, a_2 = 0, a_1 = 1, a_0 = -2$

31 $-5x + 20; n = 1;$
   $a_1 = -5, a_0 = 20$

33 $(\sqrt[3]{5}/7)x^2; n = 2;$
   $a_2 = \sqrt[3]{5}/7, a_1 = 0, a_0 = 0$

35 $-6x^5 - 5x^4 - 4x^3 + 36x; n = 5;$
   $a_5 = -6, a_4 = -5, a_3 = -4, a_2 = 0,$
   $a_1 = 36, a_0 = 0$

37 $x^3 + 3x^2 + 3x + 1; n = 3;$
   $a_3 = 1, a_2 = 3, a_1 = 3, a_0 = 1$

39 720

41 0, 4

43 Birthday/ Present: 15th/ \$800; 16th/ \$900
17th/ nothing; 18th/ \$300
19th/ \$500; 20th/ \$1200

45 2.741%

47 2

49 10

51 $n = 5$

53 $h(0) = 12$

55 Leading term is $12t^5$; constant term is -10

57 8

59 7

61 $a = 1, b = 4, c = 2, d = 9$

63 Possible answer: $2x^5 + 3x^4$ and $-2x^5 + 7x^4$

65 Impossible

67 Impossible

69 1

71 $\pm 3/2$

73 $50x^5 - 45x^4 + 19x^3 - 8x^2 + 3x - 1$

75 2, 3, 4

Section 12.3

1 $x = 3, 4, -2$

3 $x = 3, -2$

5 None

7 $1, -2, 3$

9 $0, -1, -2$

11 $0, 1, -3$

13 $0, 2, -2$

15 All u

17 $-3, -4$

19 $(x - 2)(x + 3)$

21 $(x - 5)(x + 13)^2$

23 $(x + 6)^5$ is one possibility

25 $0, 600, -600$; 0 and 600 make sense; profit zero dollars from selling these quantities

27 $(1/2)(x + 2)(x + 1)(x - 2)(x - 3)$

29 (a) $(x+1)(2x-5)$ and $-3(x+1)(2x-5)$
     (b) $2(x + 1)(2x - 5)$

31 One

33 Three

35 Four

37 None

39 $1, -1, 0$

41 $a \leq 0$

43 All a

45 All a

47 $a \leq 0$

49 $-1.414, 0.667, 1.414$

Section 12.4

1 $-(x + 2)(x - 1)$

3 $-\frac{1}{2}(x + 3)x(x - 3)$

5 $y = (-5/6)(x + 3)(x + 1)^2(x - 2)$

7 $y = (x + 2)(x - 3)$

9 $y = (x + 2)^2(x - 3)$

11 (a) II: $x = 3, -2, 1/2$
     (b) I; 6
     (c) I; positive, negative
     (d) II; three times

13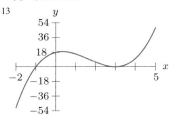

19 (a)

Chapter 12 Review

1 2

3 4

5 8

7 -1

9 0

11 Polynomial, degree 5, leading coefficient 8

13 Not a polynomial

15 Not a polynomial

17 Degree 2; leading term $-3x^2$;
    leading coefficient -3; constant term 5

19 Degree 6; leading term $4x^6$;
    leading coefficient 4; constant term 0

21 (a) 1
     (b) 4
     (c) 0
     (d) $-t^3 + t^2 - t + 1$

23 (a) 1
     (b) 5
     (c) 5
     (d) $t^4 + 3t^2 + 1$

25 $17x^6 + 4x^5 + 9x^2$

27 $2x$

29 $1, -1, -9$

31 $1, -1$

33 $y = (-3/4)(x + 1)^2(x - 2)$

35 $y = -2x(x + 1)(x - 1)(x - 2)$

37 $m, 5a$

39 $m; a$

41 a, m

43 One

45 One

47 None

51 $(x - a)(x^2 + a^2) = 0; x = a$

53 $(x + 7)(x + 2)(x - 3)/14 = x^3/14 + 3x^2/7 - 13x/14 - 3$

55 $x^6 + x^2 + 1 > 0$

59 (a)

61 $-4 < x < 3$ except $x = -1$

63 $t > a$

65 Yes; at least two

67

1		5		10		10		5		1	
	1		6		15		20		15		6		1
1		7		21		35		35		21		7		1

69 $(x + 1)^7 = x^7 + 7x^6 + 21x^5 + 35x^4 + 35x^3 + 21x^2 + 7x + 1$

71 $4, 12, 32; n \cdot 2^{n-1}$

73 $9x^2, 27x^3, 81x^4; 3^n x^n$

75 $-2, -3, -4; -n$

77 (a) \$1995.00; \$1989.95; \$1984.85
     (b) $b_2 = 2000x^2 - 25x - 25; b_3 = 2000x^3 - 25x^2 - 25x - 25; b_4 = 2000x^4 - 25x^3 - 25x^2 - 25x - 25; b_5 = 2000x^5 - 25x^4 - 25x^3 - 25x^2 - 25x - 25$
     (c) \$2025.76; balance increases

79 Initial balance: \$3000
    Payments: \$120, \$90, \$200, \$250 in months 1, 2, 5, 6, resp
    Expenditures: \$50, \$100 in months 3, 4 resp

81 $w(x) = (1/4)(x+3)(x+2)(x-2)^2(x-3)$

83 $w(x) = -0.1x(x+3)^2(x+2)(x-2)^2$

Section 13.1

1 $a(x) = 10x + 33, b(x) = 2x + 6$

3 $a(x) = x + 1, b(x) = 2x + 1, x \neq 0, x \neq -1$

5 $x = -3/5$

7 $x = -1/5, 1$

9 All real numbers except $x = -5/2$

11 All real numbers except $x = \sqrt{2}$

13 All real numbers except $x = 0$

15 All x except $x = \pm 2$

17 $r = -1, 4$

19 $x - 1$; not equivalent

21 $2/(3x^2 + 5)$; equivalent

23 $(3, 7)$

25 $(-2, -40)$

27 (a) $(10 + 0.02n)/n$ \$/poster
     (b) 500 posters
     (c) $(10/n) + 0.02$

29 (a) $R(z)$
     (b) $-R(z)$
     (c) $R(z)$
     (d) $-R(z)$

31 $x = -1/2$

33 $x = (-1 \pm \sqrt{82})/9$

35 Two

37 Two

39 $x^2((x+3)/(x+3)); (-3, 9)$

Section 13.2

1 $g = 0$

3 None

5 $q(x) = 2, r(x) = 1$

7 $q(x) = x + 2, r(x) = 7$

9 $q(x) = 1; r(x) = 1$

11 $q(x) = 2; r(x) = 3$

13 (a)

15 (a) and (c)

17 (a) and (d)

19 (a), (b), (c)

21 (a) I; $x = 1, 2$
(b) II; $r(x) \approx 1$

23 Same horizontal asymptotes, $y = 3$

25 Denominator has no zeros

27 $y = -(x + 1)/(x - 2)$

29 $y = x/((x + 2)(x - 3))$

31 $y = -(x - 3)(x + 2)/((x + 1)(x - 2))$

33 Zero: $x = -3$;
Asymptote: $x = -5$;
$y \to 0$ as $x \to \pm\infty$

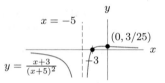

35 Zeros: $x = \pm 2$;
Asymptote: $x = 9$;
Approaches $y = x$ as
$x \to \pm\infty$

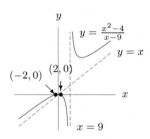

Section 13.3

1 $a(x) = 5x^2 - 23x + 16, b(x) = x - 4, q(x) = 5x - 3, r(x) = 4.$

3 $a(x) = 2x^2 - 3x + 11, b(x) = x + 7, q(x) = 2x - 17, r(x) = 130$

5 $a(x) = x^4 - 2, b(x) = x - 1, q(x) = x^3 + x^2 + x + 1, r(x) = -1$

7 $a(x) = 6x^2 + 5x + 3, b(x) = 3x - 2, q(x) = 2x + 3, r(x) = 9.$

9 $a(x) = 6x^3 - 11x^2 + 18x - 7, b(x) = 3x - 1, q(x) = 2x^2 - 3x + 5, r(x) = -2.$

11 $x - 3 + \frac{1}{x-2}$
Horiz. asym.: none
Slant asym.: $y = x - 3$

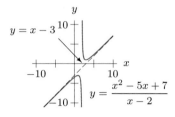

13 $6x^2 - 9x + 35 + \frac{-62}{x+2}$
Horiz. asym.: none
Slant asym.: none

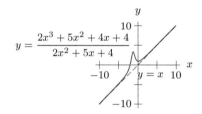

15 $x + \frac{4}{2x^2+5x+4}$
Horiz. asym.: none
Slant asym.: $y = x$

17 No

19 Yes; $2x^2 - 5x + 2$

21 $x + 2, x \neq -5$

23 $r = 15$

25 $p(4) = 7$

27 $r = 2$

29 $r = -6$

Chapter 13 Review

1 $(x - 5)/(2 - x)$ provided $x \neq -3$

3 $3x + a$ provided $x \neq a$

5 $x = 0$

7 All real numbers

9 All real numbers except $x = 2$

11 Hole at $x = 3$; zero at $x = 1$; vertical asymptote at $x = -2$; horizontal asymptote at $y = 1$; y-intercept at $y = -1/2$

13 No common factors

15 $-1/(x + 2)$; not equivalent

17 (a) $w = 0$

(b) Chances of winning approach zero

19 (a) II; $x = 3, -3$
(b) I; $9/4$
(c) II; $x = 2, -2$

21 (a) and (c)

23 (d)

25 $x^2 + 2x + 2 + \frac{11}{x-2}$
Horiz. asym.: none
Slant asym.: none

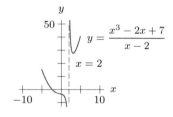

27 (a) (iii)
(b) (i)
(c) (ii)
(d) (iv)
(e) (vi)
(f) (v)

29

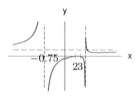

31 One possibility: $u(x) = 1/((x - 2)(x + 3))$

33 $(1/x)((x - 0.02)/(x - 0.02))$;
$(0.02, 50)$

Section 14.1

1 $\sum_{i=1}^{7} e_i$

3 $\sum_{i=1}^{30} r_i$

5 $0^2 + 1^2 + 2^2 + 3^2 + 4^2 + 5^2$

7 $1 + 4 + 7 + 10 + 13 + 16$

9 $(-1)^1 + (-1)^2 + (-1)^3 + (-1)^4$

11 20

13 30

15 5

17 Yes

19 Yes.

21 No

23 Yes

25 No

Section 15.1

1 Arithmetic

3 Not arithmetic

5 Not arithmetic

7 Arithmetic, $a_n = -2.9 - 0.1n$

9 13.8, 24.8, $2.8 + 2.2n$

11 6.9, 17.4, $-3.6 + 2.1n$

13 2, 5, 8, 11, 14

15 43, 47, 53, 61, 71

17 (a) $35,000 + 7000m$
   (b) 15

19 Arithmetic, $d = 1, a_6 = 6$

21 Arithmetic, $d = -1, a_6 = -1$

23 2, 6, 10, 14; $a_n = -2 + 4n$

25 1, 2, 4, 8; $a_n = 2^{n-1}$

27 (a) $a_{25} < a_{50}$
   (b) Fraction with incomes of \$25,000–\$50,000
   (c) No
   (d) $a_n = 1$ for large n

Section 15.2

1 125, 250

3 $a_1 = 2, d = 4$

5 360

7 36.5

9 138

11 108

13 $\sum_{n=1}^{6} 4n$

15 $\sum_{n=1}^{5} (3/2)n$

17 (a) $\sum_{i=1}^{9} 2i$
   (b) 90

19 (a) 248.7, 281.4; population in millions at census time
   (b) 22.1, 32.7; population change in millions over a decade
   (c) 2.21; average yearly population growth in millions over the 1990s

21 (a), (b), (d); (a), (b), (d)

23 (a), (b) and (d); (b) = (c) = (d) = 630

25 $F_n = \sum_{i=n-2}^{n-1} F_i$

27 Last row: 76
   Auditorium: 720

Section 15.3

1 Geometric; $8(1/2)^{n-1}$

3 Not geometric

5 Geometric; $2(0.1)^{n-1}$

7 243, 3^{n-1}

9 If $r = 3$: 1029, 3^n;
   if $r = -3$: -1029, $(-1)^{n-1}3^n$

11 Yes, $n = 9$, ratio $= 1/2$

13 Not geometric

15 9.995

17 364.494

19 6.932

21 19 terms; 1.443

23 (a) \$3125
   (b) 3905

25 (a) $h_n = 20(3/4)^n$
   (b) $D_1 = 20$ feet
      $D_2 = h_0 + 2h_1 = 50$ feet
      $D_3 = h_0 + 2h_1 + 2h_2 =$
      72.5 feet
      $D_4 = h_0 + 2h_1 + 2h_2$
      $+ 2h_3 = 89.375$ feet
   (c) $D_n =$
      $20 + 120\left(1 - (3/4)^{n-1}\right)$

Section 15.4

1 235/999

3 11/90

5 3781/4950

7 (a) 784 feet, 1024 feet, 1296 feet
   (b) 216 feet, -24 feet, -296 feet
   (c) No, stops before 8 seconds

9 (a) 7.906 sec
   (b) 11.180 sec
   (c) No

11 (a) $P_n = 250(0.016) + 250(0.016)^2 + \cdots + 250(0.016)^{n-1}$
   (b) $P_n = 4(1 - (0.016)^{n-1})/(1 - 0.016)$
   (c) $P_n = 4.065$ mg

13 (a) 6.25%.
   (b) $Q_n = (50(1 - (0.0625)^{n+1}))/(1 - 0.0625)$
   (c) $P_n = (0.0625(50)(1 - (0.0625)^n))/(1 - 0.0625)$

15 \$29,778.08

Chapter 15 Review

1 Not geometric

3 Geometric

5 Not arithmetic

7 Not arithmetic

9 $a_1 = 5, d = 3$

11 $\sum_{n=0}^{5} 3^n$

13 $\sum_{n=0}^{6} 3(5^n)$

15 $\sum_{n=0}^{10} (1000 - 100n)$

17 215; 35

19 $a_1 = 1$, ratio $= 3x$, $\sum_{n=0}^{11} (3x)^n = (1 - (3x)^{12})/(1 - 3x)$,
   for $x \neq 1/3$, $\sum_{n=0}^{11} (3x)^n = 12$ if $x = 1/3$

21 25 cans at bottom
   3 less per row
   9 rows

23 300

25 (a) After second withdrawal
   (b) \$121,350.68
   (c) \$3846.15

27 (a) $a_n = 100(0.8)^{n-1}$
   (b) $a_{42} = 100(0.8)^{42-1} = 0.011$
   (c) \$500

(d) 5

29 (a)

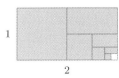

(b) $1 + 1/2 + 1/4 + \cdots + 1/2^{n-1}$
(c) 2

Section 16.1

1 2×2

3 2×2

5 6

7 -3

9 $\begin{pmatrix} 7 & 4 \\ 1 & 11 \end{pmatrix}$

11 Not possible

13 $\begin{pmatrix} -6 & 4 \\ 2 & -12 \end{pmatrix}$

15 $\begin{pmatrix} 2 & 0 & 6 \\ -11 & -1 & -5 \end{pmatrix}$

17 Not possible

19 $\begin{pmatrix} 7 & -2 \\ 5 & 9 \end{pmatrix} = \mathbf{U} + \mathbf{W} = \mathbf{W} + \mathbf{U}$

21 $\begin{pmatrix} 4 & 28 \\ 4 & 16 \end{pmatrix}$

23 $\begin{pmatrix} 6 & 2 \\ 5 & 11 \end{pmatrix}$

25 $\begin{pmatrix} 24 & 27 & 25 & 25 & 25 & 31 & 29 & 29 \\ -4 & -2 & 0 & -3 & -1 & 0 & -3 & 0 \end{pmatrix}$

27 $\begin{pmatrix} 56 & 65 & 73 \\ 127 & 76 & 69 \\ 63 & 61 & 86 \\ 108 & 97 & 110 \\ 155 & 133 & 123 \end{pmatrix}$

Section 16.2

1 $\begin{pmatrix} 4 & 15 \\ 32 & 4 \end{pmatrix}$

3 $\begin{pmatrix} -8 & 16 & -45 \\ -18 & 28 & -23 \end{pmatrix}$

5 $\begin{pmatrix} 8 & -16 & -19 \\ 32 & -32 & -52 \end{pmatrix}$

7 Not possible

9 $\begin{pmatrix} 32 & 33 \\ 80 & 44 \end{pmatrix}$

11 $\begin{pmatrix} 16 & -68 & -65 \\ 128 & -80 & -172 \end{pmatrix}$

13 No. of columns of first = no. of rows of second

15 $\begin{pmatrix} -12 & 54 \\ 21 & 3 \end{pmatrix} = 3(\mathbf{GH}) = (3\mathbf{G})\mathbf{H}$

17 (a) $\begin{pmatrix} 4080 \\ 3160 \end{pmatrix}$

   (b) The cost of all meals for Saturday and Sunday.

Section 16.3

1 $(7, 13, 21)$

3 $(4/3, 2, 16/7)$

5 $(15, 21, 39)$

7 $(5x, 7x, 13x)$

9 $(25, 40, 55)$

11 $(5y, 8y, 11y)$

13 $\begin{pmatrix} 111 \\ 38 \end{pmatrix}$

15 $\begin{pmatrix} 66 \\ 174 \end{pmatrix}$

17 Summer

19 Evening

21 $\vec{H} = (182, 1729, 3822, 1092)$; total customers each time of day in Winter

23 $(550, 640, 110)$, voters after drive

25 $(510, 610, 0)$

27 (c)

Section 16.4

1 $x = 2, y = 7$

3 $x = 3, y = -3$

5 $x = 75/2, y = 5$

7 $x = 2, y = 3$

9 $\begin{cases} 5x + 7y = 9 \\ 2x + 4y = -5 \end{cases}$

11 $\begin{cases} cx + dy + ez = 1 \\ fx + gy + hz = 2 \\ lx + my + nz = 3 \end{cases}$

13 $\begin{cases} x + 2y + 3z = 4 \\ 5x + 6y + 7z = 8 \\ 9x + 10y + 11z = 12 \end{cases}$

15 $\begin{cases} 7x + 3z + 5t = a \\ (6 + m)x + 2t = b \\ y + z + t = c \\ 5x + 7y + 9z + 11t = d \end{cases}$

17 $x = -1, y = 3$

19 $x = 5, y = 2$

21 $x = 1/2, y = 2$

23 $x = -1, y = -1$

27 (a) $\begin{pmatrix} 3 & 0 & 6 \\ -5 & 2 & 0 \\ 0 & 4 & -1 \end{pmatrix} \begin{pmatrix} x \\ y \\ z \end{pmatrix} = \begin{pmatrix} 7 \\ 8 \\ 9 \end{pmatrix}$

(b) $\left(\begin{array}{ccc|c} 3 & 0 & 6 & 7 \\ -5 & 2 & 0 & 8 \\ 0 & 4 & -1 & 9 \end{array} \right)$

29 (a) $\begin{pmatrix} 1 & -a & 1 \\ -a-1 & 1 & 3 \\ -4 & -3 & b+c \end{pmatrix} \begin{pmatrix} x \\ y \\ z \end{pmatrix} = \begin{pmatrix} 4 \\ 19 \\ -15 \end{pmatrix}$

(b) $\left(\begin{array}{ccc|c} 1 & -a & 1 & 4 \\ -a-1 & 1 & 3 & 19 \\ -4 & -3 & b+c & -15 \end{array} \right)$

31 (a) $\left(\begin{array}{cc|c} 1 & 8 & 500 \\ 1 & -2 & 10 \end{array} \right)$

(b) $Q = 108,000$ at a price of $p = \$49$

33 (a) $\begin{pmatrix} 40 & 30 & 25 \\ 50 & 115 & 60 \\ 30 & 15 & 5 \end{pmatrix} \begin{pmatrix} B \\ S \\ F \end{pmatrix} = \begin{pmatrix} 250 \\ 505 \\ 125 \end{pmatrix}$

(b) $\left(\begin{array}{ccc|c} 40 & 30 & 25 & 250 \\ 50 & 115 & 60 & 505 \\ 30 & 15 & 5 & 125 \end{array} \right)$

(c) $B = 3, S = 1, F = 4$

Chapter 16 Review

1 $\begin{pmatrix} 4 & 9 \\ -10 & 6 \end{pmatrix}$

3 $\begin{pmatrix} -1.4 & 0.9 \\ -1.6 & 3 \end{pmatrix}$

5 $\begin{pmatrix} 1.5 & -0.9 \\ 2.2 & -7.1 \end{pmatrix}$

7 Yes

9 Yes

11 No

13 No

15 $x = 7, y = 8, z = -6$

17 $x = 2, y = -3, z = 4$

19 $x = 1, y = 2, z = 3$

21 $x = 33, y = 7, z = -1$

23 (a) $\begin{pmatrix} 10 & 30 & 2 \\ 4.5 & 50 & 20 \\ 0 & 11 & 3 \end{pmatrix} \begin{pmatrix} 9 \\ 4 \\ 4 \end{pmatrix} =$ $\begin{pmatrix} 218 \\ 320.5 \\ 56 \end{pmatrix}$

(b) Sandwich

25 (a) 3×4

(b) Calories in salt

27 $\begin{pmatrix} 7774.4 \\ 625.6 \end{pmatrix}$

29 (a) $A + B + C = 1$

    $8A + 6B + 12C = 9$

    $B - \frac{1}{2}C = 0$

(b) $\left(\begin{array}{ccc|c} 1 & 1 & 1 & 1 \\ 8 & 6 & 12 & 9 \\ 0 & 1 & -1/2 & 0 \end{array} \right)$

(c) $A = 1/2, B = 1/6, C = 1/3$

Section 17.1

1 5

3 105

5 6.5

7 15.167

9 2

11 1.5

13 $54.3°F$

15 (a) (i) $240 \, \text{ft}^3/\text{sec}$

     (ii) $300 \, \text{ft}^3/\text{sec}$

(b) Week 2

(c) Yes, week 2

17 (a) 2

(b) 2.5

(c) 3

(d) 3.5

(e) $k + 1/2$

(f) $k + 1$

19 (a) $W/20$

(b) $(W - x + y)/20$

21 (a) $\$1,190,000$

(b) No

23 (a) 12

(b) Less extreme

(c) The same

25 (a) 135.3

(b) $(3x_1 + 6x_2 + 7x_3 + 5x_4 + 4x_5)/25$

(c) $(n_1x_1 + n_2x_2 + n_3x_3 + n_4x_4 + n_5x_5)/(n_1 + n_2 + n_3 + n_4 + n_5)$

27 (a) 5.2, 6.5

(b) 6

(c) No

(d) Different sample sizes

Section 17.2

1 26.43, 11.843

3 42.49, 33.657

5 0, 35.50

7 (a) 5

(b) 1.92

9 (a) 10 lizards

(b) 7 cm

(c) 5.8 cm to 8.1 cm

(d) 0.527 cm

11 (a) 10 lizards
   (b) 7 cm
   (c) 4 cm to 10 cm
   (d) 1.11 cm

13 (a) 12
   (b) Less extreme
   (c) Smaller

15 $(((x_1 - m)^2 + (x_2 - m)^2 + (x_3 - m)^2 + (x_4 - m)^2 + (x_5 - m)^2)/5)^{1/2}$

17 $((3(x_1 - m)^2 + 6(x_2 - m)^2 + 7(x_3 - m)^2 + 5(x_4 - m)^2 + 4(x_5 - m)^2)/25)^{1/2}$

Section 17.3

1 7.69%

3 50%

5 15.38%

7 (a) 0.455%
   (b) 2%
   (c) 0.741%

9 (a) 30.9%
   (b) 20.8%
   (c) 14.2%

   (d) 34.2%

11 2/52

13 2/26

15 13/26

17 0/26

19 (a) 84.6%
   (b) 0.3%

21 (a) 20.10%
   (b) 21.25%
   (c) 12.09%
   (d) 25.62%
   (e) 20.95%

25 Independent

27 Not independent

Chapter 17 Review

1 4

3 2

5 5.4

7 5.5

9 25, 12.862

11 5, 0

13 5.5, 2.872

15 1.92%

17 23.08%

19 25%

21 2/52

23 2/26

25 13/26

27 0/26

29 Store A should be higher

31 (a) $41
   (b) Increases to $46; increases to $51
   (c) Increases to $42; Increases to $44

33 (a) 42.9%
   (b) 11.1%

35 (a) 19.23%, 7.69%, 15.38%, 23.08%, 13.46%, 21.15%
   (b) 27.5%

37 (a) 40 use drugs; 960 don't
   (b) 38 positive; 2 negative
   (c) 912 negative; 48 positive
   (d) 86 positive; 38/86 use drugs
   (e) $38/86 = 44.2\%$

INDEX

Δ, delta, 95
Σ, sigma, 433, 449
σ and σ-notation
   standard deviation, 507
σ, lower-case sigma, 507
e, 330
i, the number, 286
   powers of, 289

absolute value, 68
   inequalities, 72
addition
   of complex numbers, 288
Alfa Romeo, 186, 204, 215
algebraic expression, 2
algebraic fraction, 46
alternating sequence, 441
area
   circle, 55, 79, 211
   rectangle, 7
   triangle, 7
arithmetic sequence, 441
   and linear functions, 443
arithmetic series, 447
   sum of, 447
astronaut, 191, 204, 205
astronomy
   planetary motion, 220
asymptote
   horizontal, 407, 414, 415
   rational function, 415, 416, 421
   slant, 416
   vertical, 407
augmented matrix, 483
   echelon form, 485
   row operations, 486
Austin, 309
average rate of change, 95, 96
   expression for, 95
   interpreting, 97
   units, 98

balancing equation, 19, 20
base
   common, 168
   exponent, 168
   exponential function, 299
Bhutan, 303

Bismarck, 309
bond, 5
borehole, 119, 143
Burkina Faso, 309

calculator, approximation, 374
cancelation, 46
Celsius, 232, 243
circle
   area, 79
circumference
   circle, 211
coefficient, 39, 121
   linear expression, 121
   polynomial, 380
   power function, 187
coffee, 11, 148, 165, 354
combining like terms, 32
common
   denominator, 51
   logarithm, 340
common factor, 37
completing the square, 266
   solving by, 273
   visualizing, 269
complex number, 286
   addition of, 288
   conjugate, 287, 290
   definition, 286
   division of, 290
   imaginary part, 286
   multiplication of, 288
   powers of i, 289
   real part, 286
   subtraction of, 288
composition, 232
   inside function, 233
   outside function, 233
compound interest, 458
concert pitch, 366
conjugate, 181
   complex number, 287
constant
   and variable, 86
   of proportionality, 101
      and units, 104
      solving for, 103
   term, 121, 256

polynomial, 380
constant polynomial, 380
constraint equation, 146
continuous growth rate, 330, 362
credit multiplier, 464
cube, 168
cubic equation, 283
Czech Republic, 303

decay rate, 305, 323
decimal
   repeating, 461
   writing as fraction, 461
decomposition, 233
decreasing
   linear function, 116
degree of polynomial, 378, 380
delta, Δ, 95
denominator
   rationalizing, 181
dependent
   variable, 79
difference of squares
   factoring, 43
dimensions
   matrix, 466
direct proportionality, 101
discriminant, 276
distributive law, 33
   algebraic fractions, 49
   common factor, 37
   division, 35
   expanding with, 39
   negative sign, 34
division
   of complex numbers, 290
domain, 224, 226
   and graph, 225
   exponential function, 301
   logarithm function, 345
   power function, 228
   rational function, 408
double zero, 389
doubling time, 310, 357
drug level, 459

e, 330
Earth

Photo Credits